Oceanography and Seamanship

Fletcher-class destroyer *Remey* leaps half out of water during heavy-weather fueling operations in South China Sea.

Oceanography and Seamanship

SECOND EDITION

William G. Van Dorn

Drawings by Richard Van Dorn

CORNELL MARITIME PRESS

Centreville, Maryland

Photographic Credits

Plate Number	Source
Frontispiece:	U.S. Navy
1, 5, 12, 13, 14:	University of California
2, 3, 4:	Bruce W. Halstead, World Life Research Institute, 2300 Grand Terrace, Colton, California 92324
6, 7, 8, 18, 22, 27:	Author
9, 11:	National Aeronautics and Space Administration, Houston, Texas
10:	Tourism, New Brunswick, P.O. Box 1030, New Brunswick, Canada
15, 16:	USNAS, Jacksonville, Florida
19, 20:	D. L. Inman
21:	Dr. Donald James, 10100 Culver Blvd., Culver City, California 90230
23:	© 1955 National Geographic Society
24:	Fotoboat Company, P.O. Box 426, Santa Barbara, California 93102
26:	Department of the Army, Waterways Experiment Station, Corps of Engineers, P.O. Box 631, Vicksburg, Mississippi 39180
17:	Courtesy K. Adlard Coles and John de Graff
25:	George Tiedemann
28:	Wide World Photo

Library of Congress Cataloging-in-Publication Data

Van Dorn, William G.
Oceanography and seamanship / William G. Van Dorn : drawings by Richard Van Dorn. — 2nd ed.
p. cm.
Includes bibliographical references and index.
ISBN 0-87033-434-4 :
1. Oceanography. 2. Seamanship. I. Title.
GC28.V36 1992
551.46—dc20 92-39150
CIP

Manufactured in the United States of America
First edition, 1974; second edition, 1993

To my lifelong friend, Robert S. Dorris, NA, Member, SNAME, USYRU, for sharing his encyclopedic nautical knowledge, interest, and humor.

Contents

Preface to the Second Edition

The first edition of this book (Dodd, Mead, 1974) was something of an experiment, in that it attempted to apply the new science of oceanography to traditional seamanship. The measure of its success I take to be its surprising breadth of readership. Although slanted primarily towards cruising yachtsmen, it has also found wide acceptance among commercial fishing and charter boat operators. It has become recommended reading at the maritime academies, has been distributed by American President lines to its officers and senior crew members, and has come into frequent use as a textbook in oceanography in colleges and secondary schools.

I am therefore extremely gratified that Cornell Maritime Press has elected to publish a second edition, giving me the opportunity to respond to many constructive comments received over the years, and to upgrade the work where there have been significant advances in expertise and technology. In particular, the latter half of the book has undergone extensive revision, with addition of new sections on the 1979 Fastnet storm and its influence on racing rules and yacht design; yacht construction and performance; planing powerboats and open-ocean racing; ship routing for storm avoidance; and coast search-and-rescue procedures. Lastly, I have added a section on cold-water survival, a subject that has received relatively little attention in the seagoing community.

Preface to the First Edition

This is a book about boats, ships, and the sea looked at from the vantage of a third of a century as a research oceanographer and of nearly twice that long going to sea in everything from two barrels lashed together to an aircraft carrier. It attempts to explain in practical terms what the ocean is really like, why boats behave as they do, and what the average skipper can do to make his voyages safer and more comfortable.

Most of what you will find in these pages is not covered in any other single volume. Much of it is new to the popular press. Some of it is my own invention. The principal excuse for its existence is the observation that most nautical books leave many questions about the sea unanswered, and those on oceanography rarely mention boats. It seemed to me that there was a need for a blending of the two.

Anyone who knows much about either subject will recognize this as an ambitious—perhaps audacious—undertaking. Oceanography is a rapidly growing science, whereas seamanship has a long tradition that tends to resist change. Clearly, I can expect critics on both sides, risking being outdated in the former and too novel in the latter. This I have tried to minimize by having each chapter reviewed by one or more authorities. Happily, their comments have been relatively minor.

Even so, the principal hazard I have encountered is that of sounding apostolic. Thus, the reader is advised not to accept everything here as gospel, but rather as a fresh slant of wind to those long becalmed in nautical doldrums.

PART I

Origin and Nature of the Oceans

1

History of the Oceans

THE BEGINNING AND THE END

Once in a high and far-off time when the world was new, there were no oceans. There were no lakes, no streams, no waterfalls, no mighty rivers. There was no ice, no rain, no snow. There was no water! There were no mountains, no valleys, no rocks, no sand—all of the surface of the earth was uniformly covered with unfathomably deep, impalpable dust, whirled in endless windrows and giant dunes by hot, poisonous winds of carbon dioxide and ammonia. There was no air!

This Kiplingesque description of the early earth is not necessarily "Just So," but probably represents the current best estimate of the state of things at a time when the solar system, as we know it, was nearing the end of its formative era about four and a half billion years ago. Studies of lunar rocks brought back during the Apollo flights indicate that the moon is about equally old. It probably looked much the same at that time, except that its low gravity was unable to retain an atmosphere long. However, it was then much closer to the earth and orbiting more rapidly, and the earth itself revolved in seventeen hours.

While it is quite beyond the scope of this chapter to further elaborate the possible origins of the solar system, we can state that the planets are considered to have accumulated out of cold dust and gases. The presence on the earth and moon of elements heavier than iron, which the sun is incapable of making within its internal nuclear furnace, suggests a possible earlier involvement of the sun with another, heavier star, perhaps a binary twin. This could have resulted in a stellar explosion (supernova), in which these heavy elements were manufactured, leaving the sun and its peripheral dust cloud as surviving remnants of the cataclysm.

For perhaps half a billion years after the earth agglomerated, things may have remained very much the same, while radioactive heat from decaying atomic nuclei such as uranium, potassium, and thorium slowly raised the earth's internal temperature toward the melting point. This heating process, abetted by initial heat of compression as dust accumulated, brought about great changes in the structure and composition of the earth, if not the moon as well. As the various minerals approached their respective softening and melting points, the heavier elements that were present in abundance, such as iron and nickel, sank slowly to the center of the earth, and the lighter ones, such as silicon, aluminum, magnesium, and potassium, tended to rise toward the surface.

Thus the earth gradually developed an onionlike, multilayer structure (fig. 1), consisting of a central heavy metallic core, the outer half of which behaves like a liquid to earthquake waves passing through it; a mantle of heavy minerals, hot

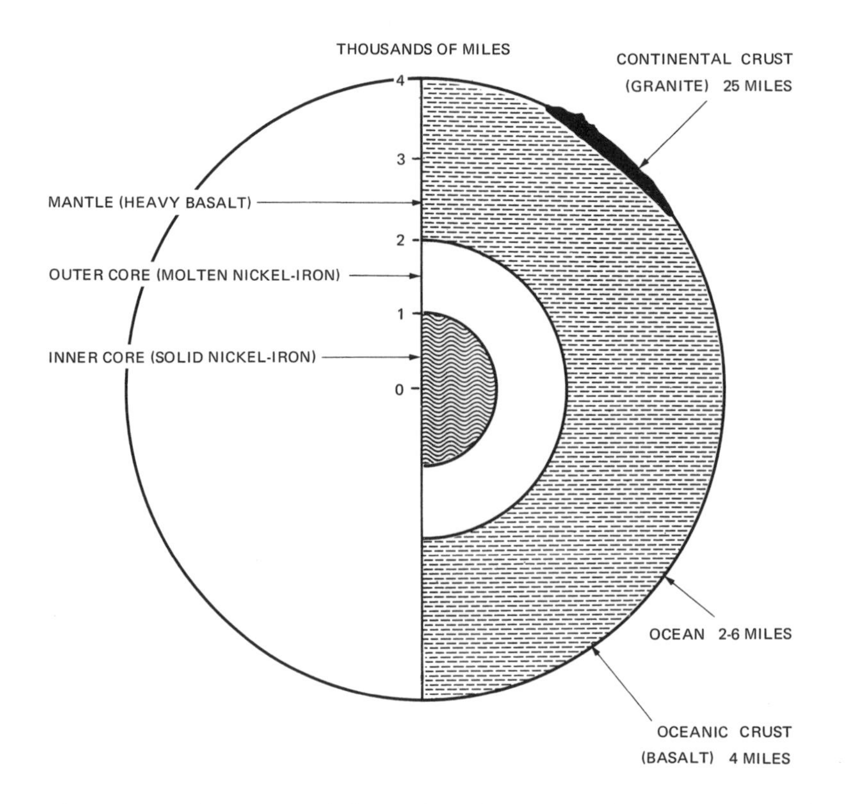

Fig. 1. Internal structure of the earth. Radioactive and/or compressional heating and melting have caused separation of constituents according to weight—iron sinking to the core, and water rising to the surface.

enough to flow sluggishly over very long periods of time; and a thin outer crust of light minerals, cool enough to behave rigidly, even under continued stress. Lastly—and most vital of all for life as we know it—the water that forms the present oceans, together with the bulk of the salts they now contain, was squeezed out of the minerals themselves by heat and pressure, and escaped to the surface. It is not clear whether our earth was ever so hot at the surface that all of this primordial water was held in the atmosphere as steam, or whether it uniformly covered the featureless earth—an endless ocean, with no continents, islands, or shoals to interrupt the march of waves and tides across it.

Born, as it were, of the radioactive fires within it, the present verdant earth may someday return to its original appearance by quite an opposite process. Gradually these fires will die down as the available radioactive materials decay. The processes of continental formation and renewal, which depend for energy upon internal heat, will also cease, and the continents themselves will be eroded away and deposited as sediments in the sea. The sea will rise and further erode the continents, until they have again disappeared below the depth of wave action.

But the sun, which for nearly five billion years has been steadily and evenly burning up its store of hydrogen fuel, will, after another five billion years or so, slowly blossom to great size and much greater heat output, following which it will shrink back to much smaller size, and cease to give out heat and light altogether. During its expanded phase, the earth's atmosphere will burn off, and the oceans will boil away, further eroding the continents as their levels are lowered, since the process of erosion is rapid compared with the time required for cosmological changes. Eventually, therefore, the earth may again become an airless, featureless sphere, covered this time by coarser sand and dust, and by salt left by the retreating oceans.

THE AGE OF THE OCEANS

A surprising thing about the oceans is that they are the oldest features on earth! This fact has only recently come to light as a result of extensive drilling into the deep sea floor, which has been revealed to be comparatively much younger than the continental rocks, which, in turn, were originally laid down as sediments in the primordial seas. We have no way of absolutely dating the age of the earth, but moon rocks and meteorites are all about 4.5 billion years old, and studies of the sun's mass and rate of burning suggest it is about the same age. But the oldest (PreCambrian) rocks presently exposed on the earth's surface are only 3.8 billion years old, whereas the oldest sea floor so far discovered dates back a mere 200 million years. This was about the time when dinosaurs ruled the tropical swamps and when much of the present land area was covered by shallow seas. Since the earth cannot be older than the sun, and since the earliest rocks were laid down as sediments on the primordial oceans, the latter are estimated to be about 4 billion years old, give or take a few hundred million years.

How do we know that all of the water was squeezed out at once? Is any of it still coming out now? These are intriguing questions, and they cannot be answered with certainty. But I know a scientist who has spent his whole career looking for "young" water—which can easily be distinguished from water that has previously been exposed at the earth's surface for any length of time. He has looked in deep wells, geysers, volcanoes—anywhere he could think of where water might be coming to the surface for the first time—but without ever finding anything younger than water freshly dipped up out of the sea.

SEA-FLOOR SPREADING AND CONTINENTAL DRIFT

These strange-sounding titles are simply different names for the same process, whereby the ocean floor is perpetually rejuvenated by replacement of surface material with lava extruded from beneath the rigid crustal layer. This surface rejuvenation is the only visible part of a very sluggish vertical circulation of hot plastic material that involves the upper part of the mantle, to a depth of four hundred miles or so. The power, of course, is supplied by radioactive heating in the upper mantle, where the material is hot enough to melt spontaneously, except for the pressure of the crustal rocks above it. The entire process somewhat resembles the cellular circulation observable when heating a very viscous liquid, such as chocolate or honey, just before it crystallizes.

Because 71 percent of the earth is covered by water, and also because the continents seem to float around as unitized blocks on the surface of the mantle, the true nature of this circulation has only become apparent through studies of the character of the sea floor. However, continental drift was suspected as early as 1910 by the German geologist Alfred Wegener, who observed that certain continental coastlines bore a remarkable resemblance to one another, even though they were in some cases several thousands of miles apart. He proposed that all of the present continents were at one time incorporated into a single large continent that later broke into pieces and drifted apart. This idea was widely ridiculed by geologists of Wegener's time. Only the recent discovery of the great midoceanic ridge systems, and evidence that new sea floor is steadily being created by hot mantle material welling up through cracks along the ridges, have vindicated his general concept of continental drift. Wegener's ideas have since been much improved upon by scientists in England and America, who have used computers to obtain a "best fit" between various continental segments. They are still somewhat divided in opinion as to whether the present continents evolved from one giant primordial continent or from two or more smaller pieces. However, because the sea floor is very young relative to the continents, and because the latter are being continually buffeted around, colliding and splitting up, the original state of affairs may now have been lost beyond identity.

While the continents are of intimate importance to us—and to all the other animals and plants who inhabit them—they are probably wholly fortuitous accidents, incidental to the general process of sea-floor rejuvenation, which proceeds somewhat as follows (fig. 2):

The earth's rigid crust, averaging four miles thick, is currently divided into about six large pieces, or *plates,* that essentially float on the hot mantle, and define the edges of the cellular circulation going on underneath. At some of the plate edges lava is oozing up, forcing the plates apart; elsewhere the mantle material is sinking, carrying one or the other of the adjacent plate edges with it. The rising zones are called *divergences,* or *spreading centers.* Here the lava cools and becomes a part of the separating plates. Along the sinking zones (*convergences*), the plate edges are consumed, either by the action of one plate edge plunging beneath the other and disappearing into the hot mantle—a process called *subduction*—or by buckling or crumpling of the colliding edges. In some places both processes (subduction and crumpling) are going on simultaneously.

It is not a matter of coincidence that most of the spreading centers are in midocean—for the ocean basins themselves are created by the spreading of oceanic

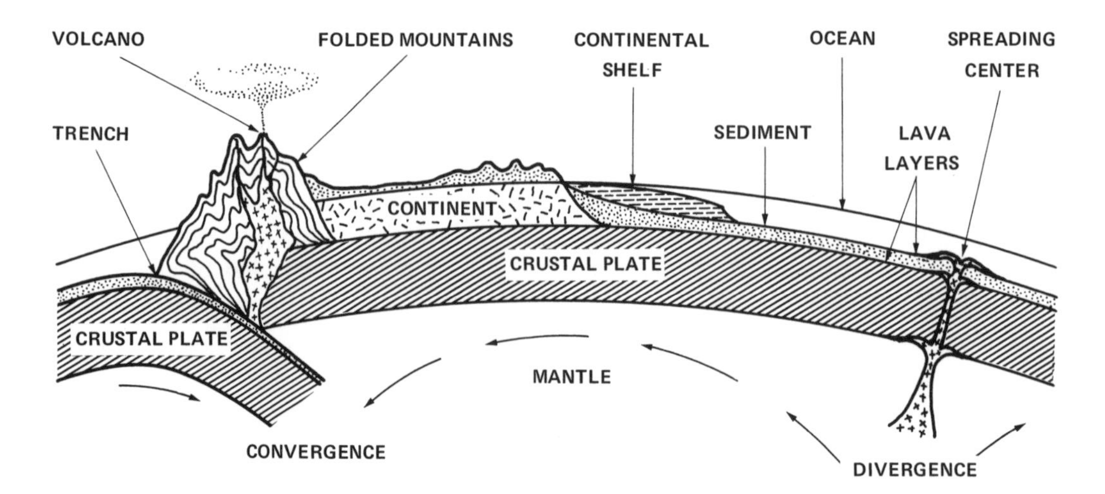

Fig. 2. Schematic section illustrating sea-floor spreading and continental drift. Crustal plates are being formed at the spreading center and consumed at the convergence. Continents are debris that collect at convergences.

plates. How, then, are the continents formed? Scientists are not universally agreed on this, but an educated guess is that they simply represent the stacking and shuffling of light debris that has accumulated at the convergences throughout geologic time. You can demonstrate the mechanism to yourself quite easily by observing the motion of sawdust sprinkled on the surface of water in a shallow pie pan, while it is being heated from beneath by a pair of candles near two opposite edges. Very soon a vertical circulation will be established with a line of convergence across the pan midway between the candles. The sawdust will be carried to—and concentrated along—this line, leaving the rest of the surface completely clear.

This convergent buckling of the plate edges is a very violent process, and explains how the high, folded mountains and their deep roots are created. Where one plate plunges under another, a great trench is formed, usually accompanied by the growth of chains of volcanoes, as lava escapes upward along the zone of weakness. Most of the world's great earthquakes occur along such convergences, of which Chile, Peru, the Aleutian Islands, and Japan are typical examples.

One must not suppose from the foregoing that the pattern of cellular circulation long remains the same, for it is continually changing. Oceans are created and destroyed, as are the plates, but the continents are usually preserved. They move from place to place, steadily accumulating light material through plate-buckling and volcanic action, and temporarily losing some of it through erosion and weathering. Thus, by and large, the continents have continued to grow since the first volcano poked its smoky head above the primordial sea, and they will continue to do so until the earth runs out of internal heat and the plates cease to move.

THE EARTH'S MAGNETIC FIELD AND ITS CHANGES

Ever since Marco Polo brought back samples of "north-seeking" iron ore from China,* the magnetic compass has been one of the principal aids to navigation. It was soon realized by navigators and mapmakers that the earth's magnetic poles do not coincide with its poles of rotation, and furthermore, that the lines of equal magnetic intensity tend to change from year to year. Since Napoleon's time, thorough study of the magnetic field and its variations has revealed that the total field can be roughly divided between two separate effects: a permanent *(dipole)* field of variable strength (equivalent to a giant bar magnet thrust off-center through the earth and tilted about 15° to its axis of rotation), and a secondary field, having quite an irregular pattern, that rotates slowly westward at about 0.18° of longitude per year (fig. 3). Unfortunately, the compass cannot distinguish between these effects. Thus, it is necessary to indicate the direction and change of the local magnetic field on charts, and to revise them periodically.

The primary field is presently presumed to arise from differential rates of rotation between the earth's molten iron-nickel core and the solid (albeit plastic) mantle. The core is a good electrical conductor. Its rotation—in a weak residual field (presumed to exist initially within the mantle)—is thought to induce a self-excited dynamo action that generates the observed field. Again, the energy to drive the dynamo is attributed to thermal differences between the hot core and the cooler

*The Chinese had long before used crude "compasses" of loadstone, suspended on a thread.

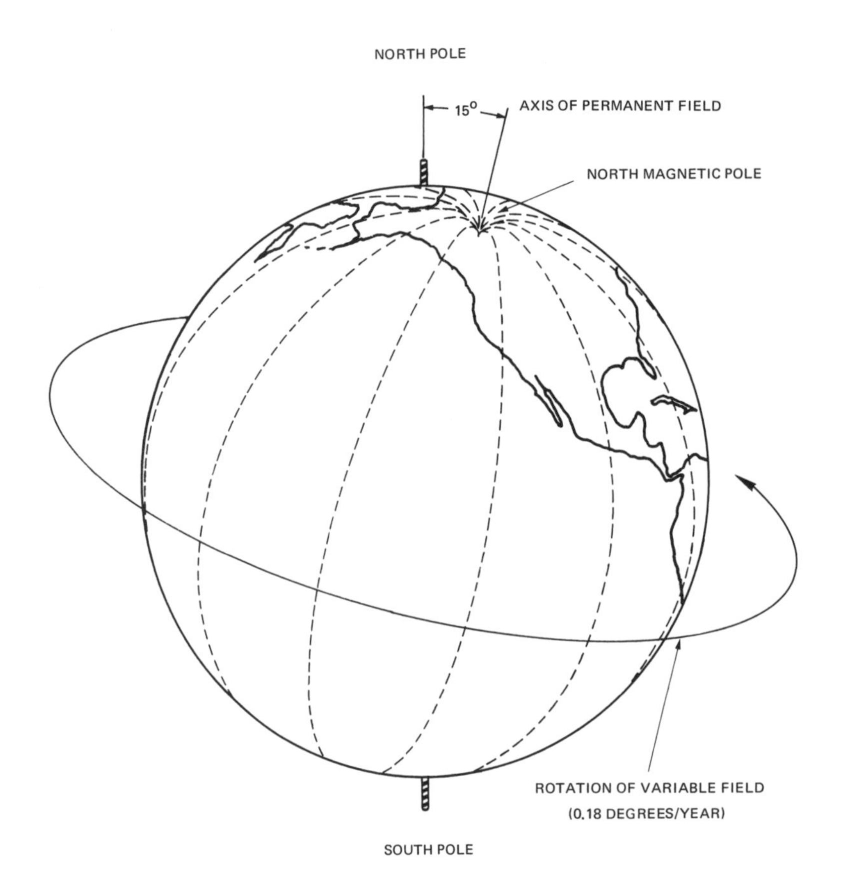

Fig. 3. The earth's magnetic field is composed of a permanent field, tilted at 15° from the pole of rotation through a meridian at 101° west longitude, and a variable field, which rotates westerly at 0.18° per year.

mantle. The cause of the secondary field is still a mystery, for nothing else we know of within the earth rotates at such a slow and uniform rate.

A ship's compass, of course, attempts to align itself with the external lines of force of the magnetic field, which arise vertically from the south magnetic pole, are essentially horizontal at the magnetic equator, and plunge back inside the earth at the north magnetic pole. As a result, the compass magnets attempt more and more to tip (dip) the compass card away from the horizontal as either pole is approached. This tendency is resisted by the small net buoyancy of the card itself, and can be further compensated for a given locality by the addition of small weights. However, there are large regions of the oceans relatively remote from the poles where the local dip angle differs from the geometric latitude by as much as 30–40 degrees. Thus, if one has his compass compensated for the northerly dip near, say, San Francisco, and attempts to sail to New Zealand, he will find his compass standing on its ear, if it is not immobilized by rubbing against its housing. Even so famous a navigator as Sir Francis Chichester found himself in trouble in this regard when sailing west from New Zealand on his globe-girdling solo voyage.

Studies of the local orientation of magnetic particles frozen into old lava flows, both on land and beneath the sea, have revealed a systematic change in direction with the age of the flows. By correlating such information from all over the world, *paleomagnetic* maps of these changes have been drawn that suggest that the magnetic poles of the earth have systematically shifted in position—relative to the poles of rotation—for the past 130 million years. However, the recent strong evidence for continental drift and sea-floor spreading has thrown the paleomagnetic picture into some confusion, since the two effects clearly must be separated if either is to be understood. The latest studies suggest that the apparent wandering of the magnetic poles can be explained solely on the basis of continental drift, the interpretation of which depends not only on magnetic data, but also upon geometric and geological similarities between the several continents.

A final—and still unexplained—peculiarity of the earth's dipole field is that it has sporadically reversed itself several dozen times within the past 70 million years, which is as far back as good paleomagnetic records go. These reversals, in which the north pole becomes the south pole for a while, and vice versa, persist from several thousand to several million years. The last reversal was about 800,000 years ago, which is a much longer time than any other reversal within the last 9 million years. The dipole field strength has been decreasing rapidly within the past 500 years, and we may all have to change the symbols on our compass cards some time in the near future—or there may be a few hundred years in which there will be no external field at all! This would be unhappy for humanity in general. The external field acts as a trap for high-energy charged particles that always are bombarding the earth. Without a magnetic trap, they would plunge straight through the atmosphere, disrupting radio communications and creating a host of other electrical disturbances.

ATMOSPHERIC EVOLUTION AND THE ORIGIN OF LIFE

It is hard to believe that our beautiful earth, with its white, fleecy clouds, its clear blue sky, and its great populations of animals and plants, could only have come about through the most complex chain of circumstances. All animal life—save for a few special kinds of bacteria—requires an oxygen atmosphere for survival, and could not have evolved without the plants, which make oxygen out of carbon dioxide. The plants, in turn, could not have originated in the world we know, but probably first appeared as single cells in a warm, slimy, brown ocean under a dark, gloomy atmosphere of water vapor, methane (marsh gas), and ammonia, with traces of carbon dioxide, nitrogen, and oxygen. If you have forgotten your high school chemistry, the present atmosphere consists of about one-fifth oxygen and four-fifths nitrogen, so you can see that things have changed quite drastically since life first appeared on earth about 3.8 billion years ago.*

The chain of events which brought about the eventual appearance of man, measles, and mosquitoes went something like a classic experiment: some years ago Stanley Miller, a graduate student at the University of Chicago, put all of the raw elements known to be ingredients of living tissue together in a test tube filled with ordinary seawater. He then exposed this mixture to ultraviolet light and to spark discharges (simulated lightning). He found that these inert chemicals had combined to form many of the very complicated compounds found in living tissue. An important feature of these experiments is that such compounds did not form if there was

*Perhaps surprisingly, the earliest protein molecules have been found associated with the earliest rocks, suggesting that life in the ocean may have predated continental formation.

free oxygen present! This is because these same elements reacted immediately with the oxygen to form other nonreactive, stable compounds that are *not* found in living tissue. Moreover, these oxidized compounds were mostly insoluble, and settled to the bottom. These and many other experiments have led us to suppose that, as the primordial oceans were squeezed out of the hot rocks of the earth's crust, they brought along with them soluble forms of life-essential elements, such as carbon, nitrogen, sulfur, phosphorus, and even iron. Many of the early lifelike compounds can be classed as "volatile," in that they could be carried up into the early atmosphere by evaporation, along with the water. There they were acted upon by sunlight and lightning discharges to form new, more complex compounds, which were carried back to the sea dissolved in raindrops. This cycle was repeated many times, until the seas were full of brown, smelly, complicated organic compounds, some of which were the harbingers of life.

Meanwhile, the water in the atmosphere was itself being dissociated into its components, oxygen and hydrogen, by ultraviolet radiation from the sun. The oxygen was rapidly removed by combining with the surface rocks to form silicates and carbonates, and with carbon to form carbon dioxide. It also combined with iron and sulfur in the sea to form insoluble ferrous sulfate, which settled out to become the present great iron-ore deposits. The hydrogen remained behind in the atmosphere and, being the lightest of all gases, it slowly escaped from the earth.

Thus the atmosphere slowly changed. The ammonia was dissolved in the sea and incorporated into the proteinlike compounds, or was dissociated into free hydrogen and nitrogen in the high atmosphere. Carbon dioxide accumulated, and began to be utilized by the newly born plants to release more oxygen until, after another billion years or so, most of the free hydrogen had disappeared from the earth, and conditions were propitious for the advent of the air-breathing animals.

The first animals, of course, developed in the sea, and could scarcely be distinguished from plants. Even today, some phytoplankton algae, which are minute, free-floating organisms that populate the sea in vast numbers, carry on both photosynthesis of carbon dioxide and respiration of oxygenlike plants. But they also possess movable cilia that propel them through the water. Little by little, over the past 3.8 billion years, the long cycle of evolution, mutation, and selection has brought about the great proliferation of life forms we know today—and the countless others that prospered in past eras, but ultimately failed to survive.

Despite their long history, the oceans and atmosphere are in still-tenuous and unstable equilibrium. Scientists are now seriously concerned that, through massive pollution of his environment, man himself, like the lowly bacterium, may produce toxins that will destroy the species or otherwise upset the delicate balance in an unpredictable manner (see Part II).

2

The Oceans Today

THE SEA FLOOR

If today we could strip away the oceans and lay bare the sea floor, we would find very little resemblance to the familiar world above sea level. Yet originally they both sprang from the same basic materials, and have merely experienced different histories. Sea-floor rock is always basalt lava no matter where you dredge it. If you then grind it and squeeze it, heat it and freeze it, soak it with fresh water and salt, expose it to sun, air, wind, and rain, and do this over and over for a few millions or billions of years, you will end up with continental granitic material. Granite is a good deal lighter than basalt, and thus resists being dragged down into the mantle along with the basaltic crustal plates that form the sea floor.

In an evolutionary sense, the sea floor is a region of quiet deposition—whether by lavas oozing up from beneath, or by sediment sifting softly down from above—whereas the continents are regions of violent uplift, folding, weathering, and erosion. Once formed, after oozing quietly forth, a submarine lava flow undergoes little change throughout its life span. It drifts slowly at the rate of 1 to 6 inches per year toward its eventual demise, as it plunges back into the mantle a few tens of millions of years later. Meanwhile, on its slow trip across the ocean, it accumulates a few hundred yards of sediment, partly silt or dust washed down from the continents, but mostly the shells of countless little organisms that live in the thin upper layer of the sea, where light can penetrate. Since the rate of wearing down of the land is some three hundred times faster than the sedimentation rate in the sea, most of the land erosion simply acts to wear down the mountains and fill in the valleys of the continents. Very little material finds its way to the deep sea by way of the rivers, or by the blowing of dust and sand by offshore desert winds. Most of the river deposits remain close to shore to form the thick sediments on the continental shelves.

But we must not suppose, because of its quiet and orderly formation, that the sea floor is a featureless desert. On the contrary, much of it appears to have a more rugged profile relief than the land (plate 1). Countless volcanoes speckle the submarine landscape—as isolated cones, small clusters, or great arcuate chains, such as the Aleutians or the Hawaiian Islands. Only a small fraction of these rise as high as present sea level, but those that do must rank with the Himalayas in splendor. Mauna Kea and Mauna Loa in Hawaii may be the highest mountains on earth, both having a total relief in excess of 30,000 feet.

Most impressive of all, if they could be seen in their entirety, are the great midocean ridges, which form a semicontinuous volcanic mountain range connect-

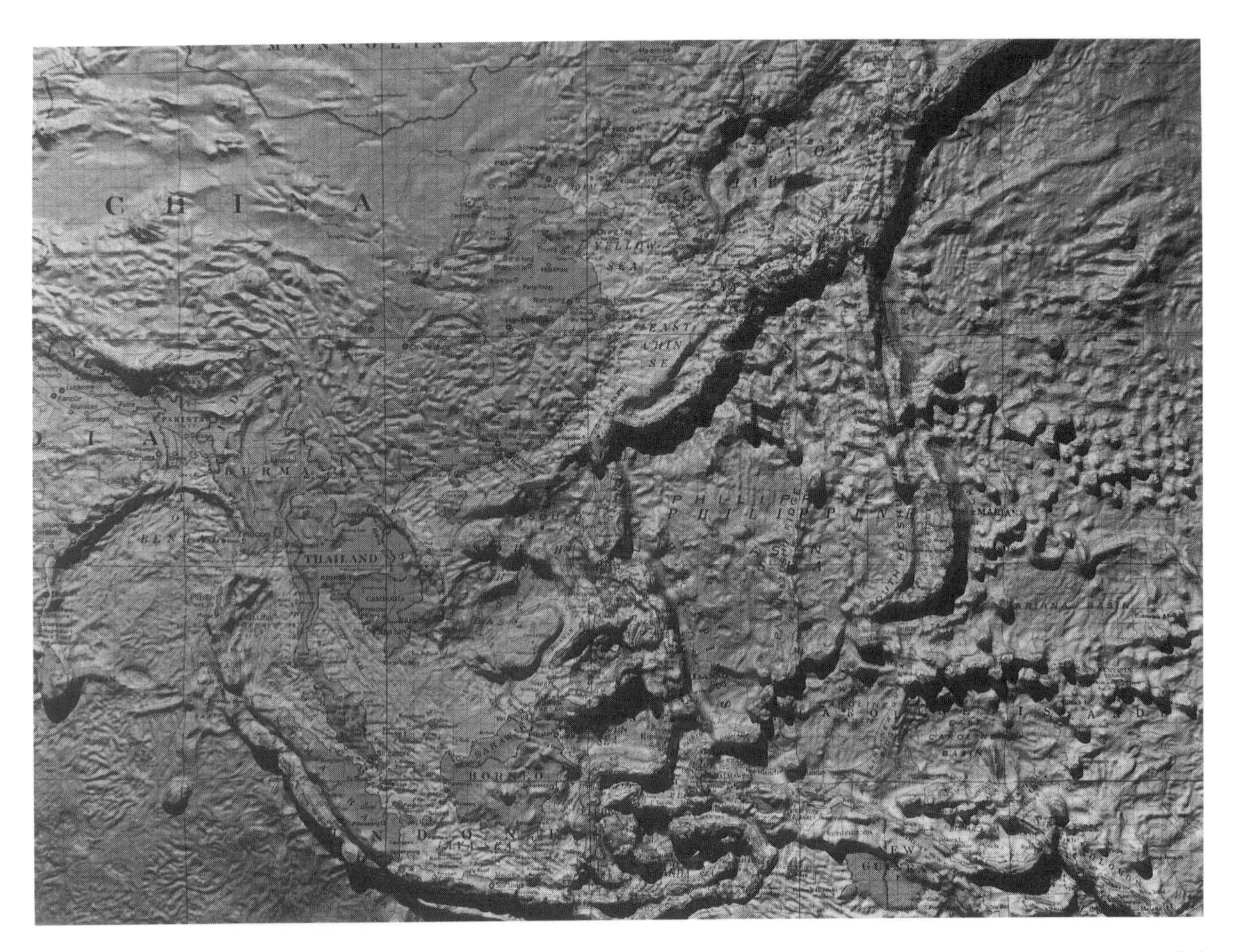

Plate 1. The dramatic contrast between the sea-floor relief in the western Pacific (*lower right*) and that of the Himalayas (*left center*) is accentuated by sidelighting a scaled topographic relief map. Note especially the great chasm of the Japan Trench (*top right*), and the cluster of volcanic atolls and seamounts.

ing all of the major ocean basins. Various sectors of this ridge system define the temporal spreading zones for the oceanic crustal plates. Although the plates appear to move as unified structures, they are continually jostling and rubbing against their neighbors, and are subject to twisting motions that produce numerous transverse cracks and fissures across the ridges, as well as local offsets that give them the appearance of being disjointed. Some of these transverse fractures are over 1,000 miles long and exhibit ridge offsets of as much as 150 miles. The resulting jigsaw puzzle is only now being resolved by patching together magnetic maps of the sea floor that clearly identify when certain discrete—but now discontinuous—lava flows were laid down.

Although the plate-forming ridge system seems to bifurcate all ocean basins, the great trenches, which mark the zones of plate destruction, are predominantly confined to the basin margins—mostly ringing the Pacific Ocean. The Atlantic has only one trench of any consequence, just north of Puerto Rico; and the Indian Ocean has two—the Java Trench, bordering the East Indies, and the Ob Trench, southwest of Australia. The depths of these trenches are, on the average, more than twice the mean depths of their respective oceans, as shown in the accompanying table.

Ocean	Mean Depth (ft)	Deepest Trench (ft)
Atlantic	12,900	28,600
Indian	13,000	23,400
Pacific	14,000	35,800*

*The Marianas Trench, south of Guam, where J. Piccard and D. Walsh set the world's deep-diving record in the bathyscaphe "Trieste," January 1960.

While the continents of Africa and America are drifting apart, recent observations indicate that the Pacific is shrinking! Perhaps in another 150 million years it will be possible to swim from San Francisco to Hong Kong across a narrow strait, just as Byron swam the Hellespont, which is all that seems to remain of the ancient Sea of Tethys that once separated Europe and Russia from Arabia and India.

THE EVOLUTION OF PRESENT SHORELINES

The geologic record, at least as far back as the beginning of the Cambrian epoch (600 million years ago), indicates that the mean level of the oceans all over the world has fluctuated up and down many times by as much as 700 feet. These fluctuations were brought about by slow climatic changes, extending over thousands of years, which altered the delicate balance between the amount of water in the oceans and that present as ice on the continents—principally in Antarctica. Ice floating in the ocean, such as that forming the Arctic ice pack, does not affect the level of the ocean.

It is now fairly well accepted that these glacial and interglacial epochs are driven by three interrelated astronomical factors: a 23,000-year wobble (*precession*) of the earth's spin axis; a 41,000-year variation in the tilt of the spin axis with respect to the earth's orbital plane; and a 100,000-year change in the eccentricity of the earth's elliptic orbit around the sun. All of these factors affect the amount of solar radiation reaching high-latitude land areas, which changes by as much as 20 percent when they act in concert or oppose one another. This change is enough to alter the balance between the advance and the retreat of polar ice. Interestingly, these epochs are not equal; it takes the ice about 90,000 years to accumulate, but only about 10,000 years to melt!

At the present time we are in the warming phase of an interglacial epoch, and the sea level stands about 550 feet above the lowest known previous level. Melting of all the remaining landlocked ice would raise the sea level by another 150 feet. Although about 18 percent of the present continental land masses are currently under water, such an increase would submerge an additional 10 percent of the land area. My home in Del Mar, near LaJolla, is situated on a hillside 180 feet above the Pacific Ocean, and thus my descendants can look forward with some assurance to having beachfront property, if they wait long enough. However, in these circumstances, much of the southeastern United States and northeastern Siberia would be submerged, and the price of the remaining real estate correspondingly elevated. Seven of the world's ten largest cities would have to be abandoned. Such calamities pose a relatively new problem for mankind, since the last major sea-level rise occurred about 100,000 years ago.

Those continental areas beneath present sea level are known as the continental shelves. They are distinguishable from the deep-sea floor in that they possess granitic basement rocks and are generally shallow, and descend rather steeply from the 100-fathom contour to depths of 1,000 fathoms or more. The shelves have breadths that vary from zero to nearly 1,000 miles, and are often covered by thick sediments carried down by rivers. In many regions sediments have been ponded behind rock dams produced by lava flows, by buckling of the sea floor, or by coral reefs growing on offshore banks. Most shelf deposition is believed to have occurred during the glacial epochs when great sheets of ice, thousands of feet thick, advanced toward the equator from the polar regions as far as the 40th parallels of latitude in both hemispheres. Not only did the glaciers level everything in their paths and carry large boulders—locked in icebergs—far out to sea, but the tropic zones were markedly cooled, and abundant rainfall occurred in present dry areas, such as the Sahara Desert. Pine trees flourished at sea level in southern California, and hardwood forests covered the continental shelf 150 miles east of New Jersey.

Thus the present shorelines appear to be ephemeral features that are continually changing. Present surveys suggest that over 70 percent of the sediments now occupying the continental shelves were laid down since the last ice age. There is good evidence that within the past five thousand years several ancient cities along the eastern Mediterranean have succumbed to the relentless rise in sea level, as the glaciers receded.

Of more immediate interest to mariners and yachtsmen are the topographical changes in shorelines that have occurred very recently—perhaps since the most recent survey from which a local chart was plotted. The sea is endlessly at war with the land, and substantial changes in the character of a coastline can take place within a few years, through the natural process of erosion; within months or weeks, through storm action; or almost instantly, as in the case of the great Alaskan earthquake of March 1964. Here, a 400-mile sector of the continental shelf of the Gulf of Alaska (as large as the state of Florida) was bodily uplifted from 6 to 50 feet within a few minutes.*

Except for the relatively slow process of coastal erosion by waves and storms, most significant changes are catastrophic. That is, little change may take place for many years, until an unusually heavy surf topples an undermined reef structure, an unusually heavy rain or early thaw swells rivers till they burst their banks and carry millions of tons of mud and sand down to the sea, or a hurricane piles water against the land until waves sweep over areas far above the normal range of surf action. While damaged areas on land are easily detected, it may be months before undersea changes can be resurveyed and rebuoyed and the appropriate notices to mariners are posted. Chart upgrading is still slower—even in the United States—and there are many areas of the world where the first report of a new nautical hazard may come from the skipper of the boat who ran afoul of it and lived to tell the tale.

The *shoreline* proper, of course, is not a line but an *intertidal zone:* the interface between the sea and the shore that includes the maximum combined range of normal wind, wave, and tidal action (fig. 4). This zone varies vertically from as little as 18 inches at Papeete, Tahiti, where the spring tide range is only 8 inches, to as much as 40–50 feet in certain semirestricted waters, such as the English Channel, Nova Scotia, and southern Alaska. In many tropical areas where the tide and surf ranges are relatively small, the largest allowance must be made for occasional

*See "Tsunamis," Part V.

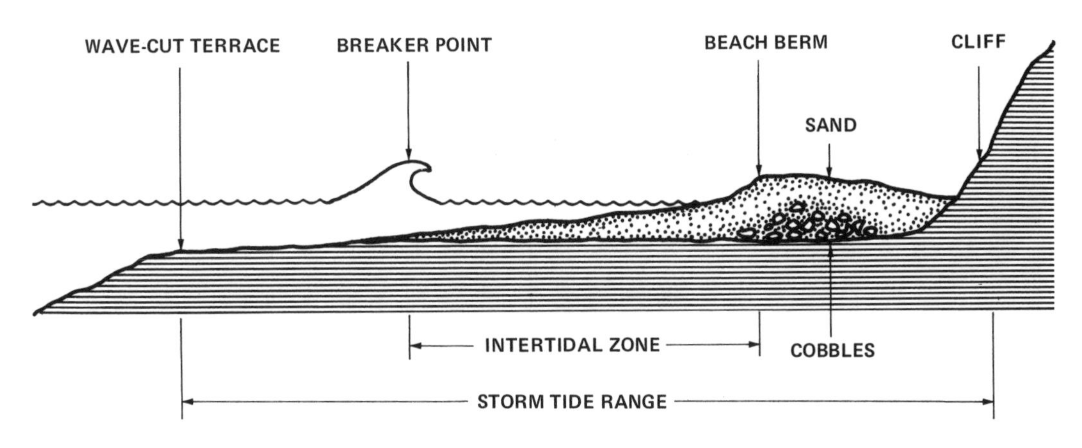

Fig. 4. Typical shoreline topography along an immature coastline. A beach can develop only after waves have cut a terrace from an existing sea cliff.

hurricanes, which can cause abnormal wind tides in excess of 20 feet. Similarly, depending upon the offshore slope, the width of the intertidal zone may vary from zero to many miles. The portion above water usually consists of a rocky cliff or a beach of some sort. While cliffs represent more-or-less durable impediments to wave action, being slowly eroded at rates varying from 1 to 100 feet per century, beaches are dynamic regimes that constantly change character, depending on prevailing sea conditions.

To most romanticists, *beach* connotes a broad strip of dazzling white sand, girdled with overhanging palms. But there are other kinds of beaches of less appeal. A beach might better be defined as any shoreline accumulation of unconsolidated debris, usually consisting of the products of local cliff erosion or of materials brought down from the hinterland by streams. Beach materials run the gamut from giant boulders to the finest silt, and often consist of interbedded layers of cobbles, well sorted according to size, with the smallest (most mobile) material at the top. This arrangement is not an accident, but part of the selective nature of wave action.

If the waves had their way about it—with no interference from the upthrusting or downwarping of the coastline—all the beaches of the world would be perfectly straight, and slope imperceptibly seaward, so that the waves would break very far out and die away completely before reaching the shore. This is what is called a *mature* shoreline, and represents the ultimate in beach stability. Although the waters may sweep inland many miles during severe storms, they tend later to drain off through local stream outlets, leaving the beaches relatively unaffected. Mature shorelines are characteristic of the southeastern Atlantic coastline of the United States and most of the Gulf of Mexico. In these areas, one can walk half a mile to sea without going over his knees, but he cannot get hurricane insurance unless he lives more than 50 miles inland.

In contrast, the western coasts of the Americas are zones of active mountain building, and thus have few mature beaches. Although the waves are steadily chopping away at the headlands—upon which their energy tends to be concentrated by the process of refraction (see Part IV)—they have not yet succeeded in their eventual task of straightening out the coastline and wearing away the cliffs. Here the products of cliff erosion are accumulated in successive *concavities* (pocket beaches) along the shoreline. Such beaches are ephemeral and precarious, and the sand can be rapidly swept away during a period of high waves, leaving the cobbles exposed. It is really astonishing how rapidly a major increase in wave intensity can cut back a stable, gently sloping beach. I have seen storm surf lower the sand level by 15 feet overnight off mile-long beaches in La Jolla. The sand is temporarily deposited in shallow water within the surf zone, which itself widens to several hundred yards during high-surf conditions. Here, wave action generally forms one or more bars that tend to coincide with the mean positions of the outer breaker lines at high and low tide. During such transitions, numerous random ridges and depressions develop within the inshore region under the combined action of breakers and the swift currents that are produced by breaker transport into the surf zone. Within a few hours, a previously safe beach becomes an extremely dangerous place for the inexpert swimmer.

Although the sand tends to be returned to shore during calmer periods, the ultimate action of the waves is to cast it above the tide limit during storms—where it can later be carried inland by prevailing winds—or to move it along the shore until it encounters a submerged channel or submarine canyon, down which it flows to depths immune to wave action. Thus immature beaches are continually being destroyed and renewed by waves, at the gradual expense of the cliffs behind them, and it is only a question of time before the sea will win the battle—but not for a few hundred million years!

The beautiful coral beaches of the tropics fall into the category of immature in the extreme. Coral sand is formed not by wave erosion of sea cliffs, but by attrition of coral rubble cast up during large storms, together with shells of marine organisms. Some beaches include a substantial contribution of ready-made coral sand produced by the slow fragmentation and ingestion of living coral by myriads of reef-feeding fishes, from whose vents issue a steady stream of sand. As with all others, the stability of tropical beaches is a delicate balance between sand supply and loss, dictated by small changes in wave intensity and direction, and by the local topography. The sand deposited below the depth of wave action (about 75 feet) sifts down to deeper water and is lost. The shallower fraction, being considerably lighter than the silica sand of granitic coastlines, is cast up on the shore by waves, where it is rapidly leached away to silt by fresh water from rains or by abrasion in the surf zone, and it must be renewed at a much more rapid rate than silica sand if a beach is to remain stable. But nothing can speed up the appetites of the reef fishes or the growth of marine organisms, and—as the people of Hawaii are sadly discovering—once sand is removed from a lovely beach, as sand was taken from Lumahae to replenish the improperly designed and wholly artificial beach at Waikiki, it may not be naturally renewed for centuries.

ISLANDS, CORAL REEFS, AND ATOLLS

Except for granitic outcrops on the continental shelves, almost all of the 5,000-odd oceanic islands are volcanoes—many of them still active. A singular exception is the Seychelles, north, northeast of Madagascar in the Indian Ocean, which are granitic and appear to represent a little chunk of continent that somehow got separated from Africa—as did Madagascar itself—during the continental migrations of the past 130 million years. Additionally, deep-sea echo soundings indicate that there are upwards of 10,000 submarine volcanoes that have never made it to the surface of the world's oceans, but which have a profile relief exceeding 3,000 feet above the sea floor.

Lastly, there are some two hundred volcanic islands that once extended above sea level, but which appear to have been eroded flat by wave action, and later submerged to various depths as great as 1,000 fathoms. This limit is far below the greatest regression attributable to previous ice ages, and is presumed to be due to the sinking of the sea floor. The depths of submersion of these flat-topped *guyots,* as they are called, do not seem to have any consistent distribution that can be simply linked to sea-floor spreading, but the deepest ones appear to be the oldest. Perhaps the Pacific deep-sea drilling program, now in progress, will provide answers to this puzzling problem. All of these islands, guyots, and sea mounts seem to be geologically very recent, that is, less than about fifty to sixty million years old, which is consistent with the premise that the entire sea floor is renewed after about twice this length of time.

A glance at your world globe will reveal two startling facts about the distribution of oceanic islands. First, the overwhelming majority are in the Pacific Ocean. There are about two hundred islands in the entire Atlantic, and perhaps twice that number in the Indian Ocean—half of them concentrated in the dense cluster of islets comprising the Laccadive and Maldive groups southwest of India—but the remainder are all in the Pacific. The Philippine archipelago alone is said to include about three thousand islands. Second, most of the Pacific islands are crowded into its southwestern half. In fact, the wedge-shaped sector of the Pacific between Santiago (Chile), Shanghai, and Djakarta probably contains 80 percent of the world's oceanic islands (fig. 5). Since a great circle from Santiago to Guam passes through the heart of this sector, it seems almost incredible that Magellan could have sailed the more than 8,000 miles between these points sighting only two uninhabited islands. Surely he must rank among the most unfortunate navigators of all time; he lost half his ships and men from thirst and

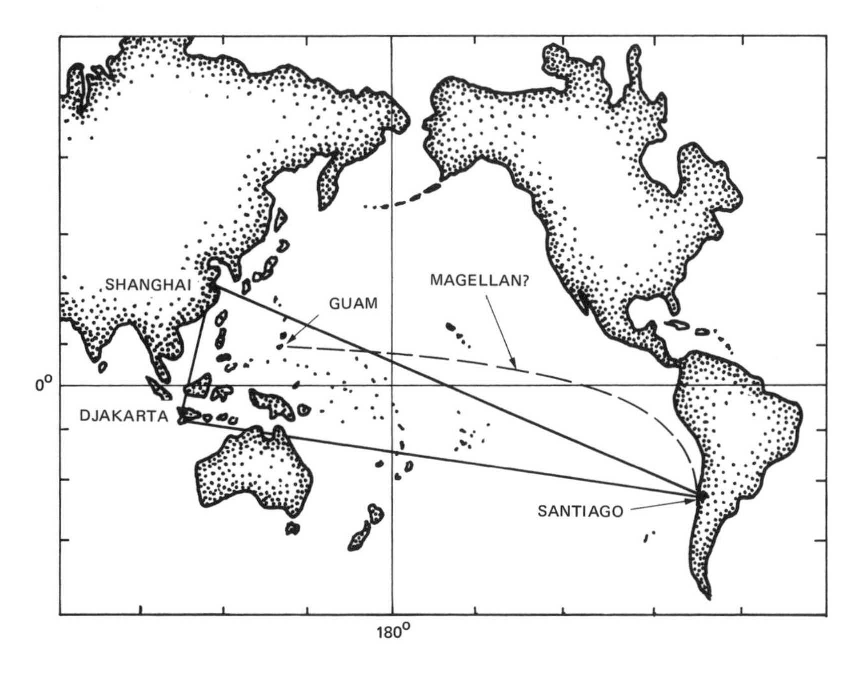

Fig. 5. Eighty percent of the world's oceanic islands are contained within a triangular sector with apexes at Santiago, Shanghai, and Djakarta. Dashed line shows probable route of Magellan.

malnutrition en route, and soon after was killed in the Philippines. Only 31 of his original complement of 240 men lived to return to Spain again.

Oceanic volcanoes are currently observable in every stage of growth, and thus the mechanism is fairly well understood. Molten lava is extruded from linear rifts in the sea floor, usually without much disturbance, because of the great pressure of water above. As the lava congeals in the cold water, it tends to block the rift passages, which may open elsewhere. Thus a long volcanic ridge is gradually built up by the concurrent or intermittent activity of a chain of vents, until one or more of these connected cones reaches the surface. Depending on the rate of lava production, activity continues quietly, as in the case of the island of Hawaii, until a great *shield* volcano is formed by successive thin layers of lava issuing over thousands of years; or explosively, as with the great eruption of Krakatoa, in the Sunda Strait, which blew its top in 1883 and almost completely disappeared below sea level. This tremendous explosion was heard 3,000 miles away. It produced a tidal wave that drowned over thirty thousand people, and filled the atmosphere with so much dust that skies were reddened around the world for several months thereafter. Explosive volcanoes appear to produce more cinders and ash than lava after reaching the surface, and they are eroded rapidly by wave action. Some new arrivals, such as Surtsee (near Iceland) and Myojin (off Japan), spend many years thundering and hissing as they laboriously thrust skyward, only to succumb between each burst of activity to the inexorable waves attacking their flanks. The island of Bora Bora (Society Islands) is a classic example of an extinct shield volcano that has been eroded almost to sea level, except for a great lava plug that formed the core of its central vent. The plug now stands alone—a 2,400-foot column of glistening basalt rock, overhanging slightly on all sides.

Irrespective of generic type, volcanic islands often occur in long, arcuate chains, such as the Hawaiian Archipelago, which extends from Hawaii 1,400 miles northwesterly to Kure Island. Through analysis of bottom heat flow measurements, this intermittent production of volcanoes has now been identified with movement of thin (2-mile) oceanic crustal plates over stationary "hot spots" in the earth's mantle, from which molten magma makes its way through fissures to the sea floor. Just what turns the spigot on and off is unknown, but another great shield volcano has risen to only 3,000 feet beneath the surface some 300 miles southeast of Hawaii, which remains the only active volcano in the Hawaiian chain. It will break the surface in another 1,000 years or so, and may help relieve the real estate problem in that overcrowded tourist paradise.

Among the most enchanting and beautiful features of cruising in tropical waters are the endless variety and color of coral formations that compete for space on the sea floor wherever they can obtain solid footing. In great contrast to the waving fields of underwater vegetation that cover the rocks of cold and temperate shorelines, the corals are rigid and immobile, although some forms exhibit great delicacy and fragility of structure. This contrast is as great as that between a tropical rain forest and a cactus garden, but is mitigated in the coral seas by the myriad brilliantly colored fishes in constant motion, gliding and whirling through the labyrinthian passages between the coral colonies. Although many coral structures look like plants, the resemblance is purely superficial. They are in fact nothing more than inanimate limestone tenements fabricated of calcium and magnesium carbonate extracted from seawater by minute marine animals called *polyps,* each of which temporarily resides in its self-made apartment, waving an array of stinging tentacles from its tiny window in hopes of capturing smaller animals or plants as they float by. Except for the relatively rare red and black Gorgonian "tree corals," most of the color of living coral is a property of the organism that inhabits it. Washed and dried in the sun, it bleaches to dazzling white, which later turns dark gray under protracted exposure to sun and rain. Differences in the shapes and structures of corals are due to genetic differences between the kinds of polyps that form the individual colonies; but the polyps in a given colony appear to be identical, and the key to the architectural plan that directs the form of the final structure—which may incorporate the labors of millions of individuals—remains one of nature's great mysteries.

A coral reef comprises a heterogeneous array of neighboring coral colonies, which may crowd each other to the point of extinction—or even build one atop another. Somewhat like humans in a big city, if things become too congested, separate colonies of a given type can be established by free-swimming or drifting larvae cast off by adult polyps, which later settle to the bottom and set up shop where rents are cheaper. The ultimate size of a coral reef is governed by the minimum temperature at which the polyps can prosper and multiply and the availability of solid substrata for attachment. The first limit seems to be a thermal minimum of about 72°F, and thus live tropical corals are not found much below 100 fathoms near the equator, nor much above 30° of latitude at the surface. This range of latitudes is considerably extended along the westerly margins of the oceans by poleward-flowing warm currents, and contracted along their easterly margins by cold currents flowing toward the equator. Suitable substrata within these temperature limits are provided by rock outcrops on the continental shelves, and by the volcanic islands rising from the deep sea.

Certainly the largest continuous living coral formation in the world today is the Great Barrier Reef, which extends for 1,100 miles along the continental margin of eastern Australia. The shallow waters of Indonesia are also noted for their beautiful reefs, which cover thousands of square miles. However, most of the south coasts of India and Asia are severely limited in coral growth because of the great amounts of silt brought down by rivers that discharge into the South China Sea and the Bay of Bengal. The same circumstances prevail all along the north coast of South America. In fact, the only region of substantial coral reef development in the Atlantic bounds the Florida peninsula and the Bahama banks offshore. Despite lurid vacationland advertisements, these waters cannot compare with the profusion of fantastic shapes and colors of the coral seas of the western Pacific.

But even where coral reefs grow in abundance, they are subject to occasional widespread devastation of the tiny polyps that produce them. It was first noticed in 1963 that large areas of Australia's Great Barrier Reef were being invaded by a rapid proliferation of the "crown-of-thorns" starfish (*Acanthaster planci*), an 18-inch echinoderm which lives principally by devouring coral polyps. Advancing at a rate of 4–6 feet a day, millions of these creatures reduced hundreds of square miles of living reef to snow-white fossil by 1970, when they first reached Hawaii. Since then, the infestation has considerably abated, having affected no more than 1 or 2 percent of all Pacific reefs.

Acanthaster is unknown in the Atlantic, but large reef areas in the Bahamas have succumbed to abnormally warm water since about 1986. Whether this temperature increase is transient or represents a long-term trend associated with the atmospheric "greenhouse effect" remains unclear (see Part II).

Atolls seem to represent a late stage of a complex chain of events in the life of volcanic islands that once rose to substantial elevations above the sea surface. Here again, clear examples exist today of atolls in all stages of development and demise. Their life cycles are intimately related to the elevation and subsidence of large areas of the sea floor, to the rate of growth and extinction of volcanoes, and to the glacial fluctuations of sea level imposed by the ice ages. All of these events have different time scales, to which the rate of growth of bulk coral reef structures must be precisely matched in order that an atoll can exist in its precarious equilibrium with present sea level.

Several conceptual stages in the development of an atoll are shown in figure 6. Stage 1 shows a mature volcano that has become extinct, with its apex several thousand feet above sea level; where wave action has eroded a terrace around it at a depth of about 5 fathoms. In stage 2, the wave terrace supports an active coral reef grown up to the low-tide level, and further wave erosion is prevented by the exceedingly hard coating of Lithothamnion algae that progressively encrusts the coral formation in the intertidal zone. This stage is well represented by the island of Oahu, although occasional lava flows and subsurface seepage of fresh water still inhibit coral growth around its periphery. In stage 3, the sea floor has subsided by a thousand feet or so at a rate commensurate with the upward growth of coral—which cannot grow actively below 50 fathoms. Because of fresh-water runoff from streams and accumulated silt, coral has grown faster around the outer margins of the fringing reef, while the inner region has lagged behind, thus forming a shallow lagoon, with numerous passes through the reef margin at places where streams come down from the island interior. This stage might be typified by Tahiti or Tahaa in the Society Islands. After another long period of subsidence—coupled with massive erosion of the central volcano by rain and weathering—stage 4 is reached. The volcano has virtually disappeared, except for vestigial peaks here and there surrounded by a wide and deep lagoon, as well represented by Truk, in the Caroline Islands. In the final stage of atoll formation, as exemplified by Bikini or Eniwetok, in the Marshall Islands, the volcanic basement has subsided to several thousand feet below sea level, and is surmounted by an equivalent column of coral, having exterior slopes as steep as 60° near the surface, and capped by a more-or-less continuous ring of coral about a thousand yards in width. The reef encloses a lagoon some 30–40 fathoms deep, from which numerous coralline columns (coral heads) rise almost vertically—some extending nearly to the surface.

Here and there along the reef rim—particularly in its windward sector—small sandy islets rise 10 or 12 feet above the high-tide berm and support a thin vegetation of hardy, salt-tolerant plants, perhaps augmented by a denser growth of imported species, such as pandanus and cocoa palms. These islets are the result of centuries of accumulation of reef debris torn loose by storm waves and thrown up on the reef flat. Some of this material is carried across and deposited within the lagoon, with the eventual result that it becomes completely filled (Christmas Island, Line Islands), or contains only a remnant depression occupied by a brackish pond (Swains Island, Samoa Group).

Certain characteristic features of atolls and coral islands are of special significance to mariners (fig. 7). First, a geologically recent, glacial sea-level lowering about

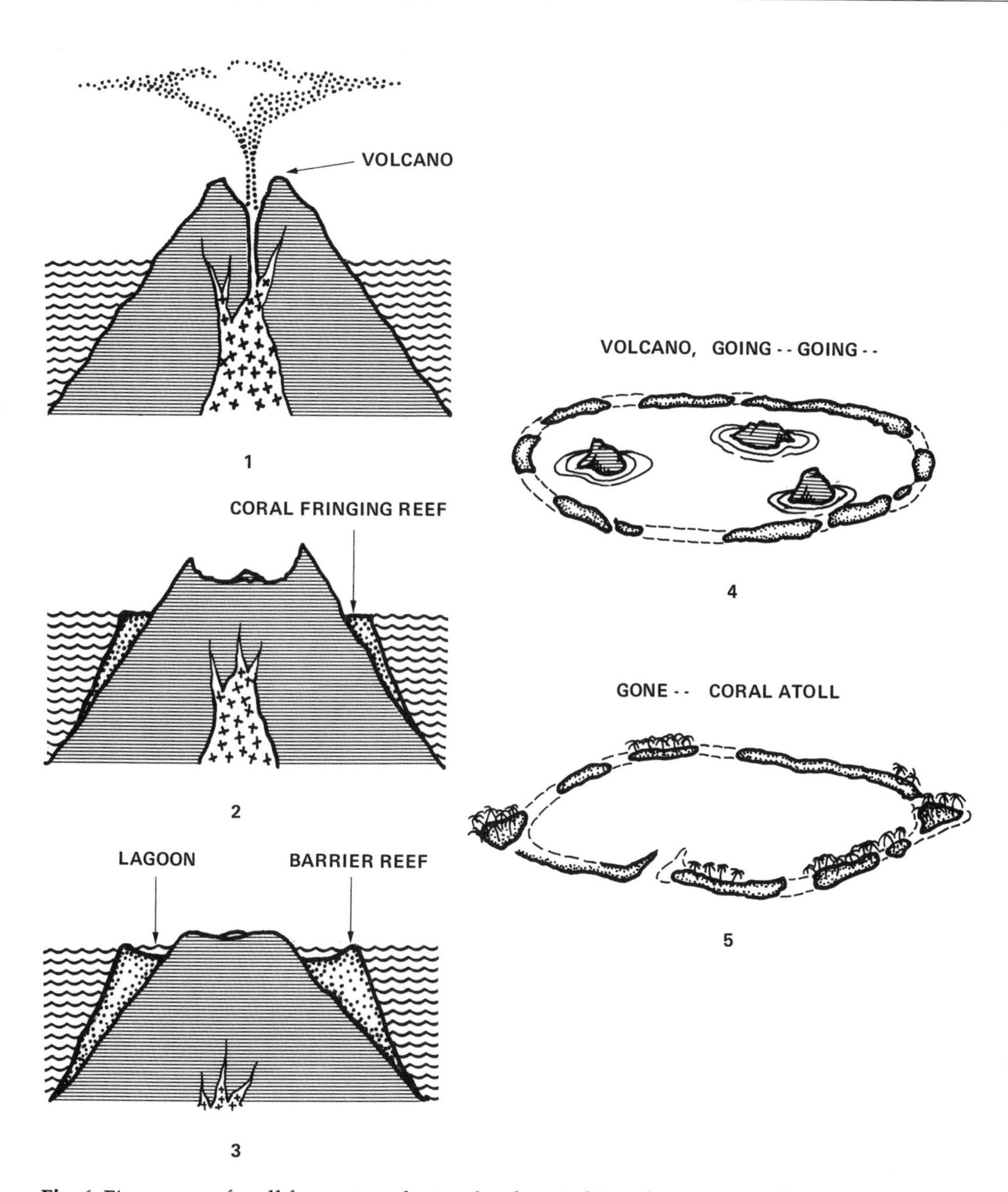

Fig. 6. Five stages of atoll formation, during the slow sinking of an extinct volcano. Final stage (5) may persist where basalt basement has sunk several thousand feet below sea level.

ten thousand years ago has resulted in a worldwide prevalence of wave-cut terraces that are now uniformly at about 10 fathoms deep. Considering that most coral islands are very steep-to, these benches provide convenient anchoring for small vessels in otherwise impossible circumstances. Such anchorages must be used with due caution, because of swift currents flowing over them during spring tides; ample swinging room must be allowed for the influence of vortex shedding in the island lee (see Part III).

The 10-fathom terrace typically slopes gently shoreward and terminates at depths of 4–6 fathoms in an abrupt rise to a flat table reef at about the elevation of

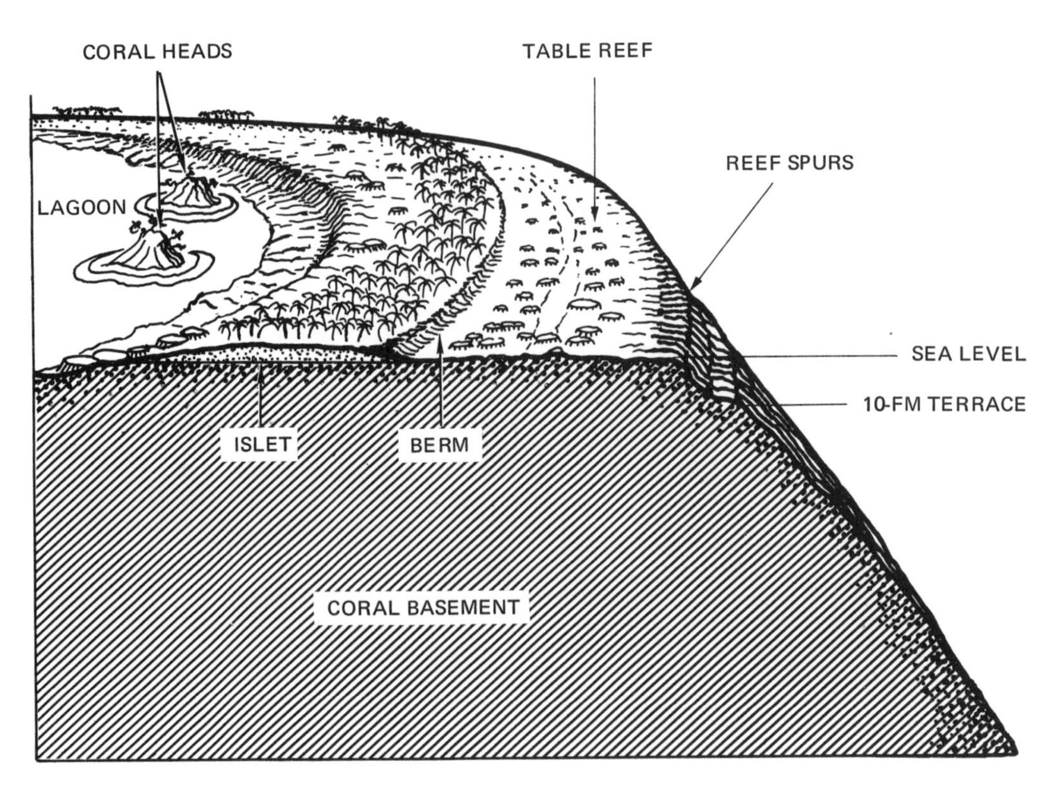

Fig. 7. Profile of an atoll, showing characteristic features.

mean low water. The seaward edge of the table reef is notched like the fingers of a hand into ridges and channels about 30–100 feet long and 10–20 feet wide, but the reef proper may extend shoreward without ridge or depression for as much as 500 yards to the coarse rubble berm that constitutes the actual high-tide shoreline. Under moderate sea conditions, such as prevail on the lee sides of atolls in spring and summer, one can walk to the edge of the table reef and jump directly into deep water when the tide is low. At high tide, when several feet of water cover the table reef, a small boat may be run right up to the rubble berm and held there for unloading. We have landed and recovered many tons of equipment in this fashion, but I recommend the use of the larger sizes of rubber boats rather than those of wood or plastic under severe surf conditions.

Where an atoll has few navigable passes, they are nearly always located in the westerly quadrant—that is, in the lee of the prevailing trade winds. I know of no more convincing explanation for this phenomenon than the observation that high-island passes usually correspond to river runoff zones. In an atoll, subject to 20-knot mean winds, a 1-inch rain within twenty-four hours can pile up 60 feet of fresh water at the leeward end by depression of the salt water under the stress of the prevailing wind. These conditions occur frequently enough in the tropics to similarly inhibit active coral growth in westerly passes.

Lastly, the bulk growth rate of coral is rapid enough to render passes unnavigable within a few years in areas where artificial channels have been dredged. Under normal circumstances, around the periphery of an atoll, growth is to some extent retarded by wave attrition, and may be limited to 1 inch per year or less. However, in protected locations with adequate circulation, I have seen cables, hoses, and supporting steel structures that had been installed six years previously become completely covered by a coarse, porous assemblage of interlacing coral fronds to a depth of several feet. Similarly, the ship channels at Johnston and Midway atolls must be blasted clean with high explosives about every ten years to keep down incipient coral heads. The channel into Palmyra, dredged to 40 feet during World War II, is now unnavigable to vessels exceeding 10-foot draft. Thus, charts are not to be trusted unless very recently upgraded. There is every advantage in utilizing the advice of native pilots, who will often come out to meet you in advance if they see you sailing into danger.

DISTRIBUTION OF SALT, TEMPERATURE, AND ICE

I have often heard visitors to La Jolla remark, "The Pacific Ocean seems so much saltier than the Atlantic. I float higher in it, and it really stings my eyes!" Not so. The oceans are so thoroughly mixed that they differ by less than 1 percent in salt content from ocean to ocean, top to bottom, and pole to pole. Seawater contains about 3.5 percent salt by weight, which is about seven times as much as the human system can tolerate for daily consumption. Only where large rivers empty into the sea is the water noticeably less salty, and a permanent feature of such coastlines as those of Venezuela, eastern Australia, and southern Indonesia is a thin layer of muddy, brackish water covering the sea surface even well beyond sight of land. Indeed, fresh water has been reported over 300 miles at sea off the Amazon.

There are also a few restricted regions, such as the Red Sea and the Persian Gulf, where low precipitation and high evaporation combine to make the water much saltier (7 percent) than the oceanic average. Because salt makes the water heavier, abnormally salty water tends to sink to the bottom without mixing, and forms a dense layer that fills the bottom of these seas, where it is prevented from escaping by shallower sills across their mouths. The all-ocean record for salt concentration (16 percent) is found in two small basins at the lowest parts of the Red Sea.

The distribution of temperature in the sea is not so uniform as that for salt, but over 90 percent of the volume of all oceans remains perpetually between 30° and 40°F. In the tropics there is a well-defined warm-surface layer about 1,000 feet thick, owing to greater absorption of solar heat at low latitudes. This layer becomes thinner in all oceans with increasing distance from the equator, and disappears altogether at about 45° latitude, above which the temperature drops by only a few degrees from top to bottom. Near the north pole and around the Antarctic continent the water is uniformly near the freezing point (30°F) at all depths. Because warm water is lighter than cold water, it is perhaps surprising that the warmest water of all (132°F) is also the saltiest, again at the bottom of the Red Sea. This is because the water is heated by geothermal action from the bottom, but is so salty that it still will not rise to the surface.

But despite the apparent uniformity of subsurface conditions, all oceans have permanent and well-defined density structures that are characterized by several layers of slightly different density, between which there is very little mixing. These density differences are almost too small to be measured directly. They arise from unique combinations of temperature and salinity that are created by special environmental conditions—usually at the sea surface.

Figure 8 shows a representative vertical slice through the Atlantic Ocean from pole to pole that illustrates its layered structure, and the source of special water making up each layer. The warm, salty water leaking from the Mediterranean Sea over the shallow sill at Gibraltar can be identified as a discrete thin layer even south of the equator. The densest oceanic water—the Antarctic bottom water—is produced in the Weddell Sea during Antarctic winter by the freezing of cold seawater. Ice thus formed is fresh, leaving the salt behind, which results in the saltiest, coldest—and thus the heaviest—of all oceanic waters; and this superdense water fills the deepest basins in all oceans—even though it is less than 0.1 percent denser than the warm surface water of the tropics!

The world's supply of ice is very unequally distributed: about 92 percent resides on the Antarctic continent, about 5 percent is on Greenland, and most of the balance is floating in the Arctic Ocean. This is not because the Antarctic climate is particularly colder, but because its mean elevation is so high. The Antarctic ice cap averages 10,000 feet in thickness, and that of Greenland is about 6,000 feet, whereas the Arctic ice pack is only some 10–12 feet thick, even in midwinter. The total amount of ice does not change significantly between summer and winter, although the Arctic pack diminishes by about 3 feet in thickness and 50–100 miles in diameter in summer.

The average annual precipitation in either the Arctic or the Antarctic is only a few feet of snow, corresponding to a few inches of rain, and thus both would qualify as arid regions in more temperate latitudes. The precipitation on Greenland and Antarctica is balanced by the slow flow and discharge of glaciers into the sea, at speeds varying from 50–1,000 yards per year, and borings indicate that the oldest ice now present in the Antarctic fell as snow about 200,000 years ago.

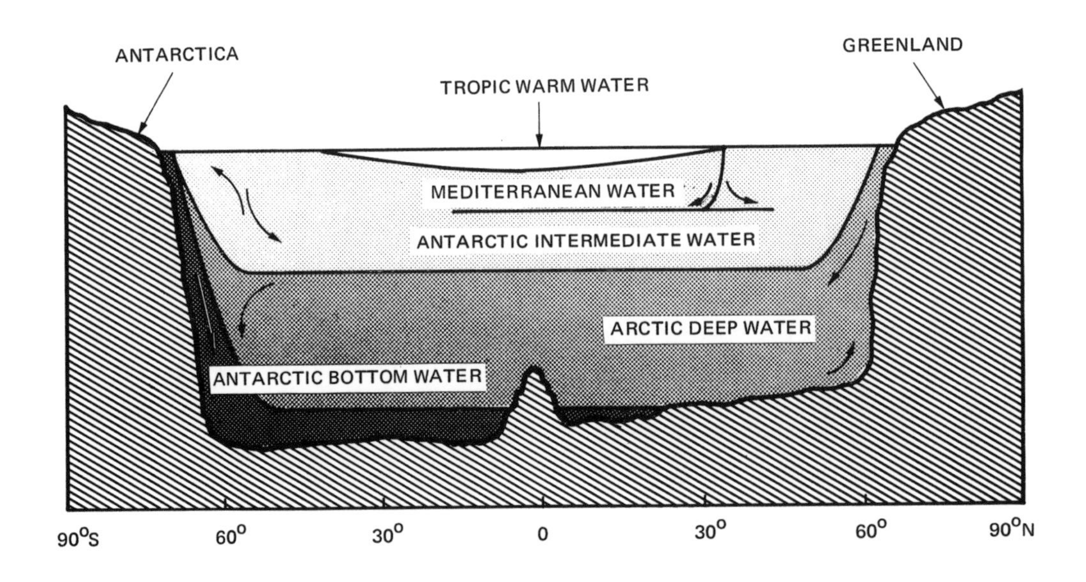

Fig. 8. Meridional cross section of the Atlantic, showing density structure and origin of different water masses. Note shallow lens of warm, salty Mediterranean water at mid-depth.

Although Antarctica discharges far more glacier ice in the form of large icebergs than does Greenland, the Antarctic icebergs are principally trapped in the subarctic circulation pattern, and remain far removed from shipping lanes. The Greenland bergs that are discharged into Baffin Bay to the west are carried far enough south by the Labrador current to pose a serious navigational hazard to transatlantic shipping (see Part III).

Until 1964, Antarctica had never been visited by a private yacht.* This interesting fact was told to me by the late Lee Quinn, who was famous for having sailed a 45-foot cutter over much of the Pacific with a female crew. Quinn learned of it in Tasmania, following the Sydney-Hobart race, whereupon he immediately sought to rectify this omission by sailing south, accompanied only by his first mate. After eight days of fine weather and favorable winds, he succeeded in reaching 60° south latitude, and had sighted the fringe of the Antarctic ice pack when a storm and accompanying seas arose of such magnitude that he was forced to remain at the helm for three days, while running west under bare poles. Finally, he fell asleep, was broached and capsized without severe damage, and eventually made his way back to Sydney without further misadventure.

*In 1970, Dr. Robert Griffith circumnavigated Antarctica in his 55-foot ferro-cement ketch, *Awahnee*.

3

Life in the Sea

DISTRIBUTION OF LIFE FORMS

It is convenient to subdivide sea life into three classifications, according to habitat, as *plankton, nekton,* or *benthos.* The plankton comprise all free-drifting species that are incapable of much locomotion; the nekton are species that swim well and are usually found in open water; and the benthos are bottom dwellers, including fish that habitually remain near or rest upon the bottom. This classification is loose, and many species (of both plants and animals) have larval or adult forms that change habitat during some stage of their evolution.

Almost all life in the sea that is of interest to mariners exists within the first 100 fathoms beneath the surface. This layer approximates the maximum depth to which light can penetrate, and includes most of the benthos, which are principally concentrated on the continental shelves and the shallow slopes of islands. Although many nektonic animals swim—and even live—at greater depths in the open sea, the majority remain near the surface, where the food is plentiful.

As on land, the broad base of the oceanic food chain rests upon the plant supply, which outweighs the animal population a thousand times. A tiny fraction of these plants (principally, the Brown Algae) resemble those on land, in that they grow upward from the sea floor and possess appendages similar to roots, branches, and leaves. Their "roots" serve no function, however, except to hold them in place, and the "leaves" are sometimes simply reproductive appendages. Of these, the common kelps are the largest, some species of which reach the sea surface from depths as great as 200 feet, being supported erect by gas-filled float bladders at the base of their fronds. Other algae resemble lichens, forming a living crust over rocks or dead coral. In very shallow water there are often abundant grasses, which are the closest oceanic relatives of flowering plants. Most of these landlike plants prefer cold water, and attain their greatest sizes and abundances where the water temperatures remain consistently below about 60°F. A notable exception is the *sargassum,* a family of tropical kelps that originated as bottom-growing, shallow-water species, but which have adapted to a floating life cycle. Because they drift freely about, under the influence of winds and currents, they tend to become concentrated in the quiet eddies of the general circulation system. So dense is the population of these plants in the tropical north Atlantic that it is a serious impediment to passage of smaller vessels. This region is shown on navigation charts as the Sargasso Sea, although sargassum is also found in the south Atlantic and Indian oceans in lesser concentrations.

The overwhelming majority of oceanic plants, however, have no terrestrial counterparts, but consist of a great variety of single cells, most of which are too small to be seen with the naked eye, and which are held in continuous suspension near the surface by water turbulence. Although these tiny plants, called *phytoplankton,* contain chlorophyll, and utilize light energy to convert dissolved carbon dioxide and water into cell tissues, they also contain animal protoplasm and respire some oxygen. The excess oxygen released by photosynthesis is redissolved in the water, where it is available for utilization by air-breathing animals.

Most of the phytoplankton also make solid skeletal structures from calcium or silicon that fall to the sea floor when they die, and in many areas these skeletal remains form the principal constituents of the sea-floor sediments. Knowing the thickness of sediments and their ages—as determined from examination of deep-sea cores—it is possible to calculate the rate of production of plant materials in the sea. This is found to be about the same as that on land, and since the latter is steadily decreasing owing to destructive farming practices and industrial pollution, we are increasingly dependent upon the phytoplankton, not only for absorbing the enormous amounts of carbon dioxide produced by burning carbon compounds, but also for the very air we breathe.

Small as they are, phytoplankton sometimes multiply so prodigiously that they render the water nearly opaque, coloring it dull green or reddish brown—hence the name *red tide*—and causing it to fluoresce brilliantly at night when disturbed by waves breaking or by fish swimming through it. Under extreme conditions, photosynthetic activity may be greatly reduced owing to competition for light and carbon dioxide, so that available oxygen is actually consumed faster than it is being produced, and fish may die in large numbers. Even in normal abundances some phytoplankton produce toxins that are concentrated by filter-feeding molluscs, such as clams and mussels, to the point where they are poisonous if eaten by humans. The old adage of avoiding mussels during months containing an "r" stems from the observation that mussel poisoning seems to occur more frequently in fall and winter. Actually, cold water appearing along the coast in these months often brings with it a resupply of nutrients that can trigger a plankton bloom almost overnight.

The phytoplankton, being nonswimmers, are at the mercy of the circulation to provide them with nutrients, carbon dioxide, and sunlight. Of these, the lack of nutrients, such as phosphorus, nitrogen, iron, sulfur, and potassium, is most apt to limit phytoplankton production in a given region, unless the nutrients are continuously resupplied by some accessory-stirring mechanism from the cold, deep, inexhaustible reservoir beneath the zone of light penetration. The warm, stable surface layer of the tropics inhibits vertical mixing in the open sea in latitudes where the layer is thicker than the effective depth of wave action (about 500 feet). Effective mixing occurs mainly in certain restricted regions where land masses interact with local currents to divert cold polar water toward the equator, or where wind blowing offshore can drive away the warm surface layer adjacent to the coast and promote "upwelling" of nutrient water from beneath.

Although there is no warm layer in high latitudes, light is severely limited during polar winter, and thus there is a strong seasonal fluctuation in phytoplankton production in polar regions. In figure 9, the areas of relative production of plant food for the world's oceans are outlined by different shadings, and it is

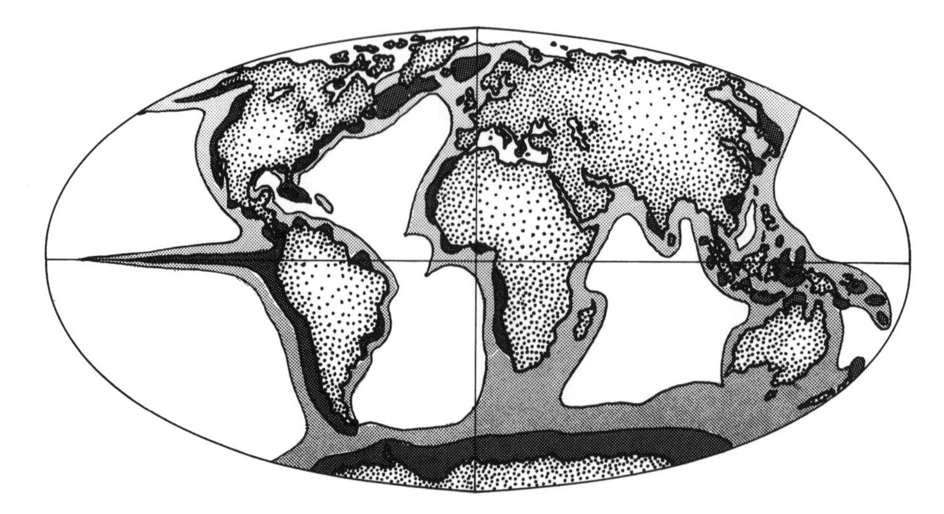

Fig. 9. Shading indicates density and oceanic distribution of phytoplankton production. Except for a tongue in the equatorial Pacific, the tropics are the "deserts" of the sea.

obvious that most production takes place near the continental margins, and increases toward the poles. The vast regions of the tropics, with their crystal-clear blue waters, are the deserts of the ocean, in which coral atolls represent little oases where nutrients can be concentrated and recycled to support a limited population of plants and animals.

Most of the animals in the sea are small zooplankton, and many are weak swimmers, barely capable of thrashing from one delectable phytoplankton morsel to the next. Some species spend the day at safe, dark depths of several hundred feet and migrate to the surface at night to feed. They, and their associated predators, frequently occur in such concentrations as to cause spurious bottom echoes on some fathometers, and have been given the generic name *deep scattering layer.* Tuna fishermen are always looking for such echoes, knowing that tuna have their own sonar detectors with which they seek out the anchovies which, in turn, are feeding on the zooplankton. The smallest and most numerous zooplankton are plant eaters—the grazers of the sea, corresponding in function to the herbivores on land. Other animals, including the larger zooplankton, eat the grazers, and so on. At the top of the ladder are the whales, which are attacked only by man—the ultimate predator.

The total number of individual animals of a given size in the sea varies roughly inversely with body weight; that is, the cube of size. There are about 1,000 phytoplankton in the sea for every zooplankton ten times as long, and about 10^{15} (one million billion) such plankton for every baleen whale thirty feet long. Nevertheless, the oceans are so large that most animals spend their lives in a perpetual search for food.

Gross phytoplankton distributions like the one shown in figure 9 can now be determined very rapidly by polar-orbiting satellites, which also report surface temperature. Because zooplankton prosper among phytoplankton, sardines and anchovies eat zooplankton, and because tuna and swordfish like to feed on the latter in clear water warmer than 70°F, fisheries specialists having real-time access to satellite data can now advise fishermen where to hang out for optimum prospects. Lacking such data, you can always fly to Cabo San Lucas and hire a Mexican fisherman to take you out in his ponga; he will know where the fish are anyway.

THE CYCLE OF LIFE IN THE SEA

Survival in the sea is rendered difficult because of the limitations of the water environment. Imagine, if you can, a world in which seeing is limited to 150 feet under exceptional conditions, and to only about 50 feet on the average; where hearing is acute, but the direction of a sound cannot be distinguished; where there is little color and everything is greenish gray a few feet below the surface. Under these circumstances animals have developed marvelously specialized organ adaptations to assist them in capturing prey—or escaping attack by other predators (fig. 10). The eyes of most fish are very large and sensitive, relative to their body weight, and are particularly adapted to sense motion, rather than color. Fish do not possess ears in the ordinary sense, but have arrayed along their sides a battery of pressure-sensitive cells with which they can detect and analyze the slight impulsive pressures generated by the swimming motions of their neighbors, thus enabling a school to move in almost perfect unison at lightning speed. The heads of sharks and rays contain special organs for sensing the weak electrical signals generated by muscle movements, which enable these scavengers to seek out a struggling victim in utter darkness. The baleen whale does not hunt at all, but simply cruises through the water with his mouth open, scooping up, separating out, and swallowing the several hundred pounds of planktonic animals required by his daily appetite. The sperm whale is capable of diving a mile beneath the surface for periods up to an hour, and somewhere in the stygian blackness of the depths he finds the giant squids that

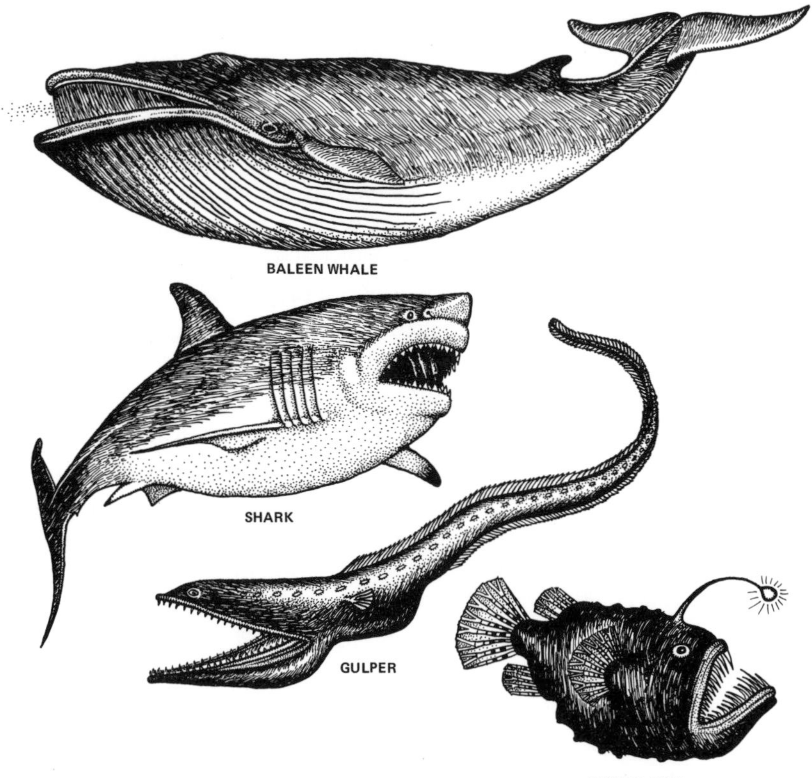

Fig. 10. Various predator adaptations for capturing food (not to scale).

constitute his principal diet. The starfish, after forcing the mussel to open by steady tension from his tentacled arms, turns his stomach inside-out, injects it into the aperture, and digests the animal within its own shell.

Each level of the sea has its own aggregation of predators and prey, with more and more fanciful adaptations as the depth increases and food becomes more scarce. At the very bottom of the deep sea live the families of marine benthic worms, tunneling endlessly through the soft ooze. Like detached segments of intestine, swallowing everything ahead and discharging it astern, they eke the last remnant of food energy from the organic silt endlessly sifting down from the unknown world above. And in the stygian darkness, along the midocean spreading centers, deep-diving submarines have discovered oases of warm water, issuing from volcanic rifts, in which live exotic lifeforms that are found nowhere else: bacteria that thrive on sulfur, brilliant red tube worms spreading like tulips from individual calcareous vases, and fragile, spiny crustaceans. One wonders how such oases, thousands of miles apart, can attract the same limited species.

As important to survival as food procurement in a hostile environment is the ability to reproduce in sufficient numbers to ensure perpetuation of the species. In contrast to the land, where the plurality of larger animals are mammalian, in the sea most of the animals are oviparous, and reproduce by casting off enormous numbers of eggs and spermatozoa to drift freely in the water, where they are widely dispersed by currents. Individuals of many species may annually produce several million eggs, of which only a few reach adulthood. Most bottom dwellers also cast off eggs that mature into larvae that swim about for a period before maturing into adult forms. At times the population of larvae and eggs exceeds that of the plankton itself, and it requires a trained eye to distinguish them. The common California lobster is believed to be represented in the plankton by between twelve and twenty separate larval stages during the first eight months of its existence, whereupon it metamorphoses into a transparent adult form that swims freely about for several years. Ultimately, its carapace grows to the point where it becomes heavy, and the animal settles down to bottom living. It may require an additional five or six years to reach "legal" size.

An interesting alternation of the lobster's reproductive cycle is that of the jellyfish *Obelia,* which casts its eggs into the water. The eggs hatch into larvae that settle to the bottom, each metamorphizing into a plantlike *hydroid,* from whose leaflike appendages buds develop into free-swimming adults again. The deep-sea angler fish *Edriolychnus* is probably the ultimate in piscean fidelity. The female has several dwarf males permanently growing from her belly. In a lightless world, where it is really difficult to find a mate, she is always certain to have a male around.

An important aspect of the balance of life in the sea is its stable equilibrium, limited only by resupply of nutrients. It is a closed cycle, in that most dead organic matter is returned to the system in usable form by the action of bacteria in the water and on the sea floor. However, despite assurances by some optimistic prognosticators, it is apparent that the sea is rapidly being depleted of food resources utilized by man. This depletion arises partly from indiscriminate overfishing of such animals as crabs, lobsters, shellfish, sardines, herring, tuna, and whales, but also by the introduction of pollutants and artificial obstacles to marine reproductive cycles. Unless international conservation agreements can be reached in the near future, the sea will not only fail to feed the starving millions of the world, but may degrade to the point where it cannot feed anyone!

MARINE LIFE OF INTEREST TO MARINERS

Edible and Inedible Species: Under duress—and with a few notable exceptions—most soft-muscle tissues of marine animals are edible. Even given wide choice, tastes vary considerably around the world. Sharks are a prized delicacy in Micronesia, although the fins are exported to China, where they command high prices for their alleged aphrodisiac properties. The Japanese, who are seafood connoisseurs, relish squid, octopus, and sea cucumber; and the Australian aborigines eagerly seek out the *holothuroidian* worms that inhabit the muddy bottoms of river estuaries—and they also like giant grasshoppers!

Just as there always seem to be one or two bad apples to the barrel, there are some things in the sea that may make you quite sick, if not dead. Aside from mussel or clam poisoning, which occurs indirectly when the flesh becomes toxic from feeding on the dinoflagellate plankton *Gonyalux catanella*, almost all seafood poisoning occurs from eating fish. Fish poisoning is not altogether understood, but apparently can arise directly, from metabolic production of toxins; indirectly, from ingestion of poisons that are only toxic to their predators; or secondarily, as a result of bacterial decay. According to a U.S. Navy–sponsored study, *pelagic* (deep-water) fish are universally edible, and 5–10 percent of near-shore species are occasionally poisonous. Since only about 2–3 percent of acute poisonings (requiring treatment) are fatal, the chances of being poisoned by eating from an utterly random sample are at most 3 per 1,000. With a little selection, and care in preparation, this probability can be greatly reduced.

Bacterial poisoning can be virtually eliminated by cooking and eating fish within an hour after catching, unless prompt deep refrigeration is available. Direct and indirect poisoning appear to be restricted to fish that live in warm, tropical waters. Unless you have sound advice to the contrary, you will do well to avoid the following, known to be occasionally toxic:

Trigger fish	Box fish
Butterfly fish	Goatfish
Porcupine fish	Trunkfish
Wrasses	Moray eels
Puffer fish	Parrot fish

or any other brightly colored or unusually shaped fish. Native advice is not always reliable, since natives constitute the majority of poisoning victims. This stems from the variability in toxicity of well-known species from place to place and time to time, depending upon their diet, size, and sexual periodicities, and from the difficulty in deciding whether a reported case of poisoning was owing to toxins, bacterial decay—which is very rapid in the tropics—or improper handling and preparation. Because most fish and bacterial toxins are undiagnosable and no antitoxins exist, treatment of fish poisoning is symptomatic.*

*Bruce Halstead's three-volume compendium *Poisonous and Venomous Marine Animals of the World* summarizes the identification of almost all harmful species, and gives appropriate diagnosis and treatment for accident victims.

Dangerous Species: The sea is fraught with fearsome and dreadful dangers, as any true-adventure book will tell you. Yet, in twenty years of leading an equal number of one- to six-month oceanographic expeditions, my closest brushes with the Unmaker have been two (widely separated) cases of severe bacillary dysentery, contracted while eating in public restaurants in Papeete, Tahiti. During this period our group's work schedules have included over five thousand diving hours, mostly in and around some fifty tropical Pacific islands and atolls, where we were engaged in making surveys, installing and retrieving instruments, and collecting specimens. All of this was accomplished without serious incident or accident, save for a single, *allegedly* unprovoked shark attack. I am therefore a great believer in the calculated risk, and am convinced that if you treat the sea with understanding and respect, you can operate with confidence in what appear to the uninitiated as truly terrifying circumstances. With this preamble, I cite the following hazards, common to the literature, with some qualifications, based on personal experience:

1. Inanimate Injuries: Nearly every exposed solid surface beneath the sea is apt to harbor sharp cutting edges. These may be the remnants of barnacle or mollusc shells, or premature species just beginning to form shells. Moreover, human skin becomes soft after even short immersion in water, and thus is much more susceptible to injury. This combination of very sharp edges and soft skin makes it possible—and even likely—that the most casual contact will result in cuts deep enough to bleed. Because blood is very soluble in seawater, and the skin somewhat anesthetized by water absorption, it is not uncommon to come out of the water dripping blood from a dozen lacerations without having been aware of any injury. Such cuts are not autoinfectious because there are ordinarily no pathogenic (man-infecting) organisms in the sea. This statement is belied in some more primitive tropical areas by missionary attempts to ensure native modesty by construction of outhouses on stilts over reef-flats or lagoons. These absurd dichotomies in a lovely tropical settings can result in fecal contamination of local waters. The only persistent skin infections we have encountered in the tropics have come from minor scratches sustained while swimming and diving in such areas.

Sea urchins are a class of marine animals whose body structure consists of a globular shell an inch or two in diameter, from which a large number of calcareous spines project radically outward (plate 2). Some species have rather blunt spines, and are not dangerous. Others possess sharp, brittle spines up to a foot long that are capable of easily penetrating a canvas shoe sole. Following penetration, the spines detach from the animal, and most are so brittle that they are difficult to withdraw without breaking them off. Aside from the pain (somewhat like a beesting) and inconvenience caused by the ensuing soreness and numbness of the tissues affected, the chief danger from urchin spines is the possibility of blood poisoning from secondary infection.* Ordinarily, the spines will be absorbed by the body within a few days. Absorption can sometimes be accelerated by soaking the injured area in dilute acetic or boric acid solution—or even urine. Sea urchins are common in all shallow waters of the world, but because they are often hidden in shallow holes or crevices with only their spines protruding, they are sometimes difficult to detect.

* Some species, such as *diomedae,* are also venomous, and require special treatment.

Plate 2. Four common varieties of venomous sea urchins. In addition to toxic slime-encoated spines, *Diodema Setosum* (*lower right*) possesses shorter venom barbs between them.

The best safeguards against inanimate injuries are protective clothing and care in movement, with the objective of preventing any direct skin contact with solid surfaces. Heavy leather work gloves and full-sole flippers, to which are vulcanized harder-grade, sponge-rubber sandal soles, permit one to wedge oneself firmly, even in very rough water. These, supplemented by a cotton undershirt and a faceplate for visibility, will provide good protection without inhibition of motion. Neither gloves nor shoes are immune to sea urchin spines. Thus, wading in shallow water is definitely risky, unless the water is perfectly clear and calm, so that each step can be planned in advance.

2. *Biting Animals:* Without doubt, sharks are the number one marine hazard to man—and then only to unprotected swimmers or divers. Figure 11 shows the thirty-year, worldwide, monthly distribution of 454 shark attacks as of 1952. Note

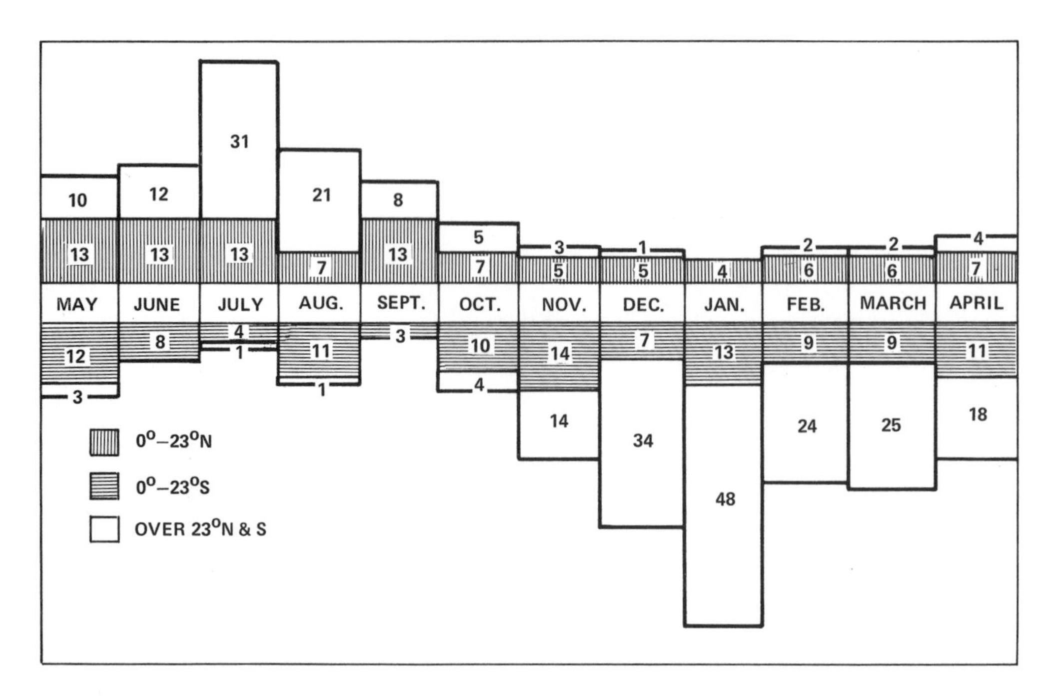

Fig. 11. Thirty-year worldwide statistics of shark attacks by month, as a function of latitude (after Coppleson).

that total attacks are about equally divided among hemispheres; low-latitude attacks are essentially independent of season, and high-latitude attacks peak in local midsummer, the peak swimming season. More recent compilations by the Smithsonian Institution's Shark Research Panel indicate a similar distribution for a total of about 20 attacks per year, until the Panel was discontinued in 1980. Of these, about 2 occur in Western and Eastern North America and 4–5 in Australia, and the remainder are scattered throughout tropical oceanic zones in all oceans.

Figure 12 shows seven species of sharks believed to have attacked humans, but 95 percent of American and Australian attacks are credited to a single species, the Great White (*Charcharodon Carcharias),* so that the remainder of all attacks is trivial. According to records along Australia's Gold Coast, frequented by millions of swimmers, Great Whites tend to be loners, and a single individual is often responsible for successive attacks at consecutive intervals of time and distance. Once isolated and killed, the menace subsides for weeks or months, until another shark develops a taste for humans. In Australia, protective nets have largely eliminated the hazard at principal bathing beaches.

In many years of tropical diving, with sharks often present in abundance, I have had the opportunity to make some observations that may be worth recording. I have never encountered a large, solitary white shark, or a tiger shark longer than 10 feet, and hope I never shall. Most of our "friends" have been the common black-tip (*C. melanopterus*), gray (*C. mennisora*), or white-tip (*P. longimanus*), none of which has seriously interfered with diving activities.* Upon the diver's first entering the water in

*On one occasion my coworkers and I had finished setting anchors into the outer reef at Canton Island at a depth of 150 feet, and were watching our workboat above. It was busy running sounding traverses, stopping every 50 feet or so to lower an 8-lb. sinker to the bottom on a steel wire. Several sharks soon appeared and began to circle the wire idly, until, just as the sinker was about to be lifted aboard, one of them closed in swiftly and nipped it off the wire. He spat it out immediately, and we retrieved it, but

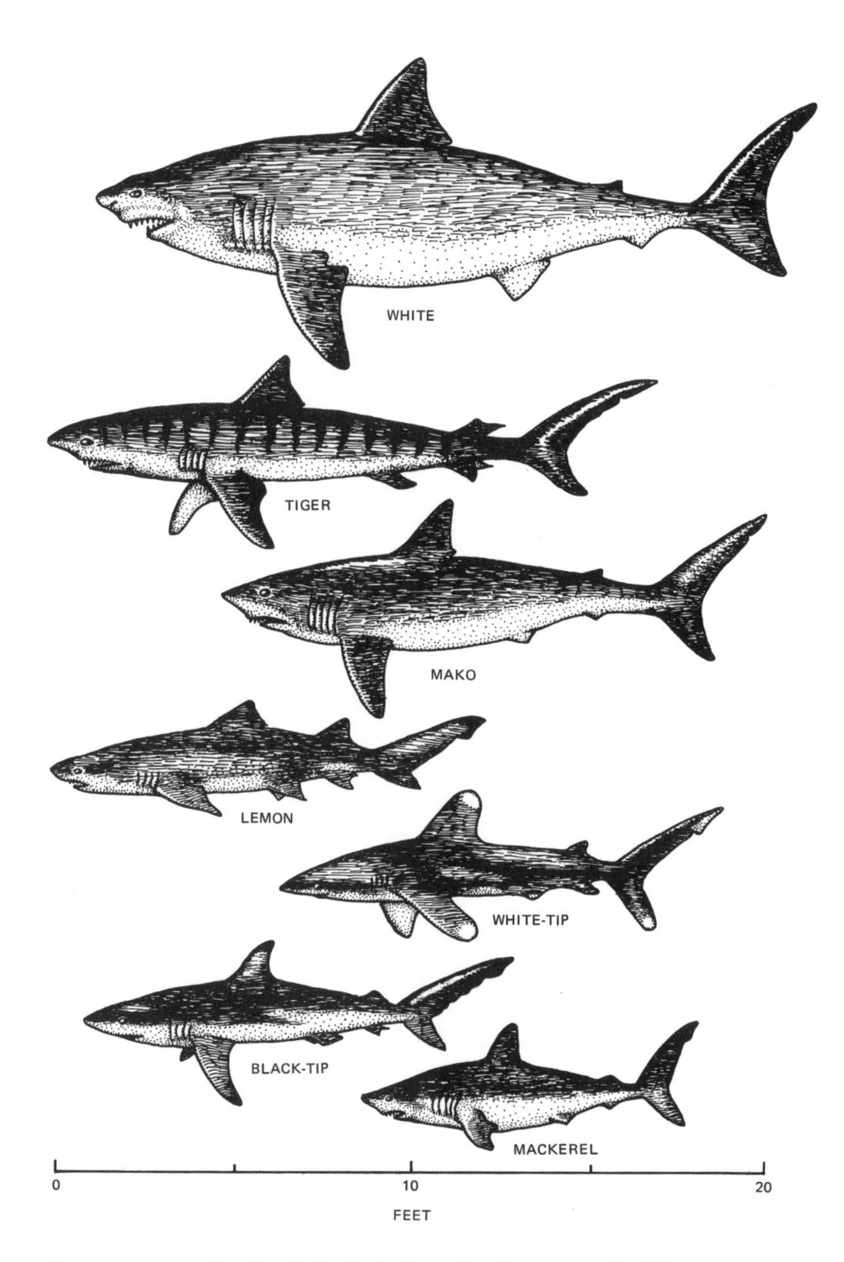

Fig. 12. Various kinds of sharks believed to have attacked humans.

a new area, the sharks soon collect, with evident curiosity about strangers in their environment, and circle rapidly at a radius of 30 to 50 feet. If studiously ignored, their attention wanders after ten or fifteen minutes, and they drift off one at a time to resume patrol. Each day thereafter fewer appear, until only an occasional sentry passes, just to keep an eye on things. These sharks appear to be indigenous, each having his own territory, and individuals can soon be recognized by markings. Except when feeding near the reef edge on a falling tide, none ever exhibited aggressive action.

---

during the course of half an hour we saw four more sinkers taken in a similar manner. Evidently there is a limit even to shark intelligence.

I do not belittle the risk of swimming where sharks are present and have adopted certain procedures in diving to minimize this risk, although I cannot vouch for their efficacy. These include:

1. Get out of the water if bleeding, or when blood is present. Sharks can sense blood and rapidly home in on its source.
2. Never enter the water without a faceplate, and unless the water is clear enough to observe potentially aggressive motions.
3. Never wade when you can swim—even in 18 inches of water.
4. Carry a stake, tool, or implement with which you can fend off an attack.
5. If caught in an aggressive situation at depth, sit with your back to something or somebody, to prevent attack from the rear.
6. Enter and leave the water rapidly, after first looking around. You are least able to cope with an attack while at the surface.
7. Always have a boat and tender available, and a medical facility alerted, when working underwater.

Barracuda are an entirely negligible risk in the Pacific—and probably in the Indian Ocean, in which the same species prevail. The Atlantic supports a larger species (*S. barracuda*), which the natives of the Caribbean are reported to fear more than sharks. Not having dived extensively in the Atlantic, I have discussed this report with Jacques Cousteau and also with Lamar Boren, both of whom have done a good deal of underwater photography in these waters, with the conclusion that there are not enough confirmed cases of barracuda attack to warrant serious concern.

Moray eels will bite viciously if provoked, but prefer to flee if encountered in the open. The largest morays in the world are found at Johnston Island, where we have dived with a Major Walker, whose principal off-duty preoccupation was hunting them with a spear gun. He has captured more than two hundred morays, the largest being over 9 feet long, many of which he has shipped to the University of Southern California for research studies. Morays live to a great age, and spend most of their time peering from holes in the coral heads, nodding and bobbing toward each other, like so many seals in a rookery.

Various authors have cited accounts of deaths due to entrapment by the giant clam *Tridacna gigas*, which, in common with smaller species of the same family, lies on the open, sandy bottom of lagoons in the west central Pacific. I frankly disbelieve these accounts. My diving group has recovered dozens of these animals—some weighing several hundred pounds—without being able to get within 20 feet of them before they sensed unusual water motion or light shadow and closed abruptly. Anyone so careless as to step into an open one while wading in the shallows along the Great Barrier Reef would probably deserve an untimely demise.

Venomous Species: There are many varieties of venomous animals in the sea that are capable of inflicting injuries ranging from minor skin rashes to convulsions, coma, and death. Most of these are not aggressive, and all can be avoided by watchful observance of one's surroundings. Fortunately, the dangerous species are relatively rare, and the chance of accidental encounter is very small.

Among the invertebrates, one can stay out of trouble by avoiding direct skin contact with all jellyfish, live corals, and a single mollusc: the *Gastropod conus* (cone shell). Most jellyfish and corals have stinging cells that produce skin irritation, the sensitivity to which varies greatly among individuals.

Of the jellyfish, particularly dangerous species are the sea wasps, *C. Quadrigatus* and *C. Chironex* (plate 3), whose sting has produced confirmed fatalities among a few victims.* Two related and painful nuisances are the *Hydroids* commonly known as fire coral and the Portuguese man-of-war (plate 4). Fire coral grows among other true corals as a smooth, plantlike structure somewhat resembling moose horns, but dull yellow in color. The man-of-war, on the other hand, resembles a large jellyfish, floating freely at the surface, where it is supported by an iridescent air bladder, and dangles a host of long, stinging tentacles beneath it. Some grow to a diameter of several feet, and have tentacles 30 to 50 feet long, which are often difficult to see under water, despite their distinctive blue color.

The cone shells are among the most common marine molluscs. Different species vary from a fraction of an inch to half a foot in length, and have a wide variety of characteristic markings, some of which are highly prized by collectors. All of the cones possess a venom apparatus (fig. 13), consisting of a poison sac and a duct leading to an extensible proboscis, through which hollow, radular darts can be ejected

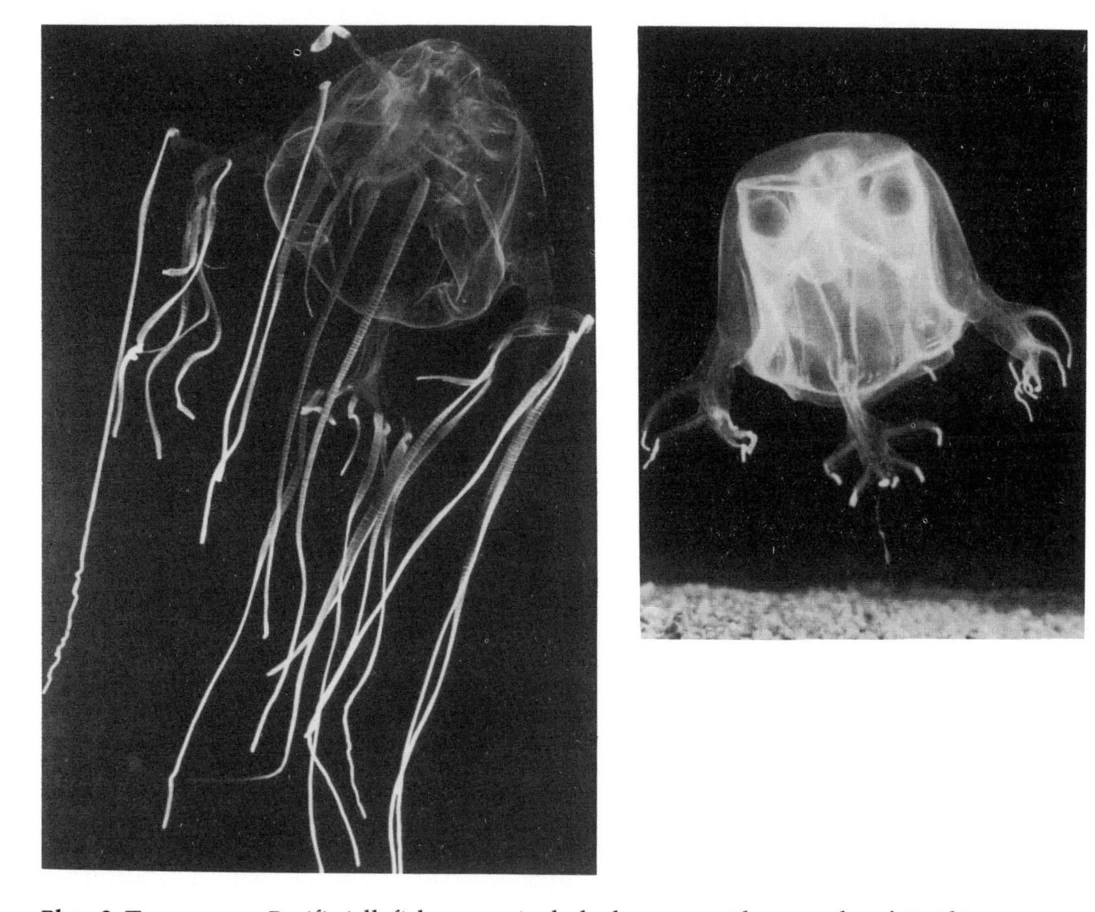

Plate 3. Two western Pacific jellyfish are particularly dangerous. The tentacles of *C. Chironex* (*left*) grow to about 2 feet, and several deaths have been reported from contact with them.

*Flooding with methyl alcohol neutralizes the poison and dissolves the stinging tentacles.

Plate 4. The Portuguese man-of-war (*left*) and fire coral are but two of some 2,700 species of marine hydroids. Both possess stinging cells that are capable of producing painful welts and rashes.

upon contact with a foreign object. Proboscis extension is a rather deliberate process, however, and easily avoided by holding the animal by its larger end. The venom is extremely toxic, and several deaths have resulted from stings by larger specimens.

Among the vertebrates, there are three related fish families that possess hollow or fluted spines, through which venom can be injected. The wounds cause intense pain, extensive inflammation, and tissue damage, and can occasionally be fatal. These are the weaver fish, scorpion fish, and toadfish, of which only the weaver fish is known to attack—and then only if disturbed. The total number and distribution of species of these families is too voluminous to include here, and interested readers are referred to Halstead's *Poisonous and Venomous Marine Animals of the World* for comprehensive descriptions. Worth mentioning, however, are two species of scorpion fish: the zebra fish, because of his beauty and prevalence, and the stonefish, because of his deadliness (plate 5).

The zebra fish (*Pterois volitans*)—also known by the common names turkey fish and lion fish—is brilliantly striped in red and white, and has long, lacelike dorsal and ventral fins, which the animal often spreads in fanciful array. It is a rather small fish (3 to 10 inches), and is widespread throughout the tropics, where it floats almost motionless in the water, often alongside pier pilings. We have found them most often upside down against the roofs of dark coral caves. Because they are essentially

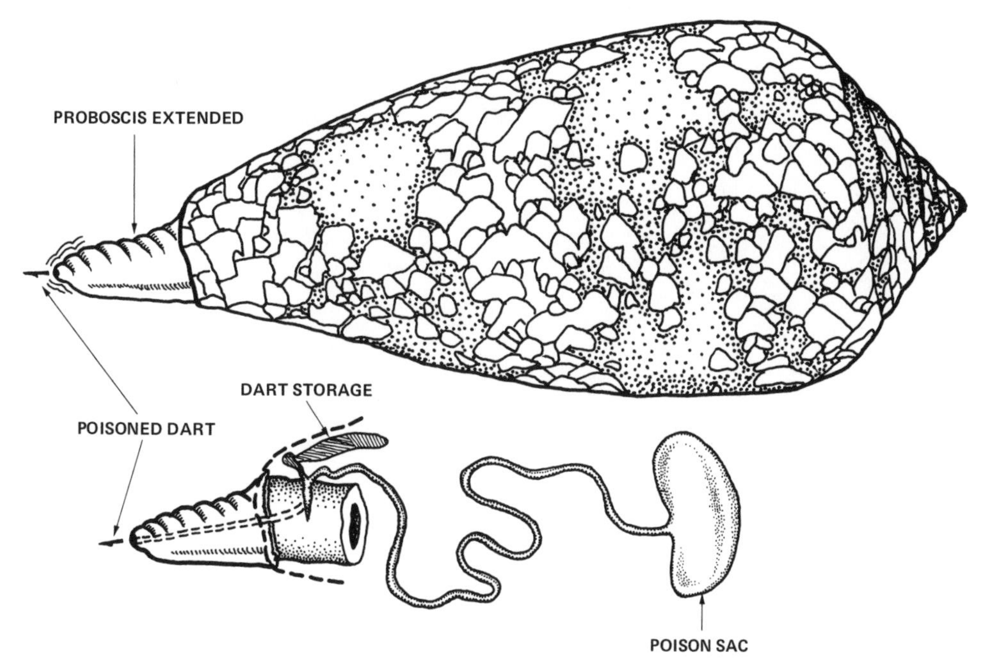

Fig. 13. Venom apparatus of the cone shell.

fearless, it is easy to brush against one unnoticed. A native swimmer was stung by a zebra fish during one of my expeditions to the Austral Islands. He was carried to the village in convulsions and great pain, unameliorated by intravenous morphine injections, but recovered after twenty-four hours. This is my only personal acquaintance with a venomous fish injury, and I have never met another person in my travels who was a firsthand witness to such an event.

The stonefish (*Sinanceja horrida*) is as ugly as the name implies, and is also the most deadly of venomous marine fishes. It is a small (6–12 inches) fish, strongly resembling a lump of coal. Its principal habitat is the tropical western Pacific, where it lies motionless in shallow water, usually in a coral crevice, or almost buried in sand. Thus, stonefish injuries are most apt to occur to waders, its dorsal spines being strong enough to easily puncture a rubber sole. The great rarity of stonefish victims implies a relatively minor risk—but a very serious one! I have seen only one specimen in twenty years, and personally know of no cases of injury.

Fouling or Boring Organisms: These marine pests have been the despair of shipowners and operators alike from the days when ships first became too large to be pulled from the water at night. Marine fouling of clean, newly immersed surfaces follows a characteristic biological and temporal pattern:

Organism	Period of Dominance
Bacteria	0–3 days
Algal slimes	3–7 days
Protozoans	1–3 weeks
Barnacles	3–10 weeks
Tunicates	10–16 weeks
Grasses	3–6 months
Mussels	6–12 months

Plate 5. The zebra fish is as beautiful as its cousin the stonefish is ugly. Both carry "hypodermic" dorsal venom spines that can inflict painful—if not deadly—injury.

The number and species vary widely depending on temperature and location, but in calm, temperate waters a clean hull can accumulate a combined growth of fouling organisms about an inch thick in two months. Contrarily, the Navy estimates that continuous motion in excess of one knot will prevent fouling beyond the slime stage.*

The first successful attempts to protect ships against both fouling and boring forms occurred in the seventeenth century, with the invention of copper sheathing. Lead sheathing, used previously, stopped borers, but was ineffective against fouling. Copper sheathing was soon found to lose effectiveness when in contact underwater with iron or other metals lower on the electrochemical scale. At the same time it was found that the sheathing was protected against slow corrosion. Evidently, protection against fouling is an electrically active process, involving release of copper ions into the water. This observation is still the basis for all effective bottom coatings.

*Yet old whales carry a considerable growth of barnacles.

With the invention of iron ships, copper sheathing became electrolytically impractical, and it has required nearly one hundred years to develop means of incorporating toxic copper compounds into paints or hot-applied coatings that are not electrolytically nullified by the iron hulls. The problem was finally solved through use of nonconducting primers, and metal ships today can expect to exceed 18 months between successive applications of antifouling coatings.

All commercially available, effective antifouling paints today contain copper oxides, although some also contain mercuric or stannous (tin) oxides.* Their toxicity depends upon continuous release of copper ions into the water at a rate of not less than 10 milligrams per square centimeter per day, either through a porous vehicle or one that dissolves away at the required rate. About 40 percent of commercial paints fail to meet minimum standards. The addition of accessory pigments inhibits the antifouling effectiveness, as does any exposure of iron, zinc, or magnesium in contact with the paint. Good bottom paint contains a minimum of six pounds of toxic copper per gallon. It will remain effective for six months under severe fouling conditions, and up to a year under less severe conditions.

Copper bottom paint does not, however, confer protection for wooden vessels against boring animals. The most common borer, *Teredo navalis,* can enter unprotected wood surfaces at a density of up to 5,000 per square foot. It has a life span of ten weeks, during which it can chew out a tube one-quarter of an inch in diameter and six inches long, with no external manifestation except for its tiny entry hole. Copper sheathing remains the most satisfactory answer to teredos, although it must be replaced periodically, as it erodes away.

The well-known immunity of fiberglass hulls to boring worms is one of their chief attractions, and some manufacturers are now adding copper to the semiporous gelcoat, so that the hull emerges from the mold with a built-in antifouling protection alleged to last 5–10 years.

* A powerful new antifoulant, *tri-butyl tin,* appeared on the market in the mid-1960s. Lethal to most marine organisms in concentrations as low as two parts per trillion, it promised a 5- to 10-year relief from annual haulout and repainting. However, by the 1970s, depletion of oyster fisheries prompted France to prohibit sale of TBT paints for vessels shorter than 30 m (100 ft.). England soon followed suit. In 1984, a California survey found that marine life was virtually disappearing from small-boat marinas, and the prohibition has now been extended to most U.S. coastal states—a victory for ecology, perhaps, but at an annual cost of about $50 million to the boating public.

PART II

A Little Meteorology

4

The Nature of the Atmosphere

A TWO-LAYER SYSTEM

The ocean and the atmosphere form a two-layer system, across whose common boundary—the sea surface—many interactions take place that are responsible for the delicate balance of heat and cold, rain or drought, calm or storm, as well as for the presence of life on earth. Although the ocean is a liquid and the air a mixture of gases, both are fluids in the sense that their molecules are free to wander about under the action of disturbing forces. Moreover, except for the presence of foreign matter, both are invisible. You might demur, saying, "But I can see water in many forms—as waves, foam, drops, etc." But what you *really* see is just the interface between air and water—and this only because the speed of light is different in the two media. Once beneath the surface, as in diving, one can no more see water than air.

Under certain circumstances, such as in violent storms, even the interface becomes smeared out. There is so much water in the air and so much air in the water that it is impossible to tell where the atmosphere stops and the sea begins. Storms are usually accompanied by high, steep waves, which can produce accelerations approaching that of gravity, and these, coupled with the spatial loss of orientation due to the absence of a recognizable interface, may literally make it impossible to distinguish up from down.

Considering that the total mass of the atmosphere corresponds to a layer of seawater only 33 feet thick, or about one-fifth of 1 percent of the mass of the oceans, it is somewhat remarkable that all of the tremendous energy represented by the waves and currents at the sea surface is imparted by the action of winds in the lower layer of the atmosphere. In turn, the energy for the wind is provided by direct heat radiation from the land and sea (about 40 percent) and by evaporation and precipitation from the sea (60 percent). Only a relatively small fraction of the incoming solar radiation—which ultimately provides all of the energy for these processes*—is directly absorbed within the atmosphere itself. This fraction is more than compensated for by radiation of heat from the atmosphere into space, leaving a deficit that must be balanced by convection of heat from the land and sea beneath.

Thus dynamic meteorology is primarily concerned with the exchange of heat between the ocean, which moves sluggishly but can store heat in illimitable quantities, and the atmosphere, which has a very limited heat capacity, but moves a thousand times faster.

*Heat flow from the hot interior of the earth is utterly negligible compared with that received at the earth's surface from solar radiation.

COMPOSITION AND PROPERTIES

Dry air, if sampled sufficiently far from a large city, is a mixture of gases, composed of 78 percent nitrogen, 21 percent oxygen, 0.93 percent argon, 0.03 percent carbon dioxide, and traces of hydrogen and ozone. The lower 5 to 10 miles of the atmosphere is sufficiently well mixed that these proportions are remarkably constant the world over. Above this level, there is little mixing, the air is stratified, and there are slight variations in the concentrations of these gases according to molecular weight. Relative to oxygen and nitrogen, the concentration of hydrogen tends to increase—and of carbon dioxide to decrease—with altitude.

The lower atmosphere also contains small amounts (0.1–4.0 percent by weight) of invisible water vapor (actually, steam), in which form it behaves like any other gas. However, water vapor is unique in that, under certain conditions, it changes *phase,* condensing to a liquid (dew, fog, clouds, rain), or freezing to a solid (frost, snow, ice crystals). These phase changes are reversible, and release or absorb large quantities of *latent* heat.* Additionally, all three phases influence the reflection and/or absorption of solar radiation, and hence significantly affect the atmospheric heat balance. Lastly, vapor condensation and precipitation play a dominant role in local weather, with all its implications for navigation and seamanship.

If all atmospheric moisture were somehow precipitated at once as water, it would cover the earth to a depth of only about two inches—about 1/80,000th the volume of the oceans. Because moisture is continuously being evaporated and precipitated, about twenty times this amount is annually recycled between oceans and atmosphere, of which roughly 10 percent falls over land and returns to the sea as river and stream runoff. Yet, in addition to making possible all life on land, this small fraction (15 inches per year) has eroded away the volume of the continents many times over in the course of the earth's history.

While the behavior of dry air can be predicted knowing only its pressure and temperature distribution, the atmosphere is only completely dry at temperatures below, say, −40°F.** In nature, such low temperatures occur only at high altitude, or at sea level near the poles in winter. Elsewhere, the moisture distribution with altitude must also be known in order to predict the course of daily weather. Because of the foregoing phase changes, pressure, temperature, and moisture content are closely interdependent, and can undergo wide local variations. Thus, in discussing their respective distributions, it should be recognized that we are referring only to average values, and not to what goes on in the centers of storms.

Moisture: Because condensed moisture does not behave like a gas and is also difficult to measure, the moisture content of air is always referred to in terms of its concentration in the vapor phase. The concentration of water vapor *by weight* in a sample of moist air is defined as the *specific humidity,* and can be thought of as the amount of available moisture in the sample. At constant temperature, dry air continuously in contact with a surplus of moisture in the liquid or solid phase will absorb evaporated vapor until it ultimately becomes *saturated,* and evaporation ceases. The specific humidity at saturation is independent of pressure over its

*A rainfall rate of ½ inch per hour releases energy equivalent to the explosion of about 20,000 tons of TNT per square mile!

**Industrial air is commonly dried by passing it over dry ice.

normal atmospheric range, and depends only on the equilibrium temperature. If the latter is increased, more moisture will be absorbed and a new equilibrium established. If it falls, moisture will condense (or freeze, if the temperature is low enough) to a lower saturation level. The saturation humidity is rather low over ice, but increases rapidly with temperature. Above the melting point (32°F), it roughly doubles for every 20°F rise, reaching a maximum value of 6.1 percent at the boiling point (212°F).

But evaporation is a relatively slow process. Even though the ocean contains an illimitable surplus of water, constant circulation of air prevents equilibrium from being established anywhere except within a thin layer at the surface, and the bulk of the atmosphere is always undersaturated. The index of undersaturation is called *relative humidity,* which is defined as the ratio of specific humidity to its saturation value at the prevailing temperature. Thus, relative humidity varies between the extremes of zero for dry air and 100 percent for saturated air, irrespective of temperature. However, because the saturation concentration increases with temperature, for the same relative humidity, there will be more water vapor in the air on a warm day than on a cold day, and the former will be sensibly more uncomfortable.

Undersaturated air can be brought to saturation by cooling to the *dew point,** which is just another name for the saturation temperature. The difference between the prevailing temperature and the dew point is called the dew-point depression. All three of these quantities can be measured directly with a wet-/dry-bulb thermometer, or its rotary equivalent, the sling psychrometer. With both, the wet bulb is maintained in a saturated state by a water-soaked wick and always indicates the dew point temperature—even when coated with ice and well below 0°F. Knowing the wet and dry-bulb temperatures, the relative humidity can be obtained from a psychrometric table, or a chart, such as shown in figure 14, which better illustrates how all the above—defined quantities are related.

The heavy curves in this figure refer to constant values of specific humidity, which are independent of temperature or pressure as long as an air sample retains its moisture. Upon cooling, any point on this chart moves upward parallel to adjacent curves until it reaches the top of the chart, where it indicates the dew-point temperature at which that moisture concentration would become saturated. The chart covers the normal range of sea-level conditions of interest to mariners, and shows that under the most humid tropical conditions, specific humidity never rises much above 3 percent by weight.

Pressure: The average pressure of the atmosphere at sea level is, by definition, one atmosphere (14.7 pounds per square inch),** equivalent to that exerted at the bottom of a column of seawater 33 feet high, or a column of mercury 29.92 inches high. This pressure is enormous compared with that exerted by hurricane winds (60–180 knots) upon a flat surface held at a right angle to the direction of air flow, which produces dynamic pressures of 14 to 130 pounds *per square foot.*

*The sweating often observed below the waterline on the inner surface of metal or fiberglass boat hulls is simply the result of warm, moist air coming in contact with a surface at or below the dew point. This effect is easily prevented by a layer of insulation.

**Meteorologists give the *normal atmosphere* as that capable of supporting a column of mercury 760 mm high at 0°C. A more common unit is the *standard atmosphere* or *bar* = 1000 millibars (mb) = 750 mm of mercury. Thus 1 atm. = 1013.3 mb.

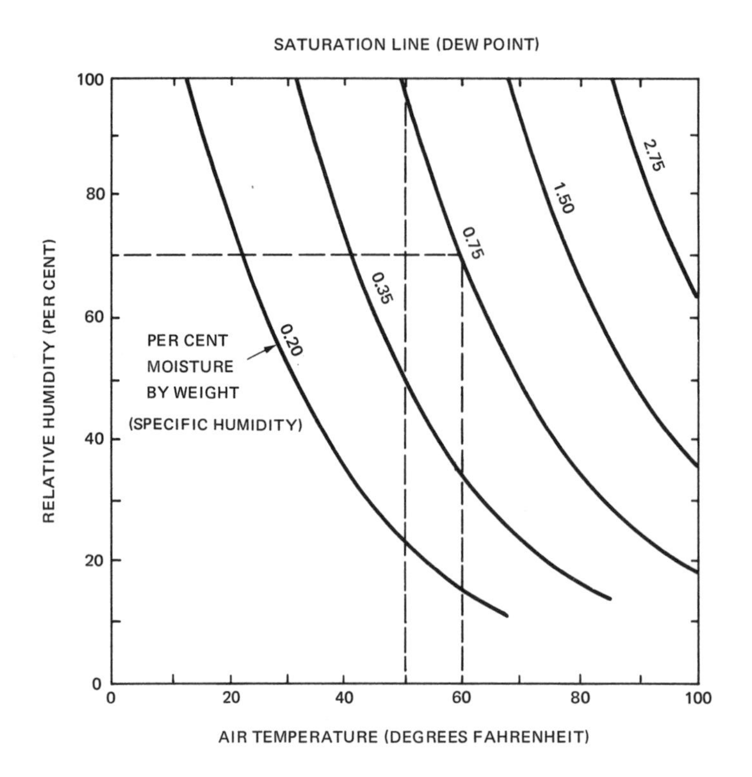

Fig. 14. Psychrometric chart for sea-level pressure. Any point on this diagram specifies an air parcel of known humidity and temperature. Cooling or heating causes point to move up or down, respectively, parallel to nearest curve. Example shows that air initially at 60°F and 70 percent relative humidity (.75 percent moisture) will become saturated if the temperature falls to 50°F, which is the wet bulb temperature under these conditions.

Differences in pressure at sea level are principally caused by thermally induced density differences within the atmosphere. Even experienced mariners may sometimes confuse the daily "barometric" tides with the familiar astronomic tides that are produced by the planetary and solar gravitational attractions. However, the former, which are characterized by twice-daily pressure maxima at 10:00 and 22:00, local apparent time, and minima at 02:00 and 14:00, have been clearly demonstrated to be caused by local solar heating as the earth rotates. Most of the effect seems to be due to twenty-four-hourly heating of the ozone layer in the stratosphere, although still unexplained is why the twelve-hour component of the barometric tide has a higher amplitude (1–2 millibars) than the twenty-four-hour component. The latter becomes smaller with increasing latitude, as expected from the reduced heating in polar regions, but the former seems also to depend upon such other factors as whether one is at sea or on land, at sea level or high in the mountains, etc. Nevertheless, in the tropics the barometric tides are extremely regular. If you break your chronometer and your radio and have to depend on your cheap dollar watch for celestial navigation, you can set it to 10:00 at the peak of the morning tide, and your longitude error will probably not exceed 200 miles.*

*Captain Joshua Slocum, the first man to sail alone around the world, jokingly reported that he reckoned longitude time by a "dollar alarm clock." In fact, he used the fairly precise, but tedious, method of lunar observations, coupled with careful dead reckoning.

Temperature: Air at sea level and at a temperature of 32°F weighs about 0.08 pounds per cubic foot, or roughly 1/840th the specific weight of water. If the atmosphere were of uniform temperature, the air density would decrease upward in proportion to altitude, and the pressure would decrease as the logarithm of altitude. However, the mean temperature at a given latitude—even averaged over a year—varies in a complex way, owing to different rates of heating and radiation at different levels. Therefore density and pressure decrease at somewhat irregular rates. Figure 15 shows the vertical distribution of temperature, density, and pressure for the International Reference Atmosphere, which is sort of a compendium average for all seasons and all latitudes. As noted later, significant short-duration departures from these mean curves occur only within the lower ten miles in the course of seasonal and daily weather fluctuations. It is of interest that about half the mass of the atmosphere lies below the top of Mt. Everest (30,000 feet), and thus high mountains pose a substantial barrier to the horizontal movement of air. At Lake Titicaca, in the Bolivian Andes (12,600 feet), the air density is only 73 percent of that at sea level. Since the dynamic pressure acting on a sail varies in proportion to the product of density × velocity squared, a sailboat would require about 15 percent higher windspeeds to attain the same hull speed at that elevation.

STRUCTURE OF THE ATMOSPHERE

Like the ocean, the atmosphere has a layered structure. While the water masses in the ocean can be identified by slight density differences owing to various combina-

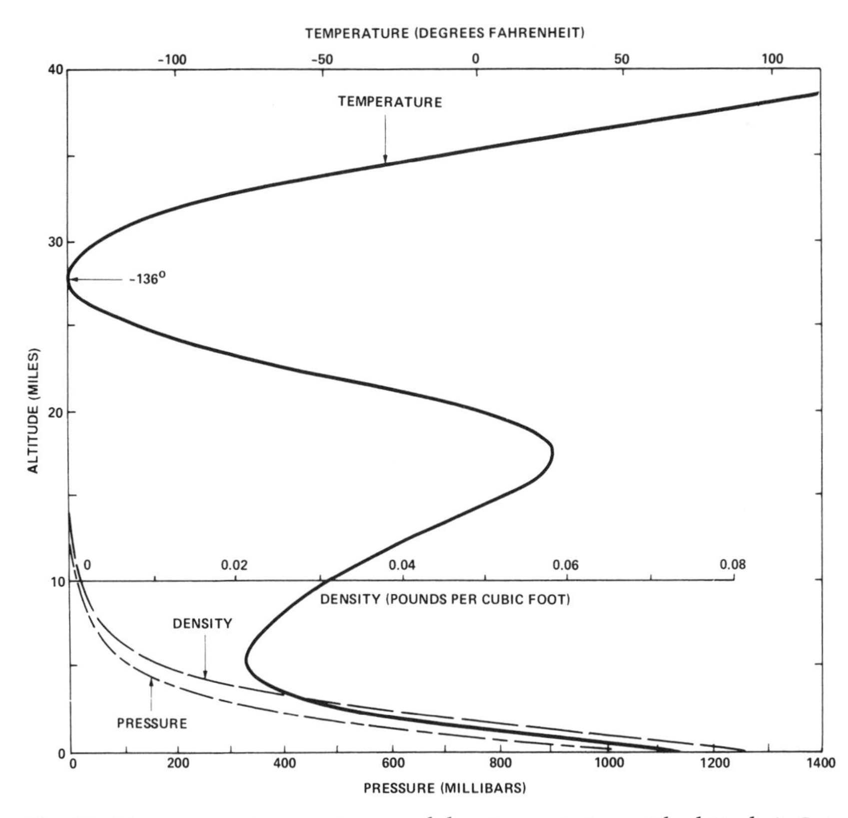

Fig. 15. Air pressure, temperature, and density variation with altitude in International Reference Atmosphere.

tions of temperature and salinity, the various atmospheric layers are distinguished by differences in *thermal* stability or instability, slightly modified (in the lowest layer only) by moisture content. Stability is conventionally defined by the subsequent behavior of a parcel of air following sudden displacement to a different elevation: if it tends to return to its original level it is in *stable* equilibrium with its surroundings; if it remains motionless it is said to be in *neutral* or *indifferent* equilibrium; but if it tends to continue in the direction displaced it is *unstable.* The degree of stability of the air depends upon the *lapse rate,* defined as the rate of *decrease* of temperature with altitude. The critical lapse rate, at which stability changes, depends chiefly upon the moisture content and mean temperature, but in general, warm, tropical air saturated with moisture over the ocean will always be stable if the lapse rate is less than 2°F per 1,000 feet. Cold Arctic air and dry air may be stable at lapse rates as high as 5°F per 1,000 feet. Air in which the lapse rate lies between 5.5°F and 19°F per 1,000 feet is in unstable equilibrium, although it may remain at rest unless displaced. At lapse rates greater than 19°F per 1,000 feet, air is always unstable, for then the density decreases with elevation. A *negative* lapse rate (temperature increasing with height) is called an *inversion,* and marks a zone of high stability.

The four atmospheric layers of principal interest to mariners (fig. 16) are defined by reversals of the thermal lapse rate, indicating varying degrees of stability. The lowest layer is the *troposphere,* which has a mean lapse rate of about 3.5°F per 1,000 feet, although it varies throughout the entire range from stable to unstable, from place to place and time to time. The troposphere might be called the *weather layer,* because it contains most of the moisture in the atmosphere, and is subject to unstable motions that determine the course of weather.

Part and parcel of the weather are the clouds, which form wherever there is sufficient moisture present and the temperature falls to the dew point. These conditions generally prevail at some elevation over the oceans, so that a completely cloudless sky is a rare event at sea. Clouds are among the mariner's most reliable guides to changing weather conditions, and we will make frequent reference to them throughout this book. By international agreement, clouds are classified according to their stability and composition, with modifying adjectives giving their distribution or indicating accompanying precipitation, etc., as follows (plates 6–8):

Name	Type	Composition	Description
Cirrus	Quasi-stable	Ice crystals	Delicate, fibrous, high-altitude clouds, resembling feathers or wisps of hair.
Stratus	Stable	Water droplets	Sheetlike, horizontal layers, usually with well-defined lower surfaces.
Cumulus	Unstable	Water droplets	Globular, puffy clouds that often visibly change shape within a few minutes.

Conditional Adjectives	
Low	Low, below 6,500 feet
Alto	Medium, 6,500–25,000 feet
Cirro	High, 25,000–60,000 feet
Scattered	Scattered (less than 60 percent sky coverage)
Fracto	Broken (60–90 percent sky coverage)
Nimbostratus Cumulonimbus	Precipitation clouds

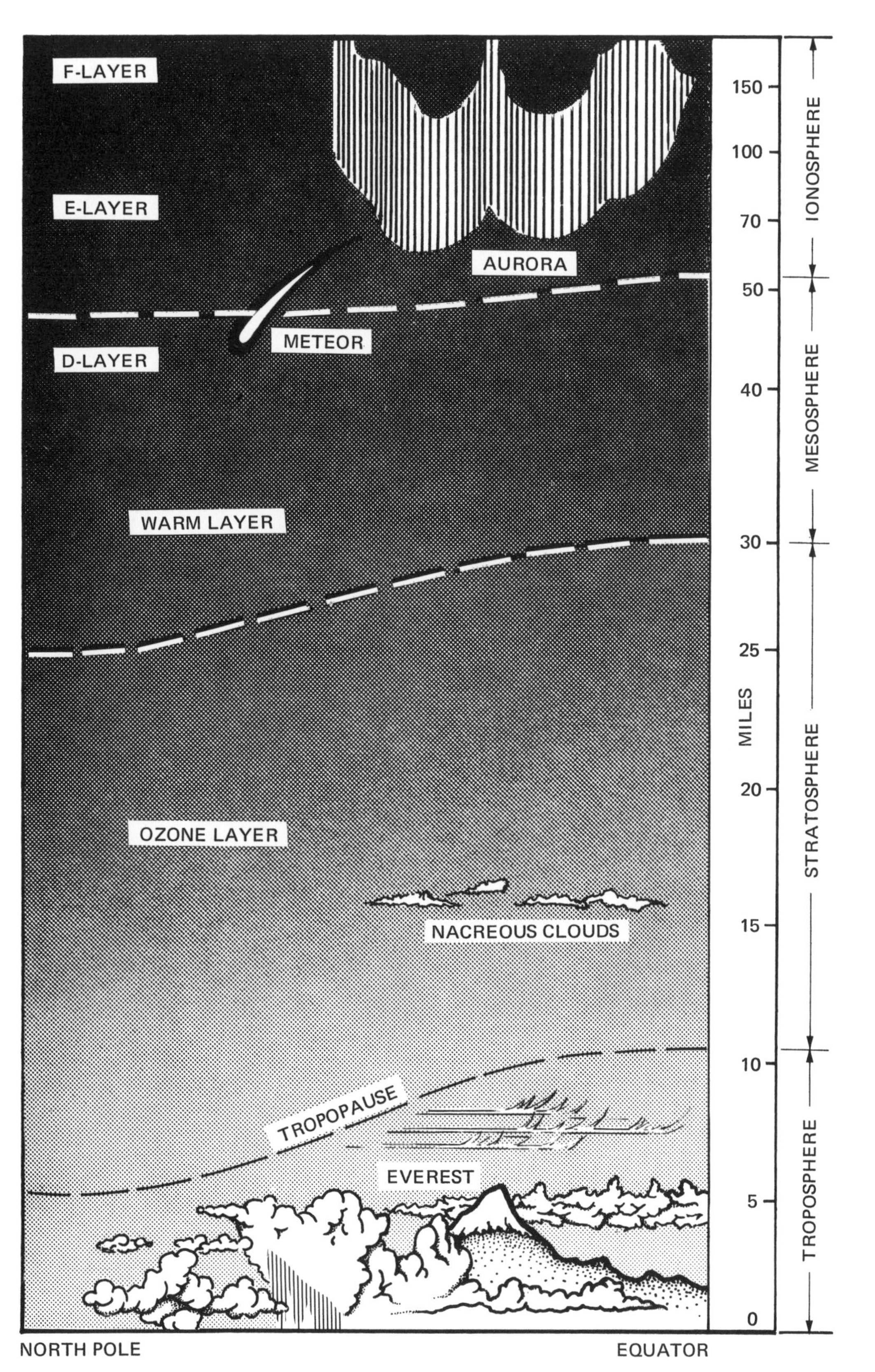

Fig. 16. Structure of the atmosphere. All weather is confined to lower 5–10 miles (troposphere).

Plate 6. This and succeeding photos show cloud patterns associated with a typical midlatitude frontal passage. (*Top*) Cirrus and cirrostratus advance from the southwest, followed by lower stratus and stratocumulus. (*Bottom*) Sky becomes obscured by stratus, with scattered cumulus beneath.

The upper bound of the troposphere is called the *tropopause* because no ordinary meteorological disturbances penetrate through it. The altitude of the tropopause varies from about 3 to 5 miles at the poles to 8 to 10 miles at the equator. This is because—unlike water—air is a compressible fluid that expands when heated. Thus the troposphere is thicker near the hot equator than near the cold poles, although the average pressure at sea level is about the same everywhere. Actually, the tropopause is not a continuous layer, but has gaps in it where the polar and subtropical jet streams flow eastward near latitudes 50° and 30° in both hemi-

Plate 7. (*Top*) Bands of cumulus increase in density. (*Bottom*) Broken cumulus gives way to solid overcast, with multilevel layers of stratocumulus and a low bank of nimbostratus.

spheres. These jet streams mark the boundaries of major convection cells in the troposphere, and consist of thin sheets of high-velocity winds that encircle the earth—now waxing, now waning, and often shifting position and altitude. They are assiduously sought out by easterly flying aircraft because of the fuel economy they afford—and carefully avoided by westerly flights for the same reason.

In addition to the horizontal layering described above, the troposphere is further subdivided in latitude by thermal convection cells, and meridionally by the oceans, continents, and mountain ranges, into characteristic air masses, the overall pattern of which we define as *climate,* and the details of which constitute the weather.

Plate 8. (*Top*) Dense rain cloud of nimbostratus blots out sky. (*Bottom*) Storm passage is marked by breakup of nimbostratus, with altocumulus and patches of clear sky.

Above the troposphere, and extending to altitudes of 25–30 miles, is the stratosphere. It is a region of clear, calm air and unlimited visibility. The high stability of the stratosphere is due to intense internal heating through absorption of solar ultraviolet radiation by oxygen molecules, some of which are split apart (ionized) and recombine to produce ozone, which also strongly absorbs ultraviolet rays. The end result is a thermal reversal at about 5 miles of altitude, and a temperature increase within the stratosphere (negative lapse rate) up to about 15 miles (fig. 16).

Terrestrial animal life as we know it is strongly dependent upon this protective ozone shield, without which many species could not exist and we humans would become increasingly susceptible to deep sunburn, skin cancers, and eye cataracts. Hence the furor over the discovery of holes in the ozone layer over the earth's poles in winter, which have been traced to ozone breakdown by chlorofluorocarbons

(CFCs) used in air conditioners and spray cans. International policy now discourages the manufacture of CFCs, but it may be a century before the tenuous equilibrium is naturally reestablished.

Above the stratosphere is a layer of about equal thickness called the *mesosphere,* of no special significance to seafarers, and chiefly distinguished by another thermal reversal, a lapse rate of about 2°F per 1,000 feet, and a pressure of less than one-millionth of an atmosphere.

The outer layer of the atmosphere is a vast shell of rarefied air known as the ionosphere, owing to its partial dissociation into electrically charged ions by ultraviolet radiation from the sun. Ultraviolet absorption also acts to heat the ionosphere, which accounts for the secondary thermal inversion. At these high altitudes, however, the word "temperature" ceases to have meaning in the ordinary sense, and refers only to the mean speed of gas molecules and ions. Although the temperature in the high ionosphere rises to thousands of degrees, a human being in an unheated space suit would quickly freeze to death if shielded from warming sunlight. The ionosphere, which extends from about 50 to 250 miles above the earth, is of great practical importance to radio communications. Both it and the mesosphere are further subdivided into several internal layers (D, E, F, and so forth), depending upon which radio wavelengths are preferentially reflected or transmitted.

These various layers descend in altitude at night, and are generally lower near the poles than at the equator. They are also subject to severe perturbations during so-called *magnetic storms,* caused by sporadic bombardment of the outer atmosphere by subatomic particles ejected from the sun during solar flares. Such flares are manifested by *sunspots* in the sun's chromosphere, and their frequency of occurrence has a pronounced 11-year cycle. Sunspot activity reached a maximum in 1981, and can be expected to reach maxima again in 1992, 2003, etc., with corresponding minima between. Magnetic storms also seem to have a weak twenty-seven-day dependency, related to the mean rotation rate of the solar sunspot belt.

A familiar by-product of magnetic storm activity within the ionosphere are the polar *auroras* (fig. 17). These brilliant-hued, diaphanous, shimmering efflorescences are produced near the magnetic poles by protons and electrons ejected in great streams from the sun, which are trapped in the earth's magnetic field. There they spiral from pole to pole, dipping briefly into the upper atmosphere, where they occasionally collide with and ionize gas molecules. These later recombine with free electrons to produce light. Because the north magnetic pole currently lies in northern Canada, auroras are seen at much lower latitudes in America than in Europe, sometimes being visible as far south as Seattle or New York.

AIR-SEA INTERACTION

The processes by which the oceans and the atmosphere exchange heat, momentum, moisture, oxygen, and carbon dioxide are only partially understood, but because of their great importance to the present and future welfare of mankind they are now receiving intensive study. The primary difficulty is the lack of sufficient data on exchange rates from all parts of the oceans to establish seasonal equilibria in different climatic zones. Such equilibria are necessary for the evaluation of slow changes that take place over many years, so that we may eventually determine whether the mean climate of the earth is changing, and to what extent such changes

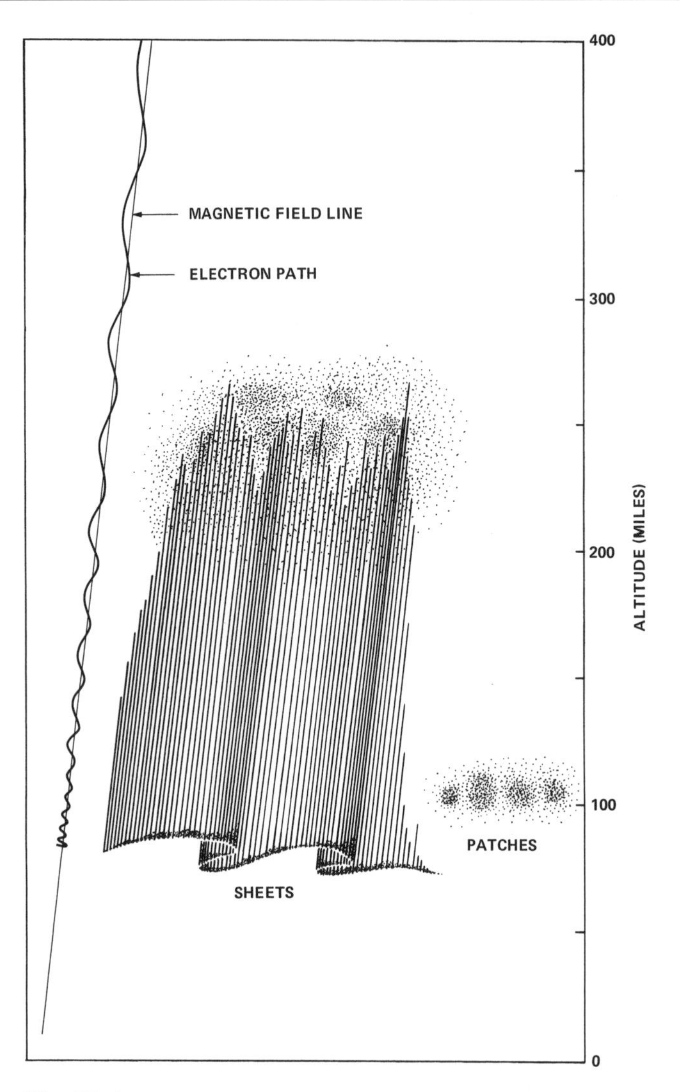

Fig. 17. Aurora structure.

are influenced by man himself. It is not at all clear at present that we are learning to control our environment at a faster rate than we are destroying it.

It is by now well established, however, that the cycling rates for all of the above processes are vastly different in the ocean and atmosphere. For example, a molecule of water evaporated from the ocean remains in the atmosphere for only a week or two, on the average, before it is precipitated back into the hydrosphere as rain, snow, or ice. Yet the mean residence time for a water molecule precipitated into the ocean, before it is again evaporated, is several thousand years! The cycling times for oxygen and carbon dioxide are thought to be similar. The cycling times for momentum in the ocean and atmosphere, however, are not so drastically different. If the supply of heat were to be cut off, the atmosphere would come to rest within a few weeks; the oceans within a year or two. Thus the oceans might be thought of as gigantic reservoirs in which all but a small fraction of the available carbon, oxygen, water, and heat are stored, to be deposited or withdrawn, depending upon external reactions. The total bank balance is very large, but what is uncertain—and perhaps critical—is the rate of deposit and withdrawal.

Of these several equilibria, the momentum balance is determined by the heat balance, which in turn depends sensitively upon the carbon dioxide balance, and to a lesser extent upon the concentrations of moisture and ozone in the atmosphere. Carbon dioxide, water vapor, and ozone, like the glass in a greenhouse, all readily transmit incoming visible radiation from the sun, but strongly absorb outgoing long-wave (infrared) radiation produced wherever visible radiation strikes the darker portions of the earth's surface and heats it up. Most of the land area and all of the oceans are relatively dark, but snow and ice reflect up to 95 percent of the visible radiation they receive back out into space. Ozone also strongly absorbs very short wave (ultraviolet) radiation from the sun.

The net effect of these selective absorption and transmission processes is that the lower atmosphere is heated from beneath, which accounts for the temperature maximum at the surface; and it is also heated in the ozone layer, which produces the second maximum at the stratopause. The temperature rise in the ionosphere does not significantly affect the heat balance, because the air is so thin.

Most of the tropospheric heating is due to carbon dioxide absorption, although its concentration is only 0.03 percent by weight in the lower atmosphere. This concentration is a delicate balance between production by combustion of organic (carbon-containing) compounds on land—plus a small increment from volcanic activity—and removal by plant photosynthesis, weathering of rocks, and solution in the oceans. Several lines of evidence suggest that the atmospheric CO_2 concentration has fluctuated markedly over geologic time, tending to decrease during glacial epochs to about 30 percent below the concentration of the shorter interglacial periods, when the mean tropospheric temperatures were 10–15 degrees warmer. Since heat production by ozone and water vapor are believed to be relatively independent of geophysical processes, scientists have speculated that the great sea-level changes associated with intermittent glaciation may have correspondingly altered the vertical ocean circulation, so as to modulate the resupply of CO_2-rich deep water to the surface—and hence to the atmosphere.

Against this background of large-scale, 100,000-year changes, we must now weigh present trends, such as CO_2 produced by man, principally by combustion of forests and fossil fuels. Figure 18 shows a recent estimate of the principal reservoirs and annual cycling rates of carbon dioxide. The total amount present in the atmosphere today is less than 2 percent of that stored in the ocean, and about 6 percent of that stored as fossil fuels underground. The annual flux from all sources is about 40 percent of the mean atmospheric concentration, of which human-produced combustion comprises about 2.5 percent. Scientists have estimated that if the present rate of increase of industrial production is maintained, the increase in carbon dioxide may raise the mean climatic temperature of the world enough (4–6°F) to melt the polar ice caps, thus raising the level of the oceans by an additional 150 feet. How long this might take will depend upon the rate of solution into the oceans—a still unknown quantity, but one that is currently receiving assiduous attention.

Momentum exchange between the atmosphere and the ocean is something of a one-way proposition—all in favor of the ocean. Wind blowing over the water exerts a stress on the surface that pushes it along, thus setting up a circulation. At the same time, wind produces waves, which transmit some momentum, but also make the surface effectively rougher, thus increasing the stress of the wind and abetting the circulation. In effect, the waves also make the wind rougher (more turbulent), but

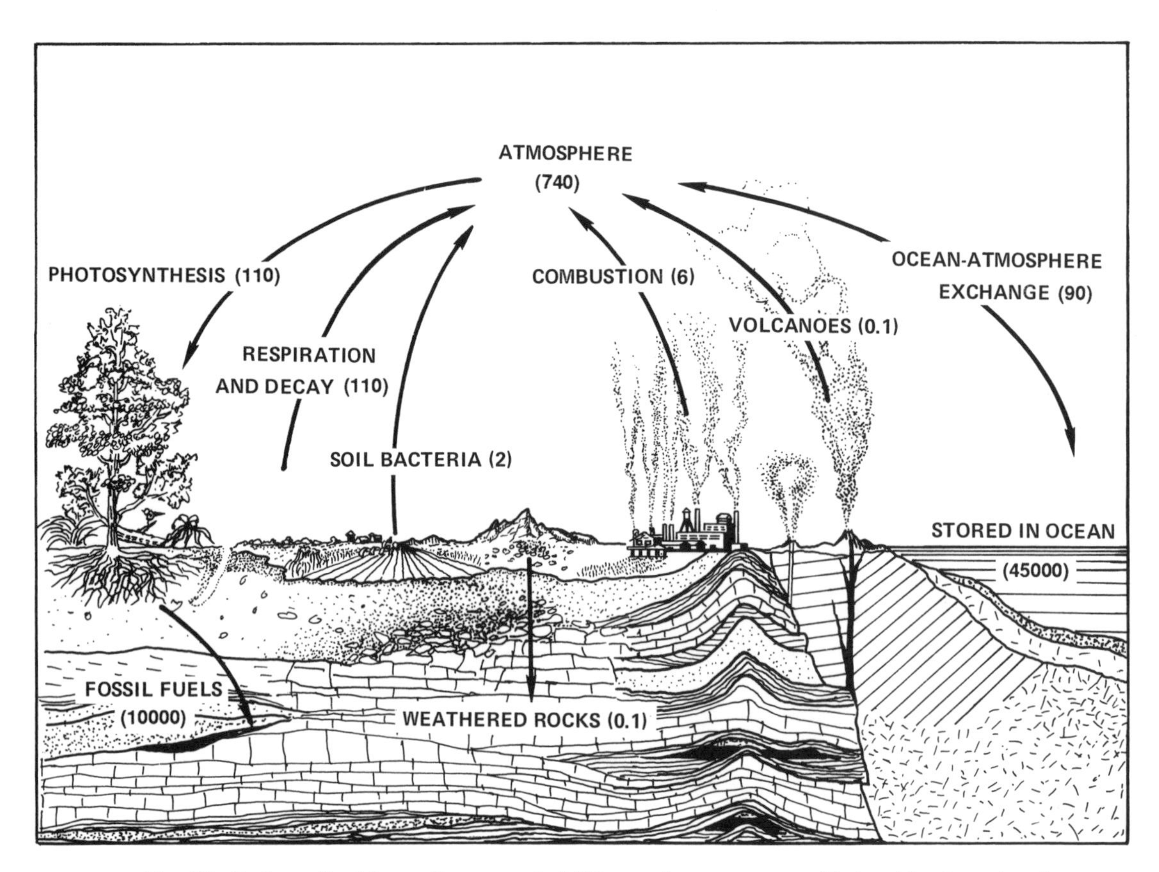

Fig. 18. Carbon dioxide cycling rates in billions of tons per year (*Scientific American*).

this turbulence usually is of fairly small scale compared with the dimensions of even a small yacht, and thus does not affect the set of sails or the heading significantly. Although very steep storm waves may shadow and momentarily becalm or back-wind a sail, this effect is difficult to distinguish from those produced by heavily rolling and pitching motions. There is no evidence that the ocean anywhere significantly transfers momentum to the atmosphere.

5

Atmospheric Circulation

THE ATMOSPHERE AS A HEAT ENGINE

The *vertical* circulation of the atmosphere might be likened to a relatively inefficient gas turbine, in the sense that air at high pressure (one atmosphere at sea level) is heated from beneath by the hot surface of the earth. The heated air expands until it becomes less dense (lighter) than the cooler air above it, whereupon it commences to rise like a hot-air balloon. As it rises it is cooled again, partly owing to further expansion under reduced pressure at higher altitudes, and partly by radiation to the cooler surrounding air. Finally, it comes into density equilibrium with neighboring air, stops rising, and spreads out to replace cooler air that has meanwhile subsided to replace the rising air.

The local, chimneylike circulation described above is called a *convection cell,* and such cells are characteristic of uniform heating of calm air over a wide area. The rapidly rising warm cores of these cells are called *thermals* by glider pilots, who can obtain free rides to high altitude by circling within them. Convection cells are often visibly manifested by puffy (cumulus) clouds at elevations of 2,000–5,000 feet, owing to condensation of surface moisture carried aloft and cooled to the dew point. Great arrays of such cumuli—known to meteorologists as "popcorn"—are often observed in satellite photographs of the relatively calm equatorial oceans. Because these are usually low-level clouds that travel with the wind, consecutive photographs provide a convenient means of estimating surface winds.

Occasionally, under appropriate warm, humid conditions, convection cells coalesce and grow to tremendous size. Rising to altitudes of 30,000 to 50,000 feet, they poke their gleaming heads, like giant cauliflowers, into the calm air of the stratosphere, where they spread out into the characteristic anvil shapes known as *cumulonimbi.* Once generated, these great thermal engines obtain their energy from condensation of water vapor and become self-perpetuating as long as the supply of warm, moist air holds out. Second only in intensity to hurricanes, their cloud chimneys contain raging thunderstorms, in which *vertical* winds can exceed 50 knots, and may be accompanied by heavy rain, snow, and even very large hailstones. There are well-documented cases of military aircraft being dismembered after penetrating such clouds, forcing their pilots to bail out. Caught in the central maelstrom, one hapless parachutist was held aloft for nearly two hours, during which time he made several vertical round trips, being successively frozen and thawed, and beaten by rain and hail. He was finally ejected, and his torn parachute lowered him to safety.

The *horizontal* circulation of the atmosphere arises from nonuniform heating rates over different portions of the earth. Globally, heating is maximized near the equator, where the sun spends most of its time and its rays are more nearly vertical. Near the poles, heating is minimal because the sun's rays have a lower average angle, and thus the net radiation per unit horizontal area is smaller than at the equator. In addition, less energy is absorbed and more is radiated by the high reflectivity (*albedo*) of snow and ice. Even in temperate latitudes there are significant differences in the heating rate over land and sea. Temperatures over land tend to fluctuate markedly—both diurnally (owing to day-night surface temperature changes) and seasonally. Ocean temperatures vary only slightly with the seasons, mainly because the heat is distributed through a relatively thick layer.

Different heating rates at any two places produce pressure differences at all corresponding altitudes, thus establishing horizontal pressure gradients that vary with altitude. With surface heating, the thermal gradient will be greatest there, and will diminish with altitude. Air, being a fluid, always tends to flow from high pressure toward lower pressure to correct the imbalance. This effect is somewhat similar to that created by opening the door between a warm room and a cold room: Cold air immediately flows along the floor toward the warm room, and hot air flows in the opposite direction near the ceiling. The system ultimately comes to rest with a stable, warm layer overlying a colder layer in both rooms. If heating is continued in the warm room, the circulation will continue. If, instead, the warm room is cooled, it will reverse direction.

In the atmosphere, such a circulation is called a *horizontal advection* cell. It is only distinguished from the vertical convection cell because its horizontal dimensions exceed the height over which vertical motions occur. This height may be only a few hundred feet under a strong inversion layer, but never exceeds the height of the tropopause (about 10 miles at the equator). The horizontal extent of the largest advection cells varies from several miles to several thousands of miles, and is usually limited by other factors yet to be discussed.

Good natural examples of horizontal advection are the familiar land and sea breezes observable along many shorelines in otherwise calm weather. During the early morning hours the land is heated by the sun and becomes warmer than the water offshore. Air warmed by the land then rises, to be replaced by an onshore flow of cool air (sea breeze). In the late afternoon, thermal equilibrium may exist for a time, and the breeze dies. At night the land temperature may drop below that of the water, air flow reverses, and a land breeze may be established until early morning. In most areas where such breezes occur, the thermal differences are greatest in the daytime, and the strength of the sea breeze greatly exceeds that of the land breeze. Both effects are usually limited to about 20 miles from the coastline.

While advection cells of all sizes exist in the atmosphere, those due to day-night temperature changes are always small, and frequently do not show up on weather maps. There are larger cells that reverse their flow directions seasonally, owing—at least in part—to thermal reversals between large ocean areas and adjacent continental areas (the monsoon winds). Still larger cells of a permanent nature, such as those involving the main circulation between the poles and the equator, only vary slightly in intensity between summer and winter. These large-scale cellular motions, which are initiated by differential heating, resisted by friction at the surface of the sea or land, and strongly altered by the effect of the earth's rotation, determine the general pattern of atmospheric circulation described below. Important modifications are provided by cellular interactions, and by the topography of the land.

THE MAJOR WIND SYSTEMS

Before undertaking the rather complicated causal description of the circulation of the atmosphere as it actually exists, consider first what the circulation would be like on a nonrotating earth. Such an Aristotelian model (fig. 19) conceives of the earth as rigidly fixed in the firmament, while the sun and stars revolve around it. The sun would thus produce a warm belt around the earth, centered on its equator and tapering off toward the colder poles. This differential heating would establish a wholly meridional circulation, as shown in the northern (upper) hemisphere of figure 19. Here, warm air rises near the equator and cold air sinks near the poles. The surface winds flow toward the equator and there is a poleward return flow aloft. Because of the convergence of air sinking near the poles, such a simple flow pattern would be inherently unstable, and would break down into two or three counterrotating *toroidal* (doughnutlike) cells, as shown in the lower hemisphere of this figure.

Returning, now, to the real earth, nothing is changed, except that the sun is standing still and the earth is rotating. Although the tilt of the earth's equator by 23.5° from its orbital plane somewhat broadens the heating pattern and introduces a seasonal variation, there is still the same general tendency to establish a cellular meridional circulation in the atmosphere. But there is one important difference that

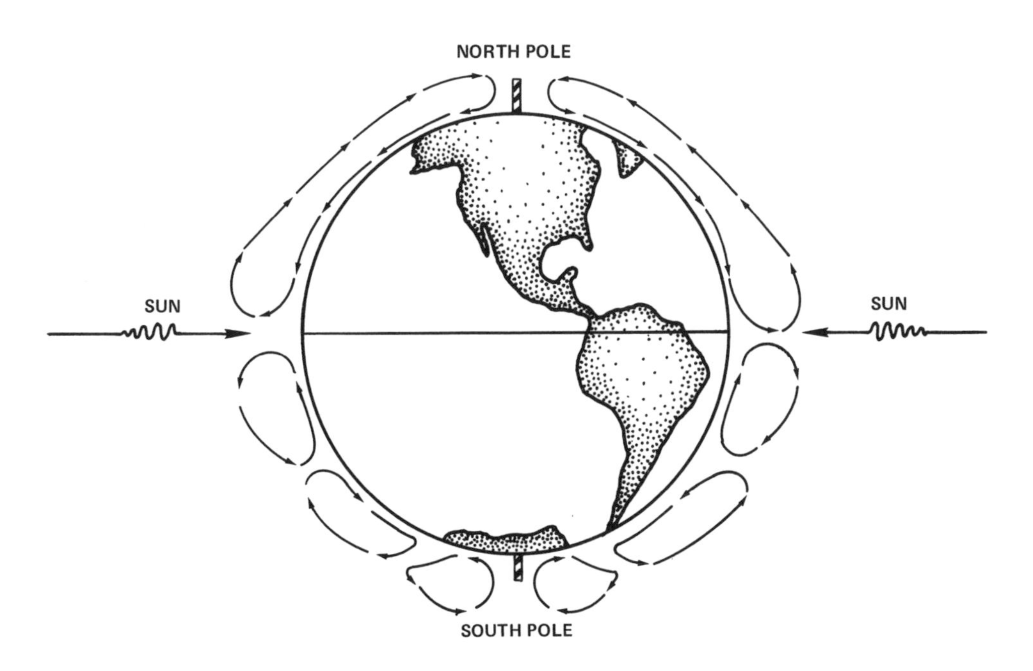

Fig. 19. Hypothetical atmospheric circulation in an Aristotelian universe, with sun circling earth. The single-cell atmospheric circulation model shown for the Northern Hemisphere would be unstable, and would probably break down into the three-cell model shown for the Southern Hemisphere.

alters the whole circulation scheme. In changing from a coordinate system fixed in space to one that rotates with the earth, we run afoul of a fundamental proposition of mechanics. Unless otherwise constrained, all moving bodies (in this case, air particles) like to continue in straight lines. Thus the orbital plane of a satellite circling the earth remains fixed in space, while the earth revolves beneath it. To an earthbound observer, no matter what the orbit,* and irrespective of geographical position, the satellite's path appears to be curved. If a person is in the Northern Hemisphere, it curves to the right; if south of the equator, to the left.

Whether satellite or air particle, if motion is reckoned with respect to the earth's rotating coordinate system, we must invoke a fictitious Coriolis** force to account for this apparent deflection from straight-line motion. As applied to atmospheric circulation, the Coriolis force acts always at right angles to the local wind direction, and its magnitude increases in direct proportion to the wind velocity V. For convenience, it is usually divided into two orthogonal components:

Horizontal component: $F_h = 2\Omega V_h \sin \phi$ (ft./sec.^{2}),
Vertical component: $F_v = 2\Omega V_e \cos \phi$ (ft./sec.^{2}),

where $\Omega = 7.3 \times 10^{-5}$ radians/sec. is the earth's angular velocity, V_h and V_e are the horizontal and easterly components of wind speed, respectively, and ϕ is the latitude in degrees. The vertical component, being vanishingly small compared with the force of gravity, is usually neglected.

Although the horizontal component is equally small, it is of the same order as the thermally induced horizontal pressure forces associated with cellular circulation. It acts to continuously deflect air flow away from meridional straight-line paths from high to low pressure—always to the right in the Northern Hemisphere, and conversely in the Southern. Thus, surface flow toward the equator is deflected westward; polar flow aloft, to the east. Air diverging from a high-pressure region spirals outward, clockwise north of the equator and counterclockwise down under. Air converging toward a low-pressure region spirals oppositely inward. To verbally simplify these reversals in direction between hemispheres, convergent spirals are called *cyclonic*, and divergent spirals *anticyclonic* in either hemisphere. Lastly, the horizontal component of the Coriolis force varies in proportion to the sine of latitude; it is maximal at the poles, and vanishes at the equator.

On a global scale, the net effect of the Coriolis force is to distort the simple toroidal model for cellular circulation within the troposphere. Although there still remain three counterrotating wind cells in each hemisphere (like the lower half of figure 19), the dominant wind directions within each cell are more east-west than north-south, and reverse direction within adjacent cells. In effect, each doughnutlike cell is a sort of spiral wind tube, with a high-velocity central core (jet stream) surrounded by concentric sheets of lower velocity winds.

Figure 20 is a somewhat idealized polar cross section of the earth, with the troposphere greatly expanded in scale.*** The left and right portions of this figure

*Except for orbits in the earth's equatorial plane.

**Named after the French physicist who first worked out the coordinate transformation. Strictly speaking, it is a force-per-unit mass, and has the units of acceleration.

***If the earth were drawn to an undistorted scale 8 feet in diameter on a chalkboard, the entire atmosphere and greatest ocean depths would be compressed to the width of the chalk line. Thus, in reality, the wind tubes are more like thin ribbons.

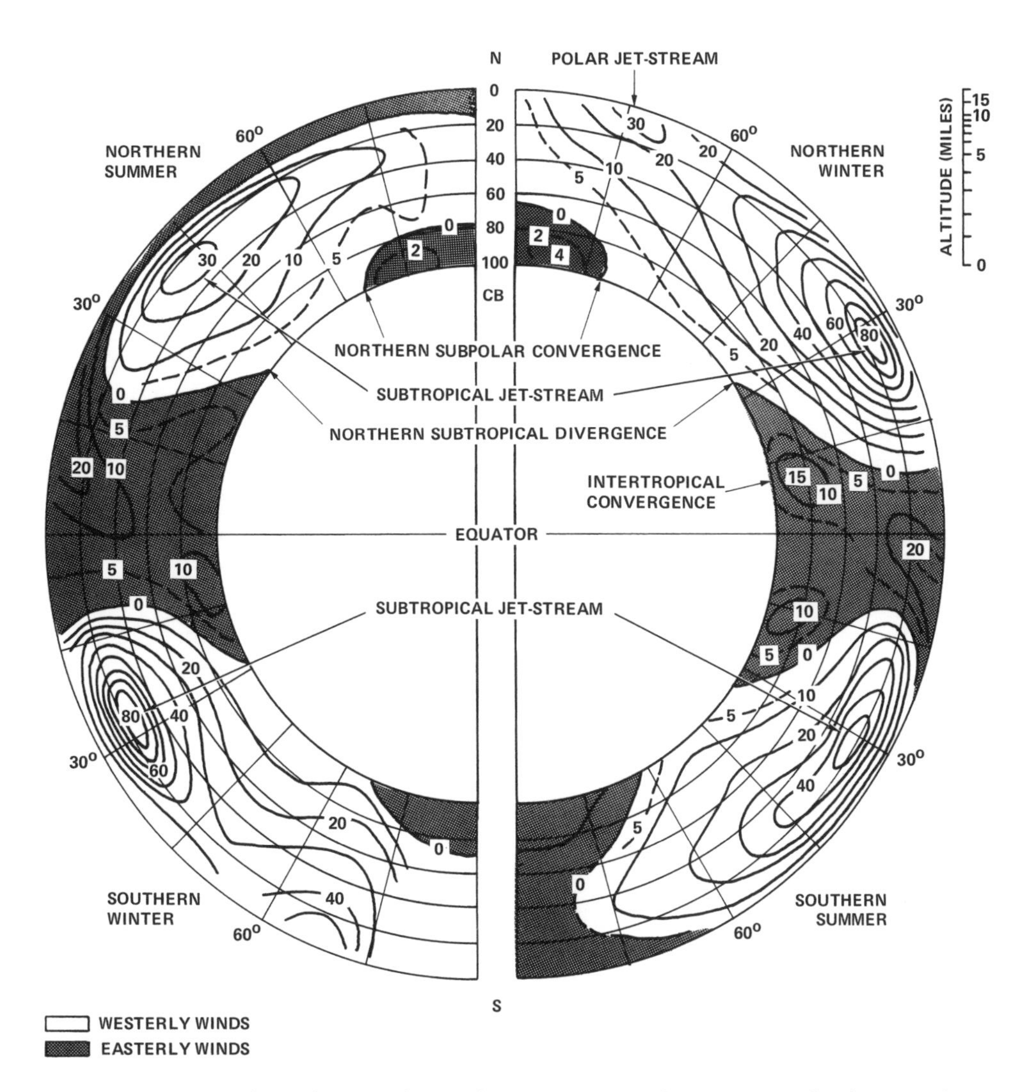

Fig. 20. Mean zonal winds over the earth in summer and winter. Wind tubes are shown as contours of equal wind/speed, in knots. Actual flow spirals around each wind tube (after Flöhn).

indicate mean wind distributions in northern summer and winter, respectively. Wind contours are shown in knots, superimposed on a grid of pressure (in centibars) vs. latitude. Places where surface air is rising are indicated as *convergences.* The intermediary descending regions, near 30° latitude in both hemispheres, are *divergences.* There are also weak divergences near the poles, but they are poorly defined and tend to wander about. All rising and sinking motions are very slow compared with the east-west flow within the wind tubes.

The general circulation scheme shown is sufficiently persistent in all oceans that its various elements have been given discrete names. It is also sufficiently altered over the continents, and by seasonal changes in the rate of heating, that some of these elements appear and disappear in certain regions, or are altogether missing in others. We shall discuss them in order of increasing north latitude, beginning at the equator, and recalling that each has its southern counterpart, although the east-west relations are the same in both hemispheres.

The wind tube nearest the equator encloses the *tropical easterlies.* Their southwesterly directed spiral components at the surface are the familiar *northeast trade winds,** so called from the obsolete word "tred," meaning "steady course." Near the equator the wind vectors swing more toward the south because of the southerly pressure gradient, surface friction, and the weakening influence of the Coriolis force. Meeting their southern counterparts, they rise steeply together in a narrow zone of low pressure called the *intertropical convergence* (more commonly known as the *doldrums*). The ITC is a region of active vertical convection cells, and is characterized by a warm, humid climate, with weak, fitful winds interspersed with strong squalls, heavy rain, and towering cumulus clouds. As the combined winds rise, they increase again in velocity toward the west, forming a strong, central jet with velocities as great as 20–30 knots at an altitude of about eight miles. Above this, near the top of the troposphere, the winds diverge. The northerly components swing to the right, passing through north toward the east. As they again descend toward the surface, they meet the counterrotating spiral wind vectors of the next great wind tube enclosing the easterly blowing *prevailing westerlies.* These combined descending wind vectors separate near the surface in a zone near 35° N called the *northern subtropical divergence* (NSD), which is a region of high pressure, clear and dry air, light winds, and generally benign climate. The NSD is otherwise called the *horse latitudes,* because ships carrying horses to the Americas from Europe frequently were becalmed there, with the result that many horses perished from thirst.

The westerlies wind tube greatly overlaps that of the easterlies at high altitude, such that its intense core of high-velocity winds, called the *northern subtropical jet stream* (NSJ), falls at about the same latitude as the NSD. The NSJ varies between 30 and 80 knots from summer to winter, with sporadic increases to as high as 150–200 knots. The westerlies spiral toward the northeast along the surface to about 70° N, where they meet and ascend with southeasterly surface winds from the polar wind tube—which rather resembles a doughnut—at the *northern subpolar convergence* (NSC). The westerlies tend also to overlie the easterlies aloft clear to the pole in winter, with the seasonal development of a secondary wind core, the *northern polar jet stream* (NPJ) almost above the subpolar convergence. The NPJ has much lower peak speeds than the NSJ (20–40 knots).

The average circulation in the southern hemisphere is essentially the same, except for one significant difference. The mean climate of the troposphere over the Antarctic continent, with its perpetual ice cover, is some 20–25°F colder than in the Arctic. This has the effect of displacing all southerly climatic zones to the north. Even the ITC, which is sometimes called the meteorological equator, lies several degrees north of the geographical equator in all oceans, except for the western Indian and Pacific oceans in late summer. This circumstance greatly reduces the frequency of hurricanes in the Southern Hemisphere, and altogether eliminates them in the south Atlantic. On the other hand, the intensity of the prevailing westerlies is about 40 percent higher in the Southern Hemisphere because of steeper thermal and pressure gradients. Strong westerly winds prevail at all altitudes over the Antarctic continent in winter, and the weak easterly "doughnut" appears only briefly in early summer.

* Through a perverse dichotomy, winds are conventionally defined by the directions from which they blow, while ocean currents are referred to in the reverse sense.

PERTURBATION OF THE ZONAL WINDS

The mean wind system described above consists of the combination of a weak, cellular, meridional circulation and a strong zonal flow (along latitude parallels) in opposite directions within adjacent cells. However, as is often the case in nature, the above idealized picture is only a rough approximation of the actual circulation, because of three disturbing influences: (1) the unequal distribution of continents and oceans, (2) the seasonal variation of solar heating with latitude occasioned by the tilt of the earth's pole of rotation with respect to the plane of its orbit, and (3) the tendency for traveling pressure waves and vortices (cyclones) to form and dissipate along the zonal convergences. All these factors affect both climate and weather in any particular place.

The continents constitute thermal and topographic barriers to the general circulation. In summer they become much hotter than the oceans in all latitudes not covered by ice and snow. Conversely, in winter and above 35° latitude, the continents are colder than the oceans. In fact, in the Arctic winter, it is much colder over Siberia and northern Canada than it is at the North Pole. This seasonal shift of the polar high-pressure zone brings about a corresponding reversal in the polar circulation, which is normally easterly in summer. Similarly, in lower latitudes, hot air rising over the continents disrupts the zonal high-pressure belts associated with descending flow in the subtropical convergences, so that they are broken up into rings of alternating high (oceanic) and low (continental) pressure areas. The anticyclonic wind flow set up around these permanent oceanic highs is abetted by the prevailing surface westerlies and easterlies of the zonal circulation. But over the continents the thermal cyclones oppose the zonal winds, resulting in hot, dry, stable cells that characterize the desert regions of Africa, America, and Australia. Elsewhere the continents are subdivided by north-south mountain chains high enough (6,000–30,000 feet) to block the surface winds, forcing them to rise above the mountains—where cooling from rapid expansion often results in heavy precipitation on their windward slopes—or to detour around through lower passes in other latitudes.

Differential heating and diversion by high mountains also produce short-term effects in the middle and upper troposphere (20,000–50,000 feet) in the form of a series of giant, horizontal waves, having wavelengths of several thousand miles (fig. 21). These waves appear to originate somewhat irregularly, as sinuous perturbations of the subtropical jet streams, and propagate slowly eastward. As they grow in amplitude and wavelength, they slow down, and may stop—or even reverse directions. After two or three weeks the wave system becomes unstable and breaks down into two bands of counterrotating eddies, which tend to drift apart as shown and die away over periods of several weeks. Because they originate along permanent discontinuity surfaces (the subtropical divergences) between cold, dry polar air and warm, moist tropical air, these successive waves and eddies represent alternating intrusions of polar air toward the equator and warm air into polar regions, temporarily establishing a "topsy-turvy" climate, during which Alaska may be warmer than Florida.

The high-level easterly waves produce corresponding pressure disturbances at the sea surface, in the form of an alternating series of high- and low-pressure cells. Centered on the subtropical divergences, these cells also propagate slowly eastward. Local convection between them generates a continuous series of smaller cyclonic

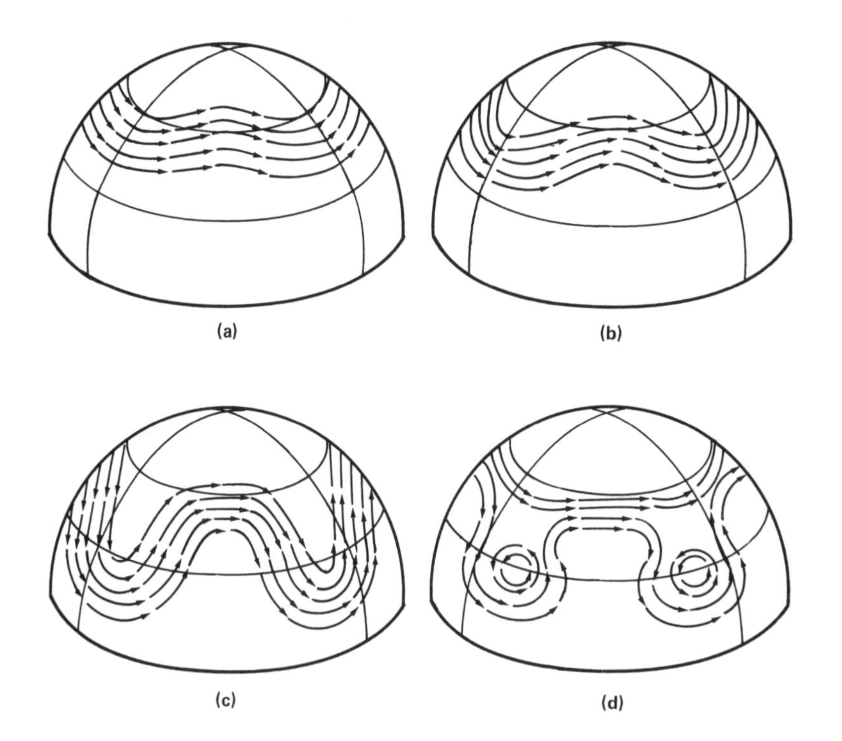

Fig. 21. Development sequence for upper-air easterly waves on the boundary between polar and tropical air masses. Cold polar anticyclones are finally cut off and drift southward.

and anticyclonic local wind vortices, the familiar meteorological *warm* and *cold fronts.* For reasons still not well understood, some of these vortices die out, while others grow rapidly in size and intensity and detach themselves from the central stream. The anticyclones drift toward the equator to merge with the subtropical high-pressure belt, while the cyclones move slowly poleward within the belts of prevailing westerlies to become the great oceanic storms for which the "roaring forties" are so aptly named. Occasionally a drifting high-pressure eddy may become stalled and effectively block the stream of westerly cyclones, such that they must detour around it. Some of these "lost storms" tend to wander toward the equator, grow greatly in size, and last for several weeks. Such *subtropical cyclones* are prevalent in the winter months at midlatitudes in all oceans, but particularly in the Pacific, where they can produce unusually strong, sustained winds from the northwest in Tahiti or from the southwest near Hawaii (Kona storms).

In somewhat the same fashion, easterly waves develop in the lower troposphere, jet over the intertropical convergence, and appear to induce westward traveling vortices in the equatorial trade winds at the surface. Again, most of them die out, merely resulting in three- to four-day shifts in the intensity and direction of the trade winds. The equatorial anticyclones tend to merge with the intertropical high-pressure zone, but an occasional cyclone drifts poleward and intensifies to become a tropical hurricane or typhoon, depending on whether you are sailing in the Atlantic or the Pacific (see page 80).

All of these effects are seasonally modified in some degree by migrations of the earth's thermal equator. In general, local surface winds are apt to be seasonally reversed in all divergence or convergence zones—wherever they are narrow compared with the extent of their annual migrations. Such reversals are called *monsoons,* the most striking example of which is that of southern Asia and the East Indies. Here, relatively cool, dry northeasterly winds prevail in fall and winter, under the influence of the Asiatic high-pressure cell over Siberia. In summer, this high is replaced by a deep low, centered over southwest Asia. As a result, the northeasterlies die away, and the intertropical convergence migrates north.* No longer opposed, the southeast trades of the Indian Ocean veer northwesterly across the equator and impinge on the massive wall of the Himalayas. Uplift and cooling produces torrential rains in India throughout the summer months. Thus, the climate of this region, as well as the lives of its inhabitants, is dominated by a seasonal wind reversal.

For reasons apparently associated with an occasional heat imbalance between the northern and southern hemispheres, very large-scale perturbations of the zonal wind systems occur at sporadic 7- to 10-year intervals that have oceanwide implications. In concordance with an extraordinarily strong winter high-pressure area over Siberia, the influence of the Asiatic dry monsoon is felt as far east as the Pacific date line, causing suppression of the easterly trades in both hemispheres in the Pacific and Indian oceans. Aside from inducing drought conditions in Asia, weakening of the equatorial driving force drastically alters the circulation along the westerly margins of both oceans, as well as the local weather (see Part III, El Niño).

MEAN SURFACE WINDS OVER THE OCEANS

The surface wind pattern is the footprint, so to speak, of all the complicated structure above. As such, it should be kept in mind that the pretty patterns of arrows showing the mean wind directions in nautical handbooks are not stream lines; that is, they are not the paths that would be followed by neutral-density particles set free in the windstream. For the air flow in the troposphere is always turbulent, and contains vertical eddies that would soon sweep such particles aloft and disperse them into the main wind tubes, where they would be carried all over the world in a matter of weeks. Instead, the wind vectors on such diagrams give only the statistically prominent speed and direction of surface winds at a particular place and season, averaged over many years. At almost any location, during the course of several days, the wind is apt to vary widely in speed and direction as consecutive minor disturbances pass through.

With these reservations, figures 22 and 23 show *average* surface wind distributions over the world's oceans, from which we can make four important generalizations:

1. Except for the north Indian monsoon, the tropical, easterly trade winds in both hemispheres—and in all oceans—are by far the most uniform and persistent. They change only superficially from summer to winter.
2. The Southern Hemisphere westerlies are the strongest prevailing winds, particularly in winter. But, owing to frequent cyclonic disturbances, they are much less persistent than the trades. Thus, mean indicated wind speeds on a surface wind chart should, rather, be interpreted as ranges, with a cycling time of three to four days, as indicated in the following table.

*In the western Pacific and Indian oceans there are occasionally two ITCs separated by a narrow belt of surface westerly winds.

Wind Speed (Knots)	Percent of Time
0–10	10
10–25	40
25–50	40
50+	10

3. In the Northern Hemisphere, the seasonal alternation of hot and cold poles over Siberia and Canada almost completely reverses the subpolar atmospheric circulation over the Pacific, and strongly alters it over the North Atlantic. In both cases, average winds are strongest in winter. Additionally, the Siberian reversal produces the monsoon changes in the western Pacific and Indian Ocean previously described.
4. Semipermanent high-pressure centers (H) near 30° latitude in both hemispheres mark the subtropical divergences that separate the easterly and westerly wind tubes (see figure 20).

More detailed information on mean surface winds can be obtained from the pilot charts issued monthly or quarterly for all oceans (except the Arctic) by the U.S. Defense Mapping Agency, which give monthly wind roses for each 5° square. These charts also give surface temperatures, compass variation, boundaries for floating ice, main shipping lanes, and a great deal of other useful information. A much more condensed compendium of weather and climatic information useful for cruise planning is the *U.S. Navy Maritime Climatic Atlas of the World,* still available in a single volume from the U.S. Government Printing Office.

LOCAL WINDS

Aside from the strong frontal winds associated with major storm systems traveling across the oceans, unusual local winds are generally confined to the weather coasts

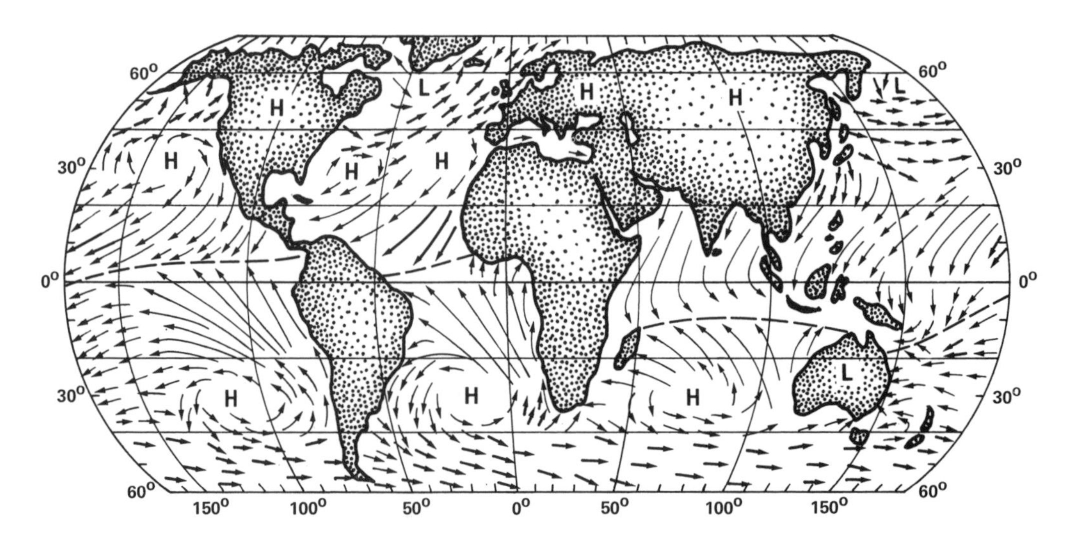

Fig. 22. Average surface winds over the oceans in January. Width of arrows indicates relative speed; length indicates persistence. Dashed line locates intertropical convergence (adapted from Dannevig).

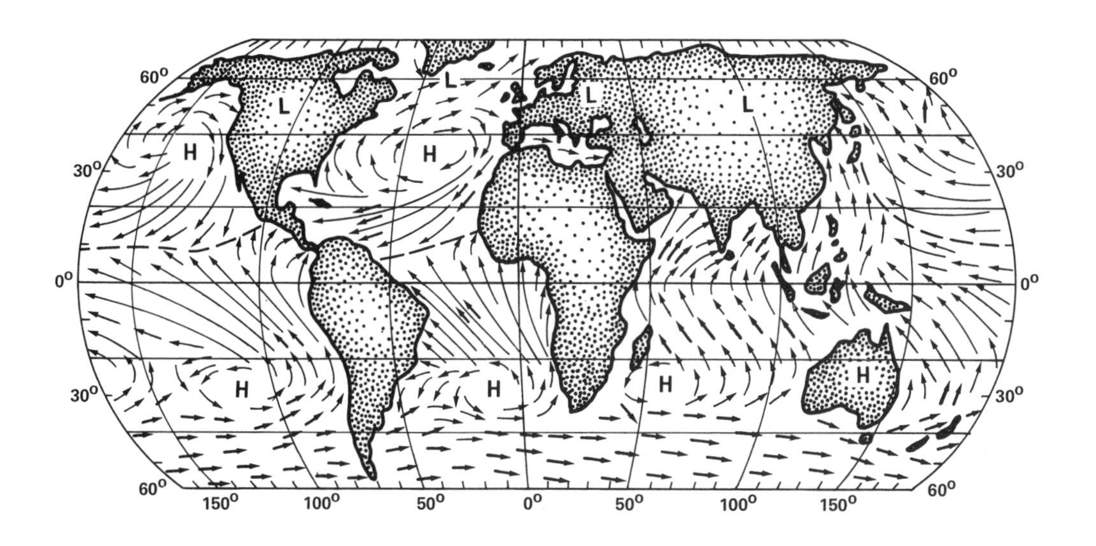

Fig. 23. Average surface winds over the oceans in August. Note monsoon reversals in north Indian Ocean; also the northwestern Pacific intertropical convergence disappears from the Indian Ocean (adapted from Dannevig).

of relatively large land masses. Such winds may result from local heating or cooling, although the region affected may be quite large and wind velocities dangerously high, or they may occur during passage of remote frontal disturbances. Local winds may arise suddenly, and as suddenly die—often without other weather symptoms. Therefore it behooves the mariner to weigh their probability in unfamiliar waters, particularly in such areas as the Gulf of California and the Mediterranean Sea, which can be beset from any direction by capricious winds of high intensity.

Local winds fall into two general categories: *katabatic* or downslope winds, caused by sinking of relatively dry, dense air into a region of lower density; and *anabatic* or upslope winds, generated by air rising in hot, low-pressure regions. Because both types are very low-level winds, surface friction and local topography exert more influence than the Coriolis force in guiding the flow, which therefore travels more nearly in the direction of lower pressure. The land and sea breezes previously discussed are small-scale versions of katabatic and anabatic winds. Katabatic winds are further divisible into *föehn* or *fall* winds, according to whether they are warmer or colder than their surroundings. Föehn winds are gradient winds that have been forced to flow upward over a mountain range before descending to a lower level. When the air is forced aloft, condensation-cooling and precipitation rob it of moisture, and subsequent pressure-heating as it descends gives it a characteristic warm, dry property. Fall winds represent the gravity flow of cold air to lower levels from sources at higher levels. Although the air is warmed during descent, it still arrives cooler than its surroundings, sometimes resulting in temperature drops as rapid as 25°F within an hour. All katabatic winds achieve their highest velocities when funneled through narrow mountain passes *(jet effect)*, and may gust to 100 knots or more, with intervening periods of relative calm.

Anabatic winds are generally steadier and less turbulent than katabatic winds, and rarely exceed speeds of 30 to 40 knots. These winds reach their greatest intensity during their onset along coastlines, where a cold current runs close inshore in tropic regions, such as west equatorial Africa, where gusts of gale proportions have been reported.

Local winds of the above types occur so frequently in many places that they have been given specific names, a few examples of which are listed in table 1. For more detailed descriptions, readers are referred to the *Coast Pilot* (U.S. and possessions) and the *Sailing Directions* (foreign coasts) obtainable from the U.S. Government Printing Office.

Table 1. Examples of strong local winds

Type	Name	Location	Direction	Max Speed (Knots)	Season
I. Katabatic					
(a) Fall winds	Bora	Adriatic	NE	40–60	Winter
	Mistral	G. of Lions	N	50–70	Winter
	Papagayo	W. Costa Rica	NE	20–40	Winter
	Williwaw	Magellan Str.	Any	50–70	Any
(b) Föehn winds	Southeaster	Cape Town	SE	50–60	Summer
	Zonda	Argentina	W	60–80	Winter
	Santa Ana	S. Calif.	NE	30–50	Fall
II. Anabatic	Sirocco	N. Medit.	SE	30–40	Spring
	Virazon	Chile	W	20–30	Summer
	Khamsin	N. Africa	SE	30–50	Spring

6

Climate and Weather

OCEANIC CLIMATE

Climate is ordinarily defined as the prevailing weather in a region. As such, it is of primary interest to persons concerned with establishing fixed facilities that are climate-oriented, as, for example, resort hotels, tracking stations, and observatories. Climate is of interest to the mariner mainly insofar as it affects operations—or, for the cruising yachtsman, comfort or discomfort. Thus I will treat climate as the background for a set of typical weather sequences to be expected at various times of the year within the principal climatic zones, defined by the oceanic wind belts and their intervening divergences and convergences.

Aside from perturbations of the zonal wind field that determine the weather, the mean climate in *midocean* is remarkably uniform compared with that over the continents. The air temperature is everywhere nearly the same as the water temperature, which decreases smoothly from the equator toward the poles, with little seasonal variation. In summer the temperature drops more slowly with latitude along the western margins of all oceans than along their eastern margins. This is a result of their uniformly anticyclonic circulation (see Part III), whereby strong, warm westerly boundary currents, such as the Gulf Stream and the Kuroshio, greatly moderate the otherwise severe climates of Scandinavia and northern Japan, respectively. In winter, air temperatures in the westerly oceans drop severely, owing to the dominant influence of cold continental westerly winds. Similarly, easterly cold currents, like the Humboldt and the Benguela, bring polar water nearly to the equator off Chile and West Africa, producing heavy fogbanks along their coasts much of the year.

The mean meridional distribution of precipitation is unlike that for temperature, despite the fact that oceanic air nearly always contains a high percentage of water vapor, which increases with temperature. Precipitation depends sensitively upon vertical air motions; ascending motions produce cooling and condensation of moisture, whereas descending motions act to heat air and reduce its relative humidity. Thus, while there is a general tendency for precipitation levels to decrease poleward in concert with surface temperature, this effect is dominated by the cellular atmospheric circulation on all scales; thus, high precipitation occurs within convergent regions (atmospheric lows), while the climate in divergent regions is relatively dry. These remarks hold not only for the semipermanent high- and low-pressure features associated with the zonal circulation, but also for the traveling, ephemeral pressure waves, cyclones, and anticyclones that compose the weather.

But precipitation is one of the most difficult weather indices to report reliably. Very subtle changes in weather patterns can drastically alter the *amount* of rain that falls in a given area. It is not unusual for exceptionally rainy regions to suddenly experience long, unseasonable periods of dry weather, or for an exceptional typhoon to dump a year's rainfall in twenty-four hours in an otherwise arid sector. The east central Pacific contains dozens of uninhabited islands in the Line and Phoenix groups that used to support abundant vegetation and substantial populations, but which have been so dry for the past one hundred years that cocoa palms no longer prosper. Therefore, one should take any generalizations regarding weather over the oceans—including the following—with a grain of salt.

TYPICAL WEATHER SEQUENCES

The word *weather* probably has had more adjectives applied to it than any other in the English language, unless it be *love.* Nevertheless, there appears to be no single volume of weather lore that, in simple terms, gives one the ordinary sequence of weather in the various oceanic climate zones. Vivid verbal pictures of typhoons have been given by Conrad and Masefield. Robinson has ultimately described the "ultimate storm." But who writes about being becalmed in the roaring forties? Yet there are intervals—albeit brief—when the number of calms in these latitudes exceeds that of gales. The probable reason is that the weather is much too variable, and the oceans are much too vast, to generalize the subject in any simple manner. Nevertheless, it may be of greater interest to devote a few remarks to the *course* of weather in a few places than to the probable state of it at any one time.

Imagine yourself bound on a hypothetical voyage from pole to pole on a median of 169.25° W. This would almost be possible without touching bottom except for tiny St. Lawrence Island in the Bering Sea and the Ross Barrier in Antarctica. For practical purposes, we will confine our voyage to the seasonal limits imposed by the polar ice packs, giving brief descriptions of representative local weather sequences in February and August along this meridian. In most latitudes these months most nearly represent midwinter and midsummer conditions because their seasonal temperature maxima lag behind the sun by a month or two. The southern limit of the solidly frozen Arctic ice pack varies in latitude from about 65 to 75 degrees between February and August, respectively. In winter, roughly once weekly, brief periods of calm, clear weather, with a mean temperature of about –5°F, are interspersed between passages of one or more cyclonic storms from the southwest. Typically, the wind will arise from the southeast, together with banks of high and then low stratus clouds, followed by snow. The wind usually increases to 30–40 knots during the next twelve hours, backing into the northeast, and may blow steadily from this direction for twelve to twenty-four hours, with heavy snow and temperatures dropping to –20° to –30°F. Depending upon the particular cyclone track—one often closely followed by another—the wind may continue to back westerly and die away or veer toward the east, with only a moderate lull before again increasing to gale proportions, which occurs about 30 percent of the time.

In summer, the situation in the Bering Sea is quite the opposite, with frequent periods of several days of calm or of light, variable winds, and temperatures in the middle 40s. Dense fogbanks are prevalent during these calms, and may even prevail during sporadic outbursts of southwest winds up to 20–30 knots that rarely persist for twenty-four hours. The skies are uniformly overcast with layers of stratus and nimbostratus clouds, and rain or drizzle can continue for weeks at a time. The conventional picture of a band of low pressure marking the northern subpolar convergence (see figure 20, chapter 5) is only poorly manifested in the Arctic by persistent troughs of low pressure that migrate slowly eastward between 40° and 60° latitude in winter, and tend to disappear in summer. Their regular recurrence in certain seasons has led to their definitions as Aleutian (Pacific) and Icelandic (Atlantic) lows, although their movements and intensities appear to be related to the development of upper troposphere waves in the subpolar jet stream, as previously discussed.

Our next weather stop is at 50° N, in the middle of the prevailing westerlies. Probably the stormiest region of the hemisphere, this belt spans 10 degrees of latitude, and has little seasonal migration, more or less oscillating above and below the Aleutian Island arc. Calms are rare in winter, and the prevailing winds are west to southwest at about 15–20 knots, with temperatures near the freezing point. This is the frequent domain of the Aleutian Low. Every two or three days wan sun and scattered clouds give way to dense stratocumuli and accompanying rain squalls. The wind veers the compass round to north, sometimes increasing in intensity to 50–60 knots; the temperature drops to −10° to −15°F; and rain turns to sleet, and then to snow or soft hail. The wind may drop abruptly with the frontal passage, and the cycle starts all over again. Summer conditions are very similar to those in the Bering Sea, with protracted periods of fog, low stratus, and drizzle, and gentle to moderate westerly breezes. A week or so of this weather may be followed by the passage of several fronts at two- or three-day intervals, bringing day-long moderate gale winds from the south with accompanying rain squalls. The weather improves in the late fall when the mean air temperature drops below that of the water, thus eliminating the stable conditions that maintain sea fog. Late September may bring a week or so of calm, clear, sunny weather—before the first autumn gale ushers in the long winter season.

Running down to 40° N in winter, the weather steadily improves. Southwest winds still prevail, alternating about equally between light breezes and moderate gales, sunshine with scattered clouds, and heavy rain squalls. The water and air temperatures hover near 50°F. Occasionally an extratropical cyclone will wander south, whomping up a full gale for a day or so, with winds from all points, and temperatures fluctuating by 20°–30°. This is the famous return route of the Spanish Manila galleons, which for two hundred years ran their easting down along this parallel, striking the American coast near San Francisco before heading south to Acapulco. In summer the westerlies shift northward and die away. The temperature rises into the 70s, and short spells of calm, sunny weather alternate with light winds from the east. Although bands of cumulus clouds pass by, there is little rain. These are the *horse latitudes,* which are found about 10 degrees farther south during the winter. However, it is only in the easterly sectors of the oceans, under the large, permanent oceanic high-pressure areas, that one encounters the weeks of flat calms and cloudless skies that were the despair of early voyagers. Because these highs migrate seasonally, and even change position from year to year, it pays to keep track of them if you depend upon the wind for power.

Below 25° N we are in the middle of the belt of northeast trade winds, which blow 5–15 knots about 95 percent of the time. Endless bands of puffy cumulus clouds sweep across the sky from the east, separated by avenues of clear sky. The low clouds appear to be uniformly flat across their bottoms, where the convection cells that produce them rise through the invisible dew-point stratum of the stable inversion layer over the tropical ocean, and moisture suddenly condenses. Here and there a larger cloud dangles dark shafts of rain—a miniature traveling thunderstorm. Behind its obscuring curtain, the wind can suddenly reverse itself and blow eastward at 40 knots for a few minutes, then again west at the same speed, before lapsing back to normal as the cloud passes over. The trades are freshest in the winter and spring, and increase in strength and become more easterly as one sails south down to about 15° N. Air and water temperature similarly increase from 70°–80°F. The summer season is much the same, except for the passage of three or four easterly waves, bringing with them multiple levels of dark stratus clouds, heavy rains, and a relaxation of the trades. Conditions may remain unsettled for a week or two, with overcast, muggy weather, before the wind returns to freshen things up.

At the midpoint of our journey we pass through the Intertropical Convergence, which shifts seasonally between the equator and 10° N. The ITC can be seen at a great distance from an airplane because of its towering cumulus, often rising to 30,000 feet. From a boat this impressive wall of clouds is hidden by increasing low scud as one sails south, and its presence is manifested by the uneasy and confused state of the swell, flukey winds that blow intensely and then subside, and intermittent showers of rain that come from nowhere in a solid, opaque overcast. Farther to the east the ITC may not exist as such, and one may make a cloudless, dry passage south, if under power, or find one morning that the northeast wind has veered into the southeast, and sail all the way down. But at 159° W the ITC is some 100 miles wide, and annually dumps about 180 inches of rain on Palmyra Atoll (7° N), while Fanning Island, at 4° N, gets less than 80 inches.

South of the ITC, and extending westward along the equator in the easterly sector of all oceans, are semipermanent tongues of calm, hot, dry air that appear to advance and retract seasonally as the ITC moves north and south. These features have not received general meteorological recognition, but are clearly visible on satellite photographs because of the lack of clouds. In the Pacific, this tongue is particularly well developed, and has received specific attention. Along the coast of central America it is 10°–25° wide, and tapers to extinction in the vicinity of the Phoenix Islands 6,000 miles to the west. This unusually dry local climate may account for the singular concentration of "filled-in" atolls in the eastern equatorial Pacific. The principal agents that counteract the natural tendency for atolls to accumulate debris that gradually fills their interior lagoons are large storm waves and abundant rainfall—to promote water solution and erosion of their coral rims—both of which are conspicuously lacking in this region.

From the equator to about 25° S the southeast trades prevail, with conditions very similar to those for their northern counterparts, except that the former are generally somewhat stronger, steadier, and cover a much wider zone of latitudes. The southern subtropical divergence, spanning the zone from 25°–30° S, is a principal shipping lane for westbound vessels between Panama and Australia because of its balmy weather and light winds, although in February there is a 5 percent probability of gale winds, usually from the northwest.

The belt of the mighty Southern Hemisphere westerlies, extending from about 35°–50° S, is probably the prime example of a zonal wind system. Except for the narrow tip of South America, the winds blow unhindered around the world, with remarkably little variation in average weather throughout the year. Sailing conditions, as succinctly described by Sir Francis Chichester, who ran down the westerlies from the Cape of Good Hope to Cape Horn, run a weekly gamut from light breezes to fresh gales, from days of bright sun to days of lowering, squally weather, with intermittent rain, with air temperatures generally in the 50s and 60s. Very large storms sweep around the South Pole along the Antarctic Circle every four to six days, producing correspondingly larger shifts in wind direction and large, confused seas as one moves south through the westerlies. But east winds are rarely encountered north of 60° S, except in the offing of Cape Horn, where the high spine of the southern Andes deflects the surface winds southerly, and thus the traveling cyclones may bring northeasterly winds.

The great Antarctic storms of the 50s are spawned in the subpolar convergence, which is well developed all around Antarctica. In winter, two semipermanent, low-pressure areas develop offshore of the Ross and Weddell seas, that act as principal foci for cyclone generation. These cyclones move westward at speeds of 15–25 knots, curving toward the south, and extinguish themselves after a week or so along the perimeter of the cold ice cap. Few of them penetrate the coastline, which is dominated by cold, strong katabatic surface winds, blowing downslope from the high interior at all times of the year.

Most of the weather patterns described above are beautifully illustrated in Plate 9, which shows the earth in bulbous phase, as seen from the Apollo 13 spacecraft in April 1970. The photograph happens to be centered on my laboratory in La Jolla, California, and the 169th meridian—along which we sailed on our hypothetical weather voyage—slices off the upper third of the earth's daylight hemisphere at an angle of about 45° NE-SW.

Although the northern polar region is almost uniformly covered with clouds, snow, and ice down to the 60th parallel, the track of the prevailing westerlies is clearly marked by three spiral storms (extratropical cyclones). The first, near the earth's northwest limb, is fully developed, and is centered over Amchitka Island in the western Aleutian chain. The second, in the Gulf of Alaska, is degrading, as shown by a clear central sector. The third, over northwest Canada, is largely dissipated. The subtropical divergence is manifested by the eastern Pacific High, a large, clear area south of storm number two. The northeast trade-wind belt runs westerly from the clear area surrounding La Jolla (of course), and is partially obscured by a large subtropical cyclone over Hawaii. To the south, the Intertropical Convergence stands out as a sharp band of clouds along latitude 5° N, beneath which the southeast trade-wind zone is wide and clear all across the Pacific. Lastly, the swirl of an Antarctic cyclone can be made out along the earth's southwest limb.

EXTREME WEATHER

Having looked at typical weather sequences, we shall now examine more extreme conditions in greater detail, bearing in mind that in some latitudes and seasons the extreme is also the usual. Ship handling in such circumstances is discussed in Part VII.

Fronts: The sloping surface of convergence between two colliding air masses having different densities is called a front. Since air density is primarily a function of temperature, fronts are thin zones of rapid temperature change, often accompanied by vertical convection and strong winds. They are mostly confined to latitudes above 30 degrees. There are fronts of all sizes and intensities, from the 5,000-mile–

Plate 9. Photograph of the earth from space shows all of the principal weather features.

long boundaries of the westerly tropospheric waves to the mild depressions that bring a touch of winter weather in late spring. Fronts may be cold or warm, dry or occluded, and moving or stationary. But in general they presage a change of weather—sometimes very sudden—and often try the survivability of ships and crew alike.

Warm fronts are distinguished from cold fronts according to whether they are moving toward a cold or a warm region, respectively. Thus, to an observer at a fixed location, the passage of easterly waves appears as a succession of alternating warm and cold fronts. Because cold air is denser than warm air, it tends to obtrude itself beneath, irrespective of the direction of frontal motion; hence frontal surfaces always slope upward in the direction of colder air. That is, a cold frontal surface slopes *rearward* of its direction of advance, while a warm front slopes *forward.* The mean slope increases with latitude (Coriolis force) and with the density difference across it, but in either case the slopes are very gentle: about 1:200 for a warm front, and 1:50 for a cold front.

Air circulation around a front is a combination of strong horizontal flow owing to cyclonic and/or anticyclonic winds blowing in the warm and cold sectors, and a vigorous, vertical, cellular circulation owing to convective mixing along the frontal surface (fig. 24). Both types of flow are unstable because converging air flows tend to become turbulent. In the horizontal flow, instability manifests itself as a series of waves, exactly like the great troposphere waves of figure 21 (p. 66), except that they are much shorter (300–500 miles). The crest of each such wave advancing into the cold sector constitutes a small warm front, and its contiguous trough moving into the warm sector a cold front (fig. 24). Because they are steeper and have higher pressure gradients across them, the cold fronts move faster and tend to overtake and spiral under the warm fronts, generating cuspated kinks, called *occlusions,* each of which is a potential cyclone. A large, polar upper-air wave, spanning half the Atlantic, may have three or four such frontal pairs in various stages of development or decay, each having a lifespan of several days, and moving along the boundary of the polar wave at speeds of 500–1,000 miles per day (20–40 knots).

The vertical circulation in a cold front differs from that in a warm front principally in the degree of turbulence, the former being much more violent. In both cases, rising warm, moist air expands and cools, and moisture condenses in the overlying warm layer to form clouds, whose characteristic shape, distribution, and degree of accompanying precipitation are determined by the frontal intensity. The release of latent heat due to cloud formation sets up local convection cells, with updrafts beneath the clouds and inflow from both sides, that may temporarily dominate the local wind pattern as the front passes through.

To an observer within the cold sector, the approach of a warm front may be preceded by a period of relatively clear, calm weather. Thin wisps of high cirrus clouds usually appear first from the west or northwest (Northern Hemisphere). These are followed by successively lower bands of stratus clouds, until after twenty-four to thirty-six hours the sky is uniformly overcast with low dark stratus, accompanied by a slowly dropping barometer, rising temperatures, and restricted visibility. Light drizzles give way to steady rain, which may last for twenty-four hours or so, gradually tapering off behind the slowly moving front (plates 6–8, pp. 52–54). There is usually little wind, and a rising barometer betokens clearing weather. The early stages of a warm front are very similar to those for a hurricane or subtropical cyclone, but the latter can be distinguished by failure of the barometer to recover after a drop of 3–5 mb,

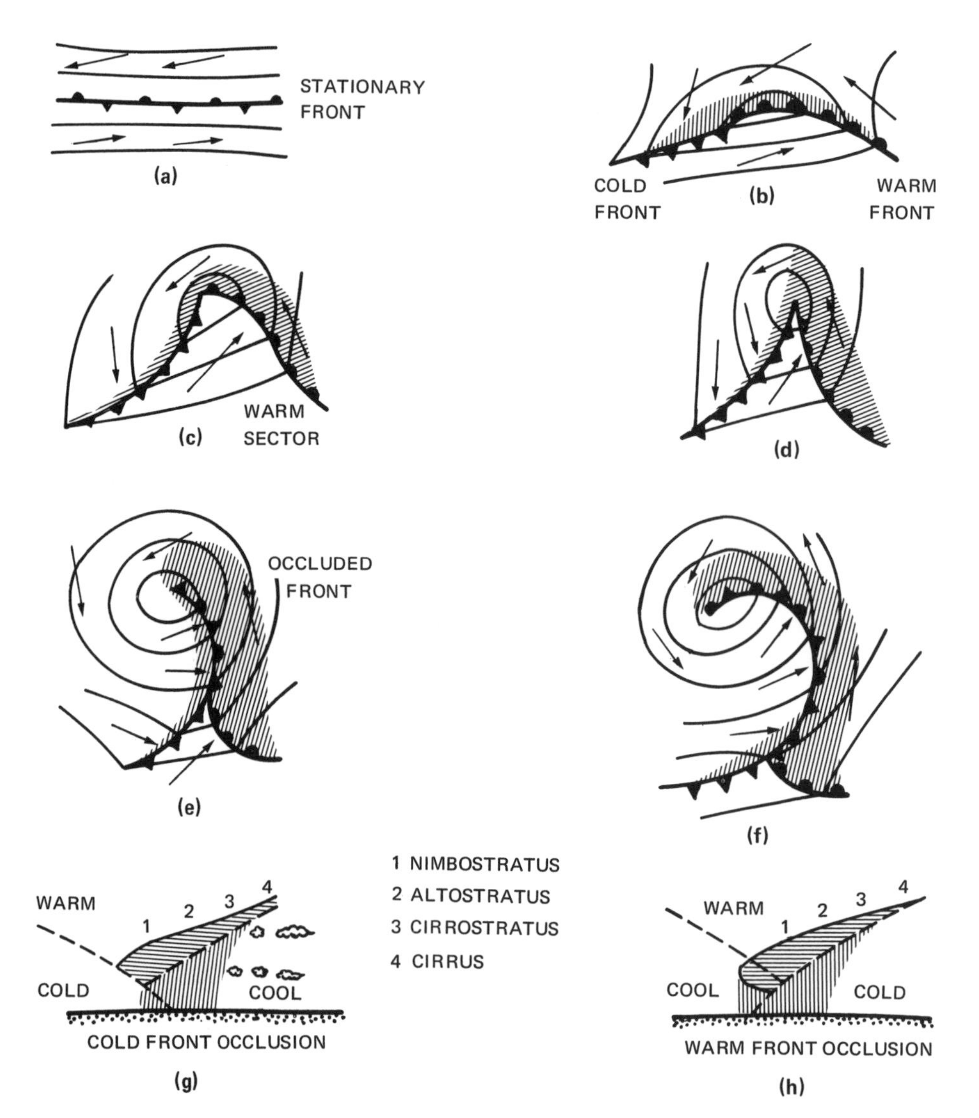

Fig. 24. Development of a frontal occlusion. Stationary front (*a*) develops a kink (*b*) that grows into a wave (*c*), bounded by local warm and cold fronts. Wave steepens (*d*) and becomes occluded (*e*) as cold front overtakes warm front. Occlusion may deepen into cyclone (*f*). Sections (*g*) and (*h*) show cloud and precipitation distributions for cold and warm occlusions, respectively.

as well as by increasing ground swell, wind, and abnormal humidity. Radio advisement should, of course, be sought if the latitude and season are appropriate.

Cold fronts usually approach much more swiftly (20–40 knots) and give less advance warning. Because they often contain strong, gusty winds they represent a serious weather hazard in middle and higher latitudes. Although virtually unknown in the tropics, in some extratropical places such as Florida an occasional *norther* sneaks south as a strong outbreak of cold continental air, and hail rattles on the windows in the Bahamas. Often, these northers are strong, cold anticyclones that develop behind occluded fronts along a larger polar front. As such, there may be no intervening warm fronts between a succession of northers, which may occur about once a week in winter along the Atlantic seaboard.

Cold fronts characteristically appear from the west or northwest as a wall of billowing cumulus clouds, sprouting from a dark, solid base of low nimbostratus, and are accompanied by heavy rain, hail, or snow, a sharp barometric rise, and falling temperatures. The prevailing wind, usually parallel to the advancing front, suddenly shifts 90° and blows violently out of the front as it passes, after which it slowly veers around the compass to its original direction as the barometer falls to normal. The fierce frontal passage is normally followed by rapidly clearing cooler weather that persists for a day or so.

An intense cold front may be preceded by an hour or two by a *line squall*—a miniature version of what is to come, but even more sudden. Line squalls have been explained as a sort of pressure shock wave produced by sudden, pistonlike advances of a cold front. Thus, they appear and disappear unpredictably, often leaving a wake of capsized boats to await further depredation when the cold front arrives. Because they often contain heavy rain or hail, line squalls are sometimes referred to as *downbursts,* and are usually visible on radar.

Occluded fronts can be either warm or cold, depending upon the density differences across the front. In either case, as an occlusion approaches, it resembles a warm front, which becomes suddenly cold as it passes over, usually accompanied by heavy, low cumulus clouds, thunderstorms, and torrential rain. Because occluded fronts are intense local depressions, they often develop into extratropical cyclones in the manner described below.

Cyclones and Vortexes, as the names imply, are rotary wind systems. This classification includes an assortment of different disturbances, ranging from tornadoes, waterspouts, and whirlwinds, through hurricanes and typhoons, up to the tremendous cyclonic depressions of the subpolar convergences, whose spiraling cloud patterns may span 3,000 miles of ocean. Most of these phenomena have been dealt with repeatedly and eloquently in the marine literature, and I will confine my remarks to certain characteristics that appear not to have received much popular attention.

First, it is no accident that they all rotate—and, except for small "dustdevils," they rotate in the same directions in each hemisphere. In the strictest sense, any two bodies in the same hemisphere revolve about each other once a day, no matter how close they are to one another. This rotation would be readily apparent to an observer in space looking at two staffs stuck in the snow near, say, the North Pole. It may not be so obvious, but is nevertheless true, at any other location, except precisely on the equator. Secondly, it follows from Newton's laws of motion that any two such bodies possess a mutual angular (or rotational) momentum equal to the product of their masses multiplied by their relative *angular velocity* and the distance between them. It turns out that this angular velocity is simply our old friend, the horizontal component of the Coriolis force (per unit mass), which varies as the sine of the latitude, and thus the angular momentum of any two bodies a constant distance apart diminishes from the poles toward the equator in the same way. Finally, for any two such bodies free to slide about on the earth's surface without friction, the angular momentum is constant. Thus, like a pirouetting ice skater who retracts his arms and legs, if the two bodies approach each other, they rotate faster and faster as their spacing diminishes.

These basic principles explain why the molecules of a fluid always tend to revolve about one another when they are mutually attracted toward some focal point, such as the region of low pressure produced by pulling the plug from a tub drain, by the tip of an oar while rowing, or by unequal heating of the atmosphere from beneath. Such rotary motions, known as *vortexes* (or vortices), can be initiated and will maintain themselves in either direction on a small scale. Contrary to popular opinion, very careful experiments have shown that even in perfectly symmetrical basins there are enough random rotations so that the drain vortices are also random. But when air converges over distances of several miles, as it does in tornadoes, the influence of the Coriolis force is sufficient to give about 80 percent of them cyclonic rotations, and all larger vortices are always cyclonic.

Waterspouts are the marine equivalent of *tornadoes,* and both occur most commonly in the region of intense shearing motions associated with or preceding a cold front. However, waterspouts sometimes precede warm fronts, and are frequently observed along the peripheries of hurricanes and typhoons. For unknown reasons, only the United States and Australia seem to be frequently afflicted with tornadoes, averaging about 150 annually, but the incidence of waterspouts must be much greater. They usually issue from the base of a particular cloud formation, and often occur in pairs, one of which may be abortive and barely visible. Sometimes six or eight such pairs may appear spontaneously along an intense front. Although they do great damage on land, encounters with ships are very rare, possibly because of their small diameters (10–100 yards) and short persistence (10–15 minutes).

The dynamics of tornadoes are fairly well understood, despite the scarcity of direct measurements. In 1904, an extreme pressure drop of 200 mb (corresponding to a vertical wind velocity of 340 knots) was recorded within a concrete structure passed over by a Minnesota tornado. Similar estimates have been obtained indirectly from the distribution and effects of flying debris. Limited evidence suggests that waterspouts can be equally dangerous: a number of small craft are known to have been sunk or dismasted by them, and the bridge deck of the liner *Pittsburgh* was severely damaged by a twister on March 30, 1923.

Hurricanes and *typhoons* are synonyms for what the weather service calls *tropical cyclones,* whose maximum winds exceed 64 knots. They represent probably the greatest natural hazard to cruising yachts because of their predilection for warm climates, although many more large ships have succumbed to winter gales in high latitudes. Hurricanes, as I shall call them, occur most frequently in the warm, western sectors of all oceans during summer and fall. They appear to materialize out of clusters of cumulonimbus clouds, often adjacent to the intertropical convergences, and frequently are the outgrowths of easterly tropospheric waves. Many such waves pass, and several tropical cyclones are generated for each hurricane that maturates. Other necessary conditions seem to be unusually high humidity, an abnormal fall of barometric pressure (3–5 mb), and water temperature above 77°F.

Hurricanes, as we have seen above, also require stored-up rotational momentum in the air. Since the ITC prevails north of the equator all year in the Atlantic and easterly Pacific, there is insufficient stored momentum between the ITC and the equator to generate and sustain a strong cyclone. Even if one should form, it cannot cross the line without encountering opposing rotations that abstract energy and snuff it out. Therefore, hurricanes are unknown in the south Atlantic and east of longitude 130° W in the south Pacific. Moreover it is generally safe to sail in any tropical sector where the ITC is well established a few degrees from the equator in the *opposite* hemisphere.

A mature hurricane is a formidable thermodynamic engine, and devoutly to be eschewed. In plan, it consists of a slightly asymmetrical array of intense line squalls, spiraling inward to a common circle of tangency some 15–30 miles in diameter, known as the *eye* (fig. 25). Surface winds blow inward along these squall lines with ever-increasing velocity, sometimes as high as 150–200 knots in the region of tangency, and gales may extend as far as 100–300 miles from the eye. In the Northern Hemisphere, wind and seas are higher on the right (dangerous) semicircle of a cyclone, viewed in its direction of travel; the local wind speed comprises the vector sum of cyclone winds and the speed of the storm as a whole. In profile, the eye has an hourglass shape, in the vertical throat of which the inward spiraling winds are diverted upward, and then again outward, now spiraling in the opposite direction. Expansion and cooling of this upward flow causes condensation of moisture with the release of the enormous quantities of heat that drive the hurricane. The resulting cloud crown spreads out for hundreds of miles, blowing off downwind, a cumulonimbus of such gigantic proportions as to be

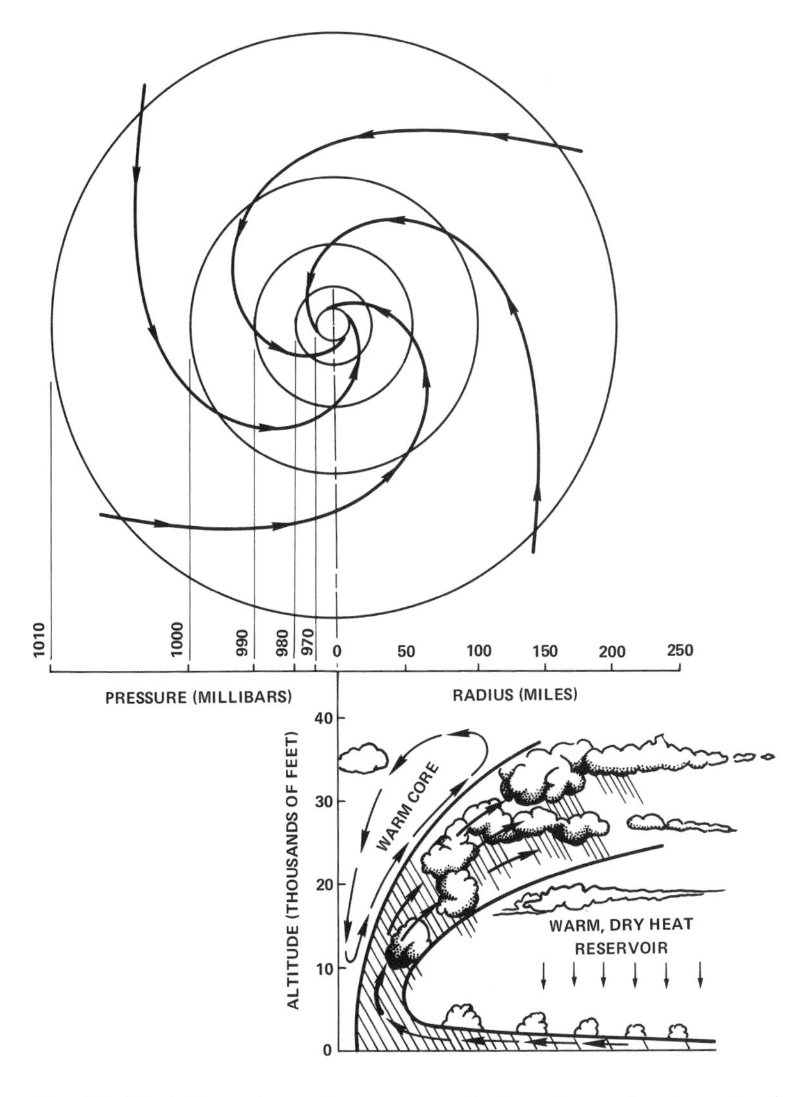

Fig. 25. (*Top*) Plan view of mature tropical cyclone, showing circular surface isobars and inward spiral wind vectors (*arrows*). (*Bottom*) Vertical section through eye, showing wet, high-velocity wind chimney separating warm core and surrounding warm heat reservoir.

visible several hundred miles away. The eye itself is a calm region of dry, warm, clear, slowly descending air, extending from the sea surface clear to the stratosphere. Pilots of high-performance aircraft who have penetrated the eyes of hurricanes report that the interior regions of whirling clouds resemble the multiple balconies of an enormous opera house, within which birds circle aimlessly without hope of escape. The extremely low central pressure within a mature hurricane—sometimes failing below 900 mb—is exceeded only within tornadoes, where pressures as low as 800 mb have been recorded. In contrast to extratropical cyclones, there is very little temperature variation across the entire depression, and thus temperature cannot be relied upon as a guide to a hurricane's progress.

Hurricanes depend for energy upon the inwardly spiraling winds, which abstract surface moisture and heat from millions of square miles of surrounding ocean. This heat is convected upward by the line squalls into the hot, warm, moist reservoir surrounding the throat, and settles slowly back as needed to keep the central furnace going. Passing over cooler water, or dry land, a hurricane will run out of energy within a day or two, as its heat reservoir runs down, and may cease to be regarded as dangerous. However, a number of these "forgotten" cyclones have reversed their erratic courses and, passing again over warm seas, have quickly been rejuvenated, to continue their destructive courses.

Hurricanes generally reach maturity in from three to five days, and have a life span of a week or two, although one five-week record holder made the entire circuit of the Atlantic before finally fizzling out near the Azores. During its early phases, a hurricane moves slowly westward with the prevailing easterlies aloft. As it grows in strength, the Coriolis force exerts a poleward influence, and the hurricane's subsequent track seems to be determined by the prevailing mean pressure gradients ("steering winds") along its path, veering northward and then eastward around the prevailing oceanic high-pressure center. Later in the fall, when equatorial waters are warmer, hurricanes' courses tend to parallel the ITC. But there are numerous exceptions to these tendencies, and no reliable long-range prediction scheme has yet been developed. Instead, hurricane tracking centers in most of the susceptible areas keep tabs on each new arrival until it has run its course.

An average of eighty tropical cyclones per year arise within the principal traffic areas shown in figure 26, whose shadings are intended to give a general idea of relative frequency and track density. Less than half of these reach hurricane intensity (central winds greater than 64 knots). Table 2 gives the monthly probability densities within these same regions.

What is the probability of encountering a hurricane? Actually, in prime recreational sailing areas, such as the West Indies and the Cook and Society Island Groups of the Pacific, it is quite small. Statistically, the densest concentration of typhoon tracks in the world lies between Manila and southern Japan, a distance of about 1,000 miles. If you were to elect a passage along this route in August—the worst possible month—at a mean speed of 6 knots, your probability of encountering gale winds or higher would be about one in twelve. Considering the excellent warning network, by making even a modest attempt to avoid their breeding seasons, this probability can be made vanishingly small.*

*For detailed descriptions of hurricanes and typhoons, see W. J. Kotch and Richard Henderson, *Heavy Weather Guide,* United States Naval Institute Press, Annapolis, Md., 1984.

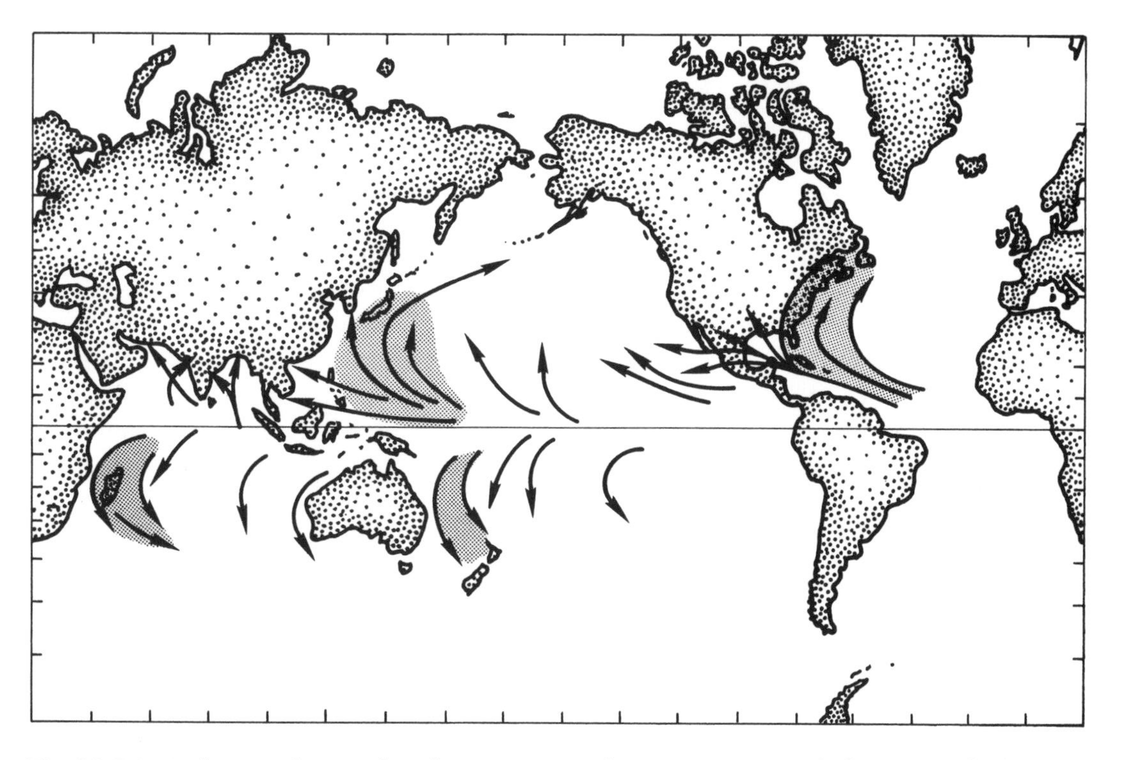

Fig. 26. Principal areas of tropical cyclone activity and representative track directions. Shading indicates increasing frequency. Cyclones do not occur in the south Atlantic.

Extratropical Cyclones are rotary wind systems that most commonly develop from occluded depressions in middle and high latitudes. They occur in all sizes and intensities, from the mild fronts associated with summer lines to the enormous, spiraling cloud systems, some of which are 3,000 miles in diameter, that are almost always visible in wintertime satellite photographs (see plate 9, page 76).

Table 2. Mean Annual Frequencies of Tropical Cyclones (after Dunn)

Month	North Atlantic	North Pacific 100E–170E	North Pacific 80W–120W	Bay of Bengal	Arabian Sea	South Indian	South Pacific 140E–140W
Jan.	0	0.4	*	0.1	0.1	1.3	8.1
Feb.	0	0.2	*	0	0	1.7	4.9
Mar.	0	0.3	*	0.2	0	1.2	7.5
Apr.	0	0.4	*	0.2	0.1	0.6	1.6
May	0.1	0.7	0.2	0.5	0.2	0.2	0.3
June	2.4	1.0	1.0	0.6	0.3	0	0
July	2.6	3.2	0.9	0.8	0.1	0	0
Aug.	2.6	4.2	1.5	0.6	0	0	0
Sept.	3.7	4.6	2.6	0.7	0.1	0	0.3
Oct.	2.5	3.2	1.5	0.9	0.2	0.1	0.3
Nov.	1.5	1.7	0.1	1.0	0.3	0.2	0.8
Dec.	0.1	1.2	0	0.4	0.1	0.8	3.2
Per year	10.9	21.1	7.9	6.0	1.5	6.1	27.0

*less than 0.1

These cyclones appear to originate—and derive their energy—from the spiraling descent of cold polar air associated with wavelike frontal outbreaks in the middle troposphere. The descending air is replaced by upward-spiraling warm surface air, as part of the normal process of heat transfer between the polar easterly and mid-latitude westerly wind tubes. The surface manifestations of these respective descending and ascending motions are the consecutive cold and warm fronts previously described, which tend to occlude as the more rapidly moving cold fronts overtake the warm fronts. Some of these occlusions dissipate with scarcely a whimper, while others intensify into broad barometric depressions, or *storms.* *

The developmental sequences for a number of severe Atlantic storms have been described in detail by Adlard Coles in his excellent book *Heavy Weather Sailing,* and I will outline here only the general features of large storms and their essential differences from hurricanes. These differences are significant because there is some seasonal and regional overlap; a few out-of-season hurricanes migrate poleward and recurve into the prevailing westerlies, and an occasional errant storm wanders into the subtropics.

In contrast to a hurricane, a major storm is a markedly unsymmetrical barometric depression, having its strongest winds concentrated along one or more lines of occlusion at some distance from their common apex (fig. 27). There is no warm, clear eye, but instead the central depression is a cold region of violent vertical convection and heavy precipitation. The storm moves in a generally easterly direction, making slightly more than a right angle with the occluded front, which divides the circulation pattern into a warm leading sector and a cold following sector. The centers of clouds and precipitation are considerably displaced from the central depression in the direction of motion—i.e., toward the warm sector—so that an advancing storm displays all the characteristics of a warm front.

The peripheral winds spiral cyclonically inward toward the central depression in much the same manner as those in a hurricane, except in the vicinity of the occlusion, which acts like a barrier or wall, distorting the wind field so that the winds blow in nearly opposite directions along it toward the storm center in the warm sector, and away from it in the cold sector. Because the front is very steep and only a few miles thick, and also because the strongest winds occur in the sector behind it, a violent and sudden wind shift usually accompanies the passage of the frontal occlusion, producing very confused sea conditions within the cold sector—and even in the warm sector, if the front is moving more slowly than the swells. Steep, crossed seas, coupled with opposing winds, make sailing within the cold sector difficult and uncomfortable, if not dangerous (see figure 82 and related discussion).

Although the maximum barometric depressions associated with large storms are much smaller than those for strong hurricanes (30–50 mb), and the sustained winds more moderate (60–70 knots), the former may extend over an area several times as large and persist for a much longer time (2–3 days) at a fixed location. However, because the maximum waves generated depend upon the persistence of the wind and the area over which it blows as well as its strength, large storms may generate more dangerous seas than a hurricane, except within a small region surrounding the eye. Thus, ground swells from great polar storms are often observable for thousands of miles, while those radiating from a major typhoon can scarcely be detected 500 miles away.

* The National Weather Service classifies such storms as extratropical cyclones.

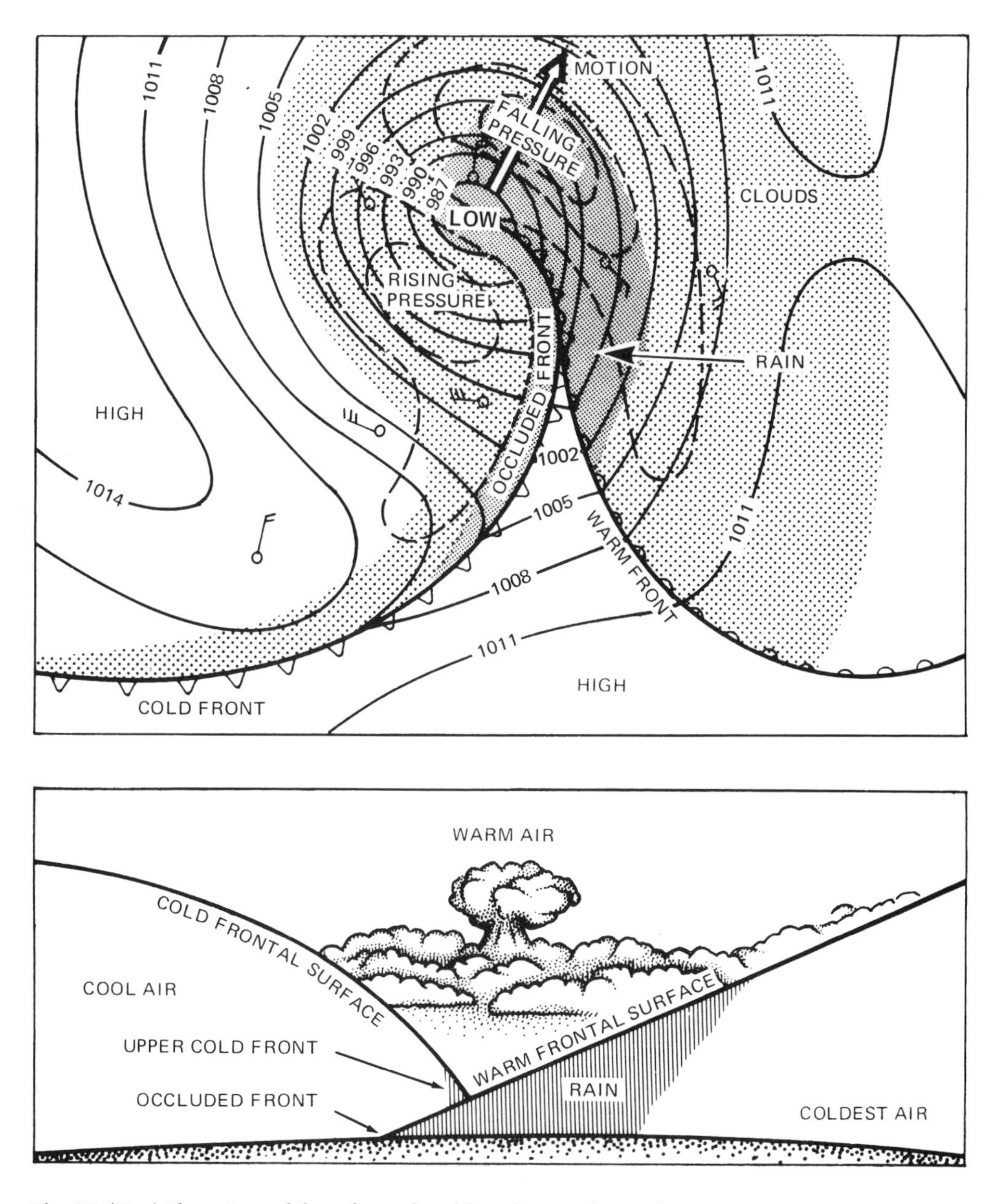

Fig. 27. (*Top*) Plan view of deep frontal, cold occlusion that is deepening into major extratropical cyclone. (*Bottom*) Vertical section perpendicular to front through central depression. Note that descending cold spiral has obtruded beneath warm front, lifting it aloft.

While the breadth and depth of a barometric depression are a good gauge of the total energy of a storm, a better index of wind intensity is provided by the rate of fall of pressure. A fall of 10 mb in three hours usually portends gale winds, and fall rates of 15 mb per hour have been observed under extreme storm conditions within the occluded sector. Large-pressure gradients are generally associated with small, intense storms, since the glass manifestly cannot long continue to plummet for long—hence the old adage, "short forecast soon past." Contrariwise, a long fall over twenty-four hours or so, accompanied by increasing cloudiness, drizzle, and then rain, adumbrates a major storm.

Fog: *Fog* does not ordinarily rank as an extreme weather condition. Yet it has been responsible for a substantial fraction of maritime disasters. Fog, in fact, is nothing more than low stratus clouds, and consists of moisture droplets so small that they are held in suspension. Like clouds, it occurs wherever the temperature drops to the dew point corresponding to the local relative humidity. This situation can arise from a variety of circumstances, such as warm, moist air blowing over colder water (advection fog), saturated air over the ocean cooling during the night hours (radiation fog), or the rapid evaporation (steaming) produced when very cold, dry air blows off land over relatively warmer water (sea smoke).

Fog almost never occurs in the tropics. In the middle latitudes, advection fog is by far the most common, and can occur wherever warm and cold water coincide, such as where the warm Gulf Stream encounters the cold waters of the north Atlantic. Along a shoreline where the daily thermal extremes over the adjacent land fall above and below the water temperature, as off western North America at different seasons of the year, the daily cycle of consecutive land and sea breezes promotes the formation of advection fog that tends to hug the coastline, advancing inland during the day, and standing out to sea at night. Advection fog is common in the polar seas during the summer months, when outbreaks of warm subtropical air move poleward over cold water.

Once formed, all fog except sea smoke is self-stabilizing, being cooled from beneath and highly reflecting warm radiation from above, and thus is apt to persist until a wind arises to sweep it away. Extremely thick (to 4,000 feet) and stable summer fog in the Bering Sea and around the Aleutian Islands sometimes withstands winds of 10–20 knots without dissipating.

7

Weather Forecasting at Sea

CIVIL FORECASTS

Weather forecasting, in general, involves the educated interpretation of relative air mass movements throughout the troposphere on a hemispheric scale. While the circulation of the atmosphere, like that of the oceans, depends upon its density distribution, density is difficult to measure directly. Instead, the density field is, in effect, determined indirectly, from measurements of pressure, temperature, and humidity, as a function of altitude, at some ten thousand reporting stations throughout the world. These data, together with observations of wind direction and velocity, and satellite images of cloud patterns, are transmitted several times daily to central-analysis headquarters in member nations of the world meteorological network, where they are used to prepare synoptic weather charts for the Northern Hemisphere.* Separate charts are prepared for the surface and several convenient pressure altitudes, the vertical spacings between which can be interpreted to give the instantaneous state of motion of all parts of the troposphere and lower stratosphere. By comparing consecutive sets of charts, and using high-speed computers, twelve-, twenty-four-, and up to seventy-two-hour weather forecasts are prepared daily, and broadcast all over the world by radio, telegraph, and television. In addition, numerous air and coastal weather-forecasting stations receive radio-facsimile copies of special regional synoptic charts, from which they prepare and broadcast local weather reports for the benefit of aircraft and shipping, respectively. Lists of radio weather station schedules and frequencies are published by the World Meteorological Organization, the National Weather Service, and numerous nautical almanacs. Lastly, in many countries, private entrepreneurs avail themselves of civil weather data banks to prepare custom forecasts and ship routing advice on demand. Such services are used routinely by large shipping companies, commercial fisherman, and major yacht racing organizations.

However, there are very large blanks on the synoptic weather charts, corresponding to ocean areas devoid of island reporting stations or weather buoys, and far removed from aircraft routes or shipping lanes. Moreover, the oceanic reporting grid is rarely dense enough so that all weather of interest to mariners is accurately foretold, and therefore it behooves anyone who goes to sea to train his eyes to recognize the preliminaries, and depend on his radio only for the main events.

*Southern Hemisphere forecasting is still on a local regional basis.

WEATHER INSTRUMENTATION

The basic weather measurements for shipboard forecasting are wind speed and direction, pressure, temperature, and humidity. Instrumentation is standard, and is described in most books on seamanship and navigation, but there are a few options worth discussing here, as well as some pointers on their employment and interpretation. Any of them may be obtained as either indicating or recording instruments. The former are much cheaper, and suffice for most ordinary purposes. But there is a certain advantage in being able to review at a glance the past few hours of a record to determine slow, average rates of change, which can be accomplished only with considerable effort and uncertainty by use of indicating instruments. As a first choice, most yachtsmen would elect a recording barograph, and as a second choice, a recording anemometer, but there is no particular necessity for recording temperature or humidity. Both of the latter can be obtained with a simple sling psychrometer, which can be whirled as often as desired.

The standard reference height for anemometers is 10 meters (33 feet) above the waterline. For a standard lapse rate, wind speed increases logarithmically with height above the sea, and any other mounting height can be corrected to standard height by using the curve of figure 28. In rough seas, a mast-mounted cup anemometer will read high by the amount of the average crosswind velocity of the mast as a ship rolls or pitches. This error increases with mounting height, but on sailing craft a masthead mounting is still preferable to spreader mounting, where a larger error is introduced through distortion of the wind field by the sails. Masthead mounting also prevents possible fouling by flopping halyards, etc., and provides a better estimate of the average wind in heavy weather, when high seas significantly perturb the wind flow.

Gustiness is another factor that affects the determination of slow changes in the mean wind. Gustiness increases with increasing wind and sea state, and can amount to instantaneous readings that differ by as much as 50 percent in speed and 10°–15° in direction from the mean values, averaged over several minutes. Studies show that five-minute averages suffice to eliminate most gust effects under moderate weather conditions, but in heavy weather twenty-minute averages may be required. Indicating anemometers are not ordinarily capable of much longer than one-minute averages, but this period can sometimes be extended by special order from the manufacturer. Time averaging with recording anemometers is readily accomplished by eye.

Mercury barometers are fast approaching extinction on shipboard, owing to their cost and fragility, and the number of corrections necessary to give a proper reading. Therefore these remarks concern only *aneroid* (bellows-actuated) pressure instruments, which are temperature-compensated and require no gravity correction. Recording barographs have no special mounting requirements, except to avoid exposure to extreme temperature changes and shock. These are best minimized by rubber-mounting in a central (midship) location, away from direct sunlight or other local heat sources. All barometers are subject to a correction for height above sea level, and should be calibrated by reference to a standard instrument at intervals best determined by experiment. Aneroid instruments normally come equipped to average out gusts, and so the principal pressure "noises" that remain to be distinguished from weather effects are the twelve-hourly barometric "tides" previously discussed. Again, the best guide to abnormal changes is the normal trend for the

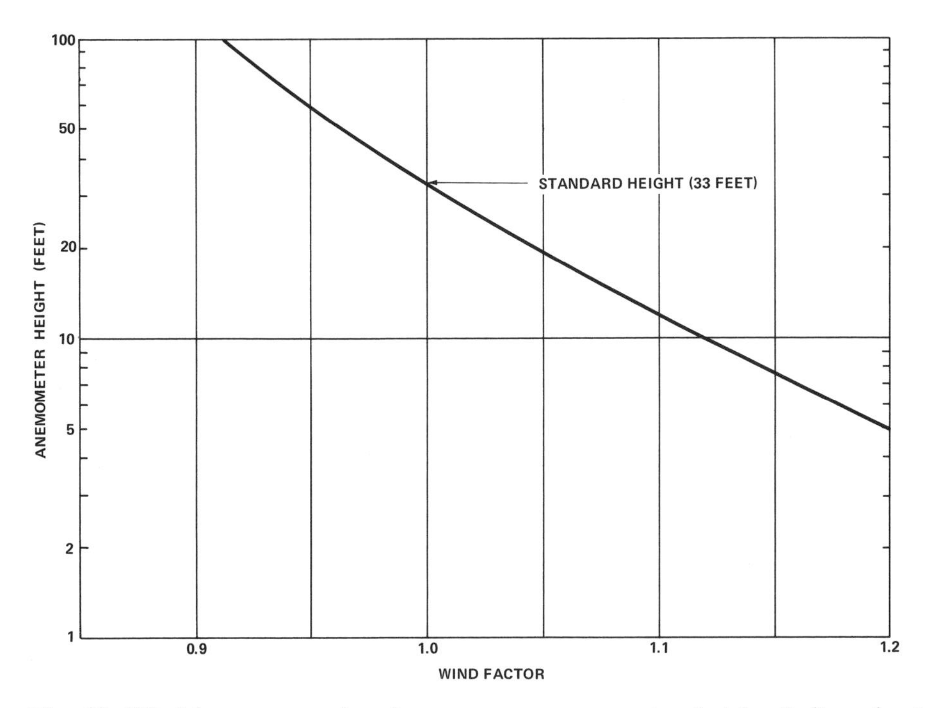

Fig. 28. Wind-factor correction for anemometer mounting height. Indicated wind speed should be multiplied by the factor corresponding to actual anemometer height.

preceding day, which is easy to judge from a continuous record, but rather difficult with an indicating instrument, unless the readings are plotted every hour or so.

Temperature and humidity indicators are conventionally supplied as round, panel-mounted instruments for cabin installation. As such, they will normally indicate higher temperature and lower humidity than the ambient values outside. This condition can be improved by mounting them on the face of a *plenum* (a thermally insulated air space having a volume of a few hundred cubic inches) that is ducted and vented to the outside, so as to ensure continuous air flow without the introduction of rain or spray. As mentioned above, a sling psychrometer, or a properly vented wet- and dry-bulb thermometer provides a more accurate measurement of temperature and humidity. In either case, you will need a psychrometric table or chart to convert from dew point to relative humidity, or vice versa (see figure 14, page 48).

SHIPBOARD FORECASTING

The essence of local forecasting is the detection and interpretation of significant changes in atmospheric conditions. If you have recording instruments, you will note that they are continuously making small excursions, from which you may deduce subtle trends that, taken together, provide a clue to coming events. Predicting the weather, somewhat like stock market analysis, can become an absorbing—if not always profitable—preoccupation.

Instruments sense only the immediate environment, whereas one's eyes can detect cloud changes tens of miles distant, and the radio provides long-range forecasts. In good weather, then, your prediction sequence should be: radio, eyes, instruments. In bad weather, you are already where the action is, and the prediction sequence is just the reverse. Shipboard forecasts are limited to about twelve hours under average circumstances, with twenty-four hours as an upper limit.

Of visual weather aids, clouds are the best indicators of change, followed by swell and sea conditions. Starting with a clear sky aloft in middle and high latitudes, the earliest visual sign of an approaching warm or occluded front are the successive appearances of high haze, cirrus, and cirrocumulus in the western sky, all moving east or northeast. If these clouds thicken and are augmented later by lower-level stratus and stratocumulus clouds, together with a falling barometer and wind blowing from an unusual direction, the front is only a few hours away. The direction of the central depression can be determined by observing the direction of clouds moving overhead, and will bear about 270° from the direction of upper-level clouds, or about 280° from the surface wind direction. Successive plots of the storm center will give you its direction and speed of travel, and the rate and extent of barometric changes will provide clues on the breadth and intensity of the depression. In midlatitudes, the barometer is always more active than in the tropics; fluctuations of 10 mb are common, and often associated with small, fast-moving depressions, whereas in the tropics, a 10-mb fall augurs a major cyclone.

In the trades, the presence of an unusually large agglomeration of low or medium cumulus, with dark rain streaks beneath, connotes a thunderstorm and accompanying strong, local winds blowing toward the central updraft below the clouds. These winds can exceed 30 knots, but blow for only 10–15 minutes, and then reverse and die away as the storm passes over. The barograph, meanwhile, writes a nice letter *M* about 3 mb high.

In hurricane season, the tropical skies may be uncommonly free of low trade-wind cloud streets for a day or so before a tropical cyclone. The first view may be the stream of cirrostratus blowing off the top of the central cloud mass above 40,000 feet. This important clue to the direction of upper-level winds may be visible several hundred miles away, and gives the direction of cyclone motion, which can be *any* direction. Other signs of a nearby cyclone are unusually warm, moist air, a small barometric rise followed by a continuing fall, and winds blowing from an unusual sector.

The appearance of a pronounced and increasing ground swell is an indicator of an approaching storm. The trained eye can readily distinguish such swells from the local seas generated by prevailing winds, and they have been used for centuries by native navigators, not only as an index of storm activity, but also as a navigation aid. If the swells are long and regular, the storm is far away; short, steep, and irregular swells betoken a nearby disturbance, whose intensity may be gauged by their size, relative to other weather symptoms. These relationships are discussed more fully in Part V.

PART III

Ocean Circulation

8

Vertical Circulation

It may come as a surprise to many that there exist any motions of consequence within the deep sea. Indeed, it was not until the first quarter of the twentieth century that the layered structure of the oceans began to be defined, and it was recognized that such layering could come about only through changes taking place at the sea surface that were slowly transmitted to all depths by some sort of molasseslike circulation along surfaces of constant density. Even as late as 1947, when I began my graduate studies at the Scripps Institution of Oceanography, the great depths were pictured as places of stygian darkness, where nothing lived except bacteria, and where only imperceptible tidal motions disturbed their infinite tranquillity. In fact, my first research project was to design a sensitive current recorder capable of free descent to the sea floor, where it was intended to remain for several days, drop its anchor, and bring back the first direct evidence of current activity. The few records I obtained, after two years of effort, indicated continuously varying, short-term oscillations as rapid as two or three-tenths of a knot. Since that time, deep photography and improved current meters have demonstrated that the sea floor is populated—albeit sparsely—with a variety of animals and fishes, and that the deep waters are never still, although in most places persistent motions are indeed very slow.

In common with the atmosphere and the mantle of the earth, the vertical circulation of the ocean is driven by regional differences in density. As in the atmosphere, these differences result principally from unequal heating between the poles and the equator. Density in the ocean, however, is also a function of salinity, which is altered by temporal and spatial differences in evaporation and precipitation, as well as by inflow from rivers. All of these factors contribute to the formation of the layered structure described in Part 1, which represents a balance between the creation of discrete water masses at the surface, shallow stirring by winds and waves, and slow internal mixing by molecular and/or eddy diffusion, all of which act to smear out the boundaries between the layers and to make the oceans homogeneous. Despite the fact that the mean vertical circulation is very sluggish, its cycling time (500-1,000 years) falls about halfway (logarithmically) between that of the hot, plastic mantle (100-300 million years) and that of the atmosphere (10-30 days).

At all depths, the ocean also experiences on a small scale other, more rapid oscillatory motions that are attributed to direct tidal action, as well as to perturbations of its quasi-stable density structure by interactions between tidal currents and the irregular sea-floor topography. Additionally, there are slow, sustained or transient flows associated with tilting of the equilibrium density layers by the wind-driven horizontal circulation described below.

But neither these latter motions nor the slow vertical circulation are of practical consequence to the ocean voyager, who has plied his trade across the sea for several millennia without being aware of their existence. Only the increasing foulness of our harbors, and the litter of unsinkable plastic trash that is accumulating in small bays and estuaries throughout the world, forces us to be aware that it is the vertical circulation that keeps the ocean clean, as well as making available all of the other essential life elements within its internal reservoir.

Lastly, it should be noted that the distinction between vertical and horizontal circulation is only categorical; they are intimately related, and the former is slower only because the ocean is stably stratified, and thus it is much easier for water to move horizontally than vertically.

9

Horizontal Circulation

WIND STRESS AND WIND DRIFT

The horizontal circulation of the oceans, also called the *wind-driven* circulation,* is much swifter than the vertical circulation. A reasonable estimate for the mean surface flow velocity is about one-quarter knot, and therefore a circuit of the north Atlantic would require about five years; of the north Pacific, about ten years. As mentioned under "Air-Sea Interaction" in chapter 4, wind blowing over water exerts a stress on the surface, such that the uppermost layer is "dragged" (actually, pushed) along. When the water is initially calm and smooth, the stress is small and proportional to the square of the wind speed (measured at some standard reference height—usually 10 meters). At higher wind speeds, wavelets develop, making the surface rougher and increasing the drag coefficient. Additionally, waves (particularly if breaking) seem to mix up the surface layer and to increase the transfer of momentum from air to water. Thus, at high wind speeds, wave momentum may contribute as much as 50 percent of the total stress. But high winds occur principally during circular cyclonic storms, such that the wave stress is applied in all directions, with only a weak net westerly momentum as a result of the mean motion of such storms. Above the threshold wind speed for significant wavelet formation (about 12 knots), the total stress increases somewhat faster than the square of wind speed. Except in very shallow water, the surface current speed amounts to roughly 3.5 percent of the (standard) wind speed, and decreases downward at a rate that depends upon the stability of the water column, the length of time the wind has been blowing, and the size of waves present.

Despite the great size of storm waves at sea, and the tremendous volumes of water transported by wind currents, the actual wind stress on the surface is some four hundred times smaller than that exerted on an equivalent area, such as a square sail, held perpendicular to the wind direction. Figure 29 shows some experimentally determined relations between wind speed and surface stress, surface current, and the normal force on a sail. At hurricane speed (64 knots), a force of about 1,700 pounds would be exerted on a square sail having an area of 100 square feet. Roughly the same force would be exerted on the bare poles and rigging of a 50-foot ketch, and would be sufficient to drive her at hull speed.

Just as in the atmosphere, the Coriolis force tends to deflect the wind-drift current to the right (Northern Hemisphere) of the direction toward which the wind

*Some strong currents, such as that through the Straits of Gibraltar, are driven primarily by density differences.

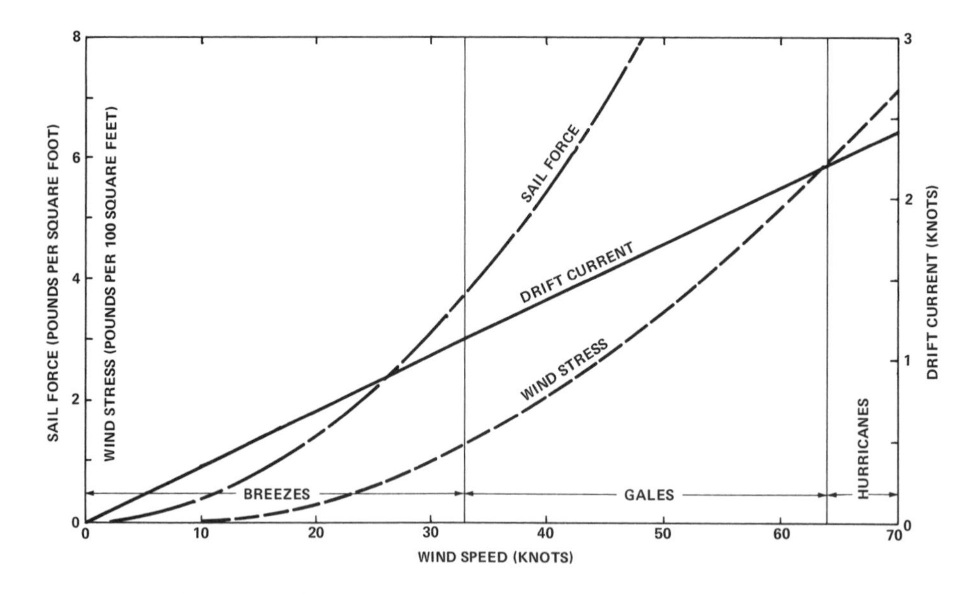

Fig. 29. Wind stress, surface current, and normal force on a sail, as functions of wind speed.

is blowing. Theoretically, for a steady wind blowing over an infinite, homogeneous ocean, with no extraneous currents owing to other factors, the surface wind drift is expected to veer the wind direction by 45° (fig. 30). The current vector continues to veer to greater angles with increasing depth, but with reduced velocity, at rates that depend upon the latitude. The mean flow, averaged over all depths, is perpendicular to the wind. Actually, the assumed conditions rarely obtain, or cannot be easily distinguished from other effects, such as tidal currents; reported deviations from the wind direction are generally less than 45°. The drift of field ice is observed to veer by 25°–35°, while floating objects that have shallow drafts and moderately high windage, such as dinghies and life rafts, drift almost directly downwind.

Where the wind drift is restricted by a lee shore, the wind stress acts to pile up water against the shoreline *(wind tide),* where the resulting pressure gradient establishes a subsurface offshore return flow to balance the onshore surface drift (fig. 31). The height of a wind tide is not simply related to wind stress, but depends also upon the water depth and bottom slope. Wind tides are not significant around small islands having steep approaches, but can rise to 20 feet or more along shallow offshore regions under hurricane conditions, and have been responsible for repeated heavy flooding in low-lying areas of the southeastern United States.*

In relatively shallow water, if the wind is offshore—or alongshore in such a direction that the Coriolis force has an offshore component—a circulation reverse to that of the wind tide will be established. In mid-latitudes, where the thermocline is shallow (10–20 fathoms), warm surface water adjacent to the shore will be blown seaward and replaced by colder water from beneath. This condition, known as *upwelling,* prevails along western subtropical coasts in summer, and produces the anomalously cold inshore water often found in such places as Baja California, northern Chile, Casablanca, and the Kalahari coast of South Africa.

*On February 1, 1953, a combination of 80- to 90-knot gales and high tides caused a storm surge that almost completely destroyed the elaborate dike system along the English and Dutch coasts, flooding 900 square miles, killing 2,000 people, and forcing evacuation of 100,000 others.

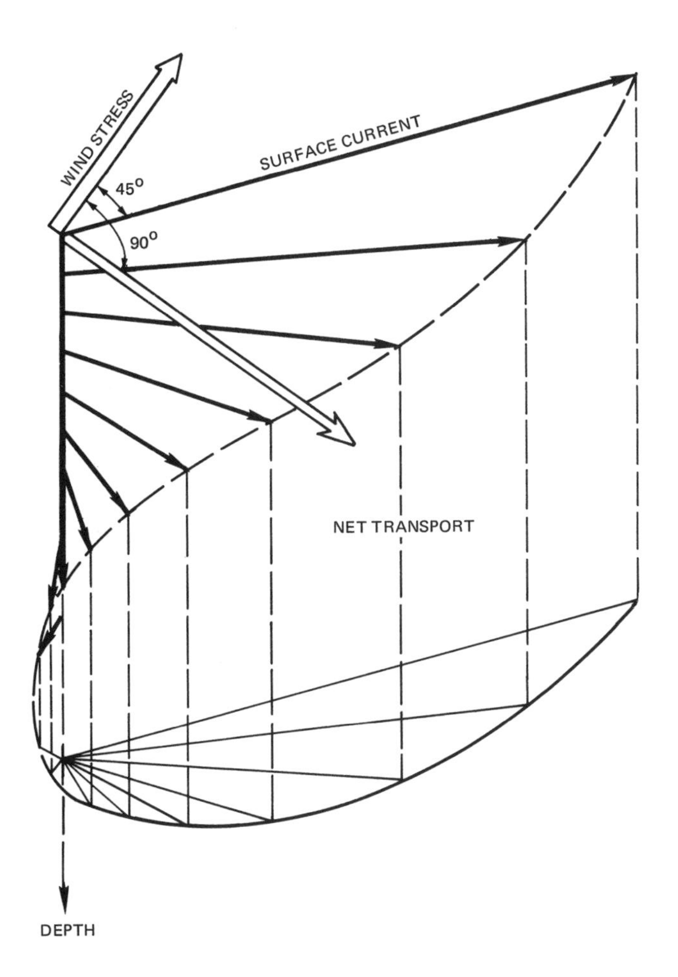

Fig. 30. Schematic changes of wind-drift current with depth in deep water. Surface current (Northern Hemisphere) veers the wind direction by 45°, and at greater angles with increasing depth. Current diminishes with depth. Net transport is at 90° from the wind. Depth of influence decreases with latitude (after Ekman).

NATURE OF THE GENERAL CIRCULATION

While the general pattern of the major ocean surface currents has been known for centuries, the bulk of ocean surface water moves so sluggishly under the combined action of wind and tides that its details are only now beginning to emerge, after 45 years of concentrated oceanographic study. Today, direct measurements of wind and water velocity from drifting and anchored buoy stations have supplemental indirect computations based upon the measured density structure, so that the general circulation now is considered to rest on a reasonably sound theoretical foundation that will be modified in detail only as more data become available.

Except for the special monsoon currents of the north Indian Ocean, the principal circulation in all oceans is anticyclonic (fig. 32), generally following the pattern of surface winds (figs. 22 and 23). There are important differences, however, because the fluid flow tends to be channelized by the Coriolis force and by the oceanic boundaries. Thus the overall circulation pattern more resembles that of the human vascular system; corresponding to the arterial network, the equatorial *"aorta"*

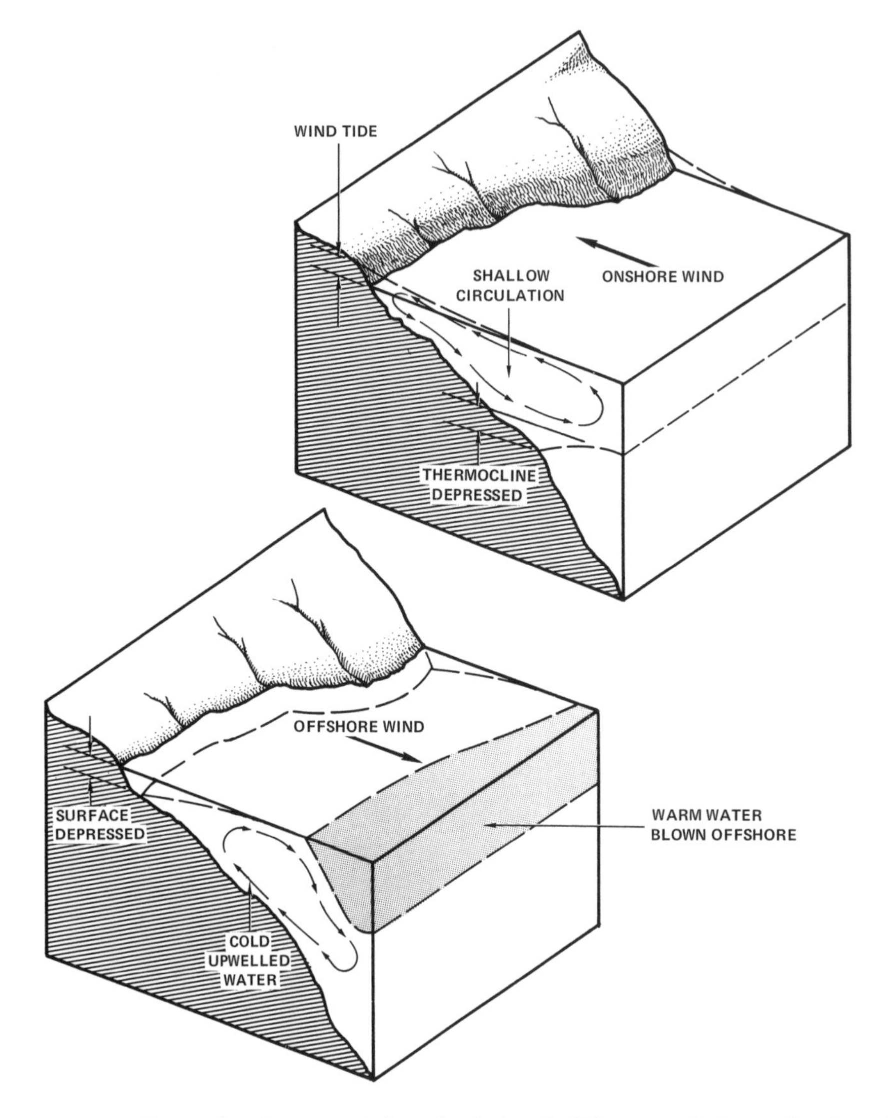

Fig. 31. Effects of onshore winds (wind tides) and offshore winds (upwelling) on shallow, warm surface layer.

branches into strong, narrow *"arteries,"* the western boundary currents, that again branch into even smaller *"capillaries."* The returning *"venous"* flow springs from weak and diffuse sources in the central and easterly sectors of the oceans, and is "pumped" back into the equatorial flow by the trade winds.

The fact that the currents usually follow the winds might seem to be intuitively obvious, but for the fact that the net wind drift is at *right angles* to the wind. Secondly, whereas the winds have prevailing anticyclonic rotation centers at latitudes of 30°–35° in the easterly sectors of the oceans (the oceanic high-pressure cells), the oceanic poles of rotation in the Northern Hemisphere are displaced to lower latitudes and far into the western sectors. These displacements are associated with strong westerly intensification of the circulation in each ocean, as manifested by the Gulf Stream in the Atlantic and the Kuroshio Current in the Pacific, and with weak, diffuse circulation in the easterly sectors of these oceans. In the Southern

Hemisphere, the poles of atmospheric and oceanic circulation largely coincide. There also is a suggestion of pairs of weak, corresponding oceanic and atmospheric poles in the western sectors of the southern oceans, but the associated poleward western boundary currents are substantially weaker than their northerly counterparts.

The above circulation pattern and its hemispheric differences are explained somewhat like this. Although the average surface-water transport is at right angles to the wind, under steady conditions this transport tends to pile up water in the central parts of the oceans, thus establishing radial pressure gradients in all directions. These gradients cause subsurface flow at all depths, such that the various density layers become slightly tilted so as to balance the unequal pressures. If there were no continents, the end result of these complicated interactions would be to force the surface currents to flow nearly in the direction of the wind—i.e., there would be bands (zones) of easterly and westerly currents beneath the corresponding wind tubes in each hemisphere. The continents, however, completely block zonal flow in the Northern Hemisphere, and nearly block it down under, so that only the *Antarctic Circumpolar Current* flows relatively unimpeded clear around the earth. Elsewhere, high-latitude eastward flow and westward flow near the equator produce the observed anticyclonic surface circulation in each ocean.

But why are the western boundary currents so much stronger than the easterly return flow toward the equator? There are several theories, none of which provides a simple explanation, and I will ask you to accept on faith that the problem is understood qualitatively—if not quantitatively. That is to say, mathematical models for a bounded, multilayer ocean with steady zonal winds, suitably adjusted for internal friction, and including the Coriolis force, give circulation patterns that reasonably resemble the observed circulation. The westerly boundary currents are

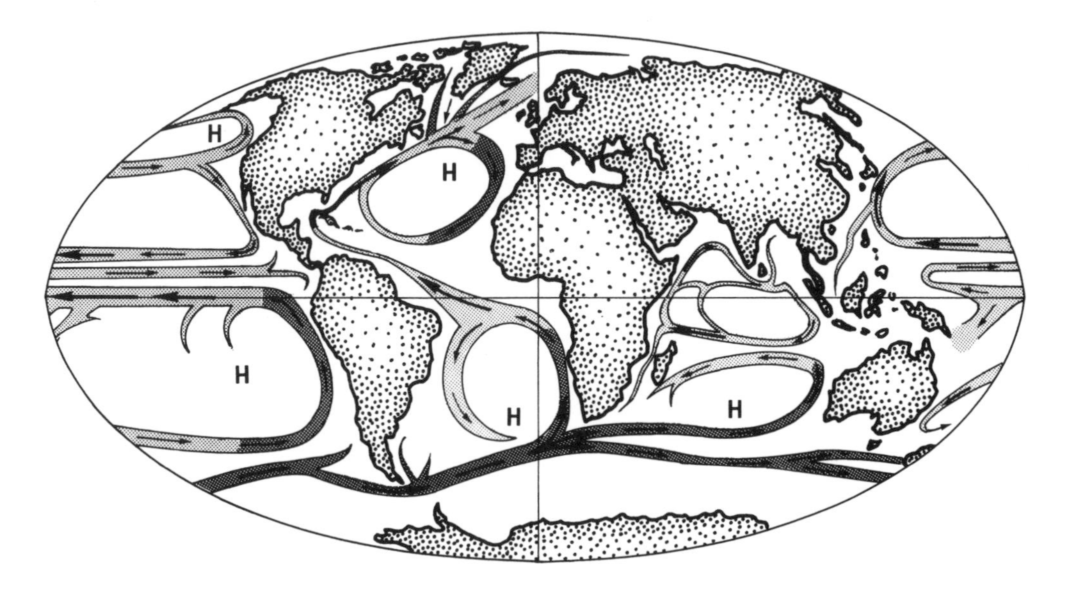

Fig. 32. General pattern of ocean circulation is anticyclonic, with circulation poles in northern oceans displaced west of center and depressed toward the equator. High-pressure poles of surface wind circulation are shown by letter *H.* Heavy arrows indicate major currents; light arrows indicate weaker, more diffuse circulation. Light shading shows warm currents; darker shading, cold currents.

weaker in the Southern Hemisphere because the circulation is not altogether blocked, and corresponds more to the unrestricted zonal model.

The navigator should keep in mind, however, that in the open sea, in all regions where he or she is not apt to be under the influence of one of the principal currents described below, and with an *increasing* wind, the wind drift will set the vessel to the right of the wind (Northern Hemisphere) by a small factor that increases with drift and latitude. This set will ordinarily be small compared with the normal leeway (3°–10° off the heading) made by a sailing vessel close on the wind.

10

Principal Ocean Currents

EQUATORIAL CURRENTS

Dominating the equatorial belt of all oceans, there is a somewhat intermittent system of several wide, shallow (100–200 fathoms) currents that nowhere extend much below the depth of the warm surface layer (fig. 33). Most commonly, two surface currents, the *North* and *South Equatorial Currents,* flow westward, separated by an eastward, surface *Equatorial Countercurrent;* while a subsurface *Equatorial Undercurrent* flows eastward beneath them. A weak secondary surface countercurrent also flows eastward in the western south Pacific. Some or all of these currents may be weak or absent in certain sectors at various seasons.

The North and South Equatorial Currents are principally driven by the trade winds. Thus they conform closely to their respective wind systems, within which there are marked seasonal and longitudinal variations in strength. The classical picture of continuous zonal bands of strong easterly trades separated by a region of light variables (the doldrums) is found only in the Atlantic and in a 2,000-mile sector of the Pacific between 120° W and 155° W (fig. 34). Elsewhere, all currents are apt to diminish wherever the trades are weak or absent, as does, for example, the South

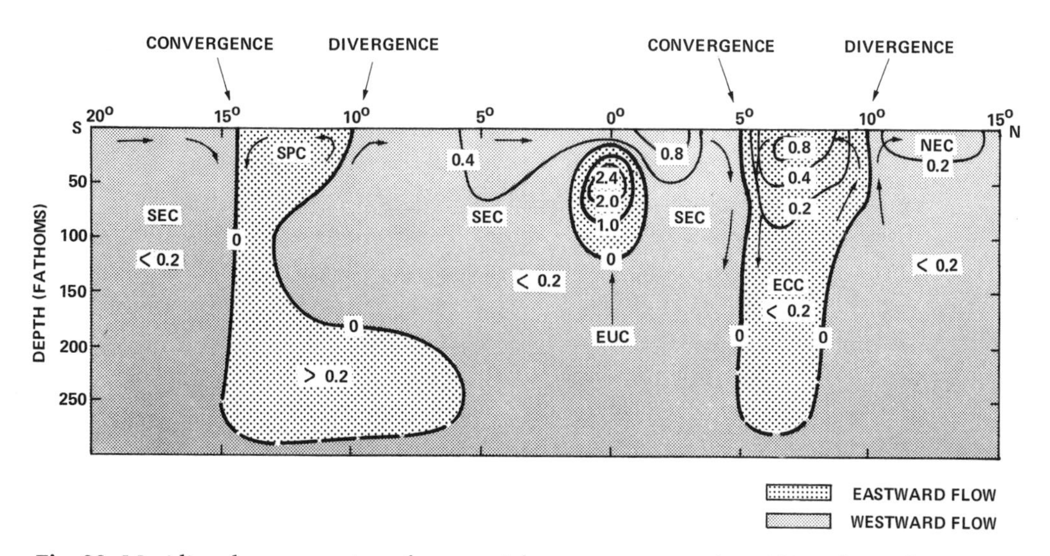

Fig. 33. Meridional cross section of equatorial current systems in mid-Pacific. Velocities are in knots. SEC: South Equatorial Current. SPC: South Pacific Countercurrent. EUC: Equatorial Undercurrent. ECC: Equatorial Countercurrent. NEC: North Equatorial Current.

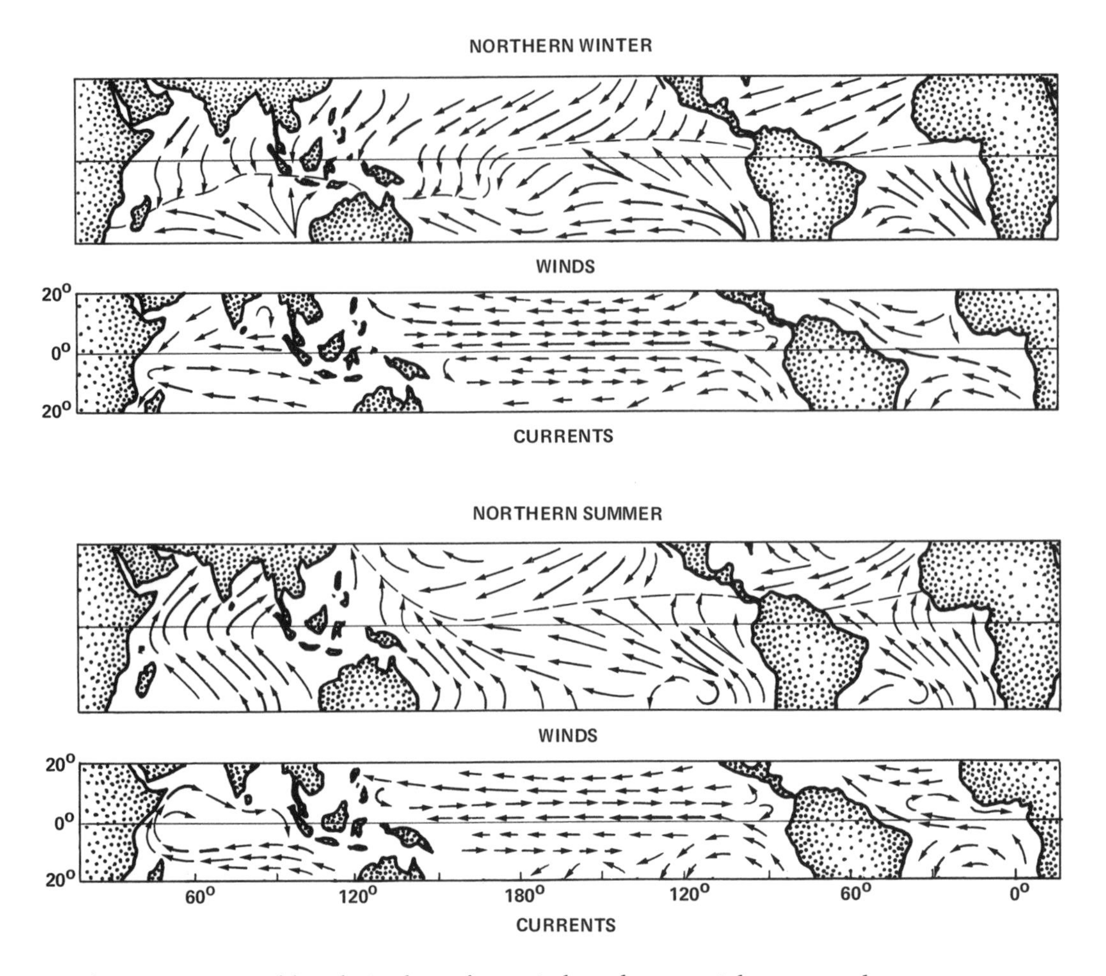

Fig. 34. Patterns of low-latitude surface winds and equatorial currents show strong resemblances in all oceans. Dashed lines in wind sections indicate positions of Intertropical Convergence.

Equatorial Current in the western Pacific all year round, and the North Equatorial Current in July and August. The latter disappears altogether in the north Indian Ocean during the southwest monsoon (May to September), being replaced by variable eastward flow.

The Equatorial Countercurrent is thought to represent the central portion of the return flow from the Equatorial Currents. Northern and southern branches of the latter form the westerly boundary currents discussed below. The Countercurrent flows eastward down the windless trade wind zone beneath the Intertropical Convergence, similarly meandering north and south with the seasons. It is best developed in the Pacific and Atlantic during northern summer, when the Southern Hemisphere winds are strongest, but is absent in the Indian Ocean. It weakens in the Pacific during the winter months, and disappears altogether in the Atlantic, except for a narrow, southeasterly coastal flow along the heel of Africa.

In concert with the weak secondary atmospheric tropical convergence near 11° S in the western Pacific, a corresponding *South Pacific Countercurrent* has been discovered at this latitude that flows eastward summer and winter from the New Hebrides to about longitude 140° W; its velocity seldom exceeds 0.2 knots.

Even in sectors where both Equatorial Currents are well developed, their outer boundaries are poorly defined, dwindling poleward toward the horse latitudes. But their respective interfaces with the Countercurrent are relatively sharp. Somewhat like air in the atmospheric wind tubes, water within these respective current tubes travels in spirals, sinking at the convergent boundary between the South Equatorial Current and the Countercurrent, and rising along the divergence between the Countercurrent and the North Equatorial Current (fig. 33).

Despite the grandeur of this massive circulation system, which transports more water than all other major ocean currents combined, there is often little manifestation of its presence. At times you can sail right across it without observing any obvious changes of color, clarity, or temperature, only to find your longitude off by 20 to 40 miles. At another time or place, the boundaries of the Countercurrent may be well defined by milky water, by foam streaks, or by a marked increase in wave steepness—and discomfort aboard—where the current sets against the wind (see chapter 17).

Although the surface currents tend to parallel the equator in all oceans, there are many temporal variations in strength and direction. In the central Pacific, both winds and currents undergo periodic three- to four-day meanders of 10°–20°, most prominent in December and June, and tending to disappear near the equinoxes. Because the Southern Hemisphere trades are stronger, the South Equatorial Current and the Countercurrent are swifter (0.5–1.5 knots) than the northern current (0.1–0.5 knots). All currents attain their peak velocities in September and October, and are weakest in March and April, when they may intermittently disappear for several days. The above figures are only average values, however, and the current speeds at any time appear to be remarkably sensitive to regional trade wind strength. During a current study near McKean Island (3° S, 174° W) in October, supposedly a month of relatively strong currents, we were chagrined to discover weak variable winds and a westerly flow of only 53 miles in 19 days!

The *Equatorial Undercurrent,* called the Cromwell Current in the Pacific, is a thin (200 yards), narrow (240 miles), high-velocity jet (2–3 knots) that seemingly ignores the peregrinations of the surface-current system, and remains glued to the equator, flowing eastward at a depth of about 50 fathoms. The Cromwell Current has been traced all the way across the Pacific from New Guinea to the Galápagos Islands, gradually decreasing in depth toward the east, where its upper boundary is occasionally detectable at the surface. Although it was only discovered in 1954, recent studies indicate that it alone transports as much water as any other equatorial current, and the reasons for its existence and remarkable stability remain one of the most puzzling problems in oceanography. Similar equatorial undercurrents have been discovered in the Atlantic and Indian oceans, but they disappear during the southwest monsoon.

WESTERN BOUNDARY CURRENTS AND THEIR EXTENSIONS

Near the western margins of all oceans, the cumulative transport of the North and South Equatorial Currents piles up deep wedges of warm water against the continental boundaries, whose effects are felt far seaward as the piled-up water attempts to escape in all directions. A portion of this water surplus returns eastward via the countercurrents and the undercurrent. But the greater fraction splits up into several branches to form the deep, narrow *western boundary currents,* which are guided poleward by the margins of the continental shelves. These currents are the strongest that flow in the open sea, sometimes achieving speeds of 3 to 5 knots, and are exceeded only by tidal flow in restricted inland waters. Because they are diversely affected by topographic conditions and by seasonal variations in coastal winds, they are best discussed as individual flow networks in each ocean.

The Gulf Stream System: Certainly the earliest and best known of the major ocean currents, the Gulf Stream appears to have been first noted in 1513 by Ponce de Leon, who could not make southing against it along the coast of Florida. While the general anticyclonic nature of the Gulf Stream system around the north Atlantic was described as early as 1663 by Isaac Vossius, the first realistic—albeit oversimplified—chart of its southerly circuit was published by Benjamin Franklin in 1770.* By this time the vicissitudes of this "ocean river" were well known to American merchants and whalers, who undertook extensive detours to avoid it when sailing from Europe to America. But with traditional, steadfast determination, English mail packets continued to plow against it until the early part of the nineteenth century.

The Gulf Stream (fig. 35) has its roots in the conflux of the North and South Equatorial Currents, which shift between 20° and 50° west longitude, reversing roughly at the fall and spring equinoxes, respectively, in concert with corresponding shifts (0°–10° N) of the Intertropical Convergence. By far the strongest contribution to the westward flow comes from the South Equatorial Current, which is deflected northwesterly along the north coast of South America (the *Guiana Current*), where it attains a peak speed of about 3 knots from April to May. East of the Antilles Arc, the central and northern parts of the flow are identified as the *Caribbean* and *North Equatorial Currents,* but the entire region exhibits more the character of a very broad, general drift, that is strongly influenced by the northeast trades, and which gradually weakens northerly until it vanishes in the Sargasso Sea.

This westward drift diffuses through the network of the Antilles into the Caribbean Basin with a mean velocity of about 1 knot, although a secondary branch (the *Antilles Current*) passes to the north of Haiti and Cuba. The Caribbean Current funnels through the Yucatan Straits, reverses direction around the east end of Cuba, and passes out northeasterly through the Florida Straits (the *Florida Current*). A left branch of the current from the Yucatan Channel sets up weak secondary circulations within the Gulf of Mexico. The Florida Current continuously accelerates from about 2 knots in the Yucatan Channel to as high as 4 knots in the Florida Straits, reaching its maximum speed on the north, or Florida, side of the channel.

Upon reentering the Atlantic, the Florida Current immediately swings northward and runs close inshore (10 to 20 miles) along the coastline of the southeastern United States as far north as Cape Hatteras, being joined en route by contributions from the Antilles Current and the North Equatorial Current, whose confluences form the true beginning of the mighty Gulf Stream. In this sector, the Gulf Stream runs between 3 and 4 knots, continuing northeasterly undiminished as far as Cape Hatteras. It sets strongest along its sharp western margin—well defined by its deep blue contrast with the green inshore water—and extends clear to the bottom across the shallow Blake Plateau connecting Florida

*The map was prepared by Timothy Folger, a Nantucket sea captain.

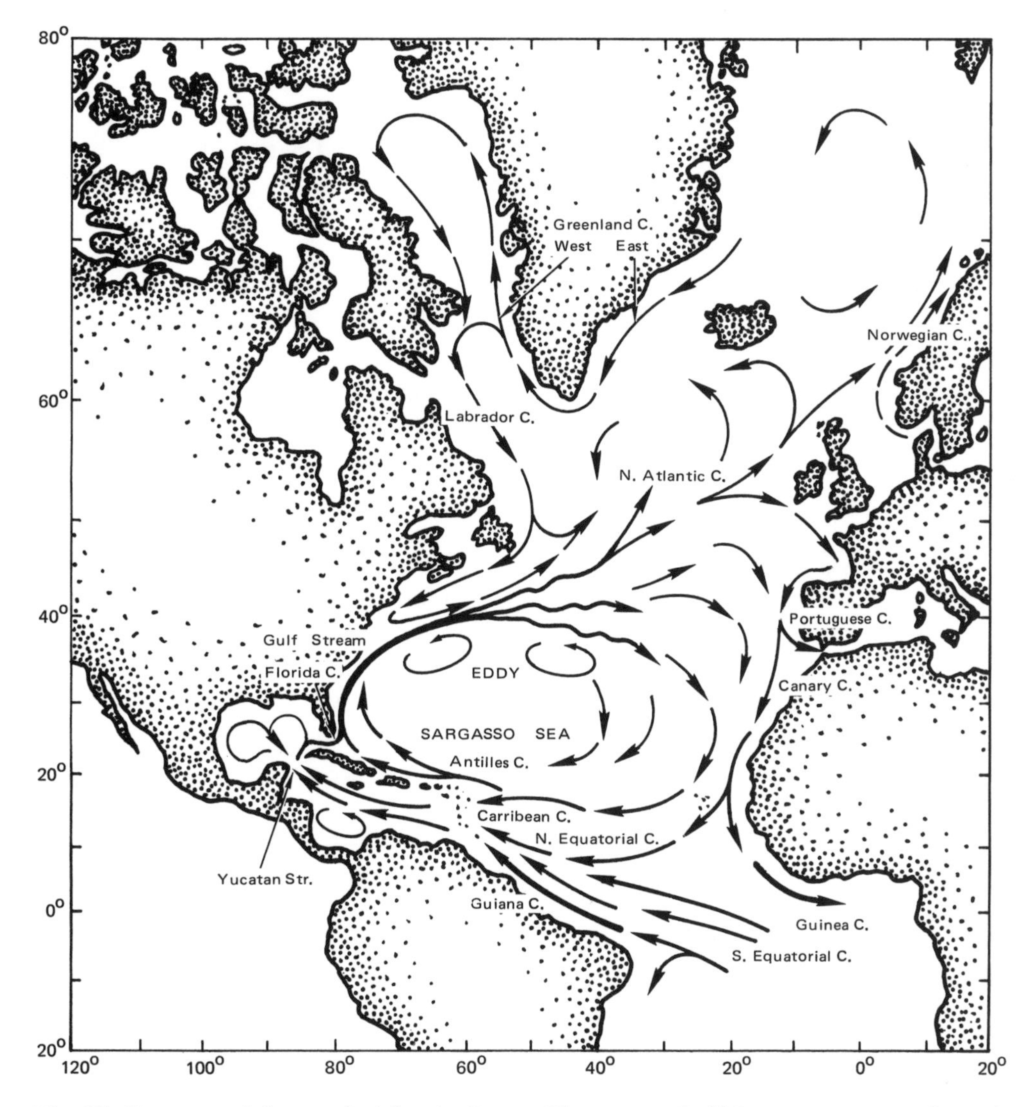

Fig. 35. Currents of the north Atlantic Ocean. The strong Gulf Stream originates from the conflux of the North and South Equatorial Currents, and diffuses in many branches throughout the northeast Atlantic.

with the Bahama Islands, in whose lee it is weak and variable. From Miami to Hatteras the flow exhibits tidal fluctuations as great as 1 to 2 knots, running strongest about two hours after mean low water in Florida ports. The tidal effect is stronger during neap tides, and stronger in February and August when the moon's declination is over the equator. These variations in the behavior of the Gulf Stream were first detailed by Lt. James Pillsbury, of the U.S. Coast and Geodetic Survey, who spent two years (1885–1887) making exhaustive current measurements from Cape Hatteras to the Yucatan Straits. Pillsbury remarks that southbound coasting vessels bound for Florida or Gulf ports can make the best time by crossing the Stream at Hatteras and running down its east side inside the Bahamas as far as Gun Cay before recrossing it near Miami.*

*Still good advice today.

After passing Cape Hatteras, the Gulf Stream—now some 100 miles wide—crosses the continental shelf and heads out across the deep water of the north Atlantic. Gradually slowing and widening, it still maintains its coherence as it curves eastward until, about 300 miles south of the Grand Banks, it breaks down into a plurality of undulating streams, vaguely defined as the *North Atlantic Current.* During this central phase of its flow, it develops unstable meanders that occasionally increase in amplitude to the point where large counterclockwise eddies break off along the right-hand side of the main stream, some of which continue to rotate independently for four to six months before they die away. During their formation, these eddies entrap central pools of colder water that may persist long after sensible eddy rotation has ceased. Here, contact of the swift left margin of the warm Gulf Stream with cold water carried down from the north by the *Labrador Current* results in an exceedingly sharp thermal gradient: the "cold wall." Condensation from warm, humid air carried north across this thermal barrier by southeasterly winds, particularly in spring and early summer, produces the dense fog banks common to the Grand Banks region. Conversely, in winter, cold, dry continental air flows down over the warm Gulf Stream, causing rapid evaporation and "sea smoke" fog, giving the sea the eerie appearance of a steaming cauldron.

North of Cape Hatteras, countercurrents occasionally develop on both sides of the Gulf Stream. There is a slow southwesterly flow over the continental shelf between Newfoundland and the Cape, while a swift, narrow flow—sometimes running at 1 to 2 knots—has been observed on the opposite, or seaward, side.

East of 45° W, the diffuse, meandering filaments of the North Atlantic Current divide into two principal branches, although the flow in this sector, perturbed by a continuous series of extratropical cyclones, more resembles a general flux of warm surface water. The southerly branch curves southeastward past the Azores, Madeira, and Cape Verde Islands, and rejoins the North Equatorial Current, being augmented en route by the wind drift of the northeast trades. The northern branch continues northeast, dividing again near 20° W. The northeast branch appears to flow only in summer, passing west of the British Isles and north along the coast of Norway (the *Norwegian Current*) until it vanishes in the weak and variable circulation of the Arctic Ocean. In winter, this flow is replaced by a cold northerly current issuing from the Baltic and out through the North Sea. The southeast branch passes south of the British Isles, circles the Bay of Biscay, and continues south along the coasts of Portugal and West Africa (the *Portuguese* and *Canary Currents*). Side branchlets flow into the English Channel and the Straits of Gibraltar. South of the Cape Verde Islands, the flow splits again, part turning west to rejoin the North Equatorial Current, and part eastward to join the Countercurrent to form the relatively swift *Guinea Current,* which runs as high as 3 knots in June.

The winds and currents of the eastern Atlantic are extremely variable, however. Sir Francis Chichester sailed from the latitude of Ascension Island to the Azores in 28 days in April and May of 1967, while Eric Hiscock required 52 days at the same time of year in 1962. This difference is only in small part due to *Gypsy Moth's* faster average speed around the world (131 versus 105 miles per day). Sailing during June and July in 1951, Alain Gerbault took 36 days from Ascension to the Cape Verde Islands (which is less than half the distance to the Azores by the route sailed), averaging only 46 miles per day. All might have done much better had they made this northerly passage in February–March, when the doldrums virtually disappear.

The Brazil Current: The feebler south Atlantic counterpart of the Gulf Stream, the *Brazil Current* is a weak, seasonal, southern branch of the South Equatorial Current that waxes strongest (0.5–1.0 knots) in September–February, running southwesterly close inshore along the coast of Brazil and Uruguay (fig. 36). It disappears from February to September, being replaced by a weaker northward flow above 20° S. The entire inshore region of the shallow shelf along this coastline is dominated by tidal currents, which in some places run as high as 4 to 5 knots. From 35° S to the Falkland Islands, the cold *Falkland Current* sets weakly north along the margin of the wide, shallow coastal shelf at certain times of the year when winds blow from the south. Its presence is easily detectable by a pronounced temperature drop, as well as by an abundance of floating kelp and other debris washed from the wooded islands to the south.

The Kuroshio and Its Extensions: The mighty *Kuroshio Current* is the Gulf Stream of the north Pacific. Indeed, these streams have much in common, as well as some significant differences. The Kuroshio is swifter (over 5 knots) at its maximum in northern summer, but its relative transport has never been determined accurately. Both have similar origins, strong central sections guided by the continental shelves, and meandering termini that extend clear across their respective oceans, and which

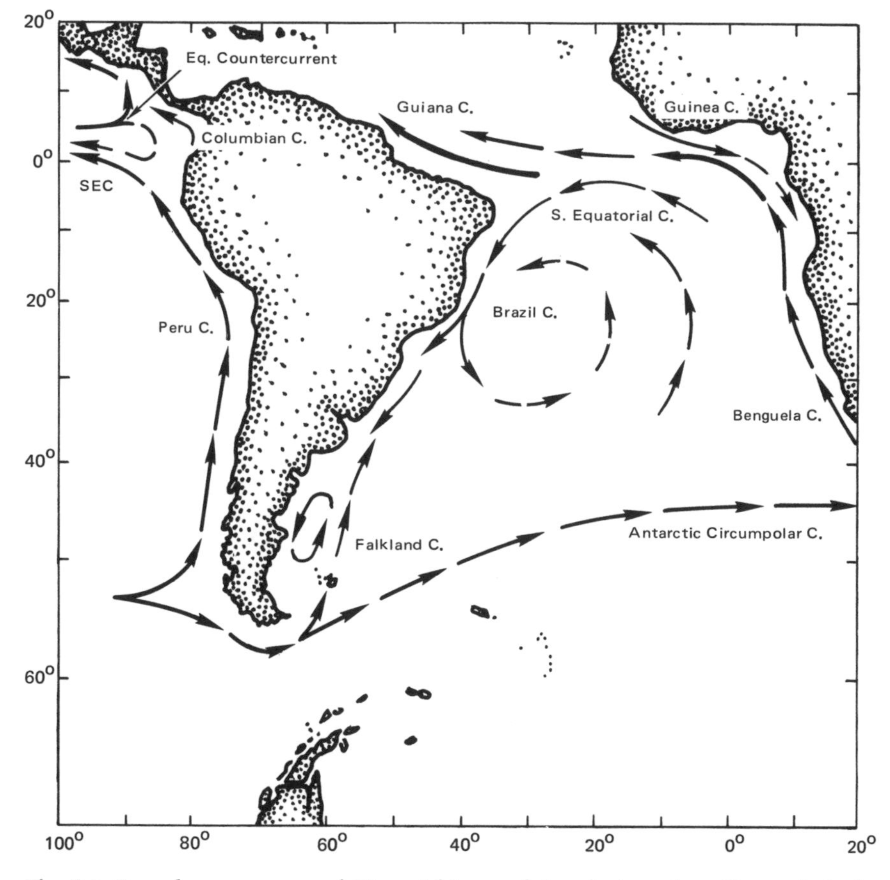

Fig. 36. Boundary currents of West Africa and South America. Short, dashed arrows show winter reversals of Equatorial Countercurrent in the Gulf of Panama, and of Benguela Current in the Gulf of Guinea, under weaker trade winds. Cold Falkland Current flows northward during southerly winds.

moderate the climates of their opposite shores. Both have northerly cold walls in deep water, where they intersect with southerly flowing Arctic currents (the Oyashio and Labrador Currents), in which regions dense fogs abound in spring and summer. Therefore, it would be pointless to further elaborate on their similarities, and I will confine the following remarks to important differences.

The Kuroshio has its headwaters near the western Philippines (fig. 37) as the north branch of the North Equatorial Current. During the northeast monsoon, a left branch flows through the Bashi Channel between Luzon and Formosa into the South China Sea. This flow reverses during the southwest monsoon. The main stream passes east of Formosa and flows northwest along the continental shelf toward Kyushu, Japan, where about 10 percent of the flow branches to the left and enters the Japan Sea through the Korean Straits (the *Tsushima Current*).

At this point the Kuroshio jogs to the right between the Ryukyu Islands and Kyushu, and enters the deep waters of the Pacific along the east coast of Japan. After crossing the coastal shelf it, like the Gulf Stream, begins to meander back and forth in much the same fashion as the free end of a hose under high pressure, except with a much longer time scale. These meanders appear to be of two types, both of which arise in the vicinity of southern Kyushu as unstable waves that amplify as they progress northeastward. Those of the first type have a lifetime of several months, but do not seem to break off as separate eddies in the manner observed in the Gulf Stream; those of the second type are similar, but occur over periods of three to ten years. They have larger amplitudes, and at least one out of three observed since 1934 broke off as a separate eddy that circulated independently for several years. In addition to the above perturbations, recent studies suggest that the main axis of the Kuroshio off the northeast coast of Japan

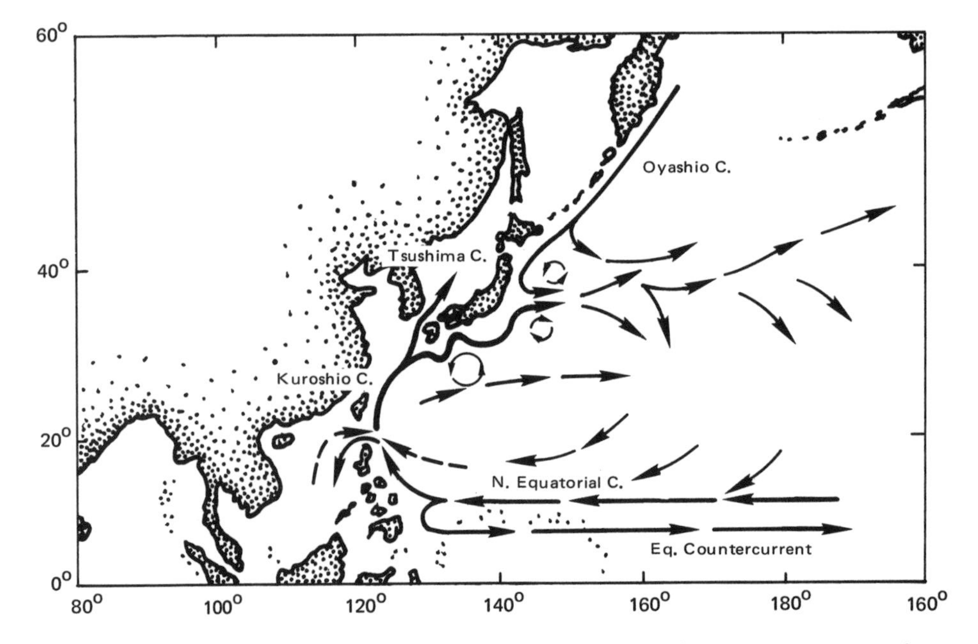

Fig. 37. The Kuroshio Current and its left branch, the Tsushima Current, arise from water piled up by the North Equatorial Current, which also feeds the South Equatorial Current. An alternate position of the Kuroshio is shown by a dashed line, as well as by circles indicating eddies near its confluence with the cold Oyashio Current. A long-persistent eddy is shown near the lower meander of the Kuroshio.

has two alternate positions of stability some 350 miles apart, between which it shifts rather suddenly at irregular intervals of several years. This shift seems to be associated with small changes in the average speed of the current, but has yet to be identified with any specific causative phenomenon, such as long-term weather changes.

The Kuroshio leaves the Japan coast near Cape Inubo-Suki, and heads eastward at about 36° N. It has been identified, although not well studied, as a well-defined meandering stream as far as 160°–170° E, beyond which it widens into a diffuse eastward drift under the influence of the prevailing westerly winds. Because of the great width of the Pacific, the influence of the Kuroshio along the west coast of North America is manifested by an abnormal warmth of the surface water, rather than by recognizable southerly coastal currents, such as the Canary Current in the Atlantic. The temperatures between San Diego and southern Alaska differ by only 20°F the year around, and only vary by 10°F between February and August. Although specific names have been given to various sectors of the general coastal circulation, the latter are dominated by local winds whose integrated effect remains much the same throughout the year. Thus the north Pacific, California, and Alaskan currents cannot realistically be described as extensions of the Kuroshio in the manner implied by some authoritative nautical references.

The New Guinea and East Australia Currents: The western boundary currents of the south Pacific are quite anomalous, owing partly to the extensive complex of islands, and partly to the seasonal reversal of the monsoon circulation (fig. 38). During the winter months (November–April), northeast monsoon winds set up a counterclockwise circulation in the East Indies, roughly centered on Borneo, with water from the South China Sea flowing through the Karimata Strait into the Java, Flores, and Banda seas, and issuing northwesterly through the Molucca and Halmahera straits. The former turns eastward, merging with the southern branch of the North Equatorial Current to strongly supplement the Equatorial Countercurrent in this region. The flow from the Halmahera Strait turns more sharply eastward and runs as the strong *New Guinea Current* along the north coast of this island. Here it joins the South Equatorial Current, which, during this season, has turned south and reversed its direction to the east. This combined flow appears to diffuse through the Solomon Island Group and continues eastward as the South Equatorial Countercurrent previously mentioned, although this connection is not firmly established.

The situation reverses during the southwest monsoon (May–September), and the general flow through the East Indies becomes clockwise around Borneo. The South Equatorial Current, now opposed by the wind, jogs southward at about 140° E and flows westward along the New Guinea coast. Just east of Halmahera, it divides; the main northern branch turns north and reverses to join the Countercurrent, while a lesser branch runs south through the Halmahera Strait and then westward through the Banda and Java seas and into the Indian Ocean, via the several passes through the Sunda Islands. In the shallow Arafura Sea south of New Guinea, a slower wind drift sets westward most of the year. Strong—and for the most part unstudied—tidal currents often oppose or reverse the flow through most of the narrower straits in these waters.

The Torres Strait, between New Guinea and Cape York, Australia, is a particularly dangerous stretch of water, abounding in uncharted coral reefs that are often hidden by turbid water from the numerous streams running down from the moun-

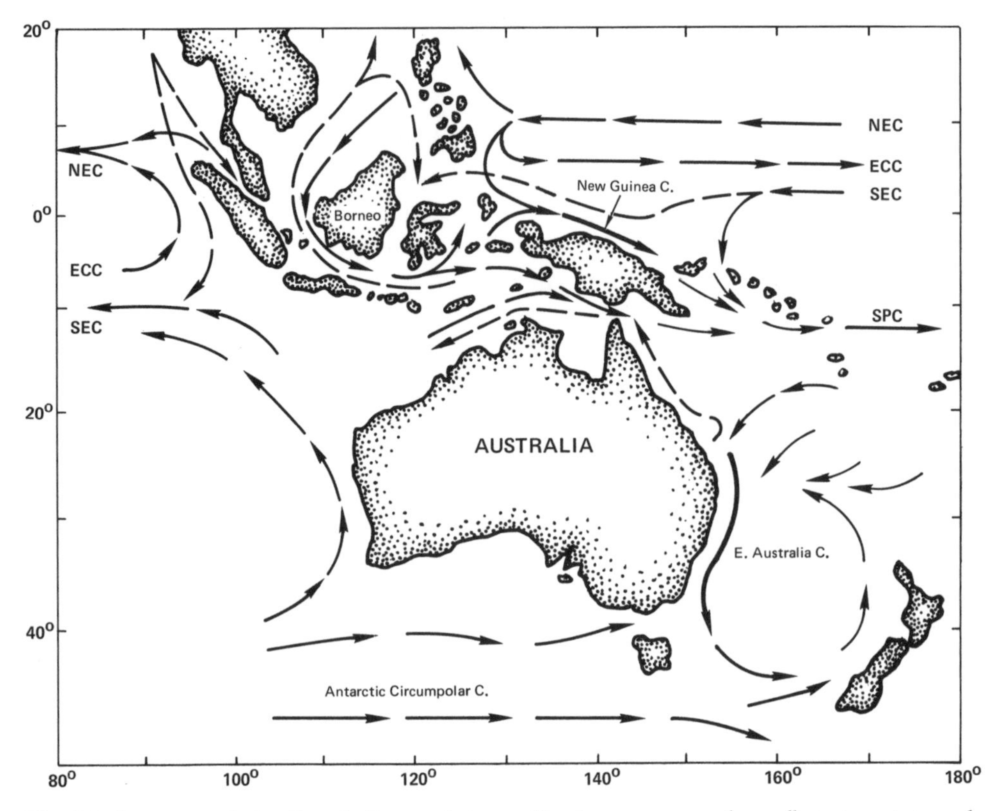

Fig. 38. Currents of the East Indies and Australia. Long arrows show flow in winter; short arrows, flow reversal in summer. Changes follow monsoon winds, as described in the text. The East Australia Current (a weaker counterpart of the Kuroshio) is not affected by monsoon winds. Legend: NEC—North Equatorial Current; ECC—Equatorial Countercurrent; SEC—South Equatorial Current; SPC—South Pacific Countercurrent.

tains of New Guinea. The current prevails to the west through this strait in southern winter, under the stress of the southeast trades. Despite these hazards, it is a preferred route for sailing yachts eastbound around the world, avoiding the much longer passage to the north of New Guinea against adverse winds. It was here that Eric Hiscock, during the second world voyage of *Wanderer III,* went fast aground at spring tide, while heeled well over under a fresh breeze. As the tide fell, the yacht laid over beam ends on the bare reef several hundred yards from the nearest navigable channel. Eric and his wife, Susan, feeling that their voyage had come to sure disaster, walked ashore to a nearby island. Fortuitously, they were later able to warp the yacht off at a lower tide stage, with the combined assistance of a Mission launch and some fifty local aborigines.

The East Australia Current: This current forms no part of the equatorial current system, but appears to be the westward intensification of a local counterclockwise circulation between Australia and New Zealand, augmented by drift from the easterly trades. It is first identifiable in the southern part of the Coral Sea east of Swain Reefs and flows south along the entire coast of Australia to Cape Howe. Here, it encounters the cold water of the Antarctic Circumpolar Current, a branch of which circles northward west of New Zealand, to complete the circuit. The East Australia Current runs strongest from 26°–32° S, close inshore along the 100-fathom contour. It attains 2 to 5 knots from February to April but generally exceeds 2 knots year-round, with frequent interruptions—and even reversals—during opposing winds. Although they both have peak velocities as high as that of the Kuroshio, the East Australia and New Guinea Currents are relatively narrow, and neither compares in width or volume transport with their northern neighbor.

Western Boundary Currents of the Indian Ocean: The circulation of the north Indian Ocean reverses seasonally, being controlled by the monsoon winds (fig. 39). The doldrums are always found south of the equator, and only the South Equatorial Current persists year-round. In northern winter, the North Equatorial Current and the Countercurrent develop in the familiar manner, except that the northeast monsoon winds extend over the entire ocean above the equator, such that the *Somali Current* sets *southwest* along the western boundary, instead of northeast, as in other northern oceans. Its southerly drift nowhere exceeds a knot or so, and at the equator it meets the northern branch of the South Equatorial Current, both of which reverse to form the Countercurrent. At this season, the Countercurrent is wider and stronger than that in the Pacific and may attain speeds of 1 to 2 knots. With the advent of the southwest monsoon—which predominates south or southwesterly north of the equator—the North Equatorial Current reverses and there is general eastward flow across the entire north Indian Ocean. Simultaneously, the greatly enhanced South Equatorial Current turns northward against the Somali coast, forcing the Somali Current to reverse itself and flow northwest at great speed (5 to 7 knots), where it continues along the African coast to 8° N, and then swings sharply eastward into the deep waters of the Arabian Sea. At the same time, strong southwest winds along the Somali coast bring about intense upwelling of cold water, with associated fog and rain, high biological productivity, and absence of coral reefs, which cannot tolerate cold water.

At all seasons of the year, the left branch of the South Equatorial Current flows south along the coast of Africa, where it is known as the *Agulhas Current,* although the section between Madagascar and Africa is also called the *Mozambique Current.* The Agulhas Current is probably the best southern counterpart of the Gulf Stream. Like all western boundary currents, it runs strongest in a narrow band at its western side, and it reaches a peak strength of 3 to 4 knots at about 30° S, off Durban, where its axis lies some 60 miles from the coast. Near the Cape of Good Hope, the flow divides; one part continues around into the Atlantic, where it joins a northerly branch of the Circumpolar Current and flows north along the Kalahari coast as the *Benguela Current.* The left branch turns eastward to join the main Circumpolar Current. As off Australia, however, this succinct description indicates only a general tendency, and the mighty Cape, with its almost continual succession of southwest and southeast gales, admits no assurance that local surface drifts are not equally variable in strength and direction. The generally ferocious weather of this region, and its uncertain winds and currents, make cape passages extremely difficult for sailing yachts.

THE ANTARCTIC CIRCUMPOLAR CURRENT AND ITS BRANCHES

Most books on navigation refer to the eastward circulation of water around Antarctica as the *West Wind Drift,* in concordance with the prevailing—if irregular—westerly

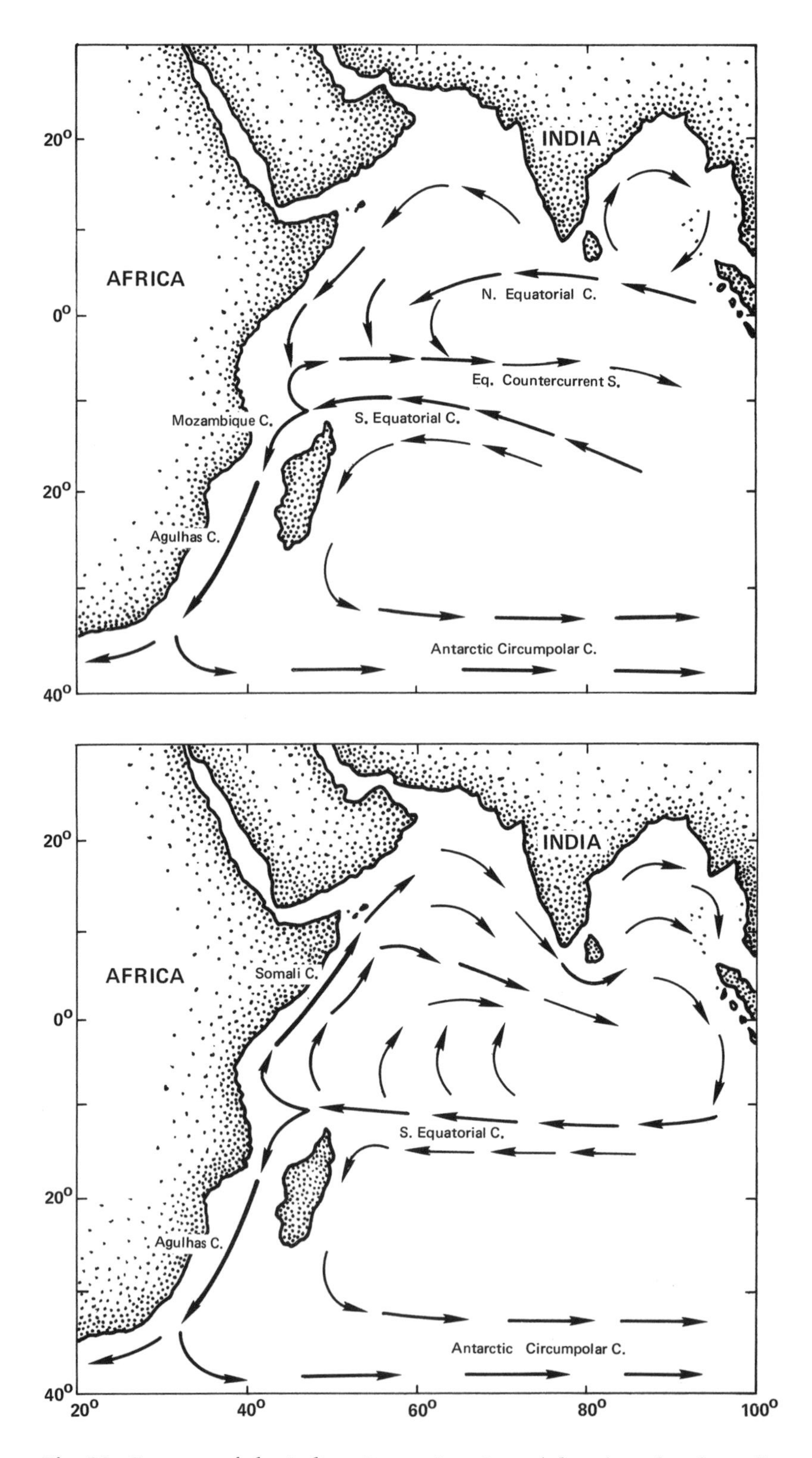

Fig. 39. Currents of the Indian Ocean. In winter (*above*), under the influence of northwest monsoon winds, the circulation is generally westward north of the equator, and the Mozambique and Agulhas currents flow south along Africa. In summer (*below*), southwest monsoon winds reverse the northern circulation, the North Equatorial Current disappears, and the strong Somali Current runs north along Africa.

winds that circle the earth between 40° S and 60° S. As pointed out in Part II, however, the *mean* westerly wind speed in the Roaring Forties is only 10 to 20 knots, and the frequent high cyclonic winds of gale force blow in circles. The strongest winds from these storms—augmented by the mean winds—blow westerly in the northern sector at about 50° S; their weakest winds, in the southern sector, blow easterly at about 70° S.

Because surface current speed increases with wind speed, the local surface currents respond more to the traveling cyclones than to the mean winds, and exhibit similar circular patterns, although inertia tends to average out the variability of wind direction. Thus, there is a wide zone of eastward wind drift at 0.3–0.5 knots centered near 50° S, and a weaker, narrow westward drift close to Antarctica with much stronger local circular flow during the passage of these extratropical cyclones (fig. 40). This entire pattern is not concentric with the South Pole but is depressed about 10° farther south in the Pacific sector than in the south Atlantic and Indian

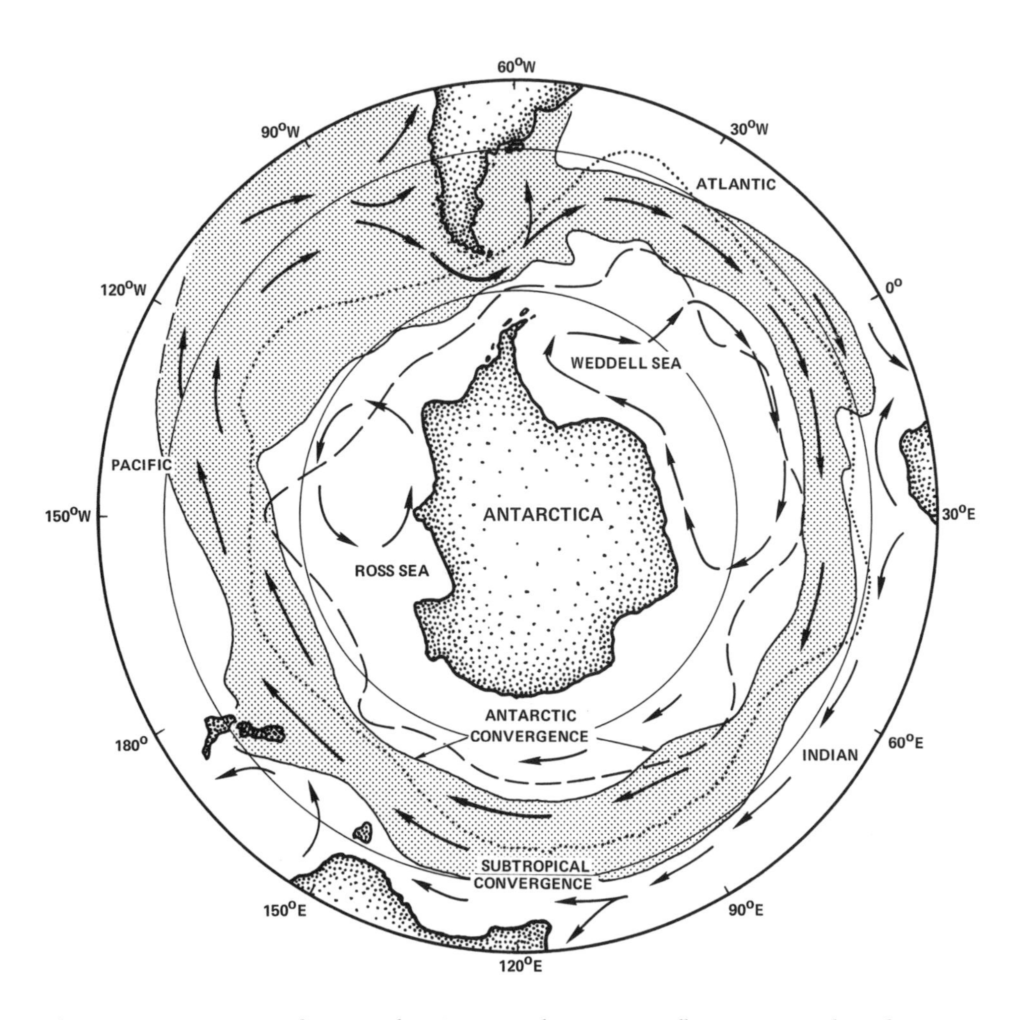

Fig. 40. Antarctic circulation. The Circumpolar Current flows eastward, and is strongest between the Antarctic and Subtropical Convergences (shaded zone). Westward flow occurs within small cells in the Ross and Weddell seas under the influence of easterly winds along the Antarctic shoreline. Dashed line shows approximate limit of pack ice in winter. Iceberg limit is given by a dotted line.

oceans, in concordance with the similar displacement of the Antarctic land mass toward 60° east longitude.

The above circulation is only the surface manifestation of a more complex vertical circulation around Antarctica. This great land mass, with its permanent ice cap, is the factory for most of the cold, deep water that fills the ocean basins; that is, the multiple layers that define the vertical density structure in midocean all outcrop at the surface near Antarctica, much as a geological formation of tilted layers often outcrops in the flat desert. In fact, these layers are *produced* by such processes as melting and freezing of ice, by evaporation and precipitation without and within the westerlies, respectively, and by mixing with adjacent waters under the action of winds and waves. The schematic of this *thermohaline* vertical circulation near Antarctica (fig. 41) illustrates how cold (36°F) water rises between 50° S and 70° S and mixes with warmer (47°–52°F), shallower subantarctic water, to form water of intermediate density and temperature that sinks along a narrow zone near 50° S, known as the *Antarctic Convergence.* Farther north (40° N) there is a weaker, secondary *Subtropical Convergence,* along which subarctic water mixes with warm, saline surface water, and sinks because of its intermediate density. As shown in figure 8, page 24, the dense Antarctic Bottom Water that pervades all ocean basins is formed adjacent to the ice shelf in the Weddell Sea, where winter freezing increases the salinity of the already cold Antarctic Deep Water.

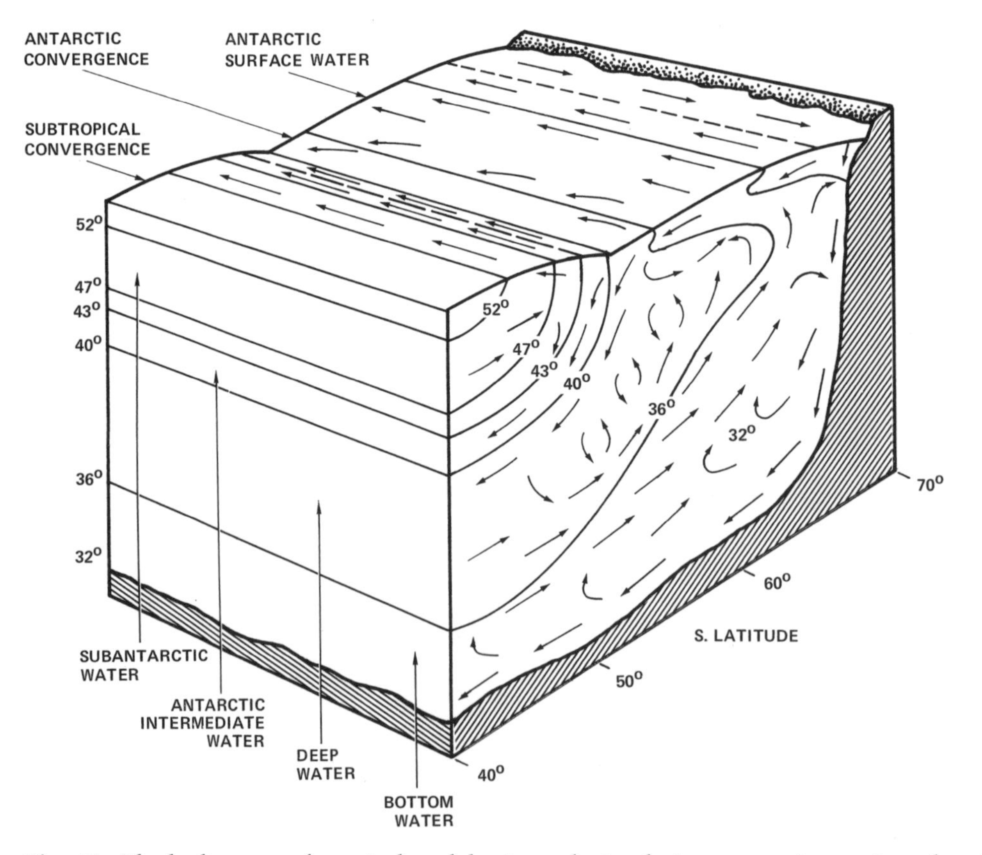

Fig. 41. Block diagram of vertical and horizontal circulations near Antarctica, showing temperature distribution (°F). The Circumpolar Current is strongest near the Antarctic Convergence.

All of these vertical motions are very slow compared with the eastward surface circulation, but the vertical differences in density are so small that even slow motions transmit horizontal momentum effectively between the several layers. Thus, the entire water mass south of the Subtropical Convergence, and from the surface to the bottom, tends to move around Antarctica as a slow, contiguous flow: the *Antarctic Circumpolar Current.* Despite its small average velocity (0.1–0.3 knots), the great width and depth of this current make its total water transport some three times larger than that of the Gulf Stream.

Wherever the wide Circumpolar Current impinges upon the southern end of a continent obtruding into the Antarctic Ocean, a portion of the flow is diverted northward to form an eastern boundary current. Because the Coriolis force acts to divert northerly flow to the west in the Southern Hemisphere, these currents are wider and slower than their western counterparts, and flow farther from the coast, although southwesterly winds, to some extent, counteract this tendency below 40° S.

Probably the best example of an eastern boundary current is the *Peru Current* (fig. 36, p. 107), which flows northward along the coasts of Chile and Peru as far as Cabo Blanco. Here, the main flow leaves the coast under the influence of the southeast trades and proceeds northwestward past the Galápagos Islands to join the South Equatorial Current. The Peru Current is characteristically cold because of the upwelling of subsurface water arising from the southerly trades and the offshore component of the Coriolis force. The combination of cold offshore water and the strong anabatic land breeze rising over the hot coastal slope accounts for the prevalence of heavy fog in this region, and produces the chilly—albeit nearly rainless—climate of the Galápagos.

Eastward, a sharp front separates the cold, salty waters of the Peru Current from the much warmer and fresher surface water along the coasts of Ecuador, Colombia, and Panama. In concert with seasonal shifts of the Intertropical Convergence in the Gulf of Panama, the North Equatorial Current swings northwest off Costa Rica in northern summer, but reverses in winter to join the South Equatorial Current. A weak inshore flow, the *Colombian Current,* persists northward all year.

Corresponding to the Peru Current, the eastern boundary current off South Africa is the *Benguela Current*—already mentioned as arising from the conflux of the right branch of the Agulhas Current, rounding the Cape of Good Hope, with a left branch of the Antarctic Circumpolar Current (fig. 36, p. 107). Off the West African Coast, its set is weak and variable (0.5–1 knot), and is strongly influenced by westerly cyclones. North of 30° S, the Benguela gradually increases in strength in the southerly trades and runs as fast as 2 knots from the Congo River to Cape Lopez, where it leaves the coast and joins the South Equatorial Current. Inshore, the Guinea Current occasionally sets farther to the south in January–April between the Cameroon River and the equator, with a deleterious effect on the marine life of the normally cold, upwelled coastal water.

There appears to be no equivalent eastern boundary current off western Australia, except for the southwest sector between Cape Leeuwin and Fremantle, where the west wind drift branches to the north at about 1 knot. To the north, as off Chile and West Africa, the combination of frequent westerly gales in southern winter and spring, and strong, local onshore wind currents, renders the entire coastline a lee shore, and considerable offing is recommended for safety.

EL NIÑO

As mentioned in Part II, a sporadic 12- to 18-month weakening of the equatorial trade winds at 7- to 10-year intervals correspondingly suppresses the equatorial circulation in the Pacific and Indian oceans. This effort, associated with exacerbation of the dry monsoons, with consequent droughts in India and Australia, often has catastrophic ecologic consequences in the southeastern Pacific. Weakening of the South Equatorial Current, particularly, causes warm, low-salinity surface water to accumulate in the Gulf of Panama, from where it spreads south along the coasts of Ecuador and Peru, overlying the colder Peru Current and inhibiting the resupply of oxygen and nutrients to the surface. This abnormal situation, known as *El Niño* ("The Christ Child," because it occurs in the Christmas season) has a profound effect on the entire food chain in one of the world's richest fisheries; fish of all sizes die of anoxia, and their bodies accumulate in windrows along the shore, poisoning the water with hydrogen sulfide to the point where it allegedly blackens the paint of ships. Concurrently, warm water brings heavy rain and disastrous floods in this otherwise arid coastal region.

A less severe but analogous effect (dubbed *La Niña*) has been noted along the southern coast of North America; here, the northward flow of warm water along the coast forces the cold California Current offshore, increasing local rainfall, but without marked harm to fisheries.

11

Secondary Currents

ARCTIC CIRCULATION

There are no currents of significance in the Arctic Ocean, which is largely covered by floating ice throughout the year. Its interest to mariners arises chiefly through its influence on the climate and circulation of the north Atlantic, as well as from the production and transport of ice and icebergs into commercial shipping lanes. While warm Gulf Stream water entering the Norwegian Sea keeps the ports of Norway always ice-free, a cold, southward return flow (the *East Greenland Current*) maintains a subarctic climate in the western north Atlantic as far south as Newfoundland (fig. 35, p. 105). Throughout its sinuous, 4,000-mile peregrination, this current wends its way along the coasts of Greenland, Baffin Bay, and Labrador, where it is successively identified as the *West Greenland* and *Labrador Currents;* it finally encounters the Gulf Stream south of the Grand Banks. Here, part turns northeastward and is recycled back into the north Atlantic; the remainder branches to the right and circulates along the continental slope.

All along its route, the East Greenland Current entrains ice and icebergs—principally from the 100-odd fjords and inlets of West Greenland. It is estimated that as many as 7,500 icebergs are formed annually, of which an average of 10 percent are carried as far south as the 48th parallel before melting. The numbers vary greatly from year to year—1,350 were sighted south of 48° N in 1929; only 2 in 1940. Icebergs have been observed as far south as Bermuda, near the Azores, and within a few hundred miles of Great Britain. The average lifetime of a berg is one to three years, and it is apt to spend one or two winters frozen into the ice pack within Baffin Bay. Ice limits in both hemispheres are shown in figure 42.

Following the sinking of the *Titanic,* which rammed an iceberg on its maiden voyage from England to New York on April 14, 1912, and went down with the loss of over 1,500 of its 2,200 passengers and crew, the International Ice Patrol was formed, under the direction of the U.S. Coast Guard. The patrol maintains survey ships and aircraft off the Grand Banks during the iceberg season, and coordinates ship sightings of bergs from all sources. Twice daily, broadcasts of ice conditions are issued by the U.S. Naval Oceanographic Office, which also selects and recommends safe sailing routes. Yachts and smaller vessels should stand well to the south of these routes, however, because the high incidence of fog and heavy ship traffic along the routes make them more dangerous than floating ice.

In addition to seasonal and yearly abundances of polar ice, recent glacial studies suggest that there have also been dramatic changes on a time scale of centuries. The

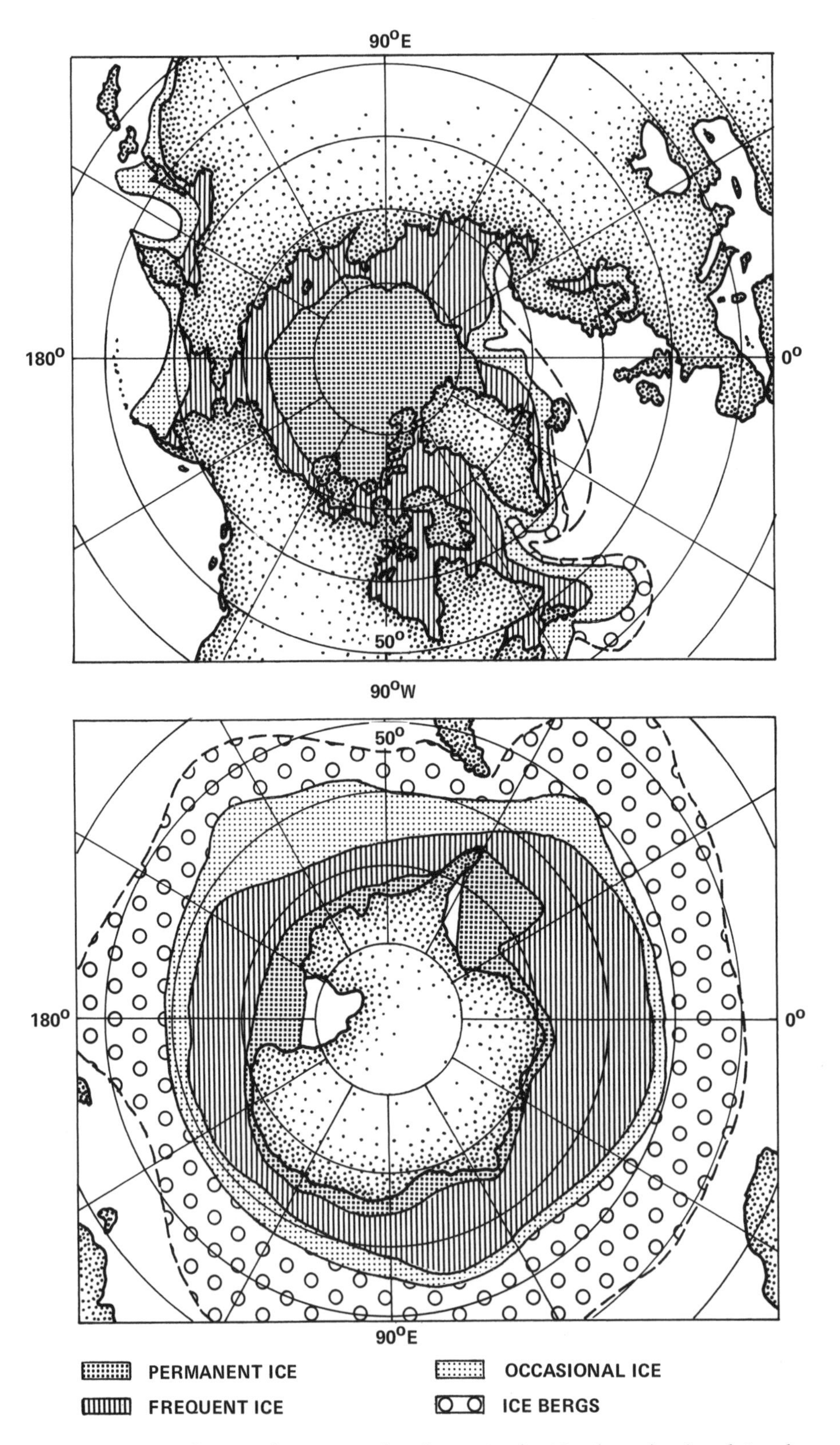

Fig. 42. Present limits of sea ice and icebergs in the Northern (*top*) and Southern (*bottom*) hemispheres. The limit of the permanent ice pack in the north Atlantic during a recent (1800) minor ice age is shown by a dashed line.

period of extensive Viking migrations (A.D. 800–1000) appears to have been even warmer than the present century is, and Iceland and southern Greenland were ice-free all year (fig. 42). By A.D. 1200-1400, ice conditions much resembled those today, and the Norse settlements in Greenland were abandoned. This minor "ice age" reached its peak about A.D. 1800; floating ice continuously blocked the north Atlantic from Greenland to the Faroe Islands, and winter icing severely restricted shipping in the English Channel. Indeed, Napoleon's only serious attempt to repossess Louisiana from Spain, which he otherwise might have easily accomplished, was aborted by the entrapment of his fleet in a late spring freeze (April 1801). While waiting for a thaw, he negotiated a trade with Spain for concessions in Italy, and decided to sell Louisiana to the United States to avoid a possible alliance with England. Thus, but for a vicissitude of climate, this book might well have been written in French.

The Pacific has access to the Arctic Ocean only through the narrow Bering Strait, through which a weak flow prevails northward all year. There is also a counterclockwise circulation of cold water in the western north Pacific and Bering Sea that somewhat resembles that of the East Greenland Current (fig. 37, p. 108). The cold, low-salinity Oyashio Current first flows southward along the coast of Kamchatka and the Kuril Islands, then turns northeastward near Hokkaido to parallel the Kuroshio, and finally turns north again, following the International Date Line around the west end of the Aleutian Arc and back into the Bering Sea. The latter, being ice-free nearly all year in its western parts, poses no iceberg problem, although the temperature contrast between the Kuroshio and the Oyashio results in frequent heavy fog when the winds are from a southerly sector. Because the Kuroshio is still a swift, coherent stream when it leaves the coast of Japan, its intersection with the Oyashio is more vigorous than that of the Gulf Stream with the Labrador Current. The subsequent northeastward meandering of these combined streams is accompanied by shedding of numerous small eddies; warm, clockwise eddies drift northward while cold, counterclockwise eddies drift to the south. These mixing processes result in high biological productivity and a fertile Japanese fishing ground.

THE MEDITERRANEAN SEA

The Mediterranean, in common with the Red Sea and the upper part of the Gulf of California, has an unusual circulation that is driven by the excess of evaporation over precipitation and runoff from rivers (fig. 43). All are connected to the oceans by straits that not only are narrow, but also have very shallow entrance sills with much deeper basins behind them. High evaporation makes the surface water abnormally salty, such that it becomes denser than the subsurface water and sinks to the bottom, filling the deep basins until they overflow their shallow sills into the deeper ocean outside. The outflow is compensated by inflow at the surface, but the total volume exchanged (inflow plus outflow) is very much greater than the difference between evaporation and precipitation, because of mixing with intermediate water during the unstable sinking process. In the Mediterranean, this vertical, salt-driven circulation is so vigorous that the entire volume of the sea is refluxed in about 75 years.

These matters would be of only academic importance to seafarers were it not for the swift prevailing currents thereby maintained through the entrance straits of

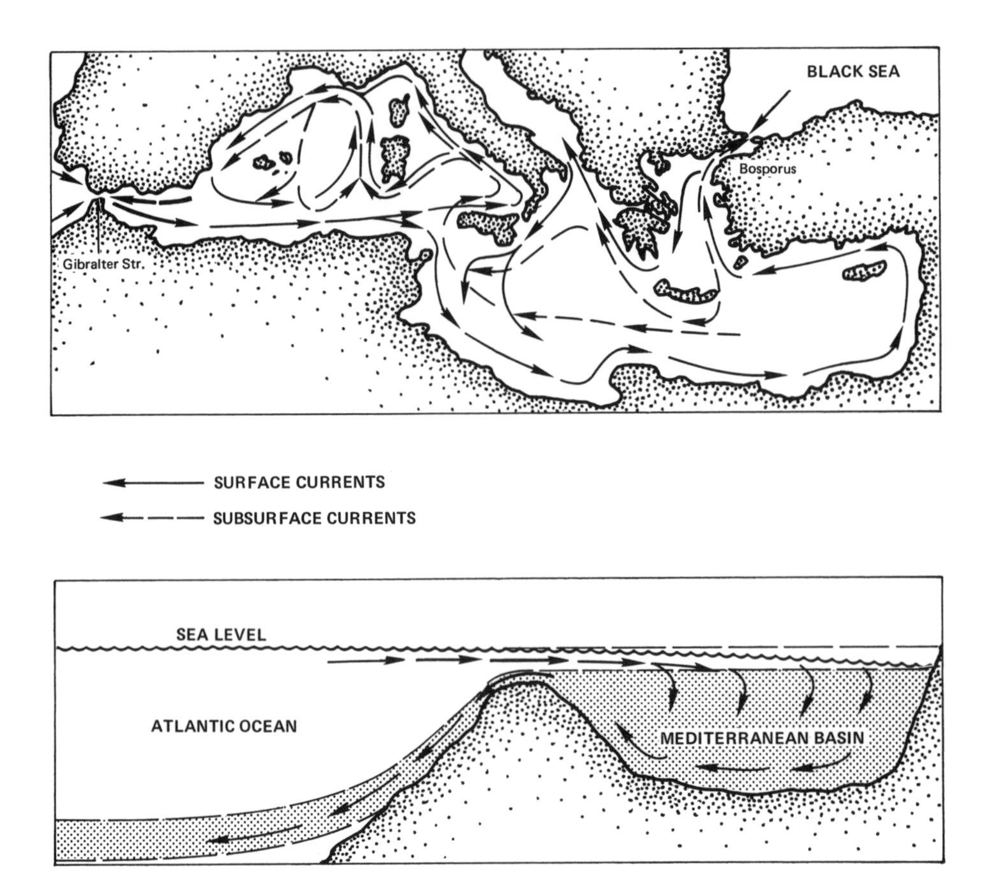

Fig. 43. Circulation of the Mediterranean Sea is driven by excessive evaporation that increases the salinity (density) of surface water, so that it sinks. Heavy water flowing out into the Atlantic over the shallow sill at Gibraltar is replaced by inflow of lighter surface water.

these seas. Through Gibraltar, a current continuously sets eastward at 2 to 4 knots, depending upon the tides; it flows strongest on the north side, and occasionally weakens—or even reverses—along the Moroccan Coast at spring tide, or under the stress of easterly Levanter winds that blow with gale force in fall and spring. Because the prevailing winds are also westerly, blowing clockwise around the Azores high-pressure center, sailing vessels have great difficulty making westing through the Straits of Gibraltar, unless they have adequate auxiliary power.*

Elsewhere within the Mediterranean, the weak, counterclockwise circulation attributable to horizontal density differences is of little consequence compared with transient currents induced by tides and local winds. However, a relatively strong current (2–4 knots) prevails westward through the Dardanelles and the Straits of Bosporus, connecting the Black Sea with the Aegean Sea. This current, which represents the balance of river runoff into the Black Sea, proved a primary obstacle to Xerxes in his war against Greece. Forced to build a two-mile pontoon bridge to transport his army of 180,000 men, he took one month to effect a crossing, giving

*Prior to the fourteenth century, when for 5,000 years the Mediterranean had been the "bathtub" of western civilization, vessels of all classes were unable to sail on the wind, and depended upon oars as their principal means of locomotion. Historians have speculated that their inability to negotiate the Gibraltar Current may have been a decisive factor in the long delay of westward expansion.

Themistocles time to build and equip several hundred galleys, which later destroyed the vastly superior Persian fleet in the narrow Straits of Salamis (480 B.C.). Unimpressed by Xerxes' example, Byron and Halliburton have since swum the Hellespont—albeit without armor or equipment.

LOCAL CURRENTS

As mentioned in Part II, the U.S. hydrographic pilot charts give mean monthly or seasonal averages of winds and surface currents for all ocean areas where sufficient data have been compiled. While these charts are useful for route selection and cruise planning, they express only probability distributions and cannot be relied upon to predict local conditions, any more than last year's weather forecast can be applied to today's weather. Only the central sectors of the western boundary currents—where they are guided by the continental shelves—flow with reasonable constancy, because they represent the integrated effect of winds over the entire ocean. Elsewhere, careful comparisons between dead-reckoning positions, navigation fixes, and local wind conditions will enable a navigator to calibrate drift on various points of sailing to the extent that reasonable projections can be made under similar situations (see Part VII).

In coastal waters, detailed descriptions of local weather and currents are given in the 100-odd looseleaf binders of the *Coast Pilot* (U.S. and possessions) and the *Sailing Directions* (foreign waters), available, respectively, from the National Ocean Service and the Defense Mapping Agency. These publications include a wealth of other information useful for coastal navigation and piloting, although the simple problem of bulk storage suggests that only those volumes pertinent to an intended voyage should be carried aboard small vessels.

Two independent oceanographic surveys of inshore current structure along the coastlines of the United States may be of interest to yachtsmen. For the East Coast, readers are referred to *Surface Circulation on the Continental Shelf*, Serial Atlas of the Marine Environment, Folio 7, published in 1965 and obtainable from the American Geographical Society, New York. For the West Coast, I include here some excerpts from a paper, "Studies of the California Current System," by Joseph L. Reid and others.

The California Current system is a part of the great clockwise circulation of the northern Pacific Ocean. At high latitudes the waters move eastward under the influence of the strong westerly winds, and near the coast of North America they divide into two branches (fig. 44). The smaller part turns northward into the Gulf of Alaska, and the larger part turns southeastward to become the California Current. In general, the temperatures in the open ocean are lower toward the north, so that the northern branch is a relatively warm current, while the California Current is relatively cooler than the water farther offshore. As the latter drifts south, at speeds generally less than half a knot, it becomes warmer under the sun's influence and by mixing with offshore water. Near latitude 25° N, it swings westward to join the North Equatorial Current.

The California Current is characterized by several persistent local circulations inshore of its main drift, which stays 40 to 100 miles off the coast. A small clockwise eddy is usually situated about 80 miles off Cape Mendocino, and a similar eddy persists between Guadalupe Island and the coast. Another permanent eddy, the

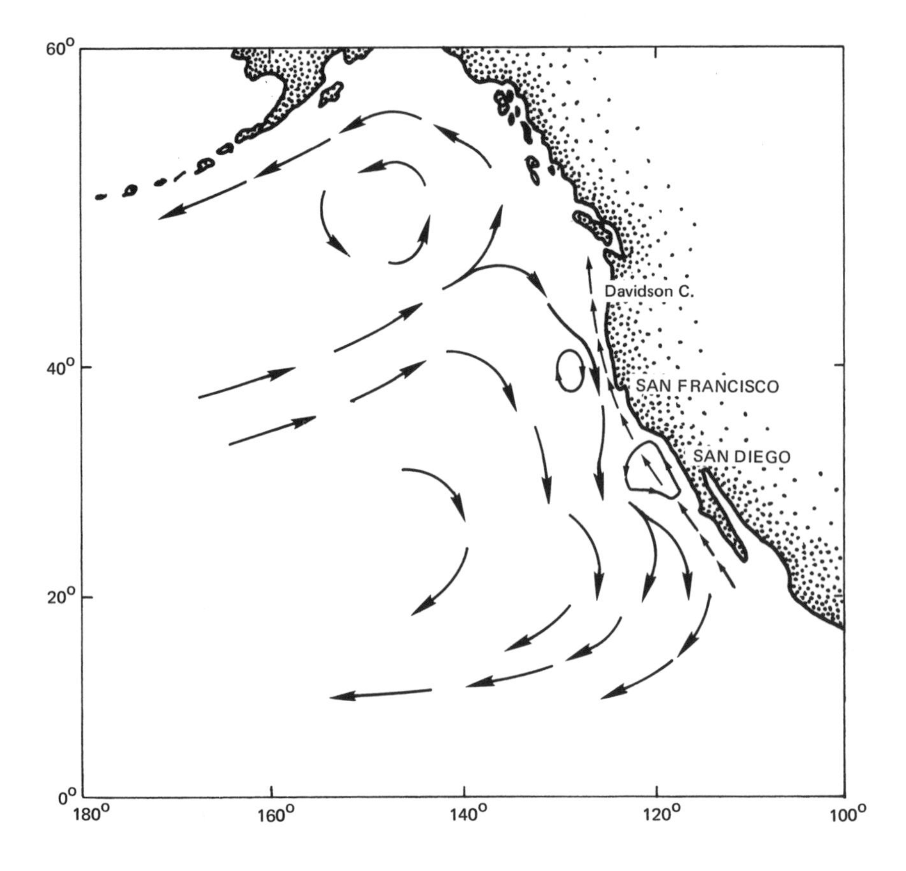

Fig. 44. Circulation off western North America generally follows the surface winds. Semipermanent eddies are found off Cape Mendocino and inshore south of Pt. Conception. Short arrows show path of cold Davidson Current in winter.

Southern California Countercurrent, is found to the south of the submerged peninsula that extends southwest from Pt. Conception to Cortez Bank, and includes Santa Rosa and San Nicolas Island. The waters southeast of Pt. Conception are protected from the prevailing northwest winds, and also from the southerly offshore current. The currents here are weaker and usually flow to the north. The waters remain off the coast of southern California for a considerable time and become much warmer than those offshore, which are constantly replaced by cooler water from the north.

A deep countercurrent, below 100 fathoms, flows to the northwest along the coast from Baja California to some point beyond Cape Mendocino. It brings warmer, more saline water great distances northward along the coast. When the north winds are weak or absent in late fall and early winter, the countercurrent appears at the surface, well on the inshore side of the main stream, and extends from the tip of Baja California to north of Cape Mendocino; it is known as the *Davidson Current.*

Lastly, some personal observations of circulation around several small atolls in the central Pacific may help explain the frequent, ubiquitous chart warnings of strong, variable currents in island waters. Even though wind or tidal currents may be negligibly small in deep water offshore, velocities as high as 3 to 4 knots running over the shallow 10-fathom coral terraces that surround such islands have been repeatedly observed. These currents appear to represent a gross enhancement of the slow general flow by the very presence of an island in the flow field. Not only do they wax and wane with tidal frequency, but alternating eddies form and break away at the leeward margins of the reef, and drift downstream (fig. 45). The sphere of influence of such currents increases with island size, and, where several islands are separated by channels whose widths do not greatly exceed the island diameters, very complex interactions may occur. For example, a strong current prevails northeast against the wind through the Alenuihaha Channel between Hawaii and Maui, thus generating notoriously steep and choppy seas. Similar conditions are found in the channels of the Lesser Antilles and Virgin Islands of the Caribbean.

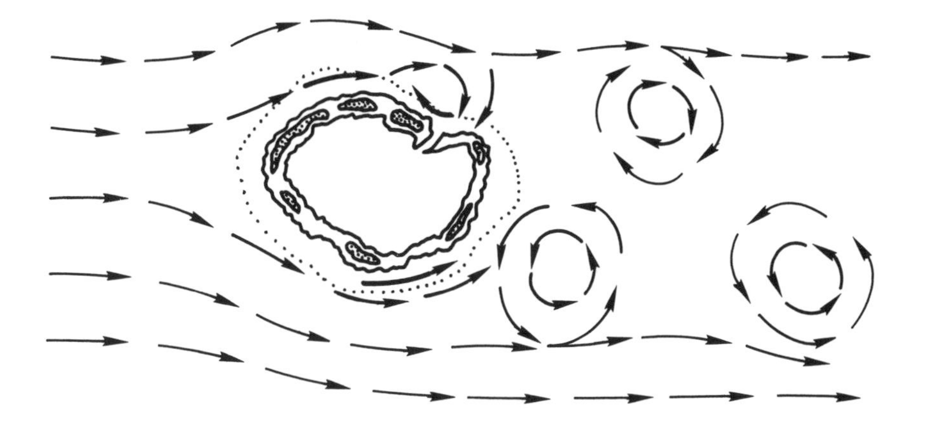

Fig. 45. Hypothetical circulation around an atoll, showing the formation of counterrotating eddies in the lee of an island, and strong currents inshore over the shallow (10-fathom) terrace.

12

Tides and Tidal Currents

PRINCIPLES OF TIDAL ACTION

One ordinarily thinks of tides as long waves traversing the oceans, and—as such—more appropriately included in Part IV. But tidal amplitudes are indistinguishable at sea and are only important in shallow coastal water insofar as they determine the safe draft for vessel passage, or the proper scope of anchoring or mooring lines. On the other hand, tidal currents are of prime importance for navigation, not only in coastal waters, but everywhere within the margins of the continental shelves (100 fathoms). Even in the open sea, their influence can be as great as that of the local winds, and all the more confusing because of their variability. Thus, I have elected to discuss them here in the context of currents, rather than waves—although the distinction is only semantic. This posture seemingly requires defense, since it is the height of the tide—as opposed to the speed and direction of the tidal current—that is continuously measured at some ten thousand places throughout the world. This precedent is due to the relative ease with which height is measured; even today there is no tidal current recording device that approaches the low-cost simplicity and survivability of the standard tide gauge. Had we as many current as tide records, our understanding of the tidal dynamics would be considerably more satisfactory than it is today.

The oceanic tides, as is well known, are caused by the combined gravitational attractions of the sun and moon, the latter being about twice the former. It is less well known that these attractions also produce tides within the elastic body of the earth, bulging it by about 1 foot from its equilibrium shape. Because the oceanic tidal bulges average about 3 feet in height, and also because these respective bulges occur at different times and places, the earth and oceanic tides are interconnected, and both must be taken into account in studying the tidal problem.*

Although detailed knowledge of tidal phenomena and their association with the sun and moon are apparent from early Greek and Roman accounts, this information was somehow lost during the Dark Ages. Records from the sixteenth century indicate that several individuals earned their livelihoods by making tide predictions for certain areas along the coast of England. But it was not until the seventeenth century that Sir Isaac Newton—who had his hand in nearly everything scientific—

*The earth tides were first measured directly by the American astronomers A. A. Michelson and H. G. Gale (1914), who observed the microscopic oscillations of water level in a very long, horizontal pipe, leading to the interesting conclusion that the earth—on the average—is more rigid than if it were made of solid steel!

published the theory of gravitation and laid the groundwork for modern tidal analysis. Since then, tidal problems have fascinated theorists in many countries.

The present status of affairs is somewhat dichotomous: while it is now possible to extrapolate from an existing set of tide records at a particular place into the indefinite future, we cannot yet make accurate predictions for places where no records exist. This is because it is not necessary to understand the causes of tidal motions in order to predict them. By this, I do not mean to imply that the principles of tidal action are not understood. Fairly accurate pictures of tidal behavior have been obtained analytically for restricted seas, such as the Mediterranean. But the extension to a global ocean clearly requires the use of large computers. There are still some tricky problems, such as the precise manner in which energy is dissipated and/or reflected at the ocean boundaries, that must be solved to bring computed tidal models into better alignment with observations.

Interestingly, we know the rate of tidal energy dissipation with accuracy; it has been calculated from the observed slowing of the earth's axial rotation, which increases the length of the day by about 0.02 second per thousand years. Reasonable estimates of the total tidal energy at any instant suggest that the two are about equal; that is, the average energy equals the daily dissipation. Thus, with removal of the tide forces, tidal motions would cease within a day or so, whereas with cessation of the winds, we recall, the major currents would continue for a year or two. The difference is attributed to the fact that the wind-driven circulation is largely confined to the surface of the oceans; tidal motions extend clear to the sea floor, where friction against the bottom—particularly in shallow seas—slows them down.

It would be redundant to reiterate here the mechanisms of the tide-producing forces, which are given in detail in standard references. Suffice it to say that, because the orbits of both the earth and the moon are elliptic and do not lie in the same plane (they are separated by an angle of about 5°), the attractive forces of the sun and moon continually vary in magnitude and direction. Moreover, the axes of these ellipses also rotate very slowly within their respective orbital planes, such that the orientations and strengths of the attractive forces reproduce themselves only roughly every year, more precisely every 18.6 years, and almost exactly every 20,940 years. These complicated motions are more easily treated mathematically by replacing the attractions of the sun and moon with a set of fictitious bodies, all orbiting the earth within different parallels of latitude and at different speeds, whose combined attractions are exactly equivalent to the sun and moon.

However, the problem of tidal analysis rests not so much on these complex planetary interactions—which are known very precisely—but upon the response of the oceans to the moving attractive forces. The relatively simpler problem of tides on an ocean of uniform depth covering the entire earth was worked out by Laplace (1774), but with little resemblance to observed conditions. Since then, the history of tidal theory includes many attempts to apply the equations of Laplace to the real oceans, introducing actual sea-floor topography, observed density stratification (which profoundly alters the character of flow in certain areas), frictional dissipation, and our old friend, the Coriolis force. Even with these concessions to realism, however, the tidal equations cannot be solved mathematically. Only recently have numerical solutions been developed that qualitatively agree with local tidal observations.

Figure 46 is a computer-generated model of what the world tides might look like if the only attractive force were due to a fictitious moon that orbited the equator in 12.42 hours. This model contains only one of ten major *constituents* ordinarily considered to represent the total gravitational effect of the sun and moon, but it serves to illustrate the principal features of tidal motion. Instead of the (perhaps) intuitive picture of a pair of tidal bulges advancing westward around the earth, the oceans are shown divided into a number of discrete cells (spider web–like structures), within each of which a local tide wave rotates in the direction indicated by the circular arrows. The successive hourly positions of these waves, reckoned from Greenwich Meridian Time, are given by the radial strands of these webs (*cotidal lines*), beginning at 0°. The dotted peripheral strands (*corange lines*) represent contours of equal tide height. The height is zero at the center of each web, which is called an *amphidromic point.* For lack of deep ocean tide measurements, no one has yet found an amphidromic point, but one is presumed to lie very near Tahiti, where the diurnal tide is only about 8 inches high. The rotary motion of the tides about their respective amphidromes can be visualized as very similar to that of the wave generated within a shallow, circular pan when it is agitated as one does when rinsing, or panning, gold.

Figure 46 also shows that most of the cotidal lines intersect the continental coastlines approximately at right angles. Thus the tide waves tend to propagate *parallel* to the coastlines, being duly recorded at adjacent tide stations as consecutive rises and falls of sea level. The additional equivalent models for the nine other

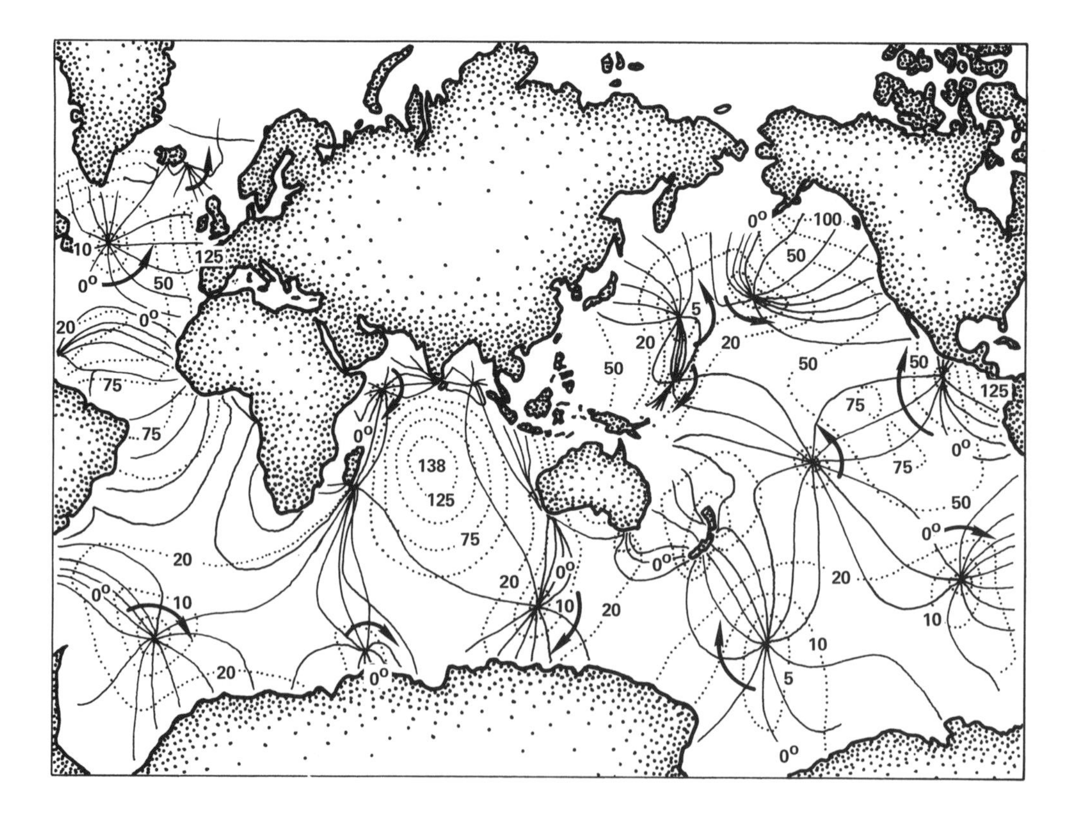

Fig. 46. Computer-generated tidal model for the principal lunar (M2) tide-producing force. Height contours (dotted) are given in centimeters (1⁄30 foot). The tide waves tend to rotate around spider web–like amphidromic points in the directions shown by the heavy arrows. Phase angles are given in 30° increments, reckoned from high tide (0°).

fictitious attractive forces considerably complicate the complete tidal picture because of interference among the many sets of waves, some of which have different sets of amphidromes and associated cotidal and corange lines. Thus, it is small wonder that the complexity of tidal motions at sea proscribes any simple description.

The test of a computer model, of course, is that it everywhere agrees with recorded heights, for it then could be used to make tide—as well as current—predictions anywhere in the oceans. But so far, agreement is only qualitative. Part of the difficulty is that the dynamics of water motion near an individual tide gauge depend upon the details of local topography to an extent that exceeds the capacity of present computers. But larger computers and computational shortcuts give promise that sufficiently accurate computer predictions are on the horizon.

TIDAL CURRENTS AND TIDAL BORES

The rise and fall of the tide at any point is associated with (largely) horizontal, oscillatory water motions known as tidal currents. In midocean, the water—from the surface to very near the sea floor—moves forward under the crest of an advancing tide wave, and backward under the following trough. At the same time, it is deflected always to the right (Northern Hemisphere) by the Coriolis force, such that its net motion, during a tide cycle, is roughly elliptical. The long (major) axis of the tidal ellipse is proportional to the tide height and duration, and inversely proportional to the square root of the water depth. Its width is proportional to the Coriolis force, which increases with the sine of latitude.

The speed of the tidal current varies along its elliptic path, usually exhibiting two maxima and two minima for each tide cycle, but rarely dropping to zero. Successive, hourly current measurements—say, from an anchored ship—when plotted in vectorial form (fig. 47), conveniently demonstrate the elliptical distribution of speed and

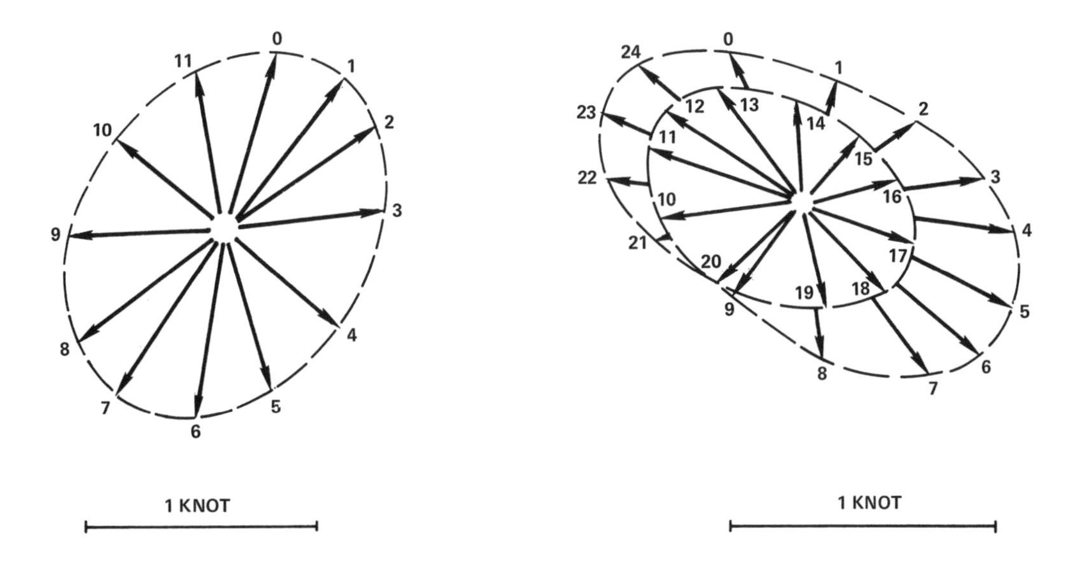

Fig. 47. Hourly tidal current vectors, showing speed and direction of current measured from an anchored vessel. Left ellipse applies to a single daily (diurnal) tide, and right, double ellipse to a twice-daily (semidiurnal) tide. The latter does not come back to its starting point after twenty-four hours because mixed tides do not repeat themselves.

direction with time. In the open sea, tidal currents rarely exceed 0.2–0.5 knots, and the length of a tidal ellipse is less than a mile. But tidal currents of 2–3 knots have been observed in depths of 100 fathoms or less along the margins of the continental shelves—which are as much as 100 miles wide in many parts of the world—and deviations of several miles from an intended course are possible in such places.

Wherever a large body of water has its access to the open sea restricted by a relatively narrow entrance channel, one is apt to encounter strong tidal currents as the level within attempts to adjust itself to the rise and fall of the tide without. In simplest aspect, this situation may be likened to a pair of open containers half full of water and interconnected by a flexible tube, the flow through which can be regulated by a valve (fig. 48a). If container *X* is raised and lowered periodically to simulate the tide in the open sea, the water will synchronously rise and fall in container *Y* by an amount that depends upon the valve setting. Obviously, if the valve is wide open, the height changes will be nearly equal; if nearly closed, very little motion takes place in *Y*. Moreover, as the valve is progressively closed, the rise and fall in container *Y* will lag further and further behind that in *X*. Ultimately, *Y* will lag to the point where its maximum rise occurs when the level in *X* is passing through its midposition. The tidal current (rate of flow through the valve) is also shown for these several situations. Maximum flow (small arrows in figure 48a) will occur in conjunction with the greatest difference in level between the two containers: this difference is greatest for the nearly closed case.

For a more realistic analogy, the stationary container is replaced by one or more U-shaped tubes (fig. 48b), within which water can oscillate from one leg to the other in a manner much like that of standing waves in a semienclosed basin (see Part IV).

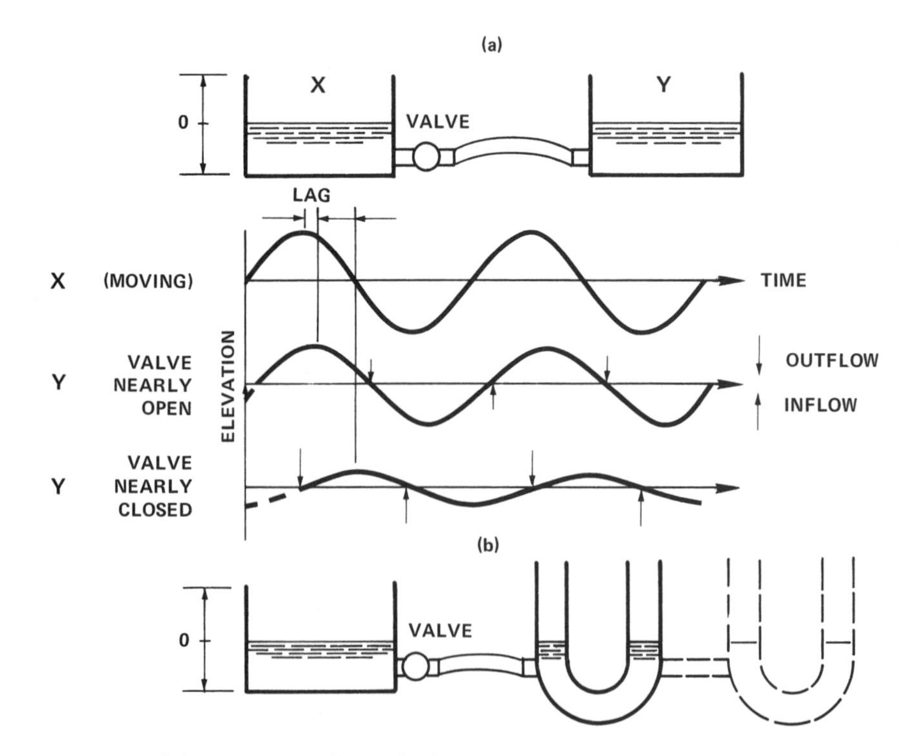

Fig. 48. Tidal response analogue for basins connecting to the sea through a narrow strait: (*a*) with frictional dissipation; (*b*) with friction and inertia (resonance).

Here, the water level within the U tube will adjust itself as above, except that, if the tidal period happens to correspond to the natural period of free oscillation in the tube, very high resonant oscillations can occur, much as a child's swing is forced to high amplitude by periodic and properly timed pushes. With other interconnected U tubes, having different natural periods, very complicated coupled oscillations result, with local water motions that have no simple relation to the tidal input. These *co-oscillating* tides are characteristic of larger seas and inland waterways that are subdivided into multiple basins by peninsulas, islands, or an irregular coastline, as are the Gulf of St. Lawrence and the English Channel. Both are notorious for their swift tidal currents and anomalously high and complicated tidal motions.

The above analogies are imperfect where the tide range is so large that water cannot flow through a shallow, convergent channel or inlet as a stable wave, but develops into a steep-fronted breaking wave (*tidal bore*) that maintains its integrity as long as the critical tidal differential persists. A bore has some analogy to the "sonic boom" shock wave produced by a supersonic airplane, in that both represent an abrupt pressure transition that propagates faster-than-normal wave motion. The speed of a bore is related to its height and the water depth ahead of it. If, as with most rivers, there is an outflow current, the bore velocity will be correspondingly reduced. Small tidal bores are common along the coasts of France and southern England, but much larger ones occur in the Bay of Fundy (plate 10) and in numerous rivers of India, Asia, and South America. A bore 25 feet high advances up the north branch of the Amazon at spring tides, in some places attaining speeds of 12–14 knots. Danger-

Plate 10. A 3-foot tidal bore in the Petitcodiac River, Bay of Fundy. Similar bores as high as 25 feet have been observed in the north branch of the Amazon River.

ous bores are well noted in the *Sailing Directions,* and are devoutly to be eschewed by small vessels.

Tides and currents in atolls (or high islands surrounded by barrier reefs) often exhibit unusual behavior, particularly if there is only one navigable entrance. Aside from the aforementioned strong tidal currents that prevail over their shallow exterior terraces, the current strength in the reef passage(s)—and its phase, relative to the tide outside—depends critically upon the extent of inflow elsewhere by wave action over the exposed portions of the table reefs. The reef elevations characteristically lie close to mean sea level; they are covered by 1 or 2 feet of water at high tide, and are laid bare at low tide. Thus, during a major part of the tide cycle, breakers transport water across the reef, maintaining the lagoon level a foot or two above the equilibrium level outside. For an atoll having a substantial fraction of its periphery so exposed, there will be a continuous outflow through the reef channel, strongest near the midpoint of the falling tide, and weakest midway of the rise, but with no period of slack water. The outflow persists even when the reef is bare, owing to the surplus built up at high water, and abetted by the trade-wind stress, which acts to pile up water by a foot or so at the western rim, where most passages are located. Such conditions are typified at Tongareva Atoll, Cook Group, and Bora Bora, in the Society Islands.*

Where islets fully occupy an atoll rim, there is no breaker inflow, and the tide phase within the lagoon may lag that outside by three to six hours, depending upon the diurnal inequality. Ebb and flood currents of 5–10 knots are not uncommon among atolls having large areas and narrow passages. Takaroa Atoll, in the Tuamotu Islands, is an extreme example of this latter type: it has an area of some 30 square miles, with a single, long, narrow pass that reduces at its inner end to only 50 yards' breadth, between two pincerlike coral fingers. The tide race through this gap reaches 15 knots. Two enormous whirlpools form alternately, during ebb and flood, in the lee of these fingers, from whose extremities one can peer down some 10–15 feet into the throat of the nearest vortex to see brightly colored fish whirling around, apparently oblivious to their predicament.

TIDE AND CURRENT PREDICTIONS

Given a year's record from any particular tide station, it is possible to make accurate future predictions without any knowledge of cause and effect. This is because tidal motions—however complicated—are highly periodic, and tend to reproduce themselves with only small variations from year to year. Such a record can be broken down mathematically, by the method of harmonic analysis, into a discrete set of *harmonic constants* (sine and cosine functions) that—once known—can then be recombined to give indefinite predictions. These constants must be determined separately for each individual station. They bear no direct relation to the above-mentioned constituents of the tide-producing forces, and usually reflect the influence of local topography, the presence of co-oscillating tides, and any frictional effects. Nonperiodic variations of sea level, such as wind tides, inflow from rivers, etc., are averaged out fairly well by harmonic analysis of a one-year record. But, if these

*While we were diving with a local native to recover an anchor in the pass at Bora Bora, he remarked with surprise that, for the first time in his recollection, water was actually flowing *into* the pass. We attributed this to several days of flat calm, with no perceptible breaking over the reef.

effects are seasonal, many years of records may be required to obtain precise predictions. Similarly, the prediction for any particular day may be in error by the amount of nonperiodic motion. However, except for abnormal local winds or floods, tide predictions are generally accurate to within a foot of the computed values.

Tidal calculations and the maintenance of tide stations within the United States and its possessions fall under the purview of the National Ocean Service, which also publishes the coastal pilot charts. Tide heights are usually referred to the same reference datum as the soundings on local charts, although this is not always the case. The several dozen datum levels in common use in various parts of the world are tabulated in the *American Practical Navigator,* H. O. Pub. No. 9, Defense Mapping Agency, Hydrographic/Topographic Center. The *Navigator* also gives a large glossary of common definitions relevant to tides, tidal cycles, and tidal currents, as well as descriptions of typical tide records and their interpretation.

Through joint cooperation between maritime nations, tables of tides and tidal currents are routinely published each year for reference stations at principal ports throughout the world. These tables also include corrections that can be applied to obtain the tide range and time difference with respect to a reference station at any of several thousand subordinate gauge locations. Tides and currents can be estimated for other places by interpolating between the closest adjacent stations given in the tables. However, this method is rather inaccurate if the station spacing is more than about 100 miles along the sea coast, or more than a few miles in restricted waters. The *Coast Pilots* and *Sailing Directions* also contain frequent references to tidal effects in strange waters, which often are simply related to meridional passages of the sun and moon. If all else fails, ask a local fisherman, or follow his boat if you do not speak the language.

PART IV

Principles of Wave Motion

13

Ideal Periodic Waves

INTRODUCTION

If clouds are the seaman's bible, the ocean waves are his apostles. Endlessly wandering the vast, untrammeled oceans, they preach the gospel of Nature's might, give warning of coming storms, and—in milder moods—provide the endless, ever-changing panorama that exalts the spirit in a manner unknown to landbound mortals.

Even in the absence of winds or currents, the sea surface is never absolutely still; there is always a residuum of subtle oscillations underlying the calmest sea. At another place or time, the unlucky or improvident seafarer may be faced with howling winds and mountainous, breaking waves that threaten his very survival. These disparate circumstances constitute but two extremes in the broad continuum of surface-wave motions normally present at sea, whose possible limits extend from the merest ripple of a falling raindrop to the gigantic waves generated by the explosion of large meteorites that impacted the primordial oceans some 3.6 billion years ago. Judging from the moon's visible impact history, some meteorites possessed explosive energies of a billion megatons—equivalent to a layer of TNT 800 feet thick covering the United States. These incredible shocks disemboweled the moon to a depth of 50–60 miles, leaving bull's-eye wave patterns frozen into the rocky lunar crust,* some of which are 20,000 feet high and 600 miles in diameter (plate 11)! The present oceans are capable of sustaining stable waves of a height equal to 70 percent of the water depth (about 14,000 feet high, in the Pacific). Fortunately, the days of meteorite bombardment are long past, and the present upper limit for waves in the open sea is governed by environmental conditions, and not by the depth of water.

Although the seafarer is principally concerned with *surface* waves, the ocean's dark interior abounds with other types of wave motion: there are *internal waves* propagating along the various density layers that fill the ocean basins, and *sound waves* radiating in all directions from a myriad of discrete sources.

While the mathematical study of waves and related flow phenomena in liquids dates back to the time of Daniel Bernoulli (1741), who is often called the father of modern fluid mechanics, it is only within the past half century that applications of wave theory have come to extensive, practical fruition. Early in World War II, the necessity for forecasting wave and surf conditions during marine military operations

*The moon's subsurface layer is presumed to have been hot enough to undergo transient melting after sufficiently energetic impact. Temporary liquefication allowed the waves to form, like ripples on a pond, but prompt (~one hour) freezing preserved them for four billion years.

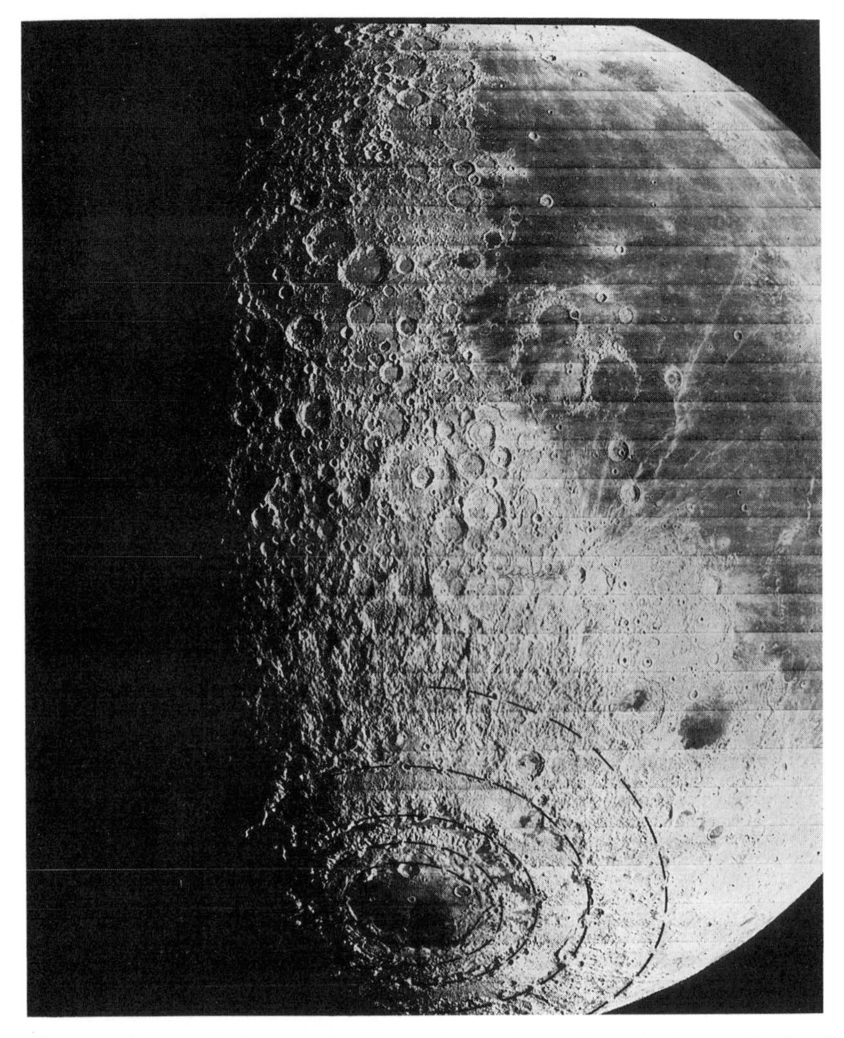

Plate 11. Orbiter IV photograph of Mare Orientale, on the moon's west limb. The central basin is surrounded by five concentric mountain rings (dashed lines) from 200 to 600 miles in diameter, and up to 20,000 feet high. These rings have been shown by the author to be gravity waves, caused by the explosive impact of a giant meteorite, and preserved by freezing of molten lava.

was brought sharply into focus by the disastrous landing of U.S. troops at Tarawa (Gilbert Islands), where, in the face of heavy enemy fire, over half of the assault craft were capsized or stranded at the reef edge.* Their personnel either drowned under the weight of their equipment or were shot while floundering in shallow water. Under urgent military sponsorship, a hastily convened group of scientists at the Scripps Institution of Oceanography developed a successful method of wind–wave prediction, based upon meteorological forecasts and local topography, that saved countless lives during subsequent landings. Since then, further studies of wave characteristics in the open sea and improved forecasting techniques have found numerous applications in the selection of shipping routes, as well as in the design of ships and a host of stationary marine structures, such as piers, breakwaters, and sea walls. Similarly, military impetus to the study of sound waves in water has led to the

*Actually, this catastrophe resulted from a tidal miscalculation, although surf effects were contributory (see U.S. Naval Institute. *Proceedings.* 88, no. 2 (February 1962): 18.

development of a variety of sound-ranging devices for measuring water depth, locating fish and icebergs, and geophysical prospecting.

Even today, however, it is recognized that substantial gaps exist between mathematical wave theories and actual waves observed at sea or in the laboratory—particularly under extreme conditions—and the former must be regarded as only qualitative guides, whose accuracy is best judged by careful experiments.

NATURE OF WAVE MOTION

All classes of water wave motion are ordinarily distinguished from other flow phenomena, as exemplified by currents, eddies, vortices, etc., in that only *energy* is propagated through the water. The water particles themselves simply oscillate back and forth, or orbit about within a restricted region, with little or no tendency to permanent displacement from their mean positions. Hence, ideal waves might be classified as oscillatory flow with no net transport. Despite the apparent complexity of the gross motions simultaneously occurring in the sea, in principle they all can be approximated by simple combinations of ideal periodic waves superimposed upon other kinds of flow. Thus, to understand waves in nature, one must first understand periodic waves. The task is not easy, and—at some risk of seeming tedious—I enjoin the reader to "drink deep, or taste not the Pierian spring . . . ."

A convenient and common device for producing and visualizing many types of wave and flow phenomena is the laboratory channel shown in figure 49. It consists of a long, closed, rectangular tunnel, with sides of plate glass, equipped at one end with a reciprocating piston for generating waves. It may include a pump for circulating water through the channel in either direction, and a blower to create wind over the water surface. Waves reaching the far end of the channel can be reflected by a vertical barrier, absorbed by a suitable arrangement of baffles, or allowed to break on "beaches" of arbitrary slope. If desirable, sand, sediment, or other beach materials can be introduced to study their movements under the action of the waves or currents. In addition to the study of waves themselves, such channels are also used for the dynamic testing of suitably scaled models of ships, piers, breakwaters, and the like under carefully simulated conditions.

Let us now conduct a series of hypothetical experiments in our channel, using only the piston to induce motions. The water is initially at rest and in equilibrium

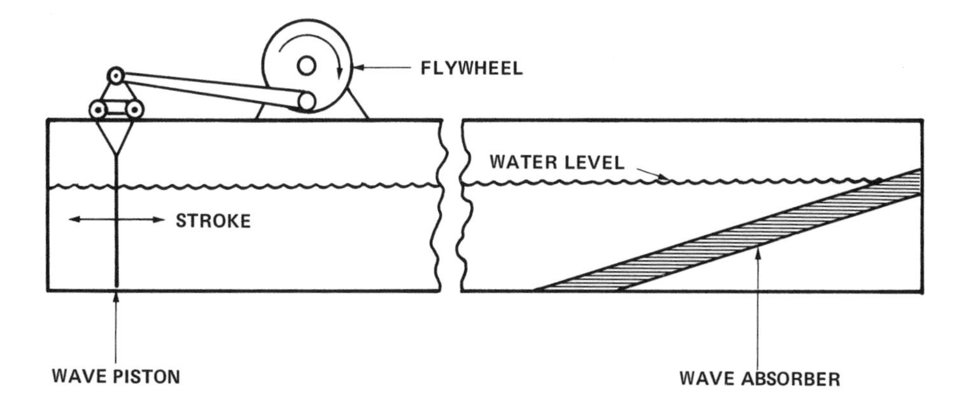

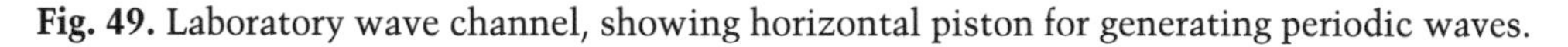

Fig. 49. Laboratory wave channel, showing horizontal piston for generating periodic waves.

under the combined forces of gravity, surface tension, and elasticity. All of these are defined as *restoring* forces because they tend to oppose any departure from equilibrium: gravity acts to keep the surface level by causing flow toward hollows and away from bumps in the surface; surface tension acts as a stretched rubber membrane that resists any surface curvature; and elasticity resists any changes in the total volume of water.*

Any motion of the piston whatever will create a disturbance from equilibrium, the nature and history of which—whether or not it is wavelike—will depend upon the character of the initial motion and the balance between the above restoring forces and the inertia of the water. Ultimately, of course, all motion will cease and equilibrium will be reestablished because of internal friction (viscosity) between water molecules, which converts the energy of motion into heat. Frictional dissipation is weakest for wavelike motions and strongest for turbulent (random), eddylike motion. Thus, the bow waves from a moving vessel are often visible for hours after its passage, whereas the turbulent propeller wake disappears within a few minutes. In our laboratory channel, most of the frictional resistance takes place within a thin (boundary) layer along the bottom and sides, which are not free to move with the water, and thus any motions will be more rapidly damped than they would be in the open sea.

PROGRESSIVE WAVES

As our first experiment (fig. 50a), we now start the generator in uniform rotation, so that the flywheel makes a single revolution in a period of T seconds. As the piston moves back and forth, a train of identical waves will appear on the water surface and travel at uniform phase (crest) velocity C down the channel, which we will suppose is equipped with a perfect absorber that reflects no energy back toward the generator. Assuming also that the wave height H is small compared with both the water depth h and the wavelength L, the wave train will closely approximate the mathematical concept of ideal *progressive* waves; that is, the surface will have the shape of a moving sine curve of amplitude $A = H/2$, symmetric about the equilibrium level. Since one wave is generated for each stroke of the piston, the time interval between consecutive crests passing a fixed point of observation is equal to that for one rotation of the flywheel, and is defined as the *wave period T*. The reciprocal of the period is the frequency f.** By convention, the dividing point is one second, with the period of slower oscillations being given in seconds, and the frequency of faster oscillations in cycles per second. The wavelength L is the distance traveled by a wave in one wave period: L = CT.

*Perhaps surprisingly, elasticity is by far the largest of these forces; the specific volume of seawater changes by only about 4 thousandths of 1 percent under a pressure change of one atmosphere. This may seem insignificant, but the Pacific Ocean would stand about 150 feet higher except for compression of the water by virtue of its own weight, or about 9 inches higher in the absence of the atmosphere. Since an atmosphere is equivalent to a column of water 33 feet high, the force of gravity is about 43 times weaker than that of elastic compression. Surface tension exerts a stress equivalent to only 74 millionths of an atmosphere *parallel* to the free surface. Its restoring force depends upon the curvature of the surface, and is still smaller. Nevertheless, it dominates the behavior of small ripples, whose presence greatly contributes to the roughness (aerodynamic drag) of the sea surface, and hence to the efficiency with which wind can generate larger waves and currents.

**A more common unit is the angular frequency, $\omega = 2\pi/T$ (radians/sec.), corresponding to the flywheel rotation rate.

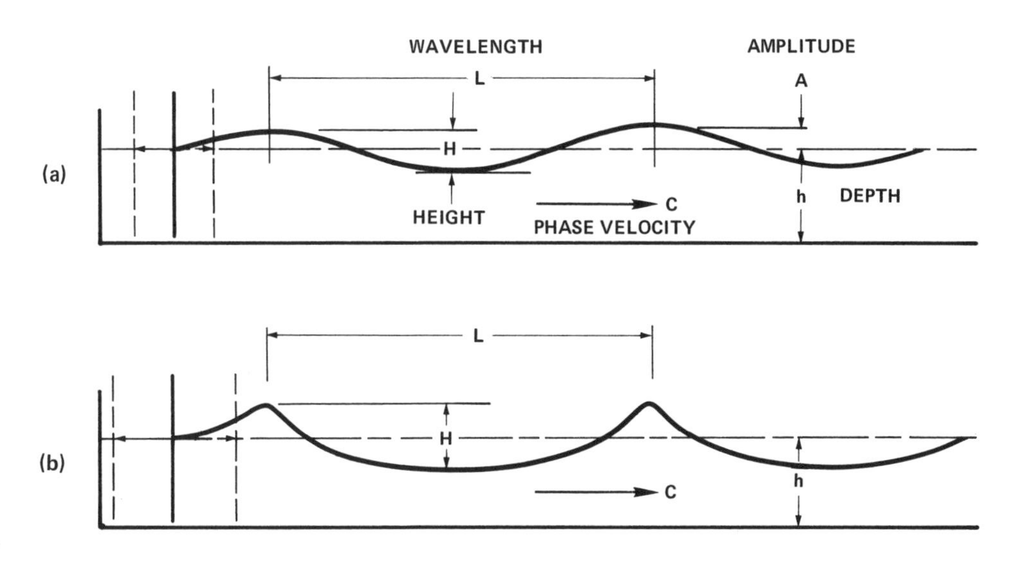

Fig. 50. (*a*) Periodic, progressive surface waves of low amplitude resemble a moving sine curve, symmetric about the equilibrium water level. (*b*) Higher periodic waves are no longer sinusoidal, nor symmetric; their crests are peaked, and extend higher above the equilibrium level than the long, smooth troughs fall below it.

As a second experiment, let us change the piston stroke, keeping the period constant. As the stroke is progressively increased, the waves become higher. Moreover, they no longer conform to a simple sine curve; their crests are more peaked and their troughs flatter, and the former rise higher above the equilibrium surface than the latter fall below (fig. 50b). If the stroke is made large enough, the waves will reach their stability limit and break. This limit is still not well defined mathematically, but can be experimentally determined to depend upon the ratio of wavelength to water depth. If $L < h/3$,* the waves are classed as *deep-water* waves, and breaking will occur when their height exceeds about one-seventh the wavelength; if $L > 11h$, they are *shallow-water* waves, and will break if their height exceeds about 80 percent of the depth. Waves having wavelengths between the limits $h/3 < L < 11h$ are classed as intermediate. The breaking criteria for intermediate waves is rather complicated, and will be discussed later.

As a third experiment, let us vary the wave period, but keep the wave height very small by suitably adjusting the piston stroke for each change of period. We will find that the relation $L = CT$ always holds, but that the phase velocity changes in a curious manner that depends upon the above-cited restoring forces.

Gravity Waves: At slow piston speeds, we are in the domain of shallow-water waves, characterized by the following approximate relations:**

Shallow water
$$C = \sqrt{gh} = 5.7\sqrt{h}\text{ ft./sec.} = 3.4\sqrt{h}\text{ knots}$$
$$T < 2\sqrt{h}\text{ sec.}$$
$$L = T\sqrt{gh} > 11h\text{ ft.}$$

* Here we introduce the inequalities *less than* (<) and *greater than* (>).

** Throughout this book, all numbers are rounded off to 5 percent error, or less, unless greater accuracy is important to the discussion.

From these relations we see that the wave parameters are functions only of the water depth and the gravitational acceleration.

If we now increase the piston speed, we will observe that phase velocity and wavelength progressively decrease, as they pass through the regime of intermediate waves, and end up related quite differently:

$$\text{Deep water}\quad \begin{aligned} &C = gT/2\pi = 5T \text{ ft./sec.} = 3T \text{ knots} \\ &T < \sqrt{h}/4 \text{ sec.} \\ &L = 5T^2 < h/3 \text{ ft.} \end{aligned}$$

Here, the wavelength (actually, particle motion) has become so small that the waves no longer "feel" the bottom; phase velocity and wavelength depend only on gravity and wave period. It should be noted that "shallow" and "deep" are only relative terms, and that phase velocity undergoes a smooth transition between them as either depth or period is diminished. Thus, ordinary wind waves in the open sea are deep-water waves that become shallow-water waves upon approaching shore. While we cannot express this transition by simple formulas like those above, it is shown graphically in figure 51 for the range of periods and depths relevant to wind waves at sea. In this figure, the shallow-water relationships above apply to the (approximately) straight portions of the various depth curves that lie to the right of the dashed lines labeled "shallow." The deep-water formulas apply to intersections of the depth curves with the curve $h = \infty$. All points between refer to intermediate depth. Some comparative values for our laboratory channel (h = 4 feet, say) and for the Pacific

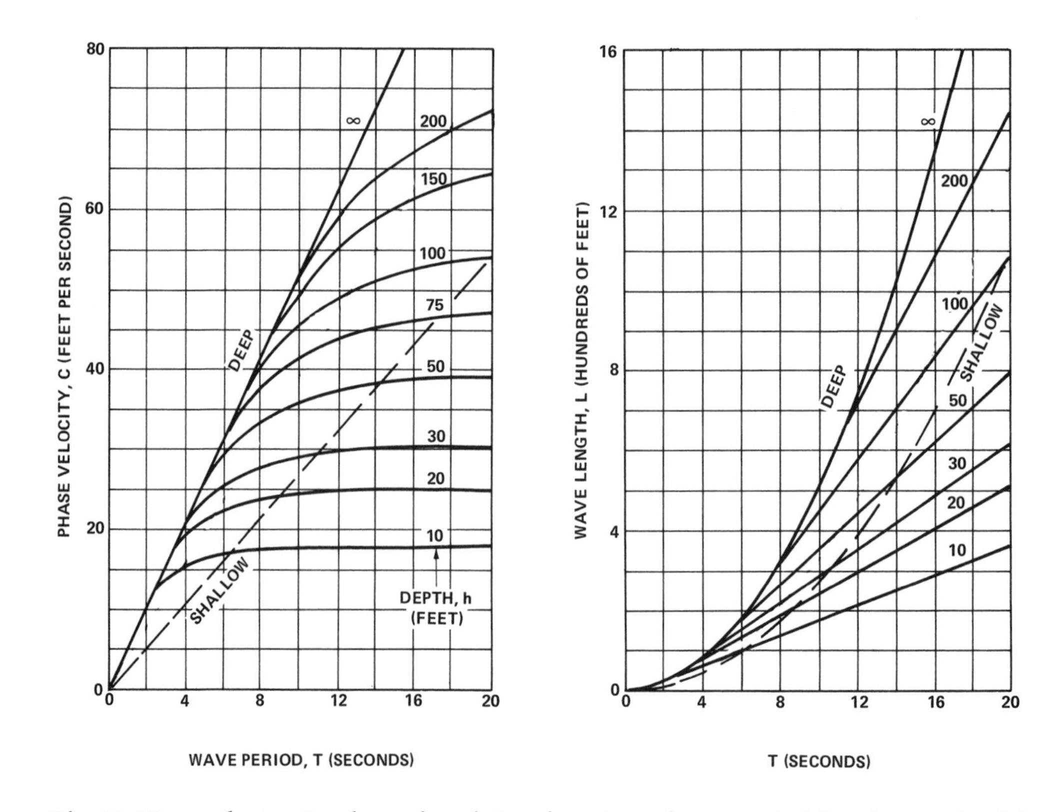

Fig. 51. Wave velocity C and wavelength L as functions of wave period T and water depth h, for periodic, progressive surface waves.

Ocean (mean depth, $h = 18{,}000$ feet), not easily obtainable from figure 51, are given in the table below:

Limiting Condition	h (ft.)	Shallow Water C (ft./sec.)	Shallow Water L (ft.)	Shallow Water T (sec.)	Deep Water C (ft./sec.)	Deep Water L (ft.)	Deep Water T (sec.)
Channel	4	11	44	4	2.5	1.3	0.5
Pacific	18,000	760	200,000	270	170	5,600	34

The high speeds (450 knots) potentially obtainable in the open sea are only realized by waves of very long period (tsunamis), such as are occasionally generated by submarine earthquakes.

Capillary Waves: Further reduction of the wave period below about $T = 0.5$ seconds ($f = 2$ cps, $L = 1.4$ ft.) brings us into the realm of *capillary waves,* which are controlled by surface tension. These are very superficial ripples, such as are familiarly generated by a puff of wind, and their influence extends only an inch or two below the surface. They are rapidly damped by the natural organic film always present on a free-water surface, and disappear within seconds after the wind ceases. In contrast to gravity waves, the speed of capillary waves *increases* as the period and wavelength decrease, in accordance with the following relations: $C = (2\pi f T'/\rho)^{1/3} = 0.25f^{1/3}$ ft./sec.; $L = CT = C/f = 0.25f^{2/3}$ ft., where $f = 1/T$ is the wave frequency, $T' = 0.005$ lb./ft. is the surface tension, and $\rho = 2$ slugs/ft.^{3} is the density of water. Thus we have the interesting result that there is also a minimum possible phase velocity for surface waves on water: $C_{min} = 0.76$ ft./sec., when $f = 13.3$ cycles/sec. and $L = 0.68$ in.

Sound Waves: In addition to capillary waves on the surface, piston frequencies higher than 5–10 cycles per second also produce accelerations large enough to do work against the strong elastic forces that hold the molecules of water together. The to-and-fro motions of the piston alternately compress and expand the water above and below its equilibrium pressure, and the resulting compressions and rarefactions propagate through the water as an invisible train of *sound waves* (fig. 52). The

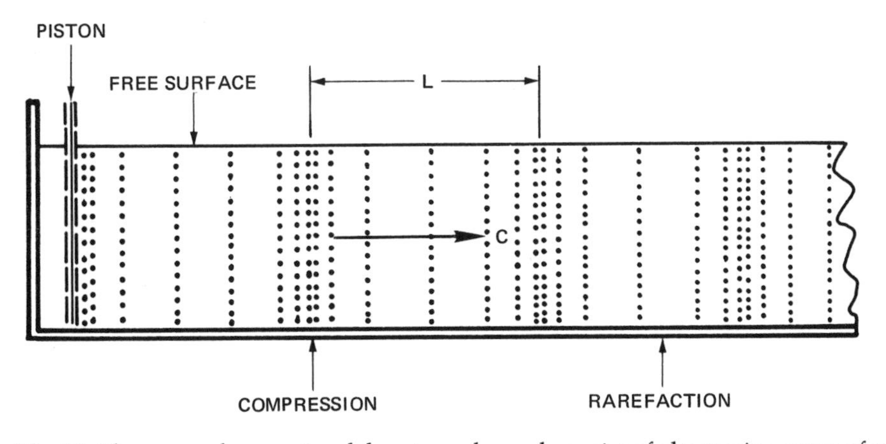

Fig. 52. Plane sound waves in a laboratory channel consist of alternating zones of compression and rarefaction that propagate at uniform speed C. The wavelength L is the distance between two adjacent compression zones.

velocity of sound waves is independent of frequency, but is related to the physical structure of water in a rather complicated way. In general, it increases with decreasing temperature and increasing mean pressure and salinity, and varies between 4,700 and 5,100 ft./sec. over the range of these variables in the open sea (fig. 53). The wavelength of sound waves obeys the same law *(L = CT)* as for surface waves, but the analogue to wave height is the *pressure intensity* (degree of compression), which increases with the square of frequency. There is no upper limit to the extent to which water can be compressed, but intensity also involves equivalent rarefactions, which are limited by the ambient hydrostatic pressure, vapor pressure, and temperature, through the mechanisms of *cavitation.* We all know that water boils at a temperature of 212°F at sea level, and at progressively lower temperatures at higher altitudes (lower atmospheric pressure). A rarefaction is a local region of low pressure which, if intense enough, will cause incipient boiling (cavitation) immediately adjacent to the piston surface during its retracting phase. The small steam bubbles produced during cavitation collapse rather violently, even at the rather moderate pressures a few feet beneath the surface, and are a common cause of erosion-pitting of screw propellers driven in excess of the critical speed at which cavitation commences. Thus, the stroke of our piston, which controls the pressure intensity at a given frequency, must be reduced in proportion to the square of frequency in order to prevent cavitation.

Perhaps the most familiar employment of sound waves at sea is the common shipboard fathometer (fig. 54), which consists of a small, electromagnetic piston that generates periodic pulses of high-frequency sound, a sensitive listening device for detecting weak echoes returning from the sea floor, and an electronic or mechanical means for accurately measuring the elapsed time between each pulse and its return echo. Such travel times can be interpreted as water depths by applying suitable corrections for temperature, pressure, and salinity from published tables. Similar, but more sophisticated, devices are used aboard military vessels for navigation and for the detection of submarines, etc.

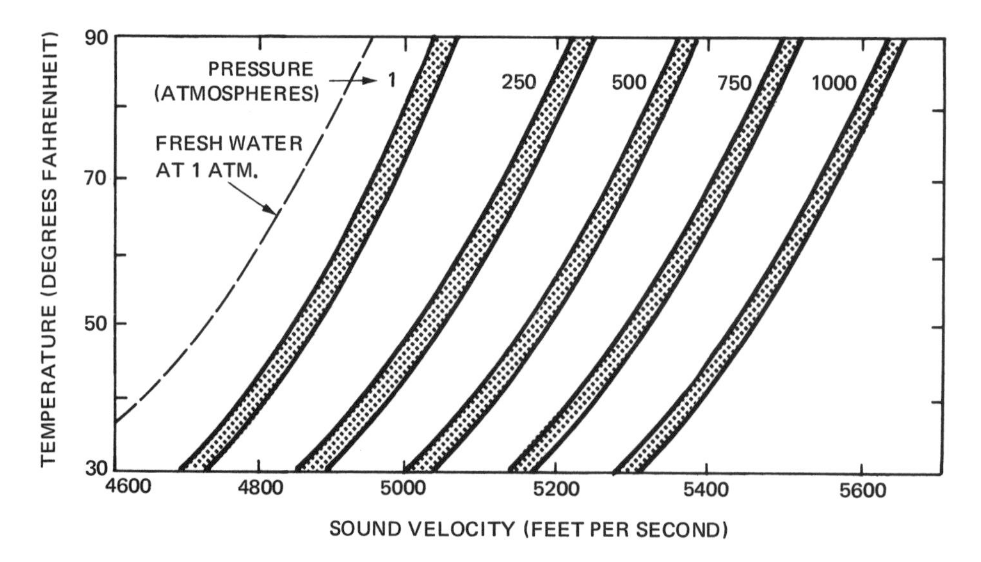

Fig. 53. Sound velocity is a function of temperature, pressure, and salinity. The width of the shaded bands gives the range of salinity effect normally present in the ocean.

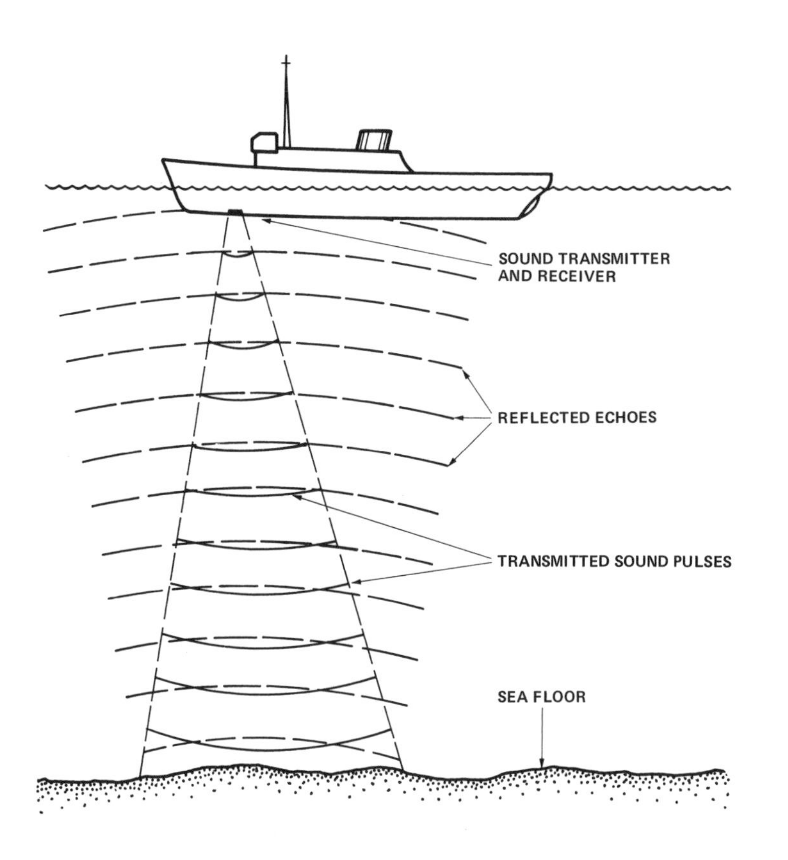

Fig. 54. The ordinary shipboard fathometer consists of a transmitter that radiates pulses of high-frequency sound in a conical pattern. Returning echoes (dashed curves) are received by an adjacent hydrophone. The elapsed-time interval, corrected for pressure, salinity, and temperature, can be interpreted as local water depth.

STANDING WAVES

Continuing our experiments, let us now interpose a vertical barrier at some distance down the channel, so that all waves are perfectly reflected (fig. 55). This situation is entirely equivalent to a second, imaginary piston, located at an equal distance behind the barrier, that generates identical waves traveling in the opposite direction, since the crests (or troughs) of both sets of waves will then arrive simultaneously at the barrier location (indicated by dashes in figure 55). After starting the generator(s), and waiting until the first wave has had time to reach the barrier and return,* it will be observed that all progressive motion of the waves has disappeared. Instead, the surface appears to be simply moving up and down like a plucked violin string. Such motions are called *standing waves* because they do not propagate along the channel. Moreover, it can be shown mathematically, by analogy to the second piston, that the superposition of two identical progressive wave trains of height H and period T, moving in opposite directions, results in a standing wave train of height $2H$; the

*We have neglected secondary reflections from the piston. For a continuous, steady-state condition to be established, the midstroke position of the piston must be distant from the barrier an exact odd multiple of quarter-wavelengths.

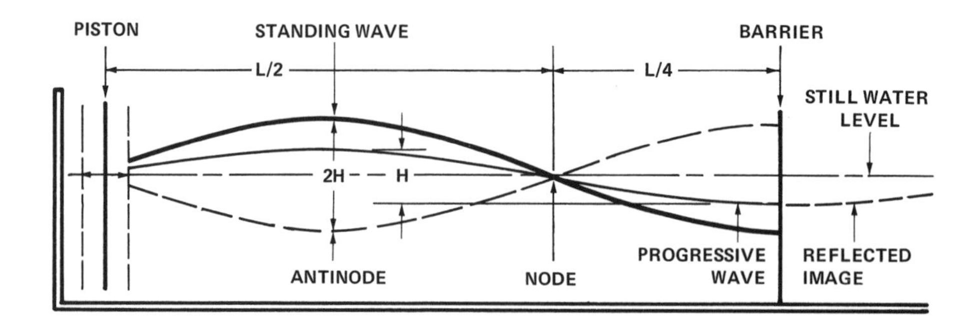

Fig. 55. Surface standing waves are produced when periodic, progressive waves are reflected by a rigid barrier. For the same period, they have the same wavelength and twice the height of their component progressive waves.

period and wavelength of the standing wave will be the same as that of its component progressive waves.

If the barrier is not a perfect reflector (for example, if it is somewhat porous), each incident progressive wave will be partly transmitted and partly reflected. In this case the wave pattern in front of the barrier will be more complicated, consisting of the superimposition of a progressive wave train of height equal to the transmitted part, and a standing wave train having twice the height of the reflected part.

The number N of standing waves between the piston and the barrier depends upon the distance X to the barrier and the phase velocity C of the component waves: $N = 2X/C$. This relation holds for any type of progressive wave discussed above—including sound waves. Thus, for very long periods, there may be only a fraction of one wave present. As the frequency of the generator is progressively increased, as if by magic, more waves will appear at the piston, and crowd into the space between it and the barrier. Irrespective of the number of waves, there will always be a quarter-wave terminated at its point of maximum height (*antinode*) by the barrier, and all waves will be separated from each other by *nodes* where the surface does not move up or down at all.

As with progressive waves, the heights of standing waves are related to the piston stroke and frequency. Similarly, higher waves become distorted and break if their stability limit is exceeded ($H > 0.1L$, in deep water). Instead of curling over, however, they become progressively more peaked, until ultimately jets of water are thrown vertically upward from each crest.

Steep, breaking standing waves are a common occurrence in many parts of the world when heavy ground swells perpendicularly approach a coastline bordered by near-vertical cliffs. The anchorage at Ascension Island, in the south central Atlantic, is particularly notorious in this regard, owing to "rollers" from distant north Atlantic storms, which make the roadstead completely untenable for days at a time.

Standing capillary waves are familiarly manifested by the stationary pattern of crispations observable on the surface of a container of water resting on a vibrating surface, or within a wine glass when the rim is stroked with a wet finger. In these examples, the patterns are two-dimensional, and result from interference between waves bouncing back and forth in several directions. In our channel, the wave crests would be straight, and parallel to the piston face. In this context, it is of interest that the stationary, V-shaped ripples formed by a fishline dependent in a moving stream are not standing waves, but forced, progressive capillary waves. Aside from their being controlled by surface tension, owing to their short wavelength, they are quite similar to the bow waves from a moving vessel.

As mentioned above, standing waves of sound can also be produced by a rigid barrier. Such waves are characterized by stationary zonal planes (antinodes) where the sound pressure rises and falls about the equilibrium pressure level, separated by alternate zones (nodes) where the pressure is constant. Again, there will be an antinode at the barrier, and the number of waves will depend upon the piston frequency and its distance from the barrier.

When progressive sound waves impinge upon the free surface from beneath, the sound pressure, instead of doubling, as it would against a rigid barrier, is instead reflected as a rarefaction; that is, there is a node at the surface. If the wave is sufficiently intense, such as the pressure shock wave radiating from an underwater explosion, the surface will cavitate, and a thin sheet of spray will be thrown upward. The same mechanism is responsible for the tissue damage sustained by animals having air voids within their body structures when exposed to strong hydrodynamic shock. Avoidance of this potential hazard is simple: stay out of the water when explosives are being detonated.

WAVE PARTICLE MOTIONS

So far, we have discussed only the nature of various wave forms, defined by the time-history of surface elevation or subsurface pressure at arbitrary points along our experimental channel. To complete this picture we need some prescription of the motions of individual water particles. Since all ideal waves are periodic, all motions exactly repeat themselves during each wave period, and a description of a single cycle of the motion suffices. In our experiments these motions can be made visible by high-speed photography of small oil droplets suspended in the water. Such photographs lead to the composite representation for particle displacements within surface waves shown in figure 56,* together with the orbits in which individual particles move during one wave period. Relative to an adjacent wave crest, the instantaneous position of each particle in its orbit is shown by a black dot; its direction of motion, by a small arrow. In general, the particle orbits in progressing waves are elliptical: particles move forward under an advancing crest and backward beneath a trough; upward before a crest and downward behind it. Particle motions in standing waves are all rectilinear: particles move vertically under antinodes, horizontally under nodes, and elsewhere at various intermediary angles, as shown. The absolute magnitudes of orbital displacements and velocities depend upon the wave properties, to which the following remarks apply:

1. For all waves, the maximum displacements occur at the surface and diminish downward.
2. For progressive waves, the maximum vertical surface displacement equals the wave height; subsurface displacements vanish at a depth of half a wavelength, or at the bottom, whichever is lesser.

*The particle motions in sound waves are confined to submicroscopic oscillations oriented in the direction of piston motion. For this reason, they are also referred to as *longitudinal* waves, as contrasted with *transverse* waves, wherein particles may also move perpendicular to this direction.

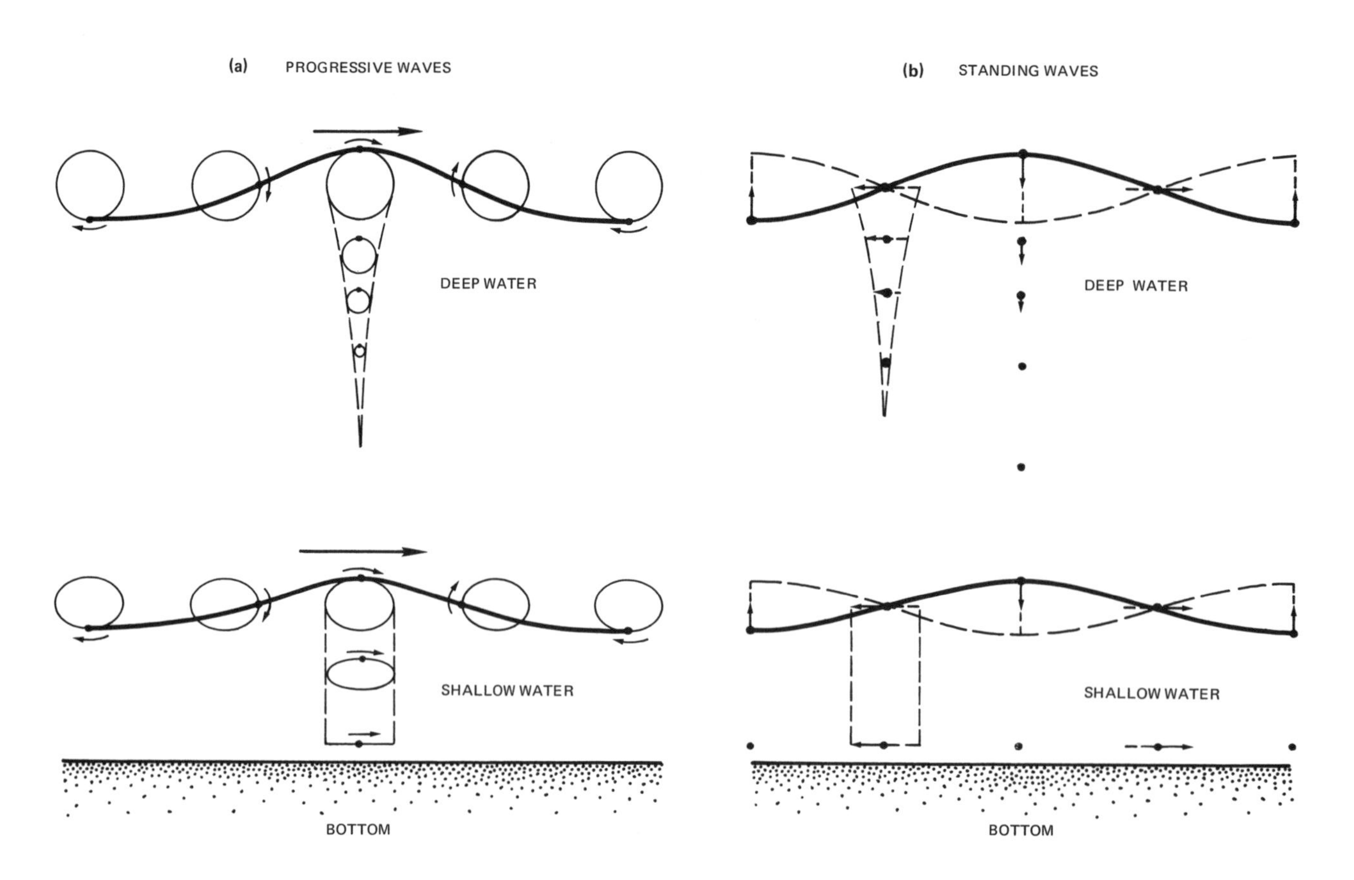

Fig. 56. Particle displacements in progressive (*a*) and standing (*b*) waves during one wave period. Small dots show particle positions with respect to an adjacent wave crest; small arrows show direction of motion.

3. The vertical surface displacement in standing waves everywhere equals that of the surface itself, being maximum at the antinodes and zero at the adjacent nodes. Subsurface displacements proportionately diminish downward in the same manner as for progressive waves.
4. Horizontal particle displacements for progressive waves in deep water are the same as the vertical displacements; that is, the particle orbits are circular. In shallow water, the former are larger in the ratio $1 : L/6.3h$, and do not decay with depth. (We have previously noted that horizontal tidal displacements in midocean can be 0.2–0.3 miles.)
5. Horizontal surface displacements in standing waves are the same as the vertical displacements, except that they are maximum at the nodes and minimum at the antinodes. Subsurface displacements decay in the same manner as for progressive waves.
6. The mean particle velocity in any orbit is equal to the distance along the excursion path divided by the wave period; for very elongate orbits, the maximum velocity will be about 70 percent higher than the mean velocity.
7. For progressive waves, the maximum particle velocity is usually very small compared with the phase (crest) velocity, except that the former approaches the latter at the crest of steep, near-breaking waves.

PLANE WAVES AND CURVED WAVES

Heretofore, we have considered only *plane waves;* that is, waves with straight crests, and within which all particle motions occur in vertical planes parallel to the axis of our wave channel. There is nothing magical about the width of the channel, which could be extended to infinity without altering any of the foregoing remarks. In nature, however, the crests of most waves are curved, and in order to study *curved waves* in the laboratory, we must use a curved piston and suitably modify our channel.

Consider a large, circular basin of uniform depth, again provided with perfectly absorbing walls, and containing a centrally located wave generator consisting of either a round piston, somehow designed to expand and contract radially, or a solid circular cone that is raised and lowered by a flywheel-and-crank system (fig. 57). Such a generator will produce annular, concentric patterns of ideal periodic waves that radiate outward in all directions. With the same stroke and frequency settings of the generator, these waves will be identical to those described in the previous experiments—with one exception: their heights will decrease as they travel outward. This is because the energy per unit crest length of periodic, progressive waves is

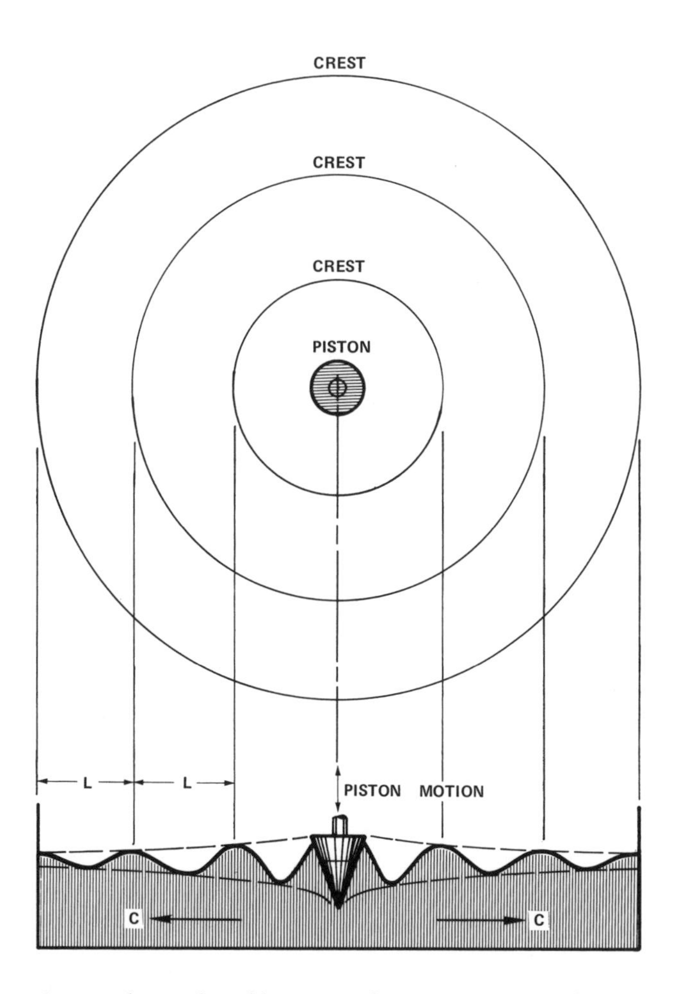

Fig. 57. Plan and profile views of progressive, periodic waves generated by a circular piston in a circular basin with absorbing walls. Wave height decreases in proportion to the square root of distance from the piston.

independent of frequency, and proportional to the *square* of the wave height. In the straight channel, the crest lengths of all waves are constant, and equal to the channel width. In the circular basin, crest length increases in proportion to the distance traveled, and thus wave height decreases in proportion to the *square root* of this distance.

By performing similar experiments with pistons of irregular shape, we will find, in general, that the outgoing waves only briefly conform to the shape of their respective pistons. Initial irregularities rapidly disappear, and the waves become circular within a radius of about ten times the maximum dimension of the piston, much as an increasing blanket of snow tends to obliterate topographic features, and assumes smoothly rounded contours.

Concentric patterns of standing waves can also be generated in the circular tank by replacing the absorbing wall with a reflecting one, and stopping the experiment just at the time when the first reflected wave returns to the generator face. As with progressive waves, all such patterns exhibit a similar decrease in wave height toward the edge of the basin. But while concentric standing waves are a phenomenon not ordinarily found in nature, it can be stated very generally that, no matter what the basin shape, a continuous, harmonic wave generator will eventually establish a permanent pattern of mixed progressive and standing waves, in which all motions periodically repeat themselves at discrete intervals. If the basin is simple and symmetrical (namely, the straight wave channel), the pattern will be simple. If it is irregular, the pattern may be very complicated, consisting of many nodal lines (straight or curved), or nodal points (amphidromes), about which progressive waves rotate like spokes of a wheel. Changing the generator period, the position, or the shape of the basin will result in a new pattern. The addition of other generators, operating at the same—or other—frequencies, will produce independent patterns, each of which will be superimposed upon the others—like the tidal patterns discussed in Part III, each of which is maintained by an independent tide-producing force.

Before considering application of the foregoing concepts to the real ocean, we can summarize the important characteristics of ideal periodic waves somewhat as follows:

1. In an unbounded basin, a harmonic generator of any shape produces corresponding patterns of periodic, progressive waves whose height and wavelength depend upon the stroke intensity and local phase velocity, respectively. Within ten generator diameters, the pattern becomes circular; all waves retain their identities, and travel radially outward to infinity. The heights of individual waves diminish in proportion to the square root of distance from the generator.
2. In a closed basin with irregular reflective walls, the outgoing wave train soon loses its identity, being replaced by an apparently random pattern of bumps and hollows caused by constructive and destructive interference between progressive waves passing through one another without interaction. Standing waves represent only special cases of interference between progressive waves traveling in opposite directions. Because the arithmetic sum of many sine waves is not a sine wave, the time-history of surface elevation at any point will be quite irregular, although it may exactly repeat itself at long intervals.
3. In a closed basin, a continuously running generator is constantly adding energy to the system. Once equilibrium is established, the wave height would grow without limit except for dissipative phenomena, such as breaking of waves that

have exceeded their stability limits internal and boundary-layer friction, and partial absorption at the walls—which are never completely reflecting.

4. With two exceptions, nothing is changed in going from a flat basin to a spherical ocean basin on the surface of a rotating earth. First, the Coriolis force now acts to displace all moving particles to the right of their intended orbits in the Northern Hemisphere, and vice versa (this effect is only important for the tides). Second, energy radiating from a single generator will spread out only halfway around the earth, and then converge again toward a point diametrically opposed (antipolar) to the generator.

Thus, on an earth completely covered by an ocean of uniform depth, the heights of individual waves in a concentric wave pattern around such a generator would decrease, not in proportion to the square root of the radius (r) from the generator, but in proportion to $H = H_0 \sqrt{r_0 / a \sin \theta}$ (fig. 58), reaching a

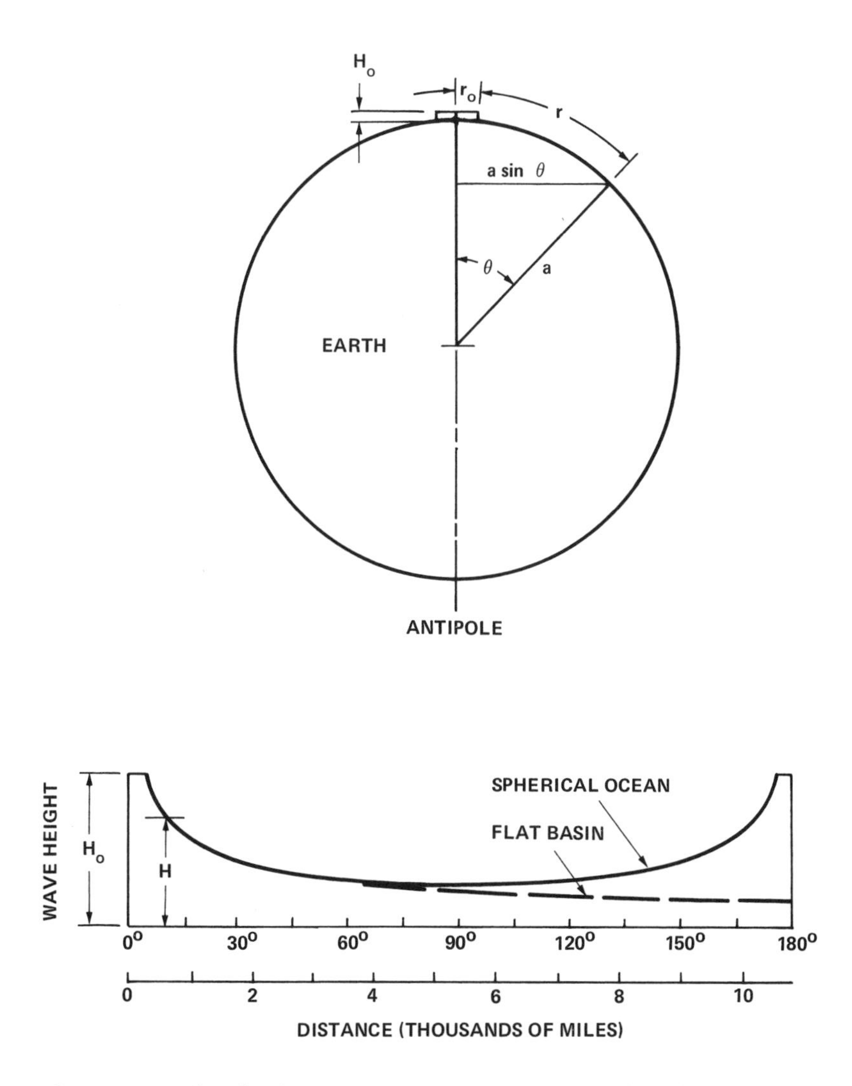

Fig. 58. Wave-height decay owing to geometric spreading for periodic waves on a spherical ocean covering the entire earth. Waves are considered to originate from a circular generator 300 miles in radius. Waves are smallest halfway round, and increase to their original height at the antipole. Dashed curve shows equivalent decay for a flat basin.

minimum height halfway around, and growing again to their original size at the antipole. This observation may seem trivial because of the continents, but there is at least one oceanic great-circle route that subtends 240° of the earth's circumference, from the Gulf of Oman to Acapulco. As we shall see later, waves from large storms in the south Indian Ocean have been tracked for over 10,000 miles northwestward along this route, and undergo considerable enhancement before reaching the coast of North America. Certain nonperiodic wave systems (tsunamis) decay somewhat faster, reaching minimum amplitude after traveling about 112° (6,700 miles), and their subsequent enhancement results in destructive effects at great distances.

14

Wave Systems

The ocean basins differ from our laboratory model in several important ways. First, they do not have uniform depths; the sea floor is highly irregular, containing deep trenches, chains of high, volcanic mountains, and long, discontinuous rifts. They are ringed by steep escarpments ascending to the shallow continental shelves, and are punctuated by numerous islands. Most submarine features do not affect the shorter wind waves* except in shallow water near shore, but the velocities and directions of long waves are significantly affected by bottom irregularities even in midocean.

Second, the ocean boundaries are neither completely absorbing nor completely reflecting. Gradually sloping beaches cause short waves to break and dissipate their energy, while vertical cliffs reflect the energy of all waves, although, because of their curvature, the reflected energy is scattered incoherently in all directions. The sloping continental shelves also act selectively to "trap" the energy of incoming waves shorter than a critical wavelength, while permitting all longer waves to escape back out to sea.

Third, the ocean basins are so large that the energy from most low-intensity sources is spread thin enough to be beyond the point of detection long before it reaches the shoreline. Only the largest storms and submarine earthquakes produce waves high enough to be readily observable at great distances.

Lastly, the oceans contain no ideal periodic wave generators. Instead, they respond to a multitude of transient disturbances, each of which generates a discrete system of waves of many frequencies (or periods), that propagates in all directions, the gross ensemble of which determines the local sea state. The size of the oceans and the brevity of these disturbances prevent the establishment of any sort of equilibrium other than the tides. Thus, sea state is a continuously varying function of time and position.

WAVE GROUPS

The principal wave systems at sea are those produced by the winds and by submarine earthquakes ("tidal waves," or *tsunamis*), the details of which are treated separately in Part V. However, any wave system can be described as an array—or train—of wave groups that results from interference between several periodic, progressive wave trains that have generally different heights and periods, but which are all traveling in nearly the same direction. As the simplest example, consider two

*The terms "long" and "short" are used interchangeably with "shallow" and "deep" in referring to wavelength, relative to the water depth.

identical plane wave trains having slightly different periods (and, hence, different velocities) propagating together in our laboratory channel (fig. 59). At any point, the instantaneous surface elevation will be the arithmetic sum of the amplitudes of the two component trains, shown by the heavy curve in figure 59. At point *A*, two component crests happen to coincide, and the wave height is doubled; at point *B*, they are opposed and cancel one another. The effect of interference is to divide the mixed wave train into groups of waves—indicated by the bounding wave envelope—that are separated by nodes where the wave height vanishes. In deep water, the velocity C_g of a wave group is one-half that of the mean component wave velocity $C_g = (C_1 + C_2)/2$, as indicated by the straight lines connecting consecutive nodes and wave crests in the figure. The average period of the mixed waves is roughly the mean of the component wave periods. Individual crests of the mixed wave system thus appear to materialize out of nowhere near a node, grow to maximum size, diminish

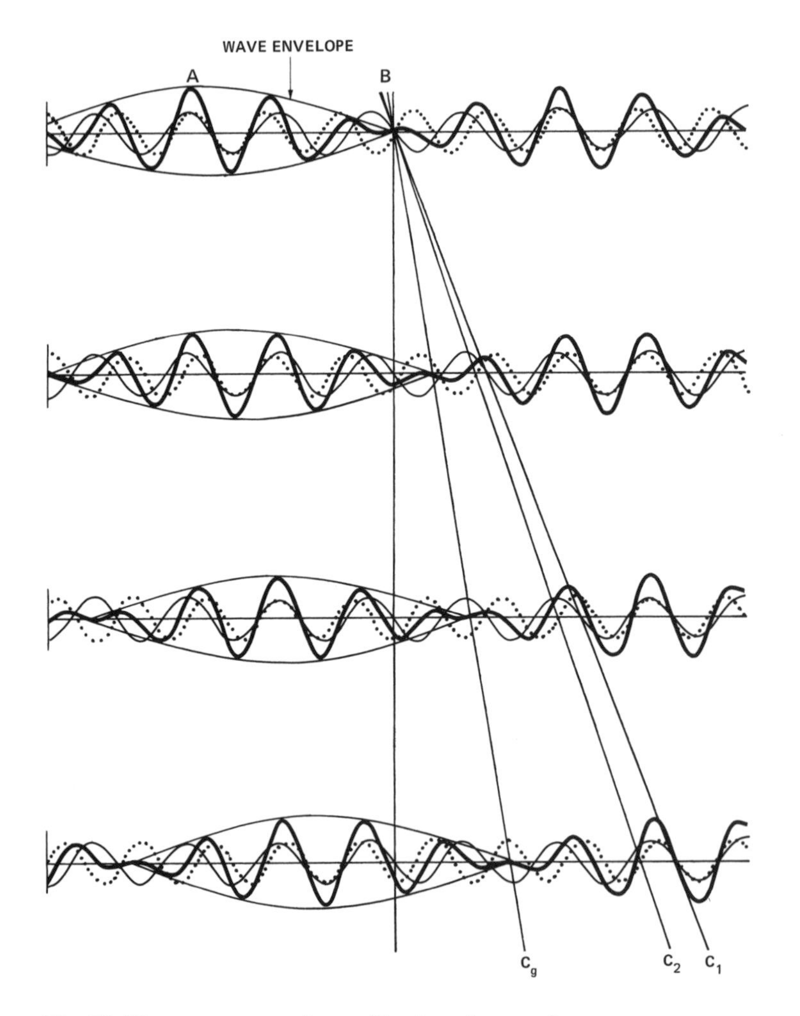

Fig. 59. Wave groups are formed by interference between two or more wave trains of different periods moving in the same direction. Here, the component trains (light continuous and dotted curves) add arithmetically to produce a composite train of variable amplitude (heavy curve). The bounding envelope defines a wave group. Its speed is half the average of the component wave velocities.

again as they advance through a wave group, and finally disappear at the succeeding node. This feature is readily apparent at sea when a well-developed swell is running; the lifetime of an individual wave crest rarely exceeds two or three wave periods. It can be shown by rather recondite mathematical arguments that the group velocity is also the rate of energy propagation for all ideal periodic, progressive waves, whether or not mixed, and thus wave energy propagates in deep water more slowly than do individual waves.

Most wave systems at sea are composed of not just two but many component wave trains, having generally different amplitudes. This does not alter the group concept, but has the effect of making the groups—and the waves within them—more irregular. Symptomatically, wave systems arise from broad, diffuse sources that simultaneously generate waves of many periods and wavelengths. In deep water, the longer waves travel faster than the shorter ones, and thus, as the wave system spreads out across the sea, like the broad beam from a poorly focused searchlight, it gradually becomes sorted out according to wavelength. This effect is somewhat analogous to a horse race; at the start, the horses are all lined up together, but as the race progresses, they become strung out in a long array, with the fastest horse in the lead and the slower ones following behind. In a wave system, this sorting effect is called *dispersion,* and occurs wherever more than one period is present in a wave ensemble.

In addition to being subdivided into groups by dispersion, the crests of individual waves within a given system are all rather short (seldom more than a few wavelengths), and are curved convexly in the direction of their travel. This is because the waves do not all travel in the same direction, but gradually spread apart at a rate that depends upon the source directivity (we recall here that all wave patterns eventually become circular, given enough space). The combination of different speeds and directions results in a spatial interference pattern, wherein wave groups are themselves staggered, and the crest intersections of individual waves have rather the appearance of a chicken-wire mesh, itself bulged up and down like an irregular waffle grid (fig. 60). This pattern is highly confused near the edge of a big storm, and becomes progressively more regular and well defined as the waves become sorted out, as if the mesh were being uniformly stretched in all directions from an initial state resembling a crumpled ball. At great distances, the long, regular swell gives little hint of its chaotic origin.

WAVE SPECTRA

A convenient, alternative method of representing such a wave system is that given in figure 61, where we have divided the system into small, discrete period intervals ΔT by harmonic analysis of a long wave record, and we have plotted each interval as a vertical *(spectral)* band whose height is a measure of its respective wave energy.* The smooth, curve bounding this aggregation of spectral bands is called the *spectrum* of the wave system. The area beneath this curve is a measure of the total energy in the system, and its shape tells us how this energy is subdivided according to period; a reverse process of analysis provides important information as to the nature of the source which produced the system. Indeed, because it is extremely difficult to obtain

* Wave energy is proportional to the square of wave height.

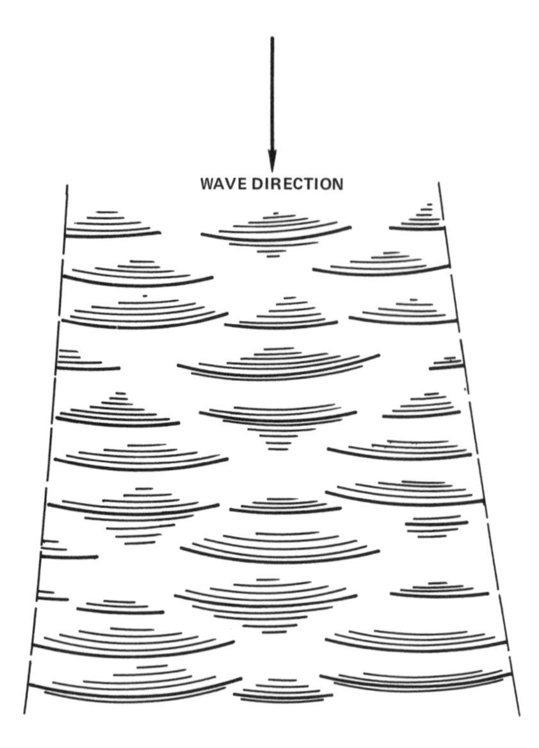

Fig. 60. Three-dimensional equivalents of wave groups are wave patches, caused by interference between swells of different periods and amplitudes traveling in nearly the same direction, as from a distant storm.

wave data from the centers of large storms, spectra from waves recorded at a distance provide the best guide for theoretical models of wave generation by wind.

While the spectrum of a wave record made at a single place gives the total energy of all component waves passing through that point during the period of observation, it says nothing about their directions of travel. Therefore, it is impossible to distinguish between two or more wave systems intersecting from different directions. This ambiguity can be resolved by analyzing simultaneous records from an array of three or more recorders. It is obvious that one detects the direction of a sound source by turning one's head until coherent sound waves—however complicated—arrive simultaneously in both ears; other sources can also be located by turning one's head. If we were equipped with three or more ears, the directions of several sources could be determined, in principle, without turning our heads, by analyzing the time difference between coherent arrivals at our several ears. For optimum resolution, our ears should be several wavelengths apart. Similarly, in the ocean, it is easier to employ a fixed array than to rotate a pair of widely spaced recorders. With such an array, the direction of a single wave system can be analyzed independently by rejecting noncoherent signals, yielding a directional spectrum. Successive spectra obtained from discrete segments of long records tell us not only the direction of a storm, but also its distance and life history. Using a linear array of recorders stretching clear across the Pacific, oceanographers have successfully tracked large storms traveling around Antarctica all the way from the Indian Ocean to Chile, and have identified individual wave systems propagating entirely across the Pacific from Antarctica to Alaska.

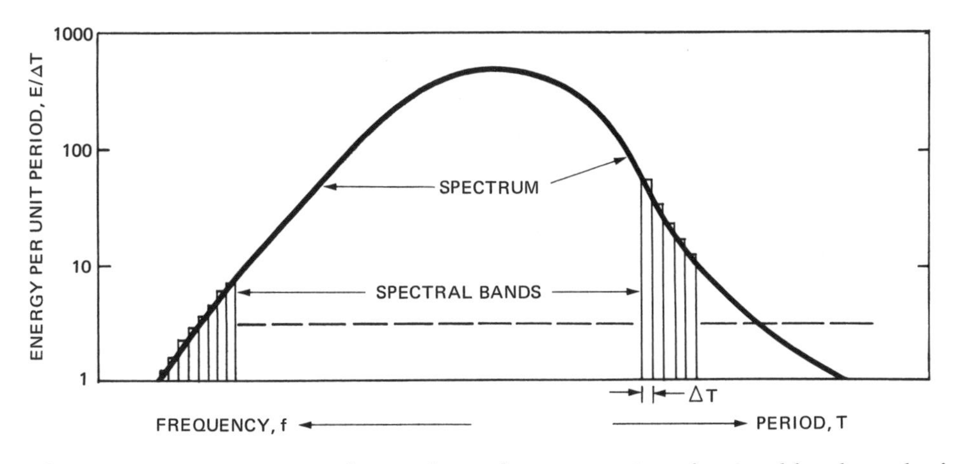

Fig. 61. A wave spectrum is the envelope of an aggregation of spectral bands, each of which represents the energy contribution from a single, component periodic wave train. Nonperiodic sources simultaneously produce wave trains of many different periods.

Figure 62 is a composite, nondirectional spectrum,* such as might be obtained from a very long (ten-year) record from several recording stations in midocean. The

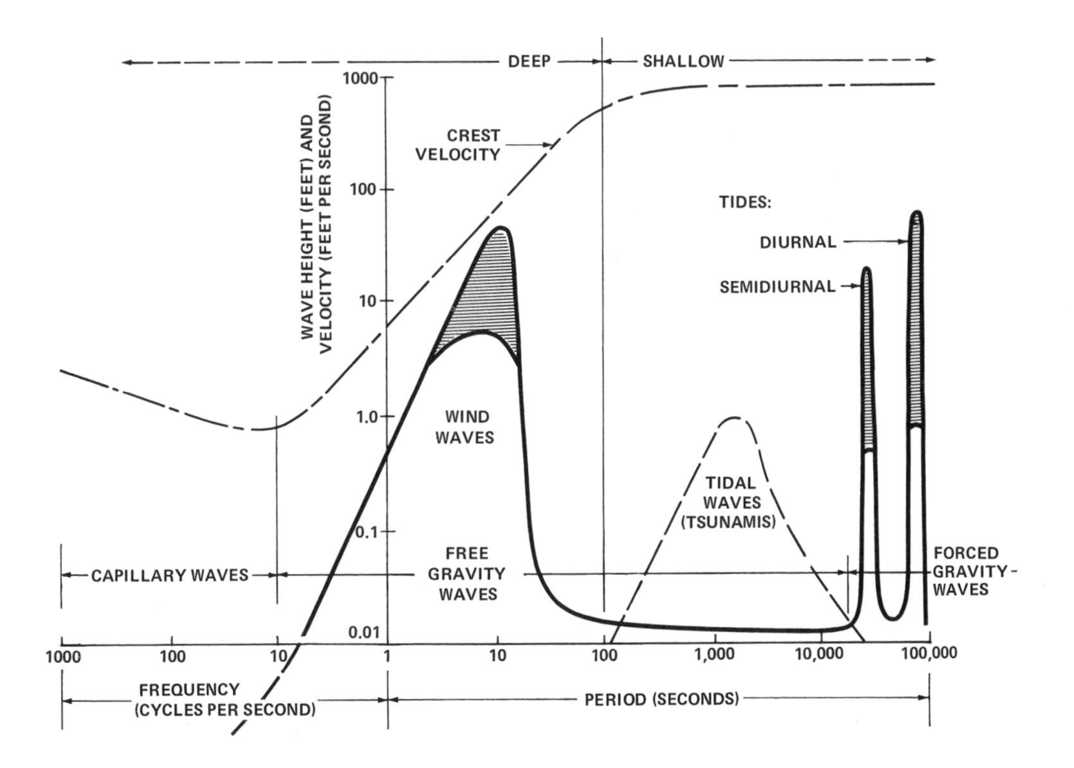

Fig. 62. Composite spectrum of ocean waves normally present at sea. The vertical height of individual spectral peaks indicates maximum probable wave height, as a function of wave period. Shaded areas denote normal range of wave height. Upper curve shows variation of phase velocity with period (or frequency).

*For convenience, the vertical scale is given in terms of wave height, instead of energy (height squared).

spectrum exhibits several peaks that can be identified with characteristically different types of disturbances that occur at sea. These peaks have shaded portions indicating the normal range of wave heights that might be observed, although their lower limits are somewhat subjective.

Beginning at the right, the tides are shown as two narrow spikes, centered around twelve- and twenty-four-hour periods. Higher resolution would show that these spikes are actually composed of many discrete spectral lines, each of which is associated with one of the periodic, harmonic constituents of the fictitious planetary driving forces. Between periods of 100 and 2,000 seconds, a dashed curve shows the spectral peak of a typical large tsunami, although its precise size and shape might vary considerably, depending upon the nature and position of its epicentral earthquake. The broad spectrum of wind waves includes contributions from a multitude of storms, the steady trades, and even the small capillary ripples that arise from the least puff of wind. This spectrum cuts off rather sharply at its right end, near 20-seconds period, and is connected to the tides by a continuous, low-amplitude residuum of long-period background oscillations of random period that arise from trapped "edge waves" leaking from the shallow continental shelves. The left end of the wind-wave spectrum is poorly defined, and trails off in the capillary region with very short waves of vanishingly small amplitude. Because storms and tsunamis contain waves of many periods, their spectra contain no discrete lines, but are said to be *continuous.*

Figure 62 also includes a curve showing the variation of wave velocity with period in midocean. As previously noted, even the great ocean depths are "shallow" for waves having periods longer than about 100 seconds.* Thus, their velocities are independent of period, and are shown here as a horizontal line, giving the mean long-wave velocity in the Pacific Ocean. Within the wind-wave spectrum, wave velocity decreases inversely with wave period, reaches a minimum in the capillary region, and then increases again to the left, in proportion to the cube root of frequency.

INTERNAL WAVES

In discussing surface waves on water, we have tacitly ignored the effect of the atmosphere. While the general theory of gravity waves treats them as periodic oscillations on the interface between two fluid layers of different density, the density of air is eight hundred times less than that of water, and can be neglected. The oceans, however, are not homogeneous, but have a layered structure within which the density varies discontinuously by small amounts. Thus, there is always an infinite variety of progressive and standing waves propagating along these various density interfaces. The phase and particle velocities of these internal waves are at most a few percent of those for surface waves, although their amplitudes may be much larger. Such ponderous, sluggish, subsurface motions would pass altogether unnoticed by seafarers, but for two unrelated phenomena: surface slicks and "dead water."

Surface slicks are a common feature at sea during light winds. They are manifested by persistent patches or streaks of calm water that stand out in sharp

*Intermediate waves are not shown. One hundred seconds is the logarithmic average between the values of 34 seconds (deep water) and 270 seconds (shallow water) given in the table on page 141.

contrast to the otherwise uniform pattern of small wind ripples. They have long been recognized as being due to localized, abnormal concentrations of organic films on the surface that strongly attenuate capillary waves, and are frequently associated with obvious film sources, such as patches of kelp. At wind speeds above 6–8 knots, the surface is covered with narrow foam streaks oriented with the wind that have no connection with internal waves. But with lighter winds, multiple arrays of slicks often appear in temperate or tropical waters as long, parallel, sinuous bands, separated by intervening bands of ripples (plate 12). In shallow water, such slicks remain nearly stationary, and tend to parallel the bottom contours. In deeper water, they move slowly (0.5 knots) at various angles to the wind. Using time-lapse photography, and simultaneous measurements of subsurface temperature, it has been shown that the organic films forming these slicks result from local surface convergences over the troughs of underlying patterns of internal waves. The slicks move with the waves, but the film is not transported; it is merely stretched thinner over the crests and compressed to greater thickness over the troughs, resulting in zones of lesser and greater ripple attenuation, respectively.

Dead water is an old nautical term applied to unusual wave resistance occasionally encountered by slowly moving vessels in stratified water. The phenomenon

Plate 12. Surface slicks that show up as bright streaks under light winds are caused by local concentrations of organic film that inhibit ripple formation.

has been studied extensively, and usually occurs where a thin layer of fresh water of thickness commensurate with the draft of the vessel overlies a denser layer of salt water. Such conditions are commonly found where a river exits into a shallow estuary or across a coastal shelf. They also may occur in high latitudes in summer, owing to the melting of ice. At low speeds, the vessel generates not only normal surface waves, but also a similar pattern of internal waves on the interfacial layer.

The net effect is to greatly impede the vessel's progress when moving at less than a critical speed, which depends upon the density ratio across the interface. If the critical speed is exceeded, internal waves are no longer generated, and the additional wave drag disappears. Figure 63 gives the critical speed, as a function of salinity and thickness of the upper layer, for vessels whose draft is roughly equal to the thickness of this layer. Vessels of shallower or deeper draft will experience similar, but less severe, effects.

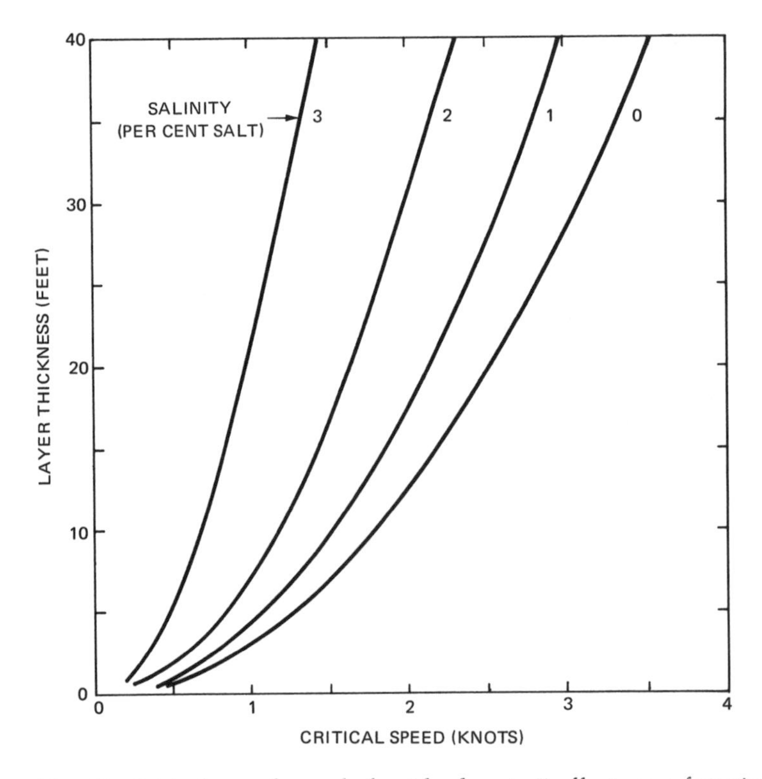

Fig. 63. Critical vessel speeds for "dead water" effect, as a function of layer thickness and salinity, relative to a sublayer salinity of 3.5 percent. Strong thermal layering can also produce a similar effect.

15

Boundary Effects

When progressive waves pass from deep to shallow water, any of several effects may occur that are related to the bottom topography and the initial angle of wave approach. Despite the recognized variability of waves at sea, these effects can best be understood by treating each wave as if it were but one of an infinite train of identical periodic waves, and later making due allowance for their differences in size and wavelength. This approximation will be fairly accurate if—as is often the case—the incoming waves have originated from a large, distant storm, because the sorting process described above arranges matters so that those waves arriving during any period of several hours will have nearly the same period and direction. However, height variation, owing to the group phenomenon, must be considered wherever it affects wave behavior.

WAVE REFRACTION

The word *refraction* is applied to all classes of periodic wave motion to describe changes in the *direction* of energy propagation owing to local differences in propagation speed. These differences are properties of the medium, although the speed is also often related to wave frequency. Familiar refraction phenomena, such as the apparent bending of a pencil standing in a glass of water, when viewed from above or below the air–water interface, or the focusing of light waves by a lens, result from the fact that the speed of light is faster in air than in water or glass (fig. 64a). Similarly, sound waves passing at shallow angles through successive layers of water having differing sound speeds (differences in pressure, temperature, or salinity) will be bent or curved (fig. 64b). These refractive effects often lead to a confusing pattern of echoes when a horizontal sonar beam encounters internal waves or tilted density surfaces before striking a reflective target. In all of the above examples, the *rays* giving the direction of energy propagation are bent in the direction of lower velocity. The change in ray angle is geometrically related to the change in propagation speed across the interface: $\sin \alpha_1/\sin \alpha_2 = s_1/s_2$. Although no refraction occurs where rays pass perpendicularly through a density interface, as previously mentioned, fathometer signals must be corrected for temperature, pressure, and salinity in order to interpret sound travel times in terms of water depth.

Progressive surface waves are not refracted in deep water, because their energy propagation speed (group velocity, C_g) depends only upon the wave period. However, upon entering shallow water ($T < 2\sqrt{h}$ sec.), group velocity becomes independent of

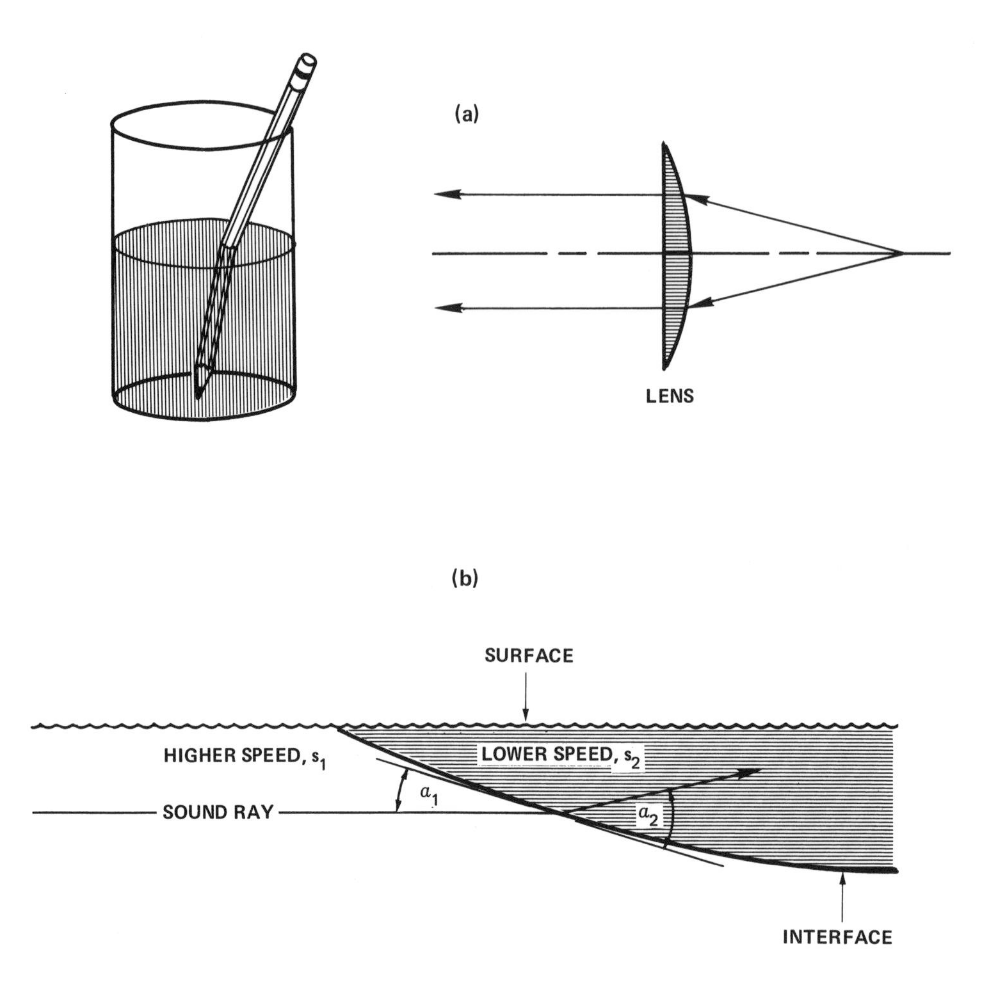

Fig. 64. (*a*) Optical refraction causes apparent bending of pencil in water glass and focusing of light rays by a lens. (*b*) Sound refraction occurs when a sound ray passes from one layer into another having a different sound speed. In either case, rays always bend in the direction of lower speed of energy propagation.

period, and is proportional to the square root of local water depth.* Thus, water wave refraction is a shallow-water phenomenon, and will occur wherever the depth is not uniform in the direction of propagation.

Consider a uniform train of plane, periodic waves—resembling a long swell from a distant storm—approaching a straight, sloping shoreline, such that their deep-water direction makes an angle α_0 with the bottom contours (fig. 65). As the waves pass over a critical bottom contour *a–a'*, where the depth is about a quarter-wavelength, their period remains the same, but their velocities and wavelengths begin to diminish in accordance with the shallow water relations: $C_g \rightarrow C \rightarrow \sqrt{gh}$, $L = CT \rightarrow T\sqrt{gh}$. Since they are approaching at an angle, that part of each crest nearest shore feels the bottom first and slows down, while the offshore portion continues straight until each point along it also passes the same contour. As the

*Waves in the gray area of "intermediate depth" are beginning to feel the bottom and will also be refracted, but this is an unnecessary complication of the present discussion.

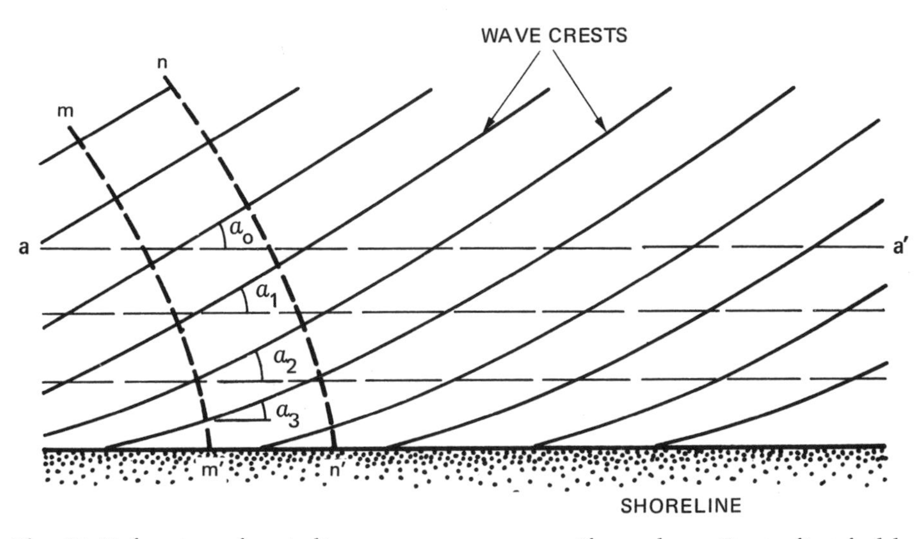

Fig. 65. Refraction of periodic water waves over uniform slope. Crests first feel bottom at contour *a–a'*, and are progressively curved as they slow down in shallower water, until they arrive nearly perpendicular to the shoreline.

depth further diminishes, each crest is bent into a curve, whose direction makes progressively larger angles, $\mathbf{a}_1$, a_2, etc., with the shoreline, finally arriving there nearly head-on, or perpendicular, to the beach.* The latter characteristic is independent of both beach slope** and the initial angle of approach, and thus the effect of refraction is to collimate all waves to the angle of normal incidence (90°) at the shore. This effect has no exact parallel in optics, because the refraction of light rays is frequency dependent. But a similar effect is achieved in a good color-corrected camera lens by employing many lens elements having special shapes and different indices of refraction; light of different colors travels by different paths, but all rays emerge from the lens at nearly the same angle.

Another important effect of refraction is an increase in the crest length of short-crested waves in shoaling water. Suppose that the incoming crests of figure 65 are bounded by the two rays *m–m'* and *n–n'*, which are everywhere perpendicular to consecutive crests at their points of intersection. In deep water, the rays are straight and parallel. In shoaling water, they trace out identical refraction paths, and both arrive perpendicular to shore. But their spacing (the local crest length) has increased during refraction, in accordance with the common observation in nature that a single breaking crest may be identifiable for a mile or more along a straight beach, whereas it is normally limited by interference to a few wavelengths in the open sea. The increase will be greater, the smaller the angle of approach.

Lastly, we note that the energy of an individual wave is constant between any two adjacent rays, and that the effect of refraction over a straight slope is to reduce the crest height in shoaling water in proportion to the square root of ray spacing.

When periodic waves approach a curved coastline, more complicated refraction patterns occur, whose character depends upon the particular situation. In general,

*Of course, we know that the waves may break before reaching shore, but breaking scarcely alters the refraction pattern described above.

**Provided that the slope extends at least two to three wavelengths offshore.

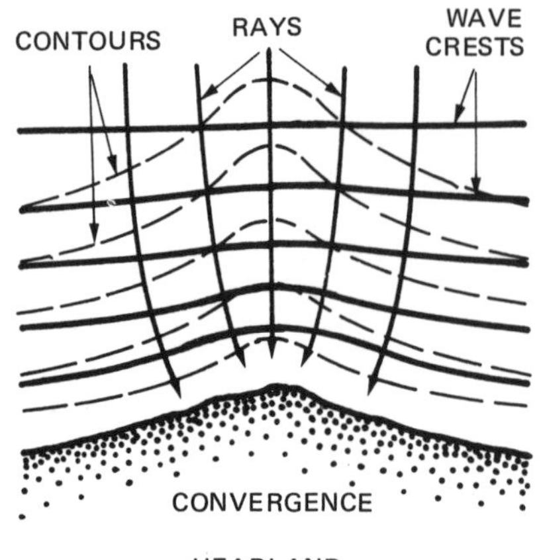

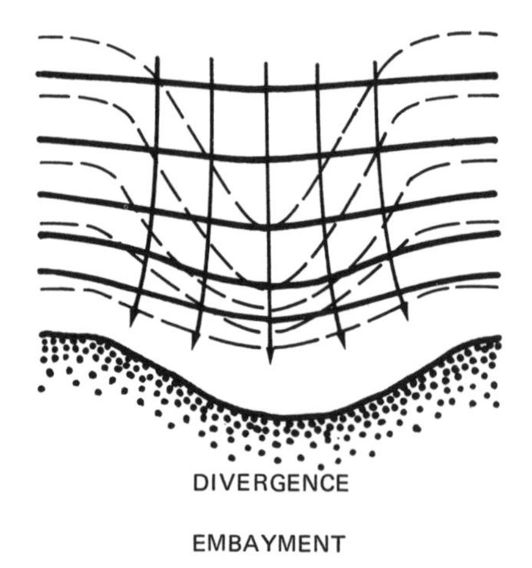

Fig. 66. Refraction causes energy divergence (lower waves) over offshore troughs, and convergence (higher waves) over ridges.

the geometric ray approximation only works where the local radius of curvature of the coastline is substantially longer than the incident wavelength. Two representative examples are of interest to mariners:

1. Refraction acts to cause wave rays to converge toward headlands terminating in offshore ridges, and to diverge from embayments associated with undersea troughs (fig. 66). This type of topography is typical of an immature, emergent coast, and results in energy concentration (high waves) on the ridges, and the converse. Given a choice in making a blind, small-boat landing at night or in a heavy fog, one is thus apt to encounter the lowest waves near the heads of bays. However, if the bay is narrow and steep-to, the reverse situation may obtain: wave energy may actually be funneled toward the bay head, and more moderate conditions prevail elsewhere.
2. Wave energy tends to be refracted around islands if their diameters exceed two to three wavelengths, and complicated wave-interference patterns are often found on their lee side. While there is usually a "shadow" zone close in the lee

of a large island, a steep, crossed sea may prevail farther offshore (fig. 67). For example, refraction studies have shown that the extraordinary surf conditions that occasionally beset the city of Long Beach, California, result from the focusing of the southern swell by Catalina Island, 14 miles offshore.

WAVE REFLECTION

In addition to refraction effects, periodic waves advancing in shoaling water toward shore undergo a continuous transformation, increasing in height and diminishing in speed and wavelength, until they either become unstable and break, or surge up the beach and get reflected back out to sea. While the characteristics of breaking waves are treated in Part V, the question of whether or not an individual wave will break is related to the maximum steepness *(H/L)* it achieves during its enhancement in shoaling water, relative to the component of bottom slope along its refracted path. Suffice it to say that if its steepness substantially exceeds the bottom slope, breaking will occur, and its energy can be regarded as totally absorbed; if not, it will be totally reflected back out to sea. As often occurs in nature, this categorical assertion is oversimplified; partial breaking and partial reflection may also occur over nonuniform slopes, or along irregular shorelines. But we will consider here only some examples of total reflection.

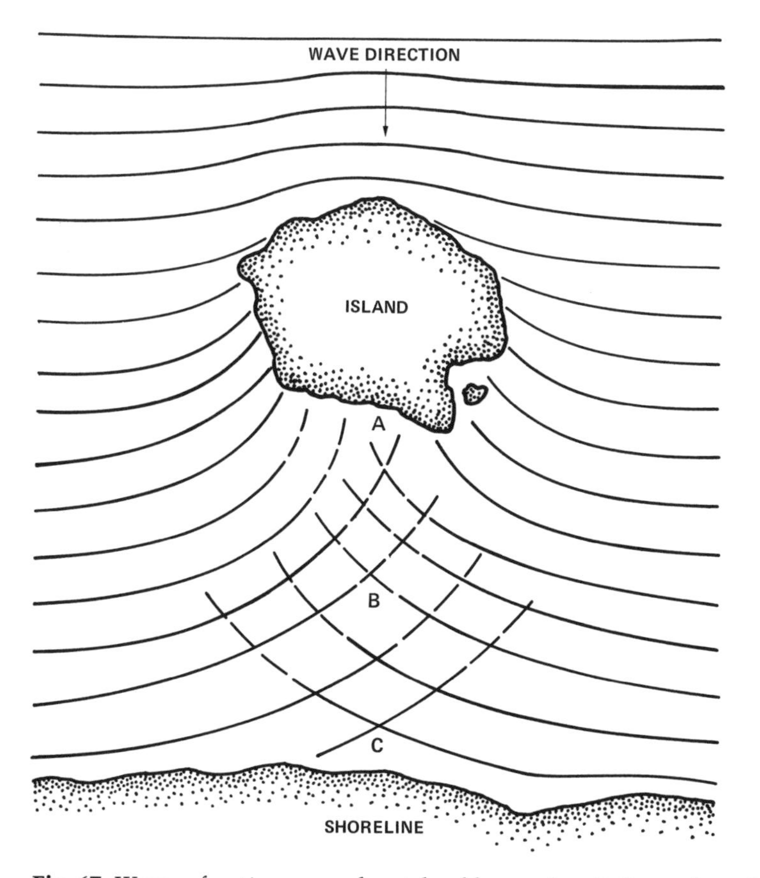

Fig. 67. Wave refraction around an island larger than 2–3 wavelengths in diameter produces a shadow zone (*A*) in the lee and confused intersecting crests about one diameter offshore (*B*), and may focus energy at some distant shoreline point (*C*).

Wave steepness and bottom slope are relative terms. All coasts slope steeply, relative to the long, low tide waves, which are always reflected without breaking, save where the tidal current exceeds a critical speed in a convergent channel. Tsunami waves are similarly reflected in most instances. Swell and wind waves generally break on slopes less than 1: 10 (a rough upper limit for sand beaches), and are reflected from steeper slopes. Even without breaking, there are turbulent energy losses attendant to most wave reflections in shallow water, and total reflection only occurs against a perfectly straight, vertical cliff extending to a depth equal to several times the height of the incident waves.

As in our laboratory channel, periodic waves perpendicularly approaching a vertical cliff will be reflected as a system of standing waves. If they approach at some other angle, they will reflect at an equal—but opposite—angle, as light from a mirror. Waves approaching a uniform slope that is steep enough to prevent breaking will first be partially refracted, then reflected, and again refracted in the opposite direction as they pass back out to sea (fig. 68a). However, natural coastlines are more commonly curved and irregular, and reflected wave energy is scattered incoherently in all directions. This produces a confused and choppy sea close inshore, which dies away within a few wavelengths from the coast. Again, these effects are relative; an irregular coastline looks much smoother to the tides than to storm waves, and the former are reflected in a more regular manner.

It has been shown both theoretically and experimentally that certain combinations of wave period, coastal curvature, and offshore slope result in the partial trapping of reflected energy by total refraction. Such a situation is shown in figure 68b, where incoming waves are first refracted and then reflected from a curving coastline. In the case of ray *m–m′*, the reflected energy is again partially refracted, and then escapes seaward. A second ray, *n–n′*, which has been reflected at a lesser angle to the shoreline, is further refracted during its outward journey to such an extent that it cannot escape, but turns inward again to complete another cycle of reflection, refraction, etc. Such ray diagrams should not be thought of as the trajectories of individual waves, but rather as the statistical flux of wave energy trapped against the coast, which shows up as long-period spectral lines when analyzing coastal wave records. The fact that tidal and tsunami wave energy dies away more slowly than can be reasonably explained by total reflection and dissipation in shallow water has been attributed to energy trapping on the shallow coastal shelves.

WAVE DIFFRACTION

Where a train of periodic waves is partially intercepted by a steep, narrow promontory, some wave energy that just clears its tip will leak into the shadow zone behind it by a process called *diffraction* (fig. 69a).* A similar effect occurs when waves pass through a breakwater gap (fig. 69b).

In both cases, interference between the incoming and reflected waves and those scattered by the end(s) of the breakwater produces distinctive diffraction patterns, whose character may be very complicated, but are related to the angle of wave approach, and to the ratio of the gap length B to the incident wavelength L. The significance of diffraction is twofold: first, for certain combinations of the above

*Light and sound waves are similarly diffracted when passing a sharp corner or through a narrow slit.

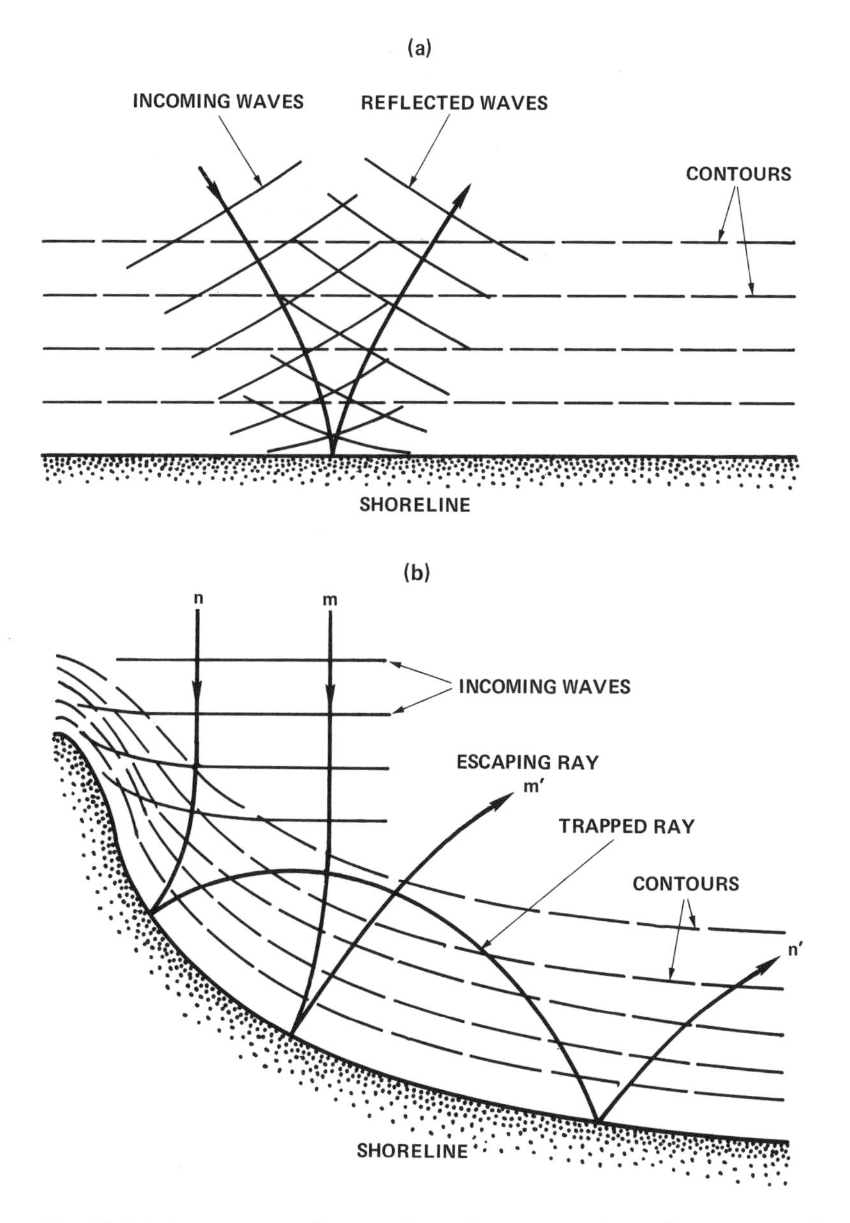

Fig. 68. (*a*) Waves approaching a relatively steep, uniform slope at an angle are first refracted, then reflected, and finally refracted again in the opposite direction. (*b*) Waves approaching a curved coastline may be totally reflected (m–m') or trapped by total refraction (n–n'), depending upon their period and the coastal curvature and bottom slope.

variables, substantial wave activity may prevail in regions within the lee of the breakwater that might seem to be well protected; second, the local wave height in such places may actually *exceed* that of the incident waves outside. Because of the many factors involved, no simple recommendations can be made to minimize wave exposure; but the provident mariner should recognize this potential hazard in selecting an anchorage, when prevailing wave conditions are apt to change during one's intended stay. A beautiful example of the complicated wave patterns produced by reflection, refraction, and diffraction is shown in plate 13.

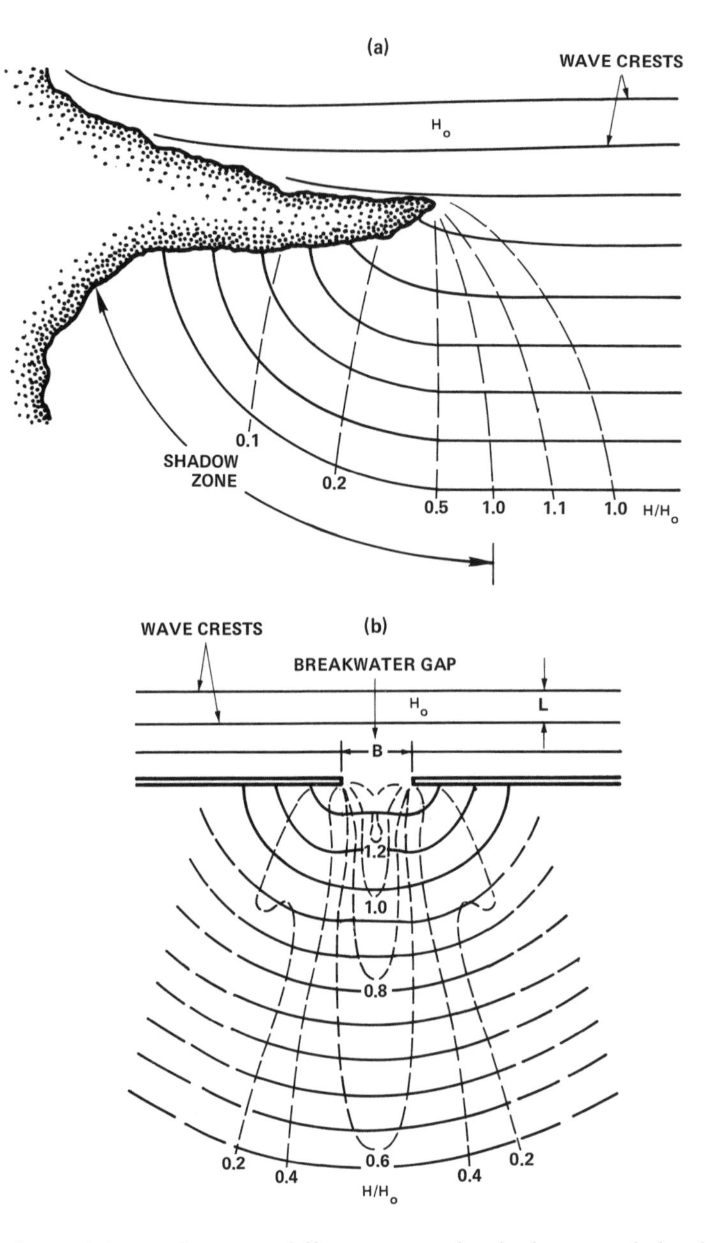

Fig. 69. (*a*) Periodic waves diffracting into the shadow zone behind a headland. Numbers give ratio of local to incident wave height. (*b*) Waves diffracting through a breakwater gap produce very complicated interference patterns, particularly where they approach at an angle. In some cases, wave height within can exceed that of the incident waves outside.

We have introduced a large number of facts and formalisms in Part IV, some of which may seem to have little application to the real ocean. But if you have persevered in beating up this tortuous channel, you will find clearer sailing in Part V, where we will consider application of these principles to natural wave systems.

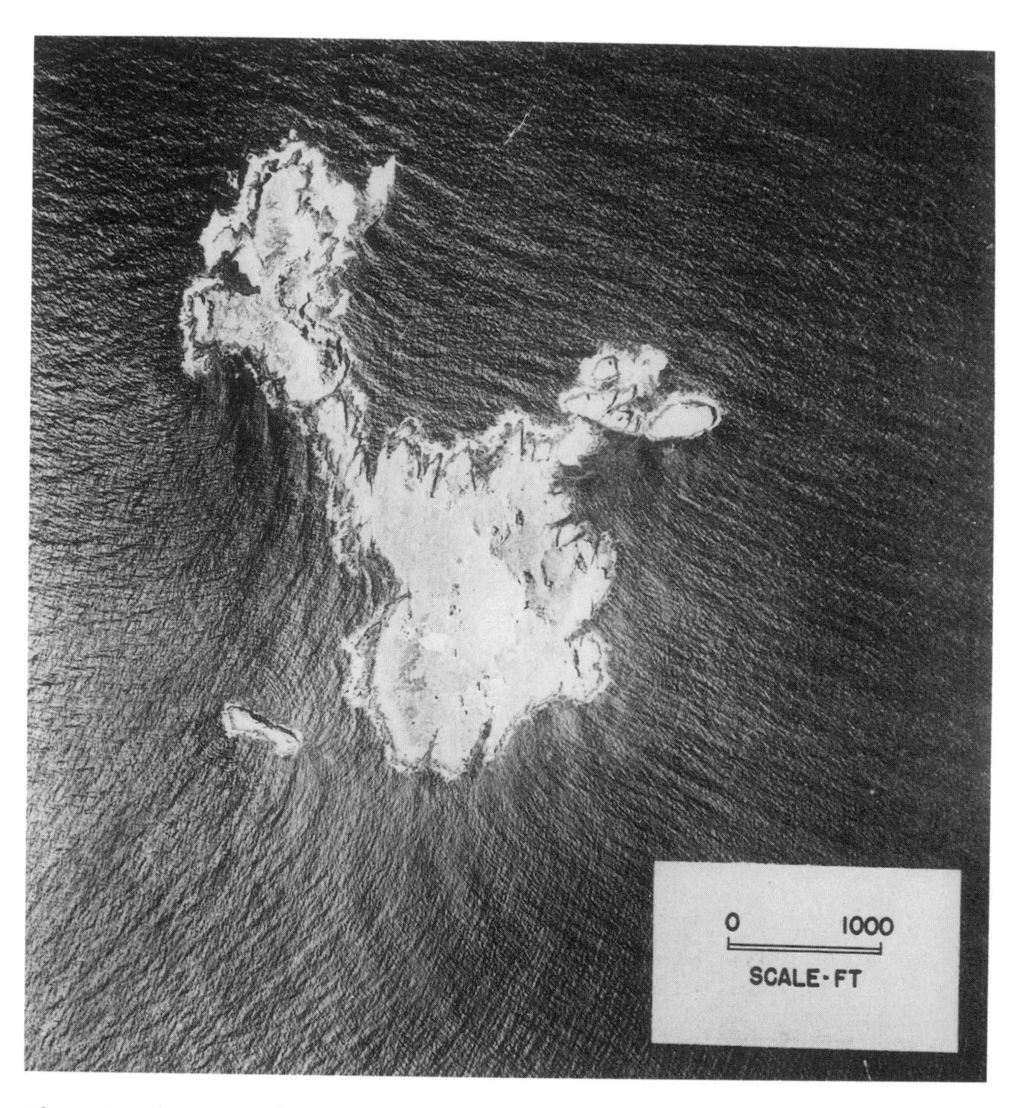

Plate 13. Refraction and diffraction of both swell and wind waves at Farallon Island, California. Note reflection from small island at lower left.

PART V

Ocean Waves

16

Wind Waves

INTRODUCTION

The importance of wind waves to mariners cannot be overstated. With increasing wind, the state of the sea to a large extent determines heading and progress, comfort or discomfort, and ultimately tests one's ability to survive. In a growing sea, wave forces increase much faster than wind forces. Under extreme conditions, pressures in excess of one ton per square foot have been measured in breaking waves, while wind pressures rarely exceed 10–12 pounds per square foot. Thus, although wind effects are not to be discounted, it is the waves that most often disable a vessel. Even the largest ships are unable to maintain steerage way under hurricane conditions, and may founder due to excessive rolling when hove to. In the typhoon of December 17–18, 1944, off the Philippines, the U.S. Navy's third fleet lost 3 destroyers, 146 aircraft, and 790 men; 18 other ships suffered major damage. An oceanographer friend, who survived this disaster as quartermaster aboard a large ammunition ship, told me that for minutes at a time he was unable to see even the superstructure of a heavy cruiser less than a quarter-mile off his starboard bow.

Seemingly, the Spanish Manila galleons—with no advance weather information—had a better batting average. During the fourteenth and fifteenth centuries, they completed more than one thousand round trips between Acapulco and Manila, losing only thirty ships. These monstrous hulks, grossing up to 2,500 tons, were solid-planked 3 feet thick at the bilges, and could not point closer than 80° to the wind, yet they sailed repeatedly through the thick of typhoon country with apparent immunity.

Despite their obvious importance, and concentrated scientific efforts to explain them, we still lack a completely satisfactory theory for the generation and growth of wind waves. Although the basic mechanisms have been reasonably well defined, wind and wave interactions are exceedingly complicated and yield only slowly to man's inquiring mind. Meanwhile, scientists have learned to forecast waves by linking wind and wave observations by a sort of statistical wave theory, and can make reasonably accurate predictions from meteorological data. While severe storms can now be identified and followed by weather satellites, we are still dependent upon ship observations for wind reports in the open sea, and there is as yet no systematic wave reporting service available to mariners.*

*Special synoptic wave forecasts are available on a contract basis from several commercial firms on both the Atlantic and Pacific coasts, which are of significant economic benefit to shipping companies in planning sailing routes.

Therefore, if you wonder why I have devoted an entire part to discussing the principles of wave motion, it is because a fundamental acquaintance with wave dynamics, coupled with familiarity with the kinds of waves one is likely to encounter, provides the best guide to good seamanship. It is ordinarily sufficient to know how waves behave, even though we cannot precisely relate cause and effect. Accordingly, the following descriptions of wind-wave development represent an interpretive blending of theory, experiment, and observation that will doubtless be improved by further study.

GENERATION OF WIND WAVES

The threshold wind speed at which ripples first appear on a calm water surface varies considerably, depending upon the vertical velocity structure of the wind, the elevation at which the wind is measured, and the degree of surface contamination. Under aseptic laboratory conditions, with triple-distilled water, ripples will appear at much lower wind speeds than they do on, say, ordinary tap water. The sea surface is always heavily contaminated, and the inextensible surface film requires substantially higher wind speeds (2–3 knots)* to deform it.

Careful experiment shows that the wind first establishes a very shallow surface current, whose velocity increases in proportion to wind speed. At some point turbulent fluctuations within the lower air layers develop to where they perturb the water surface, and a characteristic pattern of capillary ripples appears. Close inspection reveals that the earliest ripples propagate at the minimum velocity (0.76 ft./sec.) for free gravity waves, not in the direction of the wind, but as two sets of parallel crests that move in nearly opposite directions, making equal angles of 70°–80° from the wind direction (fig. 70a). These ripples are, perhaps, Nature's closest approximation to ideal periodic waves, because they are constrained to uniform velocity by the combined interactions of gravity and surface tension. Their wavelengths are less than 1 inch; their amplitudes are so low that they are nearly invisible, and can best be seen in shallow water as bright bands of focused sunlight on the bottom. That they travel at an angle to the wind is attributed to the fact that the air immediately above the surface moves faster than the waves can travel, and this angle is determined by the surface wind *component* which moves at the minimum wave speed.

As the wind increases, so do the wavelengths, heights, and periods of the ripples; moreover, the above angles progressively decrease until, at wind speeds of 4–6 knots, they travel at about 30° from the wind. This is because the growing waves begin to travel faster than the minimum speed, and their growth is selectively favored by a wind component closer to the wind direction. At this stage, the surface appears to be uniformly covered by a chicken-wire pattern of short-crested wavelets traveling *with* the wind, but each wavelet actually represents the oblique intersection of two long-crested ripples (fig. 70b). Despite its wavy appearance, the surface still looks glassy, and can be considered as hydrodynamically *smooth;* that is, it does not significantly perturb the air flow above it. If the wind should die, the wavelets would immediately disappear.

At wind speeds above 6 knots, the wavelets grow into small waves that are high enough to affect the air flow; the surface becomes hydrodynamically rough, and a

*Hereinafter, all wind speeds refer to standard anemometer height (33 feet).

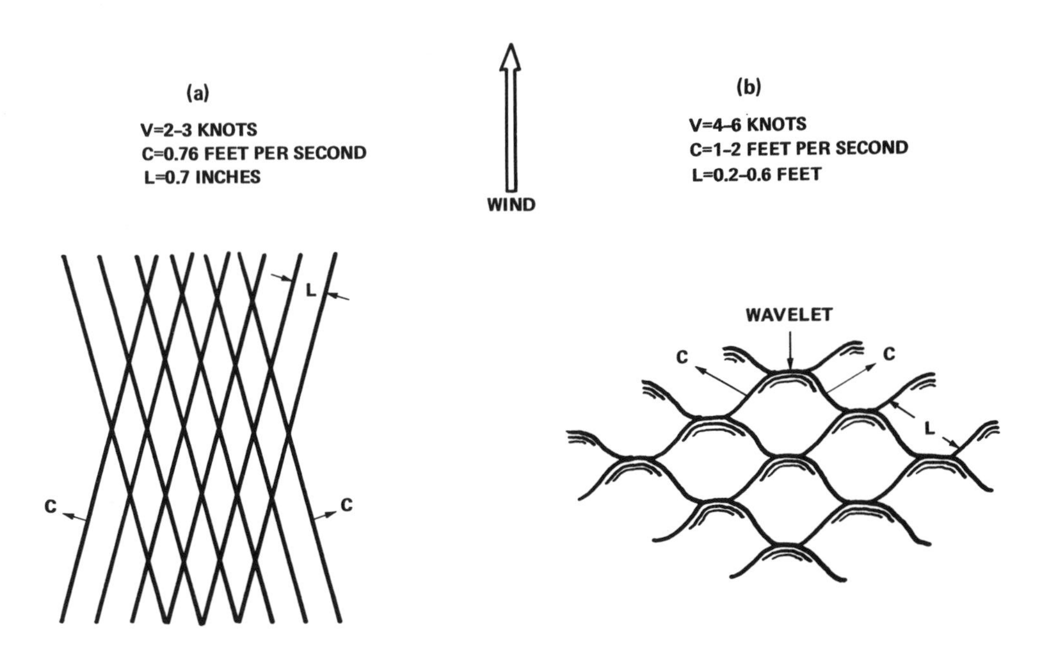

Fig. 70. Wave patterns at low speeds. (*a*) The earliest ripples travel at minimum wave velocity, nearly at right angles to the wind. (*b*) As the wind increases, this angle decreases, and the pattern becomes more irregular.

second growth mechanism comes into play. The presence of wavelets causes the air flow to become unstable, and the air-pressure distribution is altered in such a way as to increase the wave amplitude and wavelength. First, higher waves further perturb the air flow, increasing these pressure differences, etc. Second, this effect is directionally selective, enhancing most strongly those components traveling nearly in the direction of the wind—i.e., the wavelets formed by the ripple crest intersections. Thus both mechanisms (decreasing ripple angle and instability) progressively narrow the initial directional spread of wave energy. Lastly, growth by instability is also an amplitude-selective process; the shortest and steepest waves grow fastest, partly because they travel slowest and thus experience the highest relative wind speed, and partly because the steeper waves have sharper crests that more significantly perturb the air flow.

All of the above mechanisms make the wave field progressively more irregular. Even at low wind speeds, the wind is neither absolutely steady nor spatially uniform, and slight initial differences in the heights and wavelengths of individual waves are further exaggerated by unequal growth rates. Hence, for all wind speeds above 6 knots—if not lower—we can no longer follow the growth of representative, nearly identical waves, but must adjust our sights to deal with a statistical ensemble (directional spectrum) of waves moving within, roughly, 50° of the wind direction. The ensemble contains waves of all heights and periods, from a lower limit dictated by capillary dissipation to some upper limit determined by a balance between the rates at which energy is supplied by the wind and removed by dissipative processes—principally, breaking.

GROWTH AND DECAY OF A WAVE SPECTRUM

In most standard references, you will find it stated that the growth of a wave spectrum is governed by three factors: the wind *speed* (assumed to be uniform in space); its *duration* (assumed to be uniform in time); and the distance, or *fetch*, over which it blows (assumed to be finite). It should be recognized that these factors are only idealizations that are rarely realized in nature and difficult to appraise. Still, our present—and reasonably successful—techniques for describing and forecasting the state of the sea depend upon assigning numbers to them, and using statistical hydrodynamic models for spectral growth under the assumed conditions.

Wave Statistics: In a statistical model, we temporarily forget about individual waves, wave groups, and the fact that any wave system can be considered as the simple superimposition of a number of ideal, periodic wave trains, although we shall later recover these concepts in considering ocean swell after it leaves the area of wave generation. Instead, we think of sea state as a spectrum of wave energy that is constant, or slowly changing, within a given region. In any such region, we attempt to specify, on the average, the present and future probability that certain numbers of waves of any particular height, wavelength, period, and direction of motion will be present.

The statistical sea surface is represented as a quasi-random and ever-changing pattern of bumps and hollows that *never repeats itself.* Thus our previous wave properties require redefinition, since it is manifest that the distance (wavelength) and time interval (period) between any two adjacent bumps passing a fixed point of observation are unlikely to be the same as those between preceding or succeeding bumps. Yet it is bumps and hollows that we actually see, and any realistic model must be expressed in terms of quantities that we can observe and measure.

For our new definitions, consider a statistically long (20-minute) record of sea surface elevation at a fixed location in deep water (fig. 71), where the wind speed, direction, duration, and fetch are presumed known, or can be estimated from meteorological reports. We first divide the record into N equal parts at intervals of, say .5 second. In a twenty-minute record, we will have $N = 2{,}400$ intervals. Next, for each interval we tabulate the vertical distance (surface elevation) between the equilibrium (no wave) level and the wave profile, and label them $A_1, A_2, \ldots, A_N$. We now

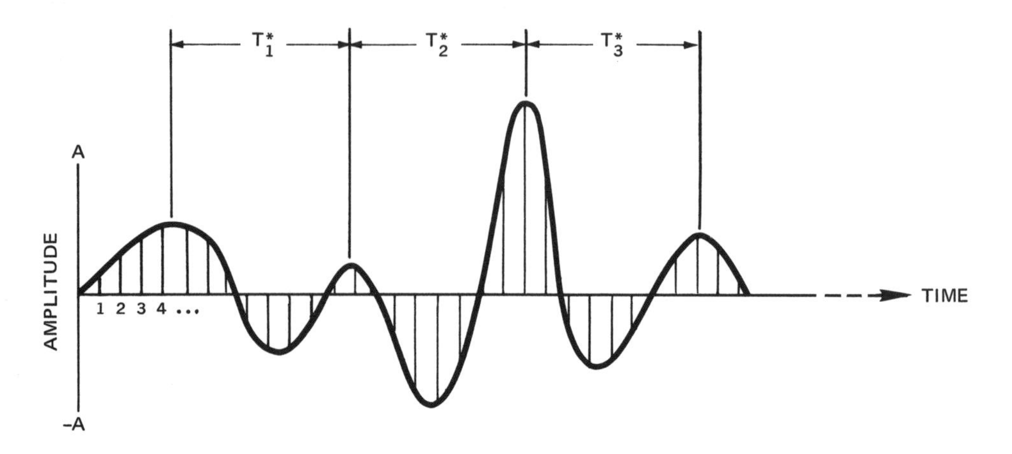

Fig. 71. Deep-water wave-record sample, showing construction for statistical analysis.

define a number E as twice the average of the sum of the squares of the individual wave amplitudes:

$$E = \frac{2}{N}\left(A_1^2 + A_2^2 + \ldots A_N^2 \right) \text{ ft.}^2$$

The number E can be thought of as a measure of the average wave energy per unit surface area of the composite sea state during the time covered by our record. Analysis of many such records has shown that, within the generating area, wave heights are randomly distributed in space and time, and that if we can forecast E, we can estimate the probability of encountering waves of any height in a slowly changing sea state. For the purposes of this book, we shall often use a more convenient wave parameter, $\sqrt{E}$ (ft.), which, for want of a better name, I shall refer to as the *sea state index*, or just *sea state*.

Certain properties of random waves allow us to say several things about their height distribution. The curve in figure 72 gives the percentage probability that all

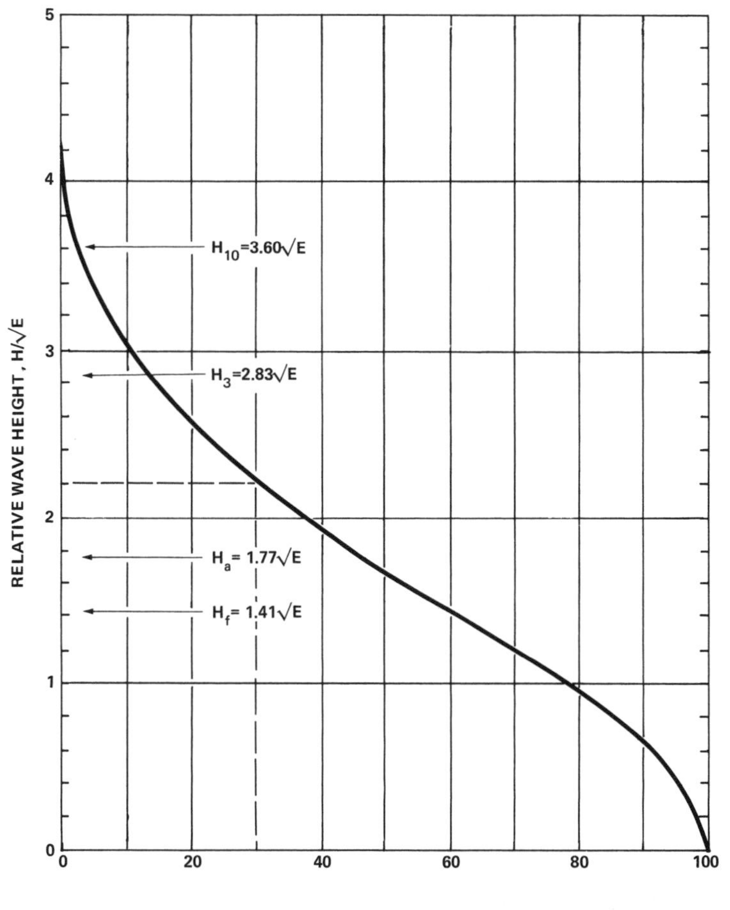

Fig. 72. Wave-height distribution as a function of probability of occurrence in a random sea. Dashed example shows that, in a long record, 30 percent of all waves will have heights greater than $2.2\sqrt{E}$.

waves in a given sea state E will be higher than the heights (in feet) obtained by multiplying the appropriate numbers along the vertical scale by the square root of E. For example, the dashed lines indicate that 30 percent of all waves will be higher than $2.2\sqrt{E}$ ft. This figure also lists four common height indices:

$H_f = 1.41\sqrt{E}$ = the most frequent probable wave height
$H_a = 1.77\sqrt{E}$ = the average height of all waves present
$H_3 = 2.83\sqrt{E}$ = the average of the highest one-third of all waves
$H_{10} = 3.60\sqrt{E}$ = the average of the highest one-tenth of all waves

The height index H_3 is often called the *significant* wave height, on the premise that mariners are chiefly interested in the larger waves present. From the distribution curve, the probability of encountering a significant wave is 15 percent, which is roughly in accord with the common saying, "Every seventh wave is highest." But we are only dealing with statistics; one is as apt to encounter two such waves in succession as one in every thirty, etc. One must flip a coin an infinite number of times to have as many heads as tails.

The shape of the distribution curve is such that very high—or very low—waves are increasingly unlikely events, but both are of special interest. In heavy seas, a group of low waves may give you time to bring a ship about, shorten sail, or take an opportune navigation sight. A group of high waves may contain a giant, breaking, "rogue" wave that can poop a vessel, or pitchpole her end for end. Because low waves are not dangerous, and because the probability of a succession of them cannot be calculated without specifying their number and size, we will say nothing more than that the chance is best following a succession of abnormally high waves. However, a single rogue wave can spell disaster, and its probability is easily estimated. Figure 73 gives the chance of encountering abnormally high waves, as a function of relative height and the total number *(N)* of waves encountered. At any sea state $\sqrt{E}$, the example (dashed lines) indicates that within any 200-wave sample there is a 5 percent chance that one wave will exceed $H = 5.8\sqrt{E}$, an equal chance that it will be lower than $H = 4.1\sqrt{E}$, and a most probable value of about $4.8\sqrt{E}$. All three curves slope upward to the right, indicating that one runs the same percentage risk of encountering progressively higher waves as more pass by. The length of time required to establish these probabilities depends upon the average wave period for any particular sea state, as defined below.*

Whereas the wave heights in a growing sea are randomly distributed, their periods and wavelengths are not. In figure 71, the time intervals between successive wave crests passing our recording station are labeled T_1^*, T_2^*, T_3^*, etc. In general, these intervals are all different, and bear no simple relation to the periods of the ideal, periodic waves that happen to combine to produce the observed wave record. Yet the theory of forecasting is based upon the statistics of observed waves and their relation to local wind conditions. Moreover, the forecasting curves, described below, assign a value of E to every ideal wave period present in a predicted spectrum, which in turn can be interpreted as the joint probability that waves of a certain height will occur at certain time intervals. Accordingly, we introduce new definitions for the observable wave "period" and "wavelength":

*There is also a physical upper limit, not well established, due to breaking.

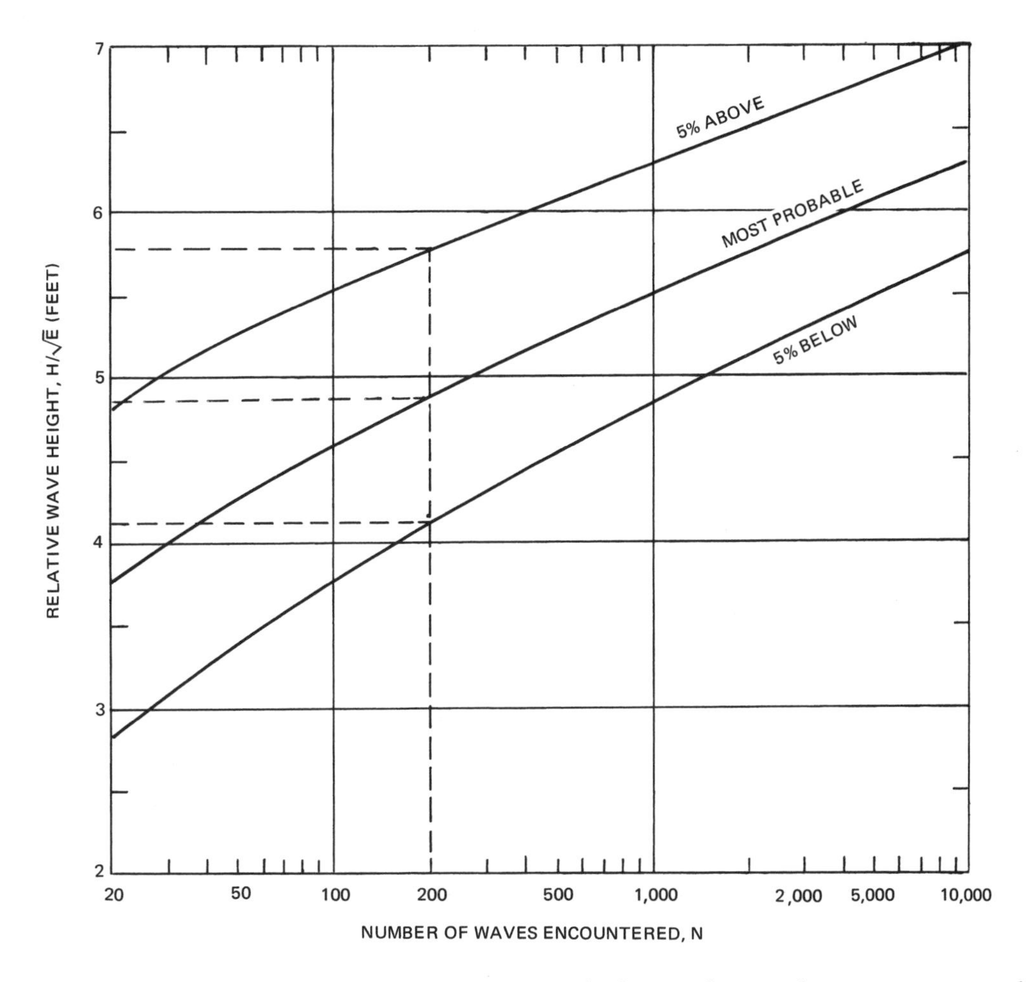

Fig. 73. Extreme wave probability increases with the number N of waves encountered. Example shows that there is a 5 percent chance that the highest of 200 successive waves will exceed $H = 5.8\sqrt{E}$, the same chance that it will be lower than $H = 4.1\sqrt{E}$, and a 90 percent probability of its being in between.

T^* = time interval between consecutive wave crests from an observed record
T_a^* = average of all of the above intervals from a given record
L^* = the distance between two adjacent crests, measured perpendicular to their direction of travel
L_a^* = the average of many consecutive measurements of L^*

The above quantities are purely statistical. For the purposes of this book, they are only used, in connection with the forecasting curves, to provide estimates of sea state and ship behavior. T^* and L^* do *not* obey the relationship $L = CT = 5T^2$ given in Part IV for periodic progressive waves in deep water. However, their respective averages obey a similar law: $L_a^* = KT_a^{*2}$, where K is a constant that depends on the stage of sea state development. For practical purposes, K varies between the limits $2.6 < K < 3.4$ from an immature to a fully developed sea. A mean value, $K = 3.0$, will be at most 12 percent in error at any sea state.

Since T_a^* is the average time interval between waves in a random sea, the total time required for N waves to pass a fixed observation point will be $NT_a^*/3,600$ hours.

This relationship can be applied in using figure 73 to estimate the time probability of extreme waves. Suppose that $T_a^* = 9$ seconds, a number easily obtained at sea by timing twenty to fifty waves. Then, for our previous example, $NT_a^*/3{,}600 = 200 \times 9/3{,}600 = 0.5$ hour; e.g., we could expect the statistics cited to apply to any half-hour sample—provided that average conditions do not change over this interval. Figure 73 applies best to a slowly decaying swell, or to the short-term expectancy of ship motions. It can provide ballpark estimates in fully developed or fetch-limited seas (see below), but will underestimate extreme-wave height in a rapidly growing sea.

However, as we shall see in Part VI, the speed of a moving vessel relative to the waves must be considered in determining the effective period of wave encounter, T_e, which would be used instead of T_a^* in estimating time probability. We will also show how similar probability estimates can be made of excessive—even catastrophic—ship motions, using the ship itself as an indicator of average conditions.

Sea State under Steady Winds: Ignoring wave energy that might be leaking into a given region from other sources, the wave spectrum described in Part IV provides all the information necessary to determine the sea state as a function of local wind conditions. Since wave energy is proportional to the square of wave amplitude A^2 (or of height, H^2), an energy spectrum can be defined as a continuous curve that gives the appropriate value of A^2 for every wave period T within the ensemble of ideal, periodic waves that combine to form the sea state. As a hypothetical case, consider that a wind of constant speed $V > 6$ knots springs up over an infinite stretch of calm water bounded at its upwind end by a straight shoreline (fig. 74a).* As described above, waves will be generated and will continue to increase in amplitude and period until, locally, a steady state is attained, the nature of which depends upon the distance offshore (fetch), and whose time of establishment increases with distance. Ultimately, after a (somewhat poorly defined) time t_m, a condition defined as a *fully developed sea* (FDS) will be achieved at all distances beyond some minimum fetch F_m, and no further changes of sea state will occur. Everywhere upwind of F_m, the average wave amplitude and range of periods present will increase with fetch, from zero at the shoreline to their constant FDS values at—and beyond—F_m.

These changes with distance are qualitatively illustrated (fig. 74b) by the spectral curves $S_1, S_2, \ldots, S_m$, corresponding to successive offshore distances (fetches) $F_1, F_2, \ldots, F_m$. The curves all arise from a common origin at the left, and initially follow the heavy curve for the fully developed spectrum S_m, but consecutively branch to the right, passing through successively higher energy maxima at successively longer periods $T_1, T_2, \ldots, T_m$, after which they turn sharply downward and cut off along the period axis. The areas under the spectral energy curves give the total wave energy present in the local sea state at the respective fetches. Since the average energy density E of the actual sea state is, by definition, the same as that for the ideal wave spectrum, we can label these areas $E_1, E_2, \ldots, E_m$. The total energy increases with offshore distance, but the fact that all the spectral curves are congruent toward the left shows that this energy increase is manifested by the progressive appearance of longer and higher waves. The rather abrupt cutoff of each spectrum

*The shoreline is not important; the upwind end of a fetch could as well be defined as the place where the wind starts to blow.

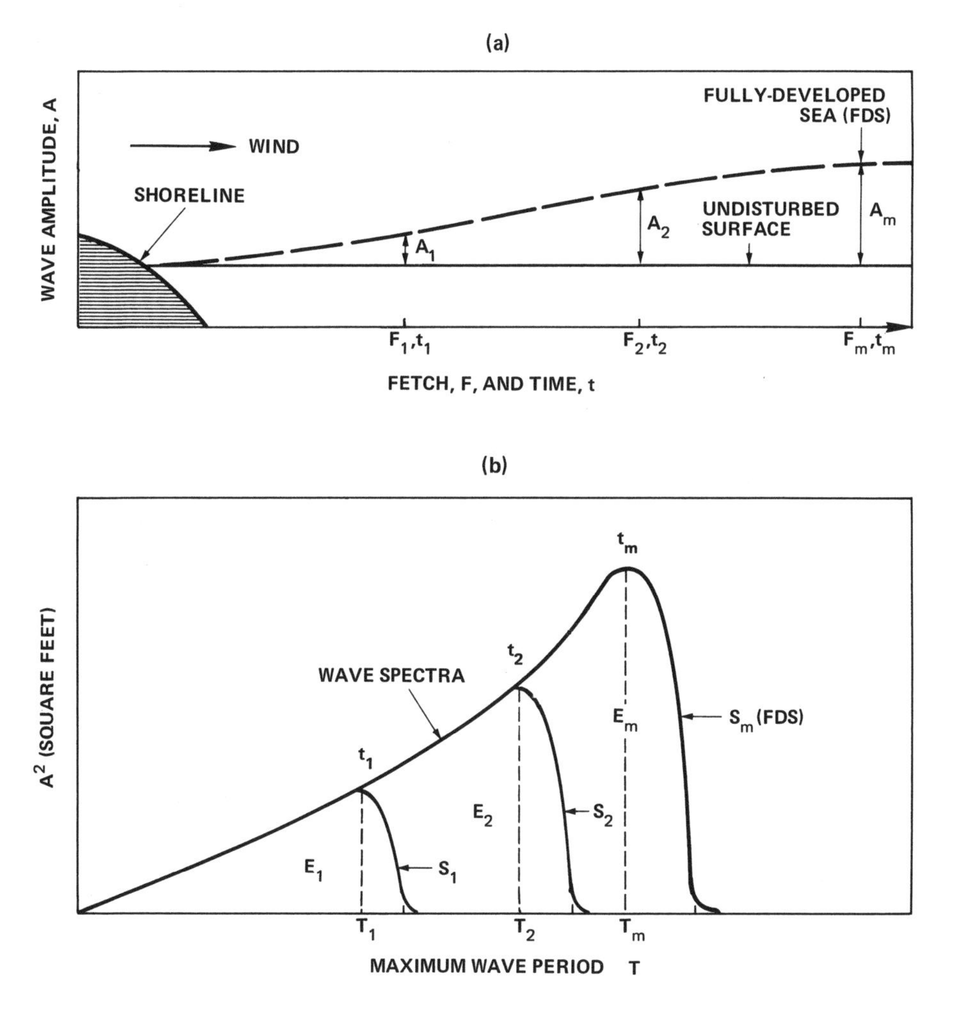

Fig. 74. (*a*) Growth of wave amplitude as a function of fetch, *F*, and time, *t*, under a steady wind. An FDS state occurs beyond F_m after a time, t_m (*b*). Growth of a wave spectrum at corresponding fetches and times.

near some maximum period indicates that there are few waves of longer period present at that particular fetch. The cutoff phenomenon is an observational fact of nature to which the forecasting theory has been adapted and, for fetches less than F_m, is attributed to the faster growth of small waves. The cutoff T_m at the minimum FDS fetch F_m—and the further observation that all spectra are constant beyond F_m—is attributed to the fact that waves of this period travel as fast as the wind, and their subsequent growth is thereby arrested. However, careful studies indicate that there actually are some longer (but lower) waves present in the spectrum of a fully developed sea, but their contribution to the total wave energy present is negligible for all practical purposes.

Should the wind cease at any time with $t < t_m$, or the available fetch be less than F_m (as, say, when restricted by a lee shore), the equilibrium sea state is said to be *duration-limited* or *fetch-limited,* respectively. Waves will develop and grow offshore as before, but their growth will be arrested at some distance or time, governed by whichever factor is limiting. If fetch-limited, the effect is merely to chop off the FDS wave spectrum at a period appropriate to the available fetch, beyond which a steady state will obtain. If duration-limited, the spectrum will be chopped off in the same way—but not necessarily at the same distance—at the moment the wind dies, and will begin to shrink again as the waves pass out of the area of generation, or are dissipated by viscosity or breaking.

To a reasonable approximation, the spectral curves of figure 74b can also be construed as giving the time history of sea state development at a fixed position. That is, curve S_1 is the spectrum for the equilibrium sea state prevailing at time t_1 at all points downwind of F_1, etc. Thus, with a steady wind of duration $t > t_m$, sea state growth at any distance $F > F_m$ will be defined by successive spectra $S_1, S_2, \ldots, S_m$ at corresponding times $t_1, t_2, \ldots, t_m$, as indicated. This important concept is the basis for our later discussion of sea state growth.

Properties of a Fully Developed Sea: Because it defines the upper limit of wave growth and the maximum range of wave periods to be expected in a steady wind field, the FDS spectrum S_m provides a convenient estimate of maximum sea conditions wherever the fetch is unlimited and the average wind can be forecast (as, for example, in the trade wind belts). This is because its relevant parameters are all simple functions of wind speed, V (knots):

$\sqrt{E_m} = 0.0068V^2$ feet = maximum sea state at fetches greater than F_m
$F_m = 3.65V^{4/3}$ miles = minimum fetch to establish an FDS*
$t_m = 6.43V^{1/3}$ hours = minimum wind duration to establish an FDS
$T_m = 0.38V$ seconds = period associated with highest waves in spectrum
$T_a^* = 0.29V$ seconds = average observable wave period
$L_a^* = 3.4T_a^{*2} = 0.28V$ feet = average wavelength in FDS spectrum.

While the theoretical FDS spectrum contains all ideal periods from zero to infinity, the waves near its ends are so small that, for practical purposes, only periods within the range $4 < T < T_m$ seconds need be considered.

The reader need not be concerned about evaluating the fractional powers of V in these equations; all of them (except L_a^*, which would not fit nicely on the same graph) are plotted for convenient reference in figure 75. For example, for a steady wind of 30 knots, the small arrows indicate the respective FDS values: $\sqrt{E_m} = 6.0$ feet, $F_m = 340$ miles, $t_m = 20$ hours, $T_m = 11.2$ seconds, and $T_a^* = 8.4$ seconds.

If the wind persists for at least 20 hours, and the available fetch is at least 340 miles, we can use figures 72 and 73 to make the following maximal sea state forecast:

$H_f = 1.41 \times 6.0 = 8.5$ feet = most frequent (most probable) wave height
$H_a = 1.77 \times 6.0 = 11$ feet = average height of all waves
$H_3 = 2.83 \times 6.0 = 17$ feet = significant wave height
$H_{10} = 3.60 \times 6.0 = 22$ feet = average height of the highest 10 percent of all waves

From figure 72, we can also determine, say, that:

10 percent of all waves will be higher than $3 \times 6.0 = 18$ feet
10 percent will be lower than about $0.6 \times 6.0 = 3.6$ feet
90 percent will be between 3.6 and 18 feet high

*If a background sea exists before the wind starts to blow, t_m may be considerably shorter (see p. 199).

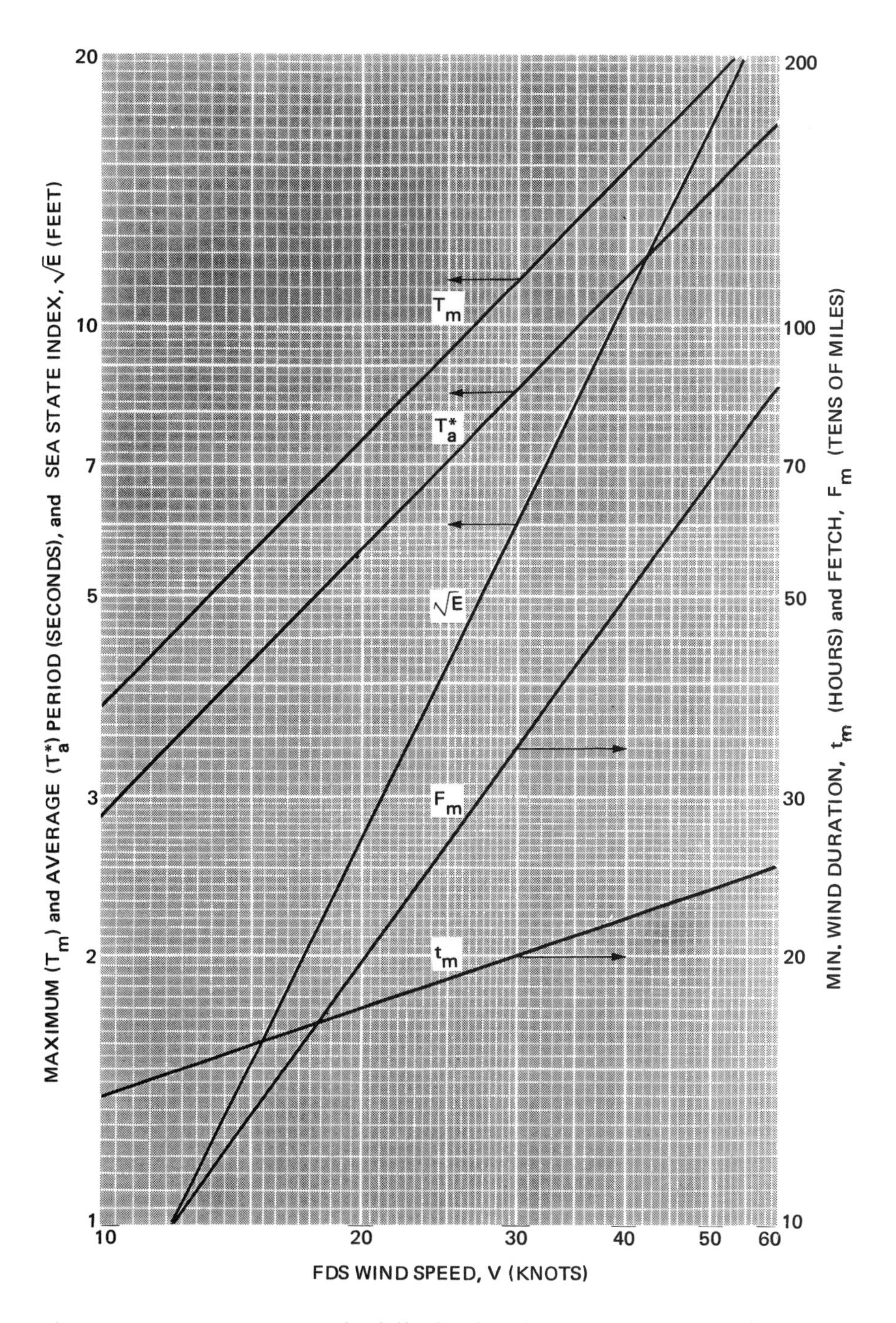

Fig. 75. Spectrum parameters for fully developed (FDS) sea states. Small arrows show sample values for a steady wind V = 30 knots.

From figure 73, and with T_a^* = 8.4 sec., there will be a 5 percent chance of encountering a single wave higher than 5.8 × 6.0 = 35 feet among every 200 waves that pass during an average interval of 200 × 8.4/3,600 = 0.47 hr. Conversely, within 5 hours, about 5 × 3,600/8.4 = 2,142 waves will pass, among which there will be a 5 percent expectancy of a single wave higher than 6.6 × 6.0 = 40 feet, or nearly four times the average wave height! Thus we see that there is a small—but real—chance of encountering a really big wave every few hours, even in a moderate sea.

Figure 75 indicates how rapidly the various FDS wave parameters increase with wind speed. Given enough time and sea room, any wave height index (H_a, H_{10}, etc.) increases as V^2; i.e., they will all be 36 times higher at $V = 60$ knots than at $V = 10$ knots. But total wave energy increases as the fourth power of wind speed, and will be *1,300* times higher at 60 knots than at 10 knots! Comparative values for other parameters are given in table 3, from which we can see that the minimum duration t_m to reach FDS conditions increases by only 80 percent, while the minimum fetch required is about eleven times larger.

Table 3. Comparative FDS wave parameters at 10 and 60 knots

V (knots)	t_m (hrs.)	F_m (mi.)	T_m (sec.)	T_a^* (sec.)	L_a^* (ft.)
10	14	80	3.7	2.9	28
60	25	870	22	17	1,000

The spectacular growth of the FDS spectrum, even between wind speeds of 20 and 40 knots, is illustrated in figure 76. The spectrum for 10 knots would be invisible at this scale, and that for 50 knots would be 44 percent larger than that for 40 knots.

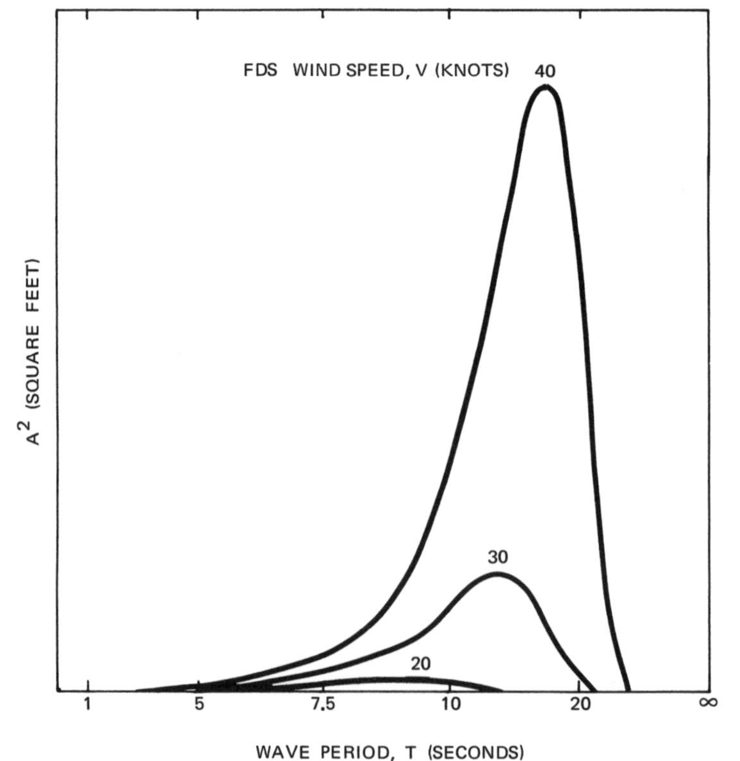

Fig. 76. The height of an FDS spectrum increases as the square of average wind speed. The area beneath it is proportional to the total energy of all waves present, and increases as the fourth power of wind speed.

At this point, we digress from statistics for the moment to consider the physical appearance of the sea surface at various stages of development. Probably the most familiar description of sea state at successively higher wind speeds is given by the misnamed* Beaufort Scale of Wind Force (table 4). The Beaufort Scale is concerned mainly with apparent wave height, and the relative prevalence of breakers, whitecaps, spray, and foam streaks. The test of its reliability is how consistently these features correlate with wind speed alone. Although we still have no quantitative data on the incidence of wave breaking in random storm seas, it should be evident from our previous discussion that only FDS conditions are reproducible functions of wind speed. With a limited fetch and high winds, the waves can be just as high (or higher) than in an FDS state at much lower wind speeds. But the fetch-limited, high-wind spectrum would have a narrower range of ideal wave periods. Hence the waves would be steeper and more likely to white-cap or break, and would be accorded a higher Beaufort rating—even though sea conditions might be more navigable. The same reasoning applies to a high-wind, duration-limited sea state, but in this case conditions will change so rapidly that one's Beaufort estimate would depend upon just when the observations were made.

Despite the ambiguities of the Beaufort Scale, there appears to be no simpler alternative description of sea state, unless having a fishing smack available, one uses it in reverse: when heeling under full sail, it is Beaufort 4; when forced to double-reef, Beaufort 6, etc.

Wave Decay: The decay of sea state from any stage of development is a balance between two factors: the rate at which the wind dies, and that rate at which wave energy leaks out of the generating fetch. In the open sea, most fetches are limited to a finite region by the extent of the wind field. Even under equilibrium conditions with a steady wind, wave energy is continuously radiating out of the fetch in the form of a *swell.* Because waves are not generated precisely in the direction of the wind, there is always some wave energy leaking out the sides of the fetch. However, close downwind, the swell spectrum resembles that within the fetch. At increasing angles off the wind, the spectra are progressively abbreviated in period and amplitude, and less than 10 percent of the total swell energy is radiated at angles greater than ± 50° from the wind direction.

With increasing distance from a stationary fetch, the swell spectrum in any direction within the 100° radiation sector will be progressively whittled away by the combined processes of dispersion, geometric spreading, and dissipation. Recall (Part IV) that dispersion is the sorting out of ideal wave components according to period, since each period travels at its own group velocity: $C_g = 1.5T$ knots. At any time or place outside the fetch, the local wave spectrum contains only those wave components that were initially generated in that direction, and which happen to arrive simultaneously at that point. Longer periods may have already passed—and shorter ones not yet arrived. Thus dispersion acts as a filter that selectively removes certain periods, the more so with increasing angle from the wind direction. Geometric spreading (fig. 58, p. 149) is just the average reduction of wave height with distance

*The Beaufort Scale increases roughly as $V^{2/3}$. Wind force, however measured, increases as the square of wind speed, as do all FDS wave height indices.

Table 4. Beaufort Scale of Wind Force

Beaufort no.	Sea Miles per Hour (Knots)	Seaman's Description	Effect at Sea	Action of Fishing Smack
0	< 1	Calm	Sea like a mirror	Makes no headway.
1	1–3	Light air	Ripples with the appearance of a scale are formed but without foam crests.	Just has headway.
2	4–6	Light breeze	Small wavelets, still short but more pronounced; crests have a glassy appearance and do not break.	Wind fills sails; makes up to 2 knots.
3	7–10	Gentle breeze	Large wavelets. Crests begin to break. Foam of glassy appearance. Perhaps scattered white horses.	Heels slightly under full canvas; makes up to 3 knots.
4	11–16	Moderate breeze	Small waves, becoming longer; fairly frequent white horses.	Good working breeze; under all sail, heels considerably.
5	17–21	Fresh breeze	Moderate waves, taking a more pronounced long form; many white horses are formed. (Chance of some spray.)	Shortens sail.
6	22–27	Strong breeze	Large waves begin to form; the white foam crests are more extensive everywhere. (Probably some spray.)	Double-reefs mainsail.
7	28–33	Moderate gale (high wind)	Sea heaps up and white foam from breaking waves begins to be blown in streaks along the direction of the wind. Spindrift begins.	Remains in harbor, or if at sea, lies to.
8	34–40	Fresh gale	Moderately high waves of greater length; edges of crests break into spindrift. The foam is blown in well-marked streaks along the direction of the wind.	Takes shelter if possible.
9	41–47	Strong gale	High waves. Dense streaks of foam along the direction of the wind. Sea begins to roll. Spray may affect visibility.	
10	48–55	Whole gale	Very high waves with long overhanging crests. The resulting foam in great patches is blown in dense white streaks along the direction of the wind. On the whole the surface of the sea takes a white appearance, The rolling of the sea becomes heavy and shocklike. Visibility is affected.	
11	56–66	Storm	Exceptionally high waves. (Small and medium-sized ships might for a long time be lost to view behind the waves.) The sea is completely covered with long white patches of foam lying along the direction of the wind. Everywhere the edges of the wave crests are blown into froth. Visibility affected.	
12	Above 66	Hurricane	The air is filled with foam and spray. Sea completely white with driving spray; visibility very seriously affected.	

as energy spreads out over larger ocean areas. Dissipation results from viscous and surface film effects, and rapidly eliminates wave periods of less than 2–3 seconds.

Figure 77 is a somewhat idealized picture of how these processes alter the initially oblong directional spectrum of a storm fetch. Their net effect is to truncate and attenuate all spectra, such that, with increasing distance, they more and more resemble isolated groups of ideal periodic waves traveling radially outward. Somewhat like a river issuing out of a narrow gorge onto a flat floodplain, at distances greater than several fetch lengths, the spreading "fan" of wave energy will have dwindled to a diffuse beam, centered on the wind direction within the fetch. This is the familiar ground swell from a distant storm that slowly decreases in height and period as the storm dies or passes by.

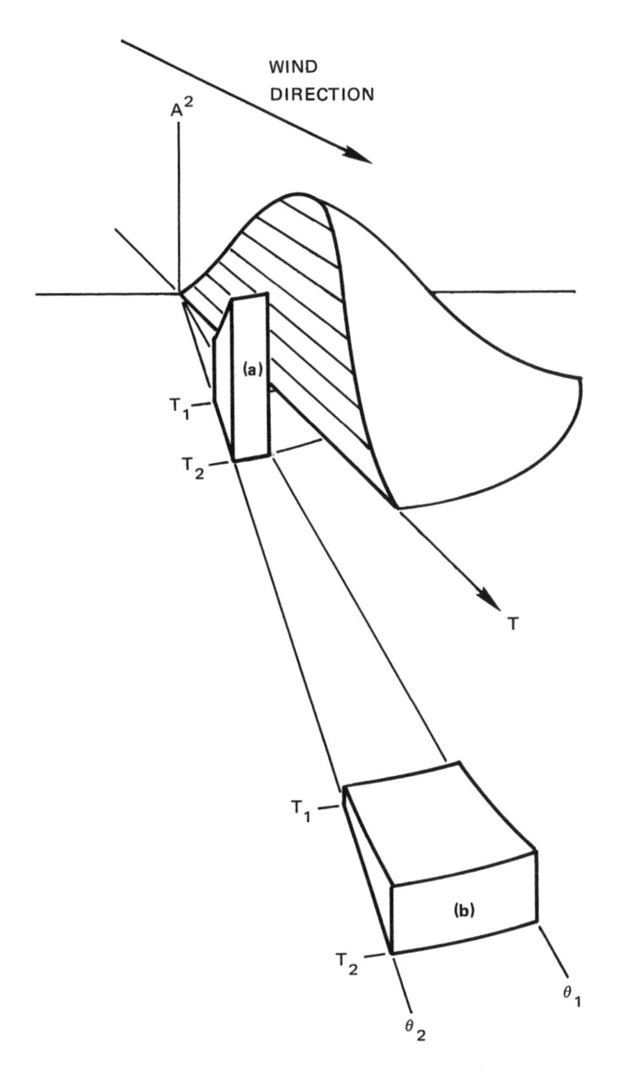

Fig. 77. The bump in this figure represents the directional energy spectrum of a storm at some instant. Its cross section is the nondirectional spectrum aligned with the wind. An energy packet (*a*) moving at some angle to the wind has a truncated spectrum, containing only component waves that happen to be moving in the same direction at that instant. At a later time (*b*), the packet contains the same energy, but lower waves, having been spread out over a larger area of the sea surface.

Because the processes that determine the rate of swell decay outside a storm fetch cannot be evaluated without more weather information than is normally contained in shipboard forecasts, there is no point in further elaboration, although we shall later point out how swell observations can provide information on movements of a distant storm. However, if we know the wind history within a fetch, we can easily estimate how long it will take for the waves to die down after the wind stops. As a first approximation, we can ignore directional spreading, and suppose that all waves are traveling in the direction of the wind. Spreading will only be important near the edge of the fetch, and will cause the waves to die a little faster.

Figure 78 illustrates three steps in the dispersive decay process within a fetch over which the wind has previously been blowing toward the right for a duration t_m, and where the sea state is everywhere in equilibrium. In the top figure (a), the heavy curve shows the variation in the sea state ($\sqrt{E}$) with distance from the upwind end of the previous fetch at the time the wind stops ($t = 0$). As in figure 74 (p. 179), at X_1 the local wave spectrum is fetch-limited to a sea state of $\sqrt{E_1}$ ft. and a maximum wave period of T_1 sec. The waves at X_1 are not all the same size, but their heights are randomly distributed in accordance with the probability curve of figure 72 (p. 175) their average height will be $H_a = 1.77\sqrt{E_1}$, and 10 percent of them will have heights greater than $3\sqrt{E_1}$, etc. The spectrum contains all ideal wave periods up to—but not larger than—T_1. Similarly, at X_2, the spectrum contains larger and longer waves up to the limits defined by $\sqrt{E_2}$, T_2. Between $X_3 = F_m$, the minimum fetch for FDS conditions, and X_4, the end of the fetch, sea state is fully developed, and the spectral limits $\sqrt{E_m}$ and T_m are everywhere uniform. Lastly, the descending portion of the

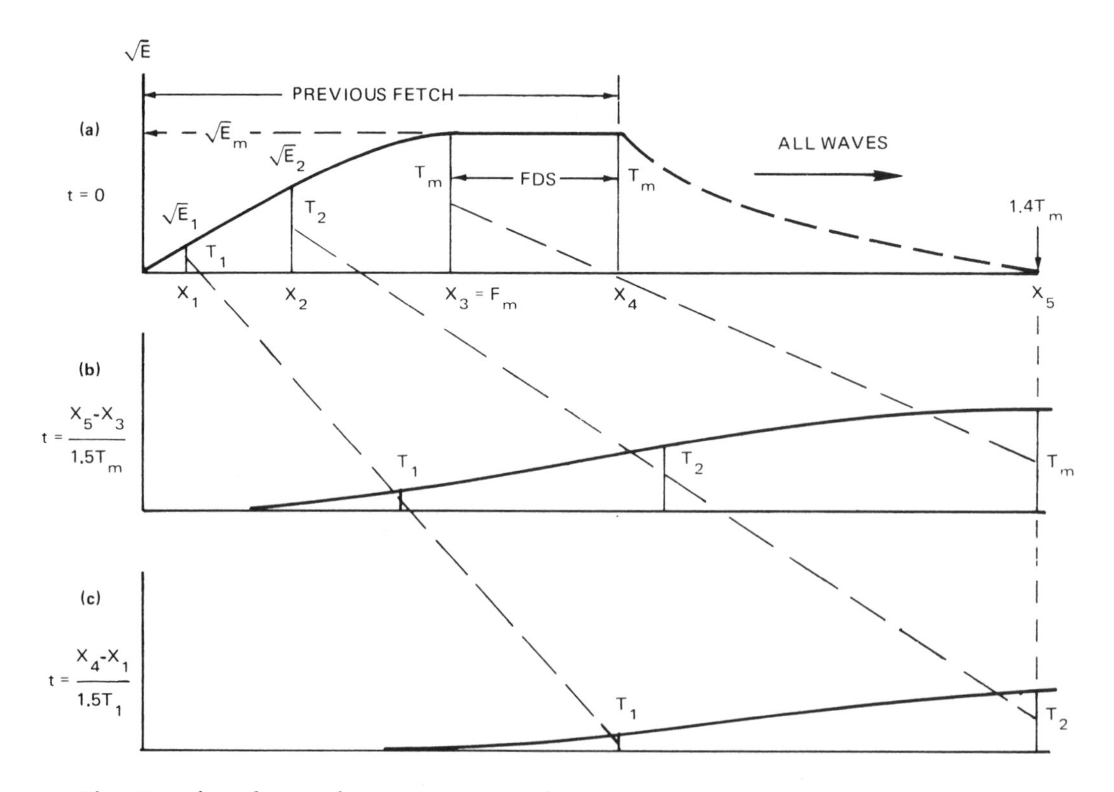

Fig. 78. After the wind stops, sea state decays at rates proportional to travel times of component periods, each of which propagates at group velocity $C_g = 1.5T$.

wave distribution curve beyond X_4 represents swell that has leaked out of the fetch while the wind was blowing. The wave height distribution in this region cannot be estimated without considering the storm's previous history as well as the effect of geometric spreading, which has been neglected here. However, we can estimate the greatest distance X_5 that could have been reached by the longest possible swell at time $t = 0$, from a storm that began t_m hours earlier. Recall (fig. 74, p. 179) that T_m is the period associated with the highest waves in an FDS spectrum. Some longer, lower waves will be generated, the longest of which will have a period of about $1.4T_m$ seconds, and will propagate at group velocity $C_g = 1.5 \times 1.4T_m = 2.1T_m$ knots. Thus, in t_m hours, these longest waves will have traveled a distance $X_5 - X_4 = C_g t_m = 2.1\ T_m t_m$ miles.

Returning to figure 78, as soon as the wind stops, waves will cease to be generated within the fetch, and those already present will travel to the right at their respective group velocities. Sea state will begin to drop everywhere within the fetch. At points beyond X_4, sea state will first increase as the higher waves from the FDS region of the previous fetch arrive, and then decay again as their smaller successors pass by.

Figure 78b shows the wave height distribution when waves of period T_m, which were formerly at X_3, have reached X_5. They have traveled a distance $X_5 - X_3$ miles, at a speed of $1.5T_m$ knots, in a time $t = (X_5 - X_3)/1.5T_m$ hours. Meanwhile, periods T_2 and T_1 have moved to new positions, progressively lagging one another, and becoming farther separated, by virtue of their lower group velocities. Because of this dispersive process, the average wave height associated with any of these periods diminishes rapidly from its maximum initial value at $t = 0$. In figure 78c, period T_2 has reached X_5, and period T_1 has moved to X_4, the end of the previous fetch. For all practical purposes, wave activity in the fetch ceases at $t = (X_4 - X_1)/1.5T_1$ hours. Thus, the problem of wave decay reduces to estimating the distribution of wave heights and periods upwind when the wind stops, and then dividing travel distance by group velocity for the shortest period of interest. Such estimates require more information about sea state growth than the upper-limit FDS values given in figure 75 (p. 181). Many storms die aborning, without establishing equilibrium wave conditions. In fact, it is a moot point whether fully developed seas are ever attained at wind speeds much over 50 knots, because strong winds require longer fetches—if not durations—than ordinarily occur in nature.* Accordingly, in the following sections, we will take a closer look at how sea state grows with time, and how the above concepts can be fit together to yield useful sea state estimates.

SEA STATE AND BREAKER FORECASTING

Accurate forecasting requires sequential weather maps, from which surface winds are inferred and fed into a computer, where they are digested and duly evacuated in the form of sequential sea state maps. However, before computers appeared and began gobbling everything up, a number of fairly successful forecasting schemes were developed, some of which are still in general use. The following is a much-abridged version of one such scheme that I have tried to adapt to the needs and

*By "duration" we mean the length of time the wind blows over the area upwind of the observer. For a stationary fetch, it is the storm duration. For a moving storm, it is the time within which the fetch can generate waves that will ultimately pass the observer.

resources of the average yachtsman or small-vessel operator. Its primary purpose here is to illustrate how sea state grows with time under steady winds to an upper limit imposed by breaking instability. Its usefulness will depend on how well the wind field can be estimated, and can be approximated by one or more uniform wind fetches. Because swell forecasting is more difficult, and only of secondary interest, we shall consider only areas directly affected by the wind, where geometric spreading can be neglected. Although our forecast model is designed for steady winds, we will give examples to show how it can be fudged to give estimates for varying winds and moving fetches.

The CSS Diagram: Where there is no fetch restriction, the growth of sea state under a steady wind can be described as the successive addition to the wave spectrum of ideal wave components having progressively higher amplitudes and longer periods. In an unlimited fetch, the growth rate is everywhere the same, and can be represented as a series of nested spectral curves, S_1, S_2, . . . (fig. 74), at times t_1, t_2, ... The respective areas, E_1, E_2, . . . , beneath them are measures of total wave energy at these times. Sea state growth is most rapid at early times, when the relative wind speed (V-C) is greatest, and progressively slows down because newer, longer-period waves travel faster and the relative wind decreases. Ultimately, at time t_m, the longest waves begin to outrun the wind, and growth stops. The spectrum S_m is fully developed, it has a total energy E_m, and the highest waves have periods near T_m. There are a few longer, lower waves with periods as long as $1.4T_m$.

At any higher wind speed, sea state grows more rapidly, and has more energy at corresponding times. However, it takes somewhat longer to establish FDS conditions, because the longest wave grows slowest, and must catch up with a higher wind speed.

Evidently, if we were to try to describe wave growth in terms of these spectral changes, we would need a complete set of spectral curves for every wind speed. But things can be greatly simplified by constructing a single *cumulative energy curve* for each wind speed, whose height, at any time, is a measure of the total wave energy (E) present in the growing spectrum. Instead of E, if we make the curve height proportional to $\sqrt{E}$, it becomes a *cumulative sea state* (CSS) curve that indicates how sea state increases with time.

With this preamble, figure 79 shows our generalized forecasting scheme for wave growth under steady winds, which I shall call a CSS diagram. It contains a set of CSS curves, a set of period curves, and a set of fetch lines, all of which are described in detail below. Although it may look complicated at first sight, it is very simple to use. For any wind speed between 20 and 60 knots, the instantaneous sea state can be read at a glance, together with the range of periods present in the local wave spectrum. With a little manipulation, we can also use the diagram to calculate wave decay after the wind stops, and to estimate the effects of varying winds and moving fetches. The CSS diagram is probably the most important single illustration in this book. I shall refer to it repeatedly, not only as regards wave forecasting, but also when discussing ship behavior, seamanship, and anchoring techniques.

Consider first the nine (heavy) CSS curves arising from the lower left corner of the diagram and spreading upward to the right. Each curve is identified by a particular wind speed, V, which increases by 5-knot increments from 20 to 60 knots. Lower winds never produce dangerous seas, and 60 knots is a reasonable upper limit

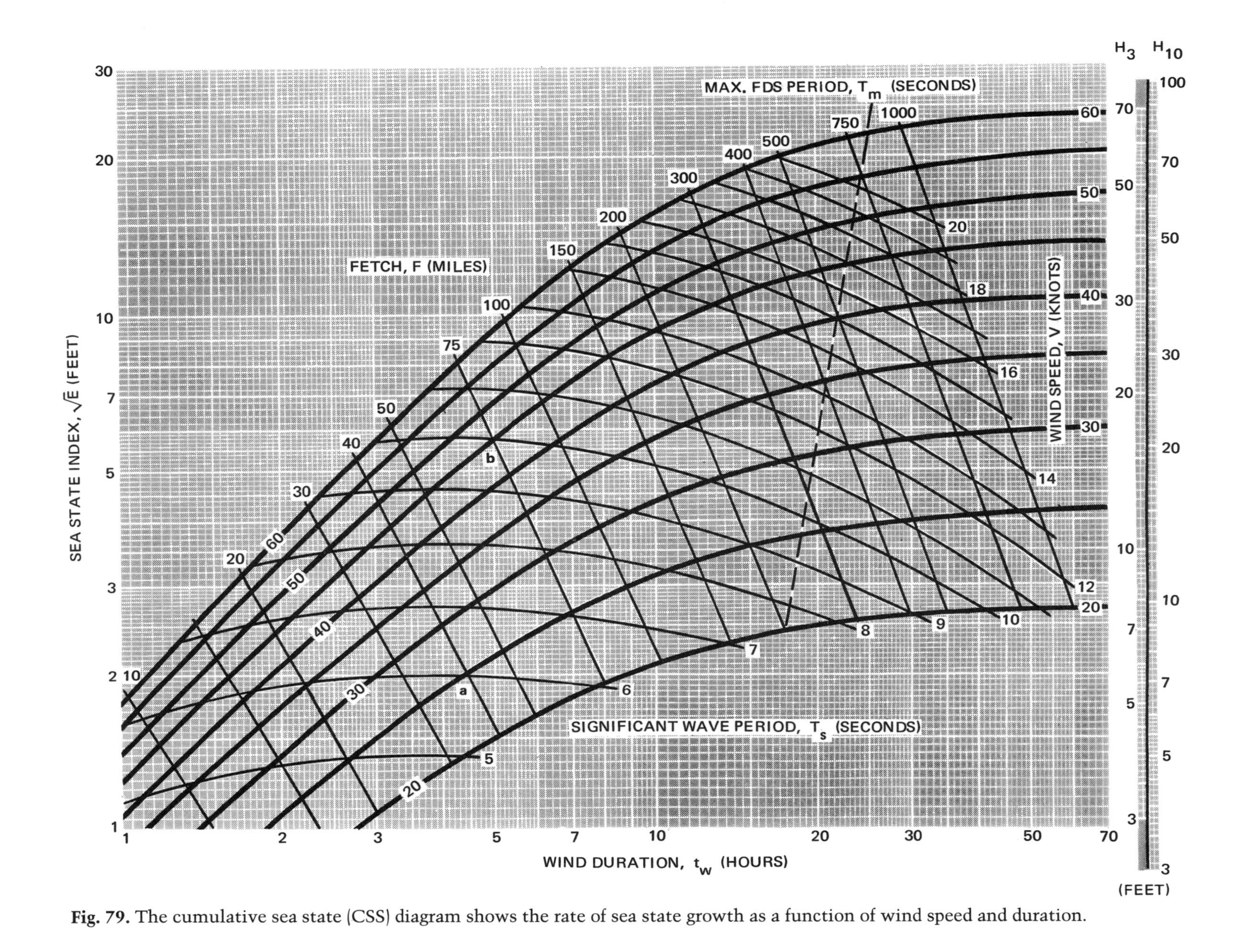

Fig. 79. The cumulative sea state (CSS) diagram shows the rate of sea state growth as a function of wind speed and duration.

for sustained winds in large storms.* Any point on one of these curves (say, *a* or *b*) represents the sea state $\sqrt{E}$ (read horizontally to the left) that would be raised everywhere on an infinite, and initially calm, ocean after a steady wind of that speed had blown over it for a duration t_w (read vertically beneath). The corresponding significant (H_3) and maximum (H_{10}) wave heights are indicated on separate scales to the right. From the general shape of these curves, it can be seen that, at all wind speeds, sea state rises fastest during the first 6 hours or so, and then much more slowly, ultimately approaching the equilibrium FDS values given by the line labeled $\sqrt{E_m}$ in figure 75.

Consider next the light curves labeled *Significant Wave Period,* T_s. Their intersections with the CSS curves represent upper limits to the important wave periods present in the sea state spectra at times read vertically beneath these intersections. Points *a* and *b* happen to fall exactly on the curves for $T_s = 6$ and $T_s = 10$ seconds, respectively, but intermediate points can be interpolated where necessary. Although any such spectrum theoretically contains all *shorter* periods, the lower limit of our diagram corresponds to a sea state of $\sqrt{E} = 1$ foot, for which the significant period is a little less than $T_s = 5$ seconds at all wind speeds. Shorter-period waves do not contribute much energy, and the latter period will be taken as the practical minimum for decay calculations. The average period present can be taken as $T_a^* = 0.75\,T_s$.

The period curves generally increase upward with increasing wind speed or duration, quantitatively illustrating how sea state grows by the addition of higher waves and longer periods to the spectrum. Note, however, that the *same* sea state can be attained by many combinations of wind speed and duration, although the range of periods present can vary considerably. At low sea states, this variation is small, because the period curves are nearly horizontal; at high speeds, they slope steeply, and the range variation is large. For example, a sea state $\sqrt{E} = 9$ feet will result from a 60-knot wind blowing for about 4.8 hours, and the spectrum will be limited to periods between 5 and 12 seconds; i.e., a short, steep sea. The same state can be produced by a 37-knot wind in 40 hours, but the longest period present has increased to 17 seconds. While the range of wave heights is the same in both cases, the latter sea is fully developed and has over twice the average wavelength. Lastly, note that $T_s = 20$ seconds is the longest significant period shown on the diagram, requiring, say, a minimum wind speed of 50 knots for 30 hours. Longer periods are meteorologically unlikely, owing to the very long fetch needed (see below).**

So far, we have discussed only uniform winds blowing over an unlimited ocean. As previously mentioned, most large storms are fetch-limited, either by land to windward, or by the finite area over which strong winds blow. The concept of fetch itself is somewhat elusive. We know that, in real cyclonic storms, the wind is

*These curves do not necessarily apply to hurricanes, for which there are still no reliable correlations between wind and sea state. Furthermore, the curves above 45 knots are extrapolations of a theoretical model that have not been experimentally verified, although they are internally consistent and agree fairly well with other forecast models. I am indebted to Dr. W. J. Pierson, New York University, for supplying the spectra from which the curves were compiled. These spectra are based on the work of Dr. Tokojiro Inuoe. Since the first edition of this book (1974), several new spectral models have appeared, differing mainly in how rapidly sea state increases. However, the present model is still quite adequate for present purposes.

** The period of swell radiating from a storm fetch increases in proportion to the cube root of travel distance. Typhoon swell periods as long as 24 seconds have been observed in Japan.

strongest near the center and tapers off toward the edges. Nevertheless, in any simple forecasting model, we are forced into the idealization of somehow approximating a real storm by one or more boxlike fetches, within which the wind is uniform, and outside of which it drops to zero. With this assumption, the limitations on sea state growth imposed by finite fetches are shown by the straight lines in figure 79. The intersection of any fetch line with a CSS curve indicates the minimum fetch required for sea state to grow to the point otherwise forecast for that wind speed and duration. If the actual fetch happened to be shorter than this minimum distance, sea state would grow with time, as indicated by that CSS curve, until it reached an intersection with the actual fetch line. It would then stop growing, and remain at this level until the wind ceased.

For example, in the CSS diagram, point *a* represents a sea state forecast for a 45-knot wind that has blown for 5 hours: $\sqrt{E}$ = 5.8 feet, H_3 = 17 feet, H_{10} = 21 feet, and T_s = 10 seconds. This point is also crossed by the 75-mile fetch line. If the actual fetch were 75 miles or longer, our forecast would be correct. Suppose, however, that a shoreline existed only 40 miles upwind. Sea state growth would then be arrested after 3.2 hours (read vertically beneath the 45-knot, 40-mile intersection), and the fetch-limited forecast would be $\sqrt{E}$ = 3.5 feet, H_3 = 11 feet, H_{10} = 15 feet, and T_s = 8.3 seconds. Conditions would remain unchanged for another 1.8 hours (t_w = 5 hrs.), and the waves would then begin to decay as described on page 186.

Thus, the influence of a finite wind fetch can be summarized by the following general rule: if the appropriate forecast fetch line crosses the forecast CSS curve *earlier* than the forecast duration, sea state growth will be arrested at the duration corresponding to the fetch intersection, and remain at that level for the remainder of the forecast duration.

Lastly, consider the dashed line labeled *Maximum FDS Period, T_m,* on the CSS diagram. Because the CSS curves approach their upper FDS states so gradually, they are difficult to define timewise. Their intersections with the T_m line indicate wind durations $t_w = t_m$, where sea state has risen to 90 percent of its FDS levels. Other intersections with the period curves define the periods $T_w = T_m$ of the highest waves in the respective FDS spectra. Periods as long as $1.4T_m$ may exist, but the waves will be lower. Similarly, T_m intersections with the fetch lines give the minimum fetches F_m necessary to achieve FDS states. Because fetches longer than 1,000 miles are meteorologically unlikely, all period curves are terminated at this fetch line. The respective FDS values T_m, t_m, and F_m indicated by these various intersections are the same as those obtainable from figure 75 at corresponding wind speeds.

Interpretation of the CSS Diagram: The CSS diagram condenses a great deal of information into a few curves. We now consider a few examples that illustrate its use and the range of sea conditions encompassed. To begin with, recall that any forecast point on this diagram represents an instantaneous sea state estimate, $\sqrt{E}$, where E is the total energy present in the local wave spectrum. The actual waves present at that moment* have heights that are randomly distributed in space and time, according to the average and extreme probability curves of figures 72 and 73, respectively. The wave periods are less random, and are roughly ordered according to height; that is, the higher waves tend to have longer periods. On the average, the

*We suppose that conditions remain constant long enough for statistics to apply.

highest waves will have periods close to those given by the T_s curve that passes through the forecast point.

Let us take a look at some extreme values. The bottom margin of our diagram corresponds to a sea state $\sqrt{E}$ = 1 foot, for which the FDS wind speed is about 12 knots (fig. 75)—just a nice sailing breeze for your Sabot. The right-hand scales on the diagram show that the average (significant) height of the highest one-third of all waves around you is H_3 = 2.9 feet; the highest 10 percent average H_{10} = 3.6 feet, occasionally brushing the horizon from where you sit at the tiller. The maximum period of these highest waves is about $T_s = T_m$ = 4.5 seconds, and the average period of all waves $T_a^* = 0.75\,T_s$ = 3.4 seconds. Every 11 minutes or so, about 200 waves will have passed (you are beam-reaching in the trough), among which the most probable height of the highest single wave is almost 4.9 feet (fig. 73). However, there is a 5 percent chance that it will exceed 5.8 feet, or be lower than 4.1 feet. About once an hour, you may encounter a wave as high as 6.3 feet (N = 3,600/3.4 = 1,060 waves). Although it took 15 hours for the 12-knot wind to raise this fully developed sea (fig. 75), the CSS diagram shows that similar conditions could have been reached in 2.8 hours with a 20-knot wind, in 1 hour at 40 knots or in 20 minutes at 60 knots! Sea state notwithstanding, higher winds would also produce heavy breaking and flying spray, making things pretty wet and uncomfortable.

Moving now to the upper right corner of the diagram, we can as easily determine what wave conditions to expect in a major storm. Assume that you are unlucky enough to get caught at sea in the worst credible situation shown: a 60-knot wind that has blown steadily for 16.5 hours over a fetch of at least 500 miles ($\sqrt{E}$ = 20 ft.)—not quite an FDS state, but bad enough. You cannot see much for the solid sheets of spray, but if you can relax your handhold long enough to snatch a glance, the diagram indicates that 33 percent of the seas hurtling by average 58 feet high; 10 percent average 72 feet. The average interval between consecutive waves is about 14 seconds, but that separating the larger waves is more like 20 seconds. According to figure 73, every hour or so 260 waves will pass by, the largest of which might exceed 5.85×20 = 117 feet! If this sounds incredible, there is at least one well-authenticated observation of a 112-foot wave under similar circumstances.*

Probably the most reliable estimate of extreme swell conditions *outside* a storm area was provided by the experience of the Scripps Institution's vessel, *FLIP* (Floating Laboratory Instrument Platform), during the great north Pacific storm of December 1–5, 1969. This unusual vessel has a tubular steel hull 350 feet long and 20 feet in diameter, with a ship's bow section appended (plate 14). She is designed to be towed from port horizontally, and erected to a vertical position in deep water by flooding her after ballast tanks, in which configuration she draws 294 feet of water and her bow is 56 feet above sea level. When erect, her 28-second heaving period is

*On February 7, 1933, the 478-foot Navy tanker, U.S.S. *Ramapo,* was running off eastward in a seven-day north Pacific storm that assertedly extended from Kamchatka to New York, with relative winds between 58 and 68 knots. The cited observation (one of several) was obtained by triangulation from a line of sight on the bridge that simultaneously intersected the crow's nest, the crest of the following wave, and the horizon, the ship's stern being then in the trough, and her bow elevated on the slope of the preceding wave. The *Ramapo* had experienced 59-knot winds for over 24 hours, and over 60-knot winds for an additional 10 hours, so we can estimate that sea state roughly coincided with the worst credible case considered above, for which there is reasonable expectancy of such a wave every hour or so. (*See* Whitemarsh, R. P., U.S. Naval Institute *Proceedings* (60), 1934.)

Plate 14. Two views of the Scripps Institution's research vessel, R/V FLIP. When erect, her specifications are: LOA = LWL = beam = 12 feet, draft 300 feet, speed under mizzen only about 0.002 knots.

much longer than that of the longest ocean swell, so that she drifts virtually immobile in any normal sea.

During the above period, she was positioned about 500 miles due north of Oahu, and about 900 miles southeast of the storm center, which comprised a 1,000-mile fetch,* with winds between 40 and 50 knots. About noon, December 1, a giant swell arose from NNW, and her crew was forced to abandon her by life raft when green water entered ventilators 28 feet above her waterline and short-circuited

*This extreme fetch is attributed to the coalescence of two major depressions. Breaking waves as high as 55 feet caused over a million dollars' damage in Hawaii, and heavily eroded the air strip at the Coast Guard Loran station on French Frigate Shoals.

her generators. Visual observations and motion pictures subsequently established that she was occasionally inundated to a maximum of 44 feet above her waterline, and exposed an equal distance below it. Taking her slight (6 inch) heaving motions into account, the considered sea state estimate over 12 hours was $\sqrt{E}$ = 17.3 feet, H_3 = 49 feet, T_m = 16–20 seconds, with an hourly extreme wave height of 80 feet. These figures are in accord with the CSS diagram prediction for a 1,000-mile FDS fetch in 50-knot winds.

The CSS diagram can also be used to make travel-time calculations for estimating wave decay (see page 187). In fact, any triple intersection of a period curve, a duration line, and a fetch line is just such a calculation. For example, at point *b*, we have T_s = 10 seconds, t_w = 5 hours, and F = 75 miles. The group velocity (energy propagation speed) for a period of 10 seconds is $C_g = 1.5T_s$ = 15 knots. Thus, the time required for this period to travel 75 miles is $t = t_w = F/1.5T_s = 75/15 = 5$ hours. This result is perfectly general, and applies to *any* point on the diagram where the necessary values can be interpolated. Knowing the decay distance for any period, simply locate the intersection of the appropriate fetch line with that period curve, and read the decay time vertically below. In many cases, you may be interested in the time for all significant waves to pass a forecast point. If the intersection of the shortest significant period (T_s = 5 sec.) with the known fetch line is off the scale, the same result can be obtained by multiplying both values by 2 or by 3. For instance, the intersection of T_s = 5 seconds with F = 200 miles is off the scale, but that for T_s = 10 seconds and F = 400 miles is not, giving t = 27 hours.

Storm Wave Breaking: Although the CSS diagram tells us nothing about the percentage of all waves that are breaking at a given sea state, any breaking effects are automatically included in the CSS curves, which were derived from actual wave height measurements at sea. Nevertheless, the distinction between breaking and nonbreaking is important, since it may ultimately determine whether or not a vessel can survive. The annals of the sea contain innumerable accounts of damage wrought by breaker impact, which run the gamut from seams started and superstructure torn away to capsizing or pitchpoling a vessel—or breaking her clean in two. In nautical literature, you will frequently find it asserted that a well-found vessel, properly handled, can take anything the sea can throw at her. But statistics would argue otherwise—that among all worst credible storms, there is a certain expectancy of encountering a wave high enough to disable any vessel—*if the wave is breaking.* It is all a matter of probability, but one that is difficult to estimate.

Do the highest storm waves ever break? Anyone who has seen them knows the answer. Doubters should look at the remarkable photographs by de Lange, in Adlard Coles' *Heavy Weather Sailing,* showing 50- to 70-foot waves breaking as heavily as any shoreline comber. Your author was among the first to demonstrate on a laboratory scale that deep-water wave breaking—as opposed to incipient toppling of wave crests by the wind—results from instability brought about by constructive interference between two or more subcritical waves so as to produce one supercritical breaker.* It was found that breaking occurs when the resulting wave attains a

*In our wave channel, interference was produced by programming the wave paddle to generate a train of steep waves of slightly different periods (hence, speeds), so that the progressively later, faster waves overtake their forerunners and all come together simultaneously as a supercritical breaker. In nature, breaking also results from lateral interaction between crests traveling in different directions.

steepness $S = H/5T^2$ within the range $0.08 < S < 0.09$. In fact, the line T_m on the CSS diagram denoting the inception of an FDS sea state is the locus of points satisfying the condition that $H_{10}/5T_m^2 = 0.086$. Furthermore, it can be shown that because of their shorter periods the highest 10 percent of all waves to the left of this line also satisfy this condition, whereas those to the right, having longer periods, do not. Therefore the T_m line effectively divides the CSS diagram into breaking and nonbreaking regions for H_{10} waves. Because their respective heights are 27 percent lower, no H_3 waves exceed the above breaking criterion.

Secondly, the experiments showed that breaking intensity increases at the rate at which the subcritical waves come together. At maximum intensity, the composite breaker speed exceeds that for lower waves ($C = 3T$ knots) by about 10 percent, and the wave crest forms a plunging jet, whose speed is about 20 percent faster than the wave speed. Thus, for a hypothetical 40-foot breaker of composite period $T_c = 12$ seconds, the jet speed would be $V_j = 1.1 \times 1.2 \times 3 \times 1.69 \times T_c = 80$ feet per second, corresponding to a dynamic pressure of about *3.4 tons per square foot!* Such pressure can cause severe damage to the deck cargo or superstructure of a large ship.

Lastly, while the foamy turbulence associated with breaking persists for the order of the wave period, the duration of the plunging jet is much shorter. Specifically, it is about the time it takes for it to fall, say, half a wave height, $t_j = (H_{10}/g)$, or about 1 second for a 32-foot breaker.

In view of its importance to ship progress and survivability, it is somewhat surprising that the statistics of deep-water breaking have received so little scientific attention. Wave growth theories eschew the question, although theoretical arguments suggest that breaking is one of the factors controlling the wave height equilibrium of fully-developed sea states. There is ample evidence that the frequency and intensity of breaking increase with wind speed, but there is still no way to specify how many waves are breaking at any instant, or how they are distributed according to height and period. The offshore oil industry, which is very much concerned about design criteria for deep-water drilling rigs, is working assiduously on computer simulations of irregular seas. Perhaps the next edition of this book will have some concrete answers.

Meanwhile, I tender the following guesstimates of breaker distributions, based on nothing more than careful examination of several dozen excellent aerial photographs, taken at uniform height and camera angle, and covering a wide range of steady wind speeds. Plates 15 and 16 show representative surface conditions at 10 and 60 knots, respectively, from which the following observations can be made:

1. At 10 knots, less than 1 percent of the higher waves are breaking.
2. At 60 knots, almost 100 percent of the high waves are breaking.
3. In both cases, as in all other photos, breaking is largely confined to the larger waves. Small, random whitecaps are plentiful at all wind speeds above 10 knots, but there is no evidence of a continuous spectrum of breakers.

After several days spent in subjectively counting waves and breakers, I have tentatively concluded that the percentage of high waves breaking in all photos examined is best represented by the linear relation $P = 2V - 20$, $0 < V < 60$ knots.

For want of wave height and wind duration data, there is no reliable way to relate the above result to sea state. However, the photography was conducted by

Plate 15. Sea surface under 10-knot winds. Less than 1 percent of higher waves are breaking.

Plate 16. Sea surface under 60-knot winds. one hundred percent of higher waves are breaking.

flying multiple traverses across large storms, so that the pictures probably include a wide range of sea state development at many different wind speeds. Thus the chance that the above result applies to any sea state is better than the chance that is applies only to immature or fully developed states. On these (admittedly shaky) premises, we can use figures 72, 73, and 79 to estimate the probability of encountering breaking waves of a given height in a random storm sea.

Because the photographic breaker estimates were restricted to "larger" waves, we will assume that they apply only to the highest 10 percent of all waves present at any sea state. From figure 72, these waves all have heights greater than $3\sqrt{E}$, and their average height is $H_{10} = 3.6\sqrt{E}$ feet. For any CSS forecast, about 10 percent × $(2V - 20)$ percent = $(0.2V - 2)$ percent of all waves present may be breaking, on the average, and their heights will also be greater than $3\sqrt{E}$, etc. The instantaneous expectancy of encountering such breakers is just the reciprocal of their probability, multiplied by the average wave period, $T_a^* = 0.75\,T_s$: $t_b = 6.3\,T_s/(V-10)$ minutes. The expectancy of encountering such a breaker in the act of plunging is the breaking expectancy multiplied by the ratio of the average wave period to the plunge duration: $t_p = t_b \times T_a^* / t_j = t_s^2/(V - 10)\sqrt{H_{10}}$ minutes.

As representative examples of breaking wave estimates, pertinent data for points a and b on the CSS diagram (fig. 79) are listed below.

	Point a	Point b
Wind speed, V (knots)	25	45
Wind duration	4.4	5.0
Sea state, $\sqrt{E}$ (feet)	2.0	5.8
Significant wave period, T_s (seconds)	6.0	10
Average period, $T_a^* = 0.75\,T_s$ (seconds)	4.5	7.5
Average breaker height, H_{10} (feet)	7.2	21
Average breaking interval, t_b (minutes)	2.5	1.8
Average plunging interval, t_p (minutes)	22.3	15.8

Note that, for roughly the same wind duration, the 45-knot gale produces breakers about three times as high as the 25-knot breeze, and at about 40 percent shorter intervals. In both cases, the probability of encountering a plunging breaker is about six times smaller than that of being exposed to subsequent breaking turbulence.

Although both the above sea states are immature and growing rapidly, the example in figure 73 tells us that among 200 average periods there is a 5 percent probability of encountering one (rogue?) breaker about 5.8/3.6 = 1.6 times as high as the H_{10} breakers considered above, so long as the sea state has not increased appreciably. Use of the CSS diagram and figure 73 to estimate catastrophic ship motions is discussed in Chapter 23.

Wave Forecasting: To the average yachtsman it may appear that large ships are ordinarily unconcerned with the vagaries of sea state, sailing as they do on prescribed schedules and detouring only to avoid major storms. But as a glance at figure 95 on page 239 will show, in order to avoid adverse motions, ships voluntarily reduce their speed such that it varies inversely with wave height, correspondingly affecting running time and fuel consumption. Within the past few years, the combined technologies of satellite data transmission, computerized weather forecasting, and spectral wave prediction models now make possible real-time ship routing by shipping companies possessing the necessary shoreside and onboard facilities. Needless to say, the naval forces of western countries have been in the forefront of this effort, albeit with somewhat different motivation.

To skippers of small vessels lacking these facilities, but to whom sea state is an operational factor, the CSS diagram can be of considerable use in determining an appropriate course of action. To anyone with experience sailing in familiar waters, this section may seem somewhat redundant, but even such readers may find it enlightening to test their intuitive computer against the factory model and see how the two compare. I have fisherman friends who have spent most of their lives at sea, to whom cause and effect are still somewhat mysterious—although they can quite accurately predict forthcoming conditions from subtle environmental factors.

Despite its many other applications, use of the CSS diagram for operational forecasting is restricted to rather simple situations, where the wind history upwind of the forecast point(s) can be reasonably estimated. Unless you are a meteorologist and have access to current weather maps, you may have to depend on local weather forecasts. These have improved remarkably since presatellite days, and often include surface wind information. If they do not, local offices of the National Weather Service or the Federal Aviation Administration Flight Service will usually respond to telephone or radio inquiries. All we can do here is to outline a few representative forecasts that illustrate how to use the diagram, assuming that the necessary wind information is otherwise obtainable.

Wave forecasting hinges around energy bookkeeping, in which one uses the CSS diagram to keep track of $\sqrt{E}$ at some fixed position in the ocean. The prevailing sea state defines a starting point on the diagram, which subsequently moves about as conditions change. Because the diagram is designed for steady winds, whereas storm winds may vary considerably in speed and direction, we try to approximate the forecast wind by one or more "average" winds for an "average" duration.

The business of averaging is one of the most subjective parts of forecasting. The best guess is to use single forecast values of wind speed, direction, and duration verbatim, and to average multiple values. That is, if the forecaster predicts winds of 30–40 knots, for 12–18 hours, veering from south to west, use 35 knots for 15 hours from the southwest, etc. However, if the wind increases slowly by more than 10 knots, or shifts by more than 90°, you will do better to make consecutive forecasts and combine them, as shown in Example II, below. All forecasts proceed by the following steps:

1. Define the fetch relative to your forecast point. This is the region over which the average wind is blowing, minus any part cut off by an upwind land mass. Your forecast point is somewhere inside, and its distance to the upwind edge of the fetch (or land mass) is the observer's fetch, F_o. If the storm is moving, the fetch will move with it.
2. Determine the wind duration. This is either the forecast duration or the time required for the fetch to move past your forecast point, whichever is shorter.
3. On your CSS diagram, move up the average wind curve to the above duration line, or to the line corresponding to your observer's fetch, whichever is shorter. This is your forecast sea state.

4. Determine the decay time, as described on page 187, using $F = F_o$, and $T_s = 5$ seconds. Add this to the forecast duration to find the time when all significant waves have passed the forecast point.

Because of the many approximations involved, if you are in doubt as how best to estimate either fetch or duration, you should err on the high side. It is better to be prepared for severe conditions that do not materialize than to be caught by more than you bargained for.

***Example I.* Stationary fetch, steady wind.**
A steady wind of 35 knots for 24 hours is forecast over a 600-mile fetch.
Required: (a) three-hourly forecasts of significant and maximum wave height at a distance $F_o = 400$ miles from the upwind end of the fetch, and
(b) decay times for these waves to pass F_o after wind stops.

Table 5. Data for Example I: $V = 35$ knots, $t_W = 24$ hours, $F_W = 600$ miles, $F_o = 400$ miles

I t(hrs.)	*II* $\sqrt{E}$(ft.)	*III* H_3 (ft.)	*IV* H_{10} (ft.)	*V* T_s (sec.)	*VI* F(mi.)	*VII* $F_o - F$(mi.)	*VIII* t_d (hrs.)	*IX* t_o (hrs.)
0	0	0	0	0	0	400	–	–
3	2.5	7	9	6.7	30	370	37.0	58
6	4.2	12	15	8.8	80	320	24.0	42
9	5.5	16	20	10.2	140	260	17.0	32
12	6.2	18	23	11.2	200	200	12.0	24
15	6.8	20	25	12.0	260	140	7.8	17
18	7.2	21	26	13.0	330	70	3.6	10
21*	7.4	22	27	13.3	400	0	0	3
24	7.4	22	27	13.3	–	–	0	0

*Sea state fetch-limited at F_o. $t_d = (F_o - F)/(1.5\,T_s)$; $t_o = 24 + t_d - t$

This is a practice exercise in picking numbers off the CSS diagram, and the answers are listed in table 5. The three-hourly heights are fetch-limited at consecutive distances F (col. VI). After the wind stops, these waves must travel the respective distances $F_o - F$ (col. VII) to pass F_o. Their decay times (col. VIII) can be calculated, as shown, or read directly from the diagram, as described on page 194. At F_o, the duration of waves of a given height is calculated from the wind duration, plus the decay time, minus the sea state rise time: $t_o = 24 + t_d - t$ (col. IX). From the table, we can see that significant waves 7 feet high will appear at F_o within 3 hours after the wind starts, and last for 58 hours; 20-foot waves will appear in 15 hours, and last for 17 more.* You should recognize that each row in table 5 is an independent forecast, and that it includes a lot of detail for the purpose of illustration. For instance, it might be sufficient merely to know that sea state would peak at about 22 feet after 23 hours, and decay in roughly 400/7.5 = 53 hours. These values can be estimated by one glance at the CSS diagram.

The above example assumes that the sea was calm when the wind started to blow. Suppose, instead, that a 5-foot swell from a preceding storm was present. This swell can be accounted for by adding to the forecast duration a *fictitious* duration, t_f,

*Here, we have neglected dispersion and geometric spreading. The average wave height will actually begin to drop as soon as the wind stops, and will decrease a little faster than shown.

equal to the time it takes for the forecast wind to raise the swell. On the CSS diagram, a significant swell H_3 = 5 feet would appear in 2 hours under a 35-knot wind. Our forecast would then proceed exactly as before, except that we would add 2 hours to each value in column I. We would find that sea state would rise a little faster, and become fetch-limited at F_o in 19 hours, instead of 21 hours.

Lastly, we note that, for a 35-knot wind, about 2 × 35 – 20 = 50 percent of the H_{10} waves may be breaking, with an encounter expectancy $t_b = 6.3T_s / (V–10) = t_b$ = 3.36 minutes.

***Example II.* Stationary fetch, variable winds.**

A stationary fetch F_w = 600 miles is developing between the high and low pressure centers shown in figure 80a. The weather forecast calls for southerly winds, commencing at 10/0000 (local time, day/hour), increasing to 40 knots by 11/0800, and dropping to zero at 13/0700 (heavy curve in figure 80b).

Required: (a) significant and maximum sea states and recommended action for ketch *A* (F_o = 360 miles), 100 miles south of a lee coast, and

(b) same for sloop *B* (F_o = 150 miles), at poor anchorage on weather side of small island, whose skipper wishes to estimate possibility of making port at *B′*, 120 miles away and 120° off the wind.

Slowly increasing winds require that we move from one CSS curve to another as the wind rises. This can be done in the same way as we accounted for a preexisting sea state in Example I. As with navigation fixes, variable wind forecasts are best made according to a set procedure:

1. On graph paper, plot a curve of forecast wind speed versus time to any convenient scale.
2. Approximate the *increasing* wind phase by one or more steps; terminate the *decreasing* phase abruptly when the forecast wind has dropped to 90 percent of its peak value.
3. Look up and tabulate the necessary values from the CSS diagram.
4. Plot sea state and/or wave height to the same scale on the wind graph.
5. Keep a log of estimated sea state for comparison with your prediction. In this way you will accumulate experience, and learn to juggle things a little to improve subsequent forecasts.

The foregoing forecast is worked out in detail below. All data are listed in table 6, in which subsequent steps are keyed by similar numbers, 1–10, to the wind–sea state graph (fig. 80b) and the CSS diagram (fig. 81).

The heavy curve in figure 80b shows our interpretation of the wind forecast, the increasing phase of which is approximated by the stepped, broken line. By our approximation, the wind rises suddenly to 20 knots at 10/0600, and blows at constant speed for 10 hours. At 10/1600, it jumps to 30 knots for 8 hours, and then increases to 40 knots at 11/0000 for 12 hours. The step wind is terminated abruptly at 11/1200, when the forecast wind has dropped to 90 percent × 40 = 36 knots. These data appear in cols. I–III of table 6.

The step wind is shown on the CSS diagram (fig. 81) as an ascending, zigzag curve (pts. 1–6). The smooth, descending curve (6–10) refers to wave decay after the wind stops. Since the initial sea state is assumed to be zero (cols. VI and VII), there is no fictitious duration for the initial 20-knot wind step (col. IV), and its effective

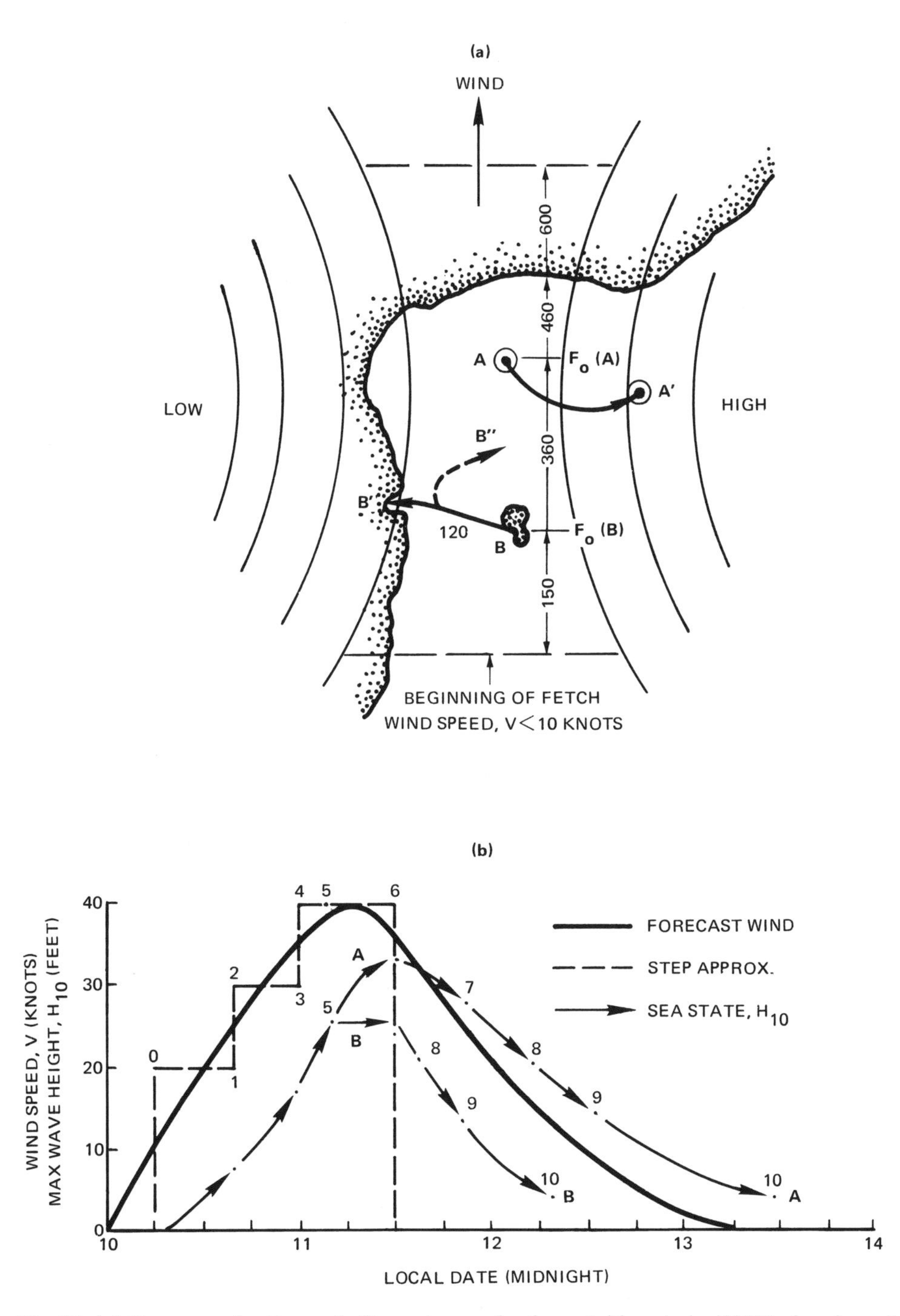

Fig. 80. (*a*) Geometry for Example II, stationary fetch, variable winds, (*b*) Wind and maximum wave history for above example.

duration is 10 + 0 = 10 hours (col. V). Accordingly, we enter figure 81 at the bottom, and move up the 20-knot CSS curve to its intersection with the 10-hour duration line (pt. 1). Read $\sqrt{E}$ = 2.1 ft. and H_{10} = 7.6 feet from the left and right scales, respectively, and enter these values in cols. VI and VII. This is the estimated sea state at 10/1600, when the wind jumps from 20 to 30 knots. The jump is made by moving left, horizontally (constant sea state), to the 30-knot CSS curve (pt. 2).

Table 6. Data for Example II

INCREASING WIND

Key	Wind	I Local time (day/hr.)	II Wind speed V (knots)	III Step duration t_w (hrs.)	IV Fictitious duration t_f (hrs.)	V Effective duration t_e (hrs.)	VI Sea state $\sqrt{E}$ (ft.)	VII Maximum height H_{10} (ft.)
0	jumps	10/0600	0–20				0	0
0–1	steady		20	10	0	10.0		
1–2	jumps	10/1600	20–30				2.1	7.6
2–3	steady		30	8	3.3	11.3		
3–4	jumps	11/1000	30–40				4.6	17.0
4–5	steady		40	4	5.0	9.0		
5		11/0400	sea state fetch-limited at B (F_o = 150 mi.)				7.0	25.0
4-6	steady		40	12	5.0	17.0		
6	drops	11/1200	40–0	maximum sea state at A			9.0	33.0

DECREASING WIND

Key		Local time (day/hr.)	Fetch F (mi.)	Travel distance $F_o - F$ (mi.)	Sig. period T_s (sec.)	Decay Time $t_d = \frac{F_o - F}{1.5T_s}$ (hrs.)	Sea state $\sqrt{E}$ (ft.)	Maximum height H_{10} (ft.)
7	Point A	11/2100	200	160	12.0	9	6.8	28.0
8	F_o = 360 mi.	12/0500	100	260	10.0	17	5.6	21.0
9		12/1300	50	310	8.3	25	3.9	14.0
10		13/1100	9	351	5.0	47	1.2	4.5
8	Point B	11/1500	100	50	10.0	3	5.6	21.0
9	F_o = 150 mi.	11/2000	50	100	8.3	8	3.9	14.0
10		12/0700	9	141	5.0	19	1.2	4.5

Vertically beneath, read 3.3 hours, the fictitious duration for a 30-knot wind to raise this sea. Add this figure to the next step duration (8 hours) to obtain the effective duration (11.3 hours) for the 30-knot wind, and enter these values in cols. III, IV, and V. We now move up the 30-knot CSS curve to its intersection with the 11.3-hour duration line (pt. 3) to obtain the sea state estimate at 11/0000: $\sqrt{E}$ = 4.6 feet, H_{10} = 17 feet. For our final wind jump, we move left again to pt. 4 on the 40-knot CSS curve, and read the new fictitious duration (5 hrs.) vertically beneath.

So far, our forecasts for vessels A and B (fig. 80a) have been identical, but now we come to a parting of the ways. The step duration at 40 knots is 12 hours, but after moving up the 40-knot CSS curve for only 4 hours we reach the limiting fetch (F_o = 150 mi.) at which sea state stops growing for vessel B (pt. 5), and will remain constant for the remaining 8 hours that the wind blows.* At this time (11/0400), the fictitious wind duration is 4 + 5 = 9 hours, and the fetch-limited sea state at B is: $\sqrt{E}$ = 7 feet, H_{10} = 25 feet, T_s = 11.2 seconds. At A, however, sea state continues to increase for the full 12-hour step duration (t_e = 5 + 12 = 17 hours), reaching pt. 6 on the 40-knot CSS curve at 11/1200, where $\sqrt{E}$ = 9 feet, H_{10} = 33 feet, and T_s = 14 seconds.

*Here, we note that pt. 3 (fig. 81) also lies to the right of the 150-mile fetch line. But if we had approximated the forecast wind by smaller steps, using the 25- and 35-knot CSS curves, this would not have happened.

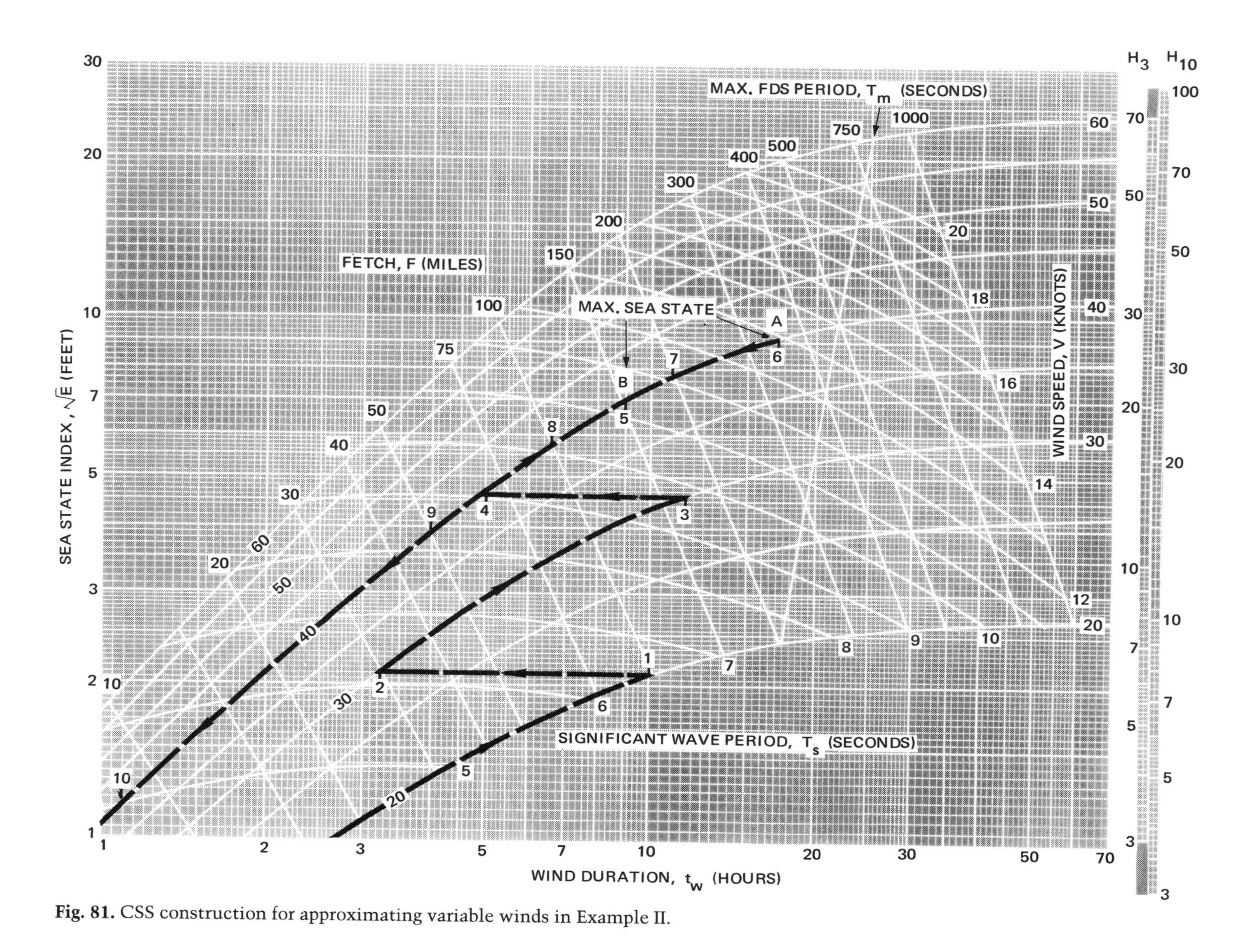

Fig. 81. CSS construction for approximating variable winds in Example II.

The lower part of table 6 outlines the wave decay calculations at *A* and *B* after the wind drops. These must be done separately because the wave heights and decay distances are different. The procedure is identical to that followed in Example I. At 11/1200, various fetch-line intersections with the heavy, descending portion of the 40-knot CSS curve show the wave height distribution within the fetch upwind of *A*. To find the time required for waves at any distance *F* from the upwind end of the fetch to pass pt. *A*, we divide their travel distance $(F_o - F)$ by the appropriate group velocity $(1.5 T_S)$, where T_s is the interpolated period corresponding to the intersection of the fetch-line *F* with the 40-knot CSS curve. The decay time is then added to 11/1200 to obtain the local time when the waves will have passed.

Since all our calculations are only rough approximations, there is no reason to do more than is necessary to plot a smooth decay curve. Table 6 lists four decay steps for point *A* and three for point *B*; the relevant wave height data for these fetch-line intersections are included in cols. VI and VII. If we were only interested in how long it would take for all significant waves to pass, we could have jumped directly to the last row of the table (pt. 10)

Using only the data from columns I and VII, the curves defined by consecutive arrows in figure 80b show how maximum wave height rises and decays with time in the vicinity of vessels *A* and *B*, and how these rates compare with the wind history. Sea state rises at the same rate at *A* and *B* until 11/0400, 28 hours after the wind starts to blow. It then becomes fetch-limited at *B*, but continues to rise at *A* until 11/1200, 4 hours after the wind has peaked out and begun to drop. This is a consequence of our step approximation, which assumes that the wind continues at 40 knots until 11/1200, and then drops abruptly to zero. It would have been slightly more accurate to have limited the last step to 38 knots for the same duration. We can now make two important observations about variable-wind forecasts:

1. Had we instead forecast a 40-knot wind beginning at 10/0600, we would have expected the same sea states some 16 hours earlier. Thus, a step forecast has given us about 16 hours more operational leeway.
2. If we had used the peak wind (40 knots) for the average storm duration (32/2 = 16 hours), we would have expected the same sea state about 4 hours earlier. This is close enough for most practical situations, and can be read directly from the CSS diagram, but we could not have known it without first going through the detailed step calculations. Such shortcuts can save a lot of arithmetic, once you are familiar with how things work.

At 11/1200, the step wind terminates, and sea state begins to decay simultaneously at *A* and *B*. But it takes more than twice as long (47 vs. 19 hours) for all significant waves to pass the former location. This is partly because the waves are higher at *A*, and partly because the decay distance is more than twice as long (360 vs. 150 miles).

Lastly, we note that the forecast wind actually persists for 44 hours after the step wind has dropped to zero, and we need to ask how much our abrupt-termination forecast will be in error through neglect of the waves generated during this period. This is a rather complicated problem, but the details need not concern us, because it has a simple answer:

1. If the wave decay time computed, as above, for an abrupt wind cutoff is longer than that required for the forecast wind to drop below 12 knots (the lowest FDS sea state on the CSS diagram), the cutoff forecast will be correct.
2. If the wind decay is longer, as may be the case for a short observer's fetch, sea state will decay in equilibrium with the wind.

In Example II, the decay time for vessel *A* is obviously longer than the forecast wind duration, so our forecast is correct. However, at *B*, we forecast that all significant waves would pass by 12/0700, whereas the wind does not drop below 12 knots until 12/0900 (fig. 80b). Properly, we should add two hours to the forecast, and recognize that the wave heights will be a little higher than calculated. Practically, the error is small and can be neglected.

Having determined the sea state histories at *A* and *B*, we return now to the question of recommended action for our two vessels. Ketch *A* is in a rather touchy situation with a lee shore 100 miles to the north. With a stationary fetch, there may be no visible signs of weather change. If its skipper fails to receive the forecast, or waits while wind and sea increase to the point where he is forced to heave-to, he could be driven ashore during the course of the gale. Even if he receives it, he cannot beat his way 260 miles upwind to port at *B′* within the 24 hours or so during which he can still make headway. His best course of action is to put on all provident sail, and make as much southeasting as possible, before heaving-to on a starboard tack, after which he will drift northeasterly without further problem—other than 33-foot breaking seas. It should be noted, however, that a significant swell, H_3 = 11.5 feet, will still be running at 12/1300, or 26 hours after the wind has begun to decrease. As many skippers have observed, a sailing yacht has great difficulty making way in a heavy swell with light winds, owing to constant jibing and backfilling of sails as the yacht pitches and rolls. Secondly, piloting is difficult due to obscuration of lights, buoys, etc., and breakers up to twice the swell height can be expected over shoals. Thus the rate of swell abatement is important to decision making.

If vessel *B* leaves its anchorage at 10/0800, as soon as the wind has freshened, and can average 6–8 knots on a broad reach, it should make port at *B′* by midnight. By then, the wind will be averaging 35 knots and gusting to 50, and the vessel will be encountering breaking seas as high as 17 feet. With any doubts about entering and anchoring, it would do better to heave-to offshore (B-B″), and ride it out under somewhat better sea conditions than vessel *A*.

Despite our lengthy discussion of this example, with a little familiarity, and by making use of the shortcuts mentioned, one can obtain the necessary operational information directly from the CSS diagram within a minute or two. Even a step forecast can be made in one's head, if the only things desired are maximum sea state and its rise and decay times.

***Example III.* Moving fetches, steady wind.**

Quantitative wave forecasts for moving storms are quite beyond the scope of our simple forecast model, and also require more information about storm movements than is ordinarily contained in radio weather broadcasts. All we can hope to do here is to describe the typical sequence of events, and show how stationary forecasts are modified by fetch motion and how the waves from a single storm system can interact to confuse sea state and forecaster alike. If this description doesn't answer all your questions, it should at least give you a feeling for what goes on in the mysterious world of wind and waves.

As discussed in Part II, a fully developed cyclonic storm consists of a roughly circular wind field blowing around a moving low-pressure center. Depending upon one's position with respect to the storm path, the wind may blow for protracted periods from any direction. However, at a fixed location, the sea knows nothing about storm motion, and senses only how long the wind blows. At any instant, the local sea state will reflect the energy of waves locally generated, as well as that of waves generated elsewhere which happen to be passing through.

Figure 82 shows a storm moving east at speed V_f, whose strong-wind field isobar is approximated by four uniform fetches, $F_1 - F_4$. Within each fetch, the wind is blowing in a different direction and at a different speed, as indicated by the respective arrow lengths, $V_1 - V_4$. Assuming that the wind has blown for some time, and that its subsequent duration is longer than the time required for the storm to move completely past stationary observers at A and B, these observers will experience quite different wind and sea state histories.

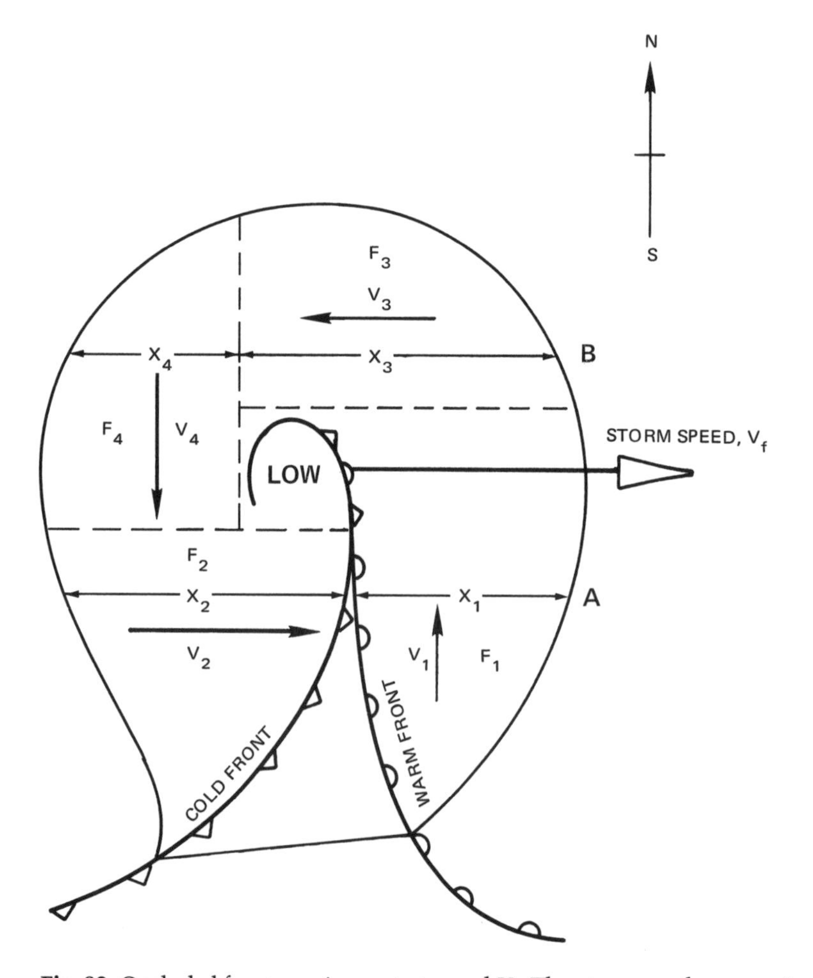

Fig. 82. Occluded front moving east at speed V_f. The storm can be approximated by four adjacent fetches, in which wind blows in different directions (arrows). Wave interactions often produce confused seas (Example III).

Observer A will first encounter light, steady southerly winds, and a short chop from the same direction that grows steadily in height and period at the same rate as it would if he were traversing a stationary fetch of this shape (say, a lake) at speed V_f, and the wind sprang up at the moment of departure. Thus, at any instant during the passage of fetch F_1, the sea state could be determined from the CSS diagram. This situation is characteristic of fetches moving at right angles to the wind.

As the warm front approaches, Observer A may become conscious of an underlying westerly ground swell radiating ahead of fetch F_2. But if the storm is moving faster than the group velocity ($1.5\,T_s$) of the longest waves in F_2, the observer may see nothing more than the lowering cloud cover and increasingly heavy rain that typically precede a cold occlusion (fig. 27, p. 85).

About $t = X_1/V_f$ hours after the storm first passes Observer A,* the front arrives, and all hell breaks loose. Fetch F_2 contains the highest winds in the storm, and they are blowing in the same direction as F_2 is moving. Although the effective wind duration in F_2 is $t_e = X_2/V_f$, to Observer A the sea state will appear to rise *faster* than indicated by the CSS diagram for a stationary fetch of the same length, wind speed, and duration. This is because the waves are also moving in the same direction, so that they remain longer under the influence of strong winds. Where the equivalent stationary forecast would indicate an immature sea, the moving fetch may achieve a higher—even a fully developed—sea state. In extremis, if F_2 is moving at just the group velocity of the highest waves, as the front moves past him, Observer A will suddenly encounter a 90° wind shift, rising to gale force, and fully developed seas from the west. To make matters worse, there will still be a residual cross-sea running from the south, soon followed by a strong north swell radiating into F_2 from fetch F_4. All this makes for a very chaotic sea condition, which may persist long after F_2 has passed on, and the wind has dropped to a whisper. The last thing Observer A sees is the tail end of the short-period swells left behind F_2, because it is moving faster than the swells can travel.

Observer B makes rather better weather of it. His first encounter with the storm is fetch F_3, within which wind and waves are traveling in the opposite direction. Because F_3 is losing energy rapidly in the form of large waves radiating from its downwind (westerly) end, the sea state at B rises more slowly and equilibrates at a lower level than would be forecast for an equivalent stationary CSS fetch. Initially, there is a moderate cross-swell coming in from F_1, to the south, but this fades away before the arrival of fetch F_4, in which wind and waves are moving southward. Unlike Observer A, Observer B encounters no sudden change in either wind or sea state, but only a smooth transition as both back around into the north.

Although the wind is stronger in F_4 than in F_3, B's fetch is very short. Sea state may rise a little and decay again, as the storm moves on, leaving behind it only swell from F_3. Whether significant swell activity continues longer at A than at B depends on the previous—as opposed to the subsequent—history of the storm, which has not been specified for either situation. But if the storm was about midway in its active life when the front passed through, the decay times will be about equal.

From this brief description, it is easy to see what is meant by the "dangerous" and "safe" sectors of a revolving storm. Both A and B experience the same *average* winds and total wind durations, but A catches it in the neck, while B gets by rather

*The time for F_1 to move its own width, X_1. This is also the maximum wind duration within F_1.

lightly only a hundred miles or so to the north. This difference is due partly to the fact that wave height increases with the square of wind speed, and partly to the much greater growth rate and maximum wave height in fetches traveling downwind.

How different would things be in a stationary or slowly moving (10- to 15-knot) storm? The highest waves would still occur in fetch F_2, but instead of being trapped within it, a strong, long-period swell would be radiated ahead of it, passing through F_1 and far beyond. This advance swell is one of the best indicators of the approach of a large, slowly moving storm (see below). We could also make fairly reliable CSS sea state estimates in all storm quadrants, instead of only for F_1 and F_4. These should be upgraded for cross-swell effects by adding sea state energies:

$$E_t = E_1 + E_2, \text{ or } \sqrt{E_t} = \sqrt{E_1 + E_2}$$

Although we have not discussed wave forecasts outside the generating area, swell behavior often provides important clues to storm movement in areas not covered by radio forecasts. Swell observations should, of course, be correlated with other weather signs described in Part II.

1. Winds from most oceanic storms tend to blow cyclonically around a moving low-pressure center, radiating swell in all directions like a garden sprinkler.
2. Swells are longest and highest in the right forward quadrant of a slowly moving storm (Northern Hemisphere), and shortest and lowest in the left rear quadrant. However, the advance of a fast-moving cloud front may precede fully developed storm seas with little or no prior swell warning.
3. If the swell direction veers with time, the storm center is passing poleward of your position, and conversely. If the direction remains constant, the storm is approaching, stationary, or receding.
4. An increase in swell height and a decrease in period denote an approaching storm. If the swell reaches a maximum and then decreases, the storm is either dying or passing to the side.
5. Only storms of gale proportions develop significant swells having periods longer than about 14 seconds, but periods longer than 20 seconds rarely occur in the largest storms. Hence, the arrival of swell with periods within the above range, and whose direction remains unchanged, indicates that you are in the dangerous sector of a major storm. Swell longer than 20 seconds implies a very large, but remote, storm, that probably will not affect you.
6. Another clue to storm distance is swell regularity; if the swells are very regular, and of relatively constant period over, say, 6 hours, the storm is remote, and conversely.

Wave Hindcasting—The Fastnet Storm: *Hindcasting* is the reconstruction of a probable scenario from data recovered after the fact. Like hindsight, it has the advantage of 20-20 vision. But it is often useful to test hypotheses, plan future strategies, and avoid future perils. In this section, we re-examine the circumstances leading to an international sailing catastrophe, the Fastnet Race of August 10–16, 1979, where a tardy meteorological warning resulted in a fleet of over 300 boats being caught in rapidly developing wind and sea conditions of near-hurricane proportions. Only 85 boats finished, while many were sunk (5), abandoned (19), or damaged (190), with a high toll of injuries (220?) and loss of lives (15).

Foregoing the details of that race, which have been set forth in several books and dozens of articles, we concentrate here on the local wind history, from which the corresponding sea state can be derived by methods previously outlined; later (chapter 23) we shall attempt to explain observed vessel responses to that catastrophic environment.

The biennial Fastnet race* is the culmination of Cowes Week at the Isle of Wight, which since 1975 has become the testing ground for the Admiral's Cup Class (ACC), an informal association of maxi-goldplaters that sponsors races all over the world under the International Offshore Rule (IOR). The Fastnet is open to anyone qualifying under IOR rules, and the qualifying formula length is only indirectly related to a yacht's actual length. In 1979, there were six qualifying classes, having minimum IOR lengths of 28, 33, 34, 39, 44, and 55 feet, respectively. Of 303 starters, 57 were Admiral's Cup contenders, ranging from 38 to 79 feet; the remainder, averaging only 34 feet in IOR length, comprised a multinational group of sailing enthusiasts with no predominate affiliation, and only the desire to pit their skills against the ACC.

Since 1925, the Fastnet has followed an invariant course (fig. 83): starting from Cowes, west along the underside of England to Land's End, thence straight across the

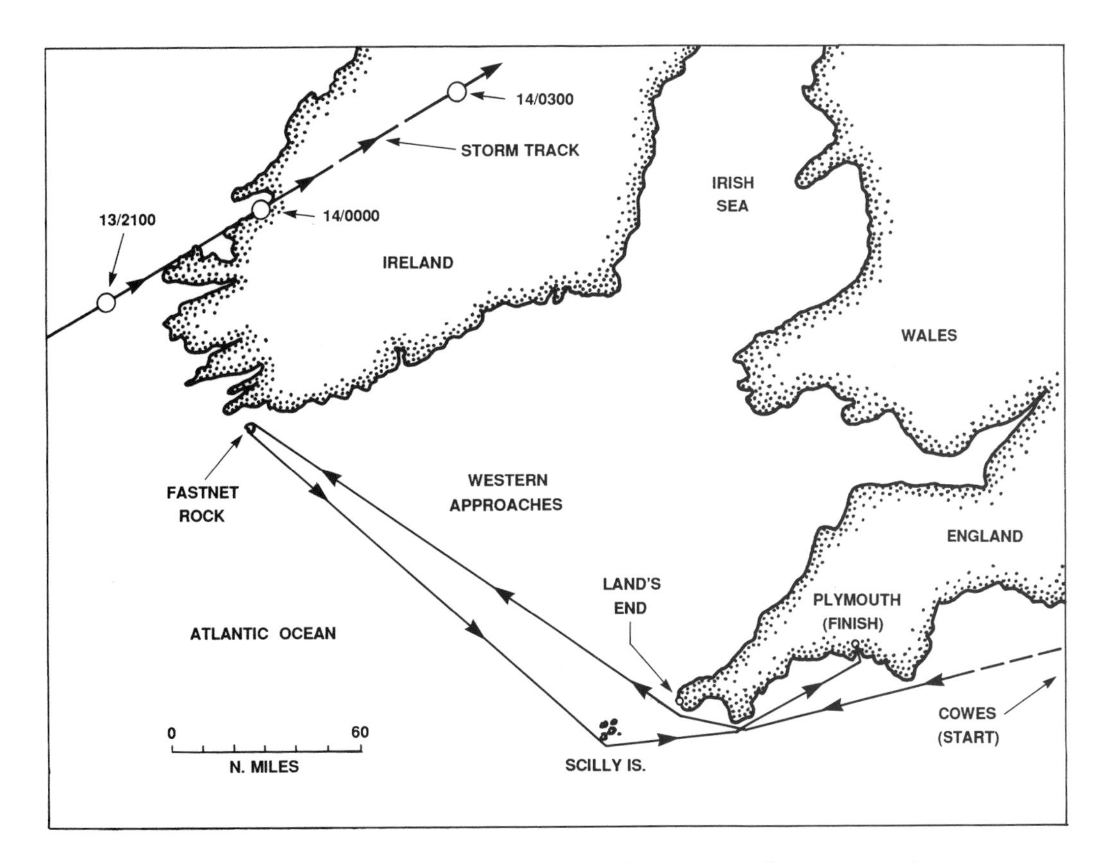

Fig. 83. Tracks of the Fastnet race (*bottom*) and critical portions of the storm (*top*).

*The first Fastnet race was planned and conducted in 1925 by E. G. Martin (UK), who also won it in the 44-ton pilot cutter, *Jolie Brise.*

Western Approaches to Fastnet Rock, south of Ireland, and reciprocally back to end at Plymouth. The distance is 605 nautical miles, of which about 60 percent is unprotected north Atlantic racing, subject in August to breezes, gales, and calms in regular succession.

The 1979 race was remarkable only because of a weak, rapidly moving meteorological depression that sneaked in almost undetected from the southwest, slowed off Ireland, and ramified into a major storm in which the wind increased from 30 to 60 knots and shifted 135° in 11 hours. The storm caught the Fastnet fleet strung out across the Western Approaches at night, with little meteorological warning, and only the fastest boats, which were able to round Fastnet Rock before the wind veered to dead ahead, were able to complete the race. The remainder hove-to or ran off downwind, many to be broached, capsized, and rolled over by steep, confused, breaking seas, which it is our present objective to reconstruct.

In most respects, the Fastnet storm was similar to that experienced by Observer A, figure 82, except that the region between the warm and cold fronts had diverged to include a separate, additional fetch containing strong southwesterly winds. The wind speed history midway of the Western Approaches (synthesized from a variety of published sources) is shown as the heavy, continuous line in figure 84. The horizontal axis gives the date and Greenwich Mean Time, with corresponding wind directions at the top.

At 0600 on Monday, August 13, the sea was calm, except for a low southwest groundswell rolling in from remote, prevailing westerlies, and the wind was so light that there was concern the race might be aborted. Because of light winds for the preceding 48 hours, the fleet was fairly bunched west of Land's End. But by 0600 Tuesday, the wind had increased almost linearly to 52 knots, and veered continuously around to due west. At 0600, the passage of the cold trough was marked by a wind jump to 60 knots, and an abrupt wind shift to northwest. Thereafter, the wind dropped just as steadily to about 10 knots by 0500 Wednesday. Since wind gustiness at sea increases with induced wave height, the highest (60-knot) winds most probably incorporated gusts to 80 knots.

Because of the moving storm and the continuously varying wind speed and direction, sea state cannot be approximated this time by a continuous line on the CSS diagram. Instead, we divide the wind history into steps, as before, and then consider each step to be a separate mini-storm of 6-hour duration (letters A–H, fig. 84), for each of which we determine the sea state $\sqrt{E}$ from the diagram, together with the significant wave period, T_S (table 7). To obtain the cumulative sea state after each wind step, we must subtract from the current value of E the wave energy from previous steps that has been lost by breaking, or that has leaked away by waves traveling out of the fetch as previously described. With a moving storm, we have no accurate gauge of these losses, except educated trial and error. After several computer trials,* I have assumed that each preceding step loses 50 percent of its energy every 6 hours; that is, the cumulative energy after step D, say, would be

$$E_C = E_D + 0.5E_C + 0.25E_B + E_A$$

*The loss index is quite sensitive; the assumption of slightly smaller losses leads to much higher wave maxima and longer storm wave persistence than were reported, and conversely.

where E_D is the CSS value from table 7, and E_A is the low southwest background that persists throughout the storm. Values of the maximum wave height H_{10} (assumed to be breaking) were then looked up on the CSS diagram, corresponding to values of $\sqrt{E_C}$. Lastly, the values for breaking and plunging expectancy intervals in table 7 were calculated by formulae given on page 197.

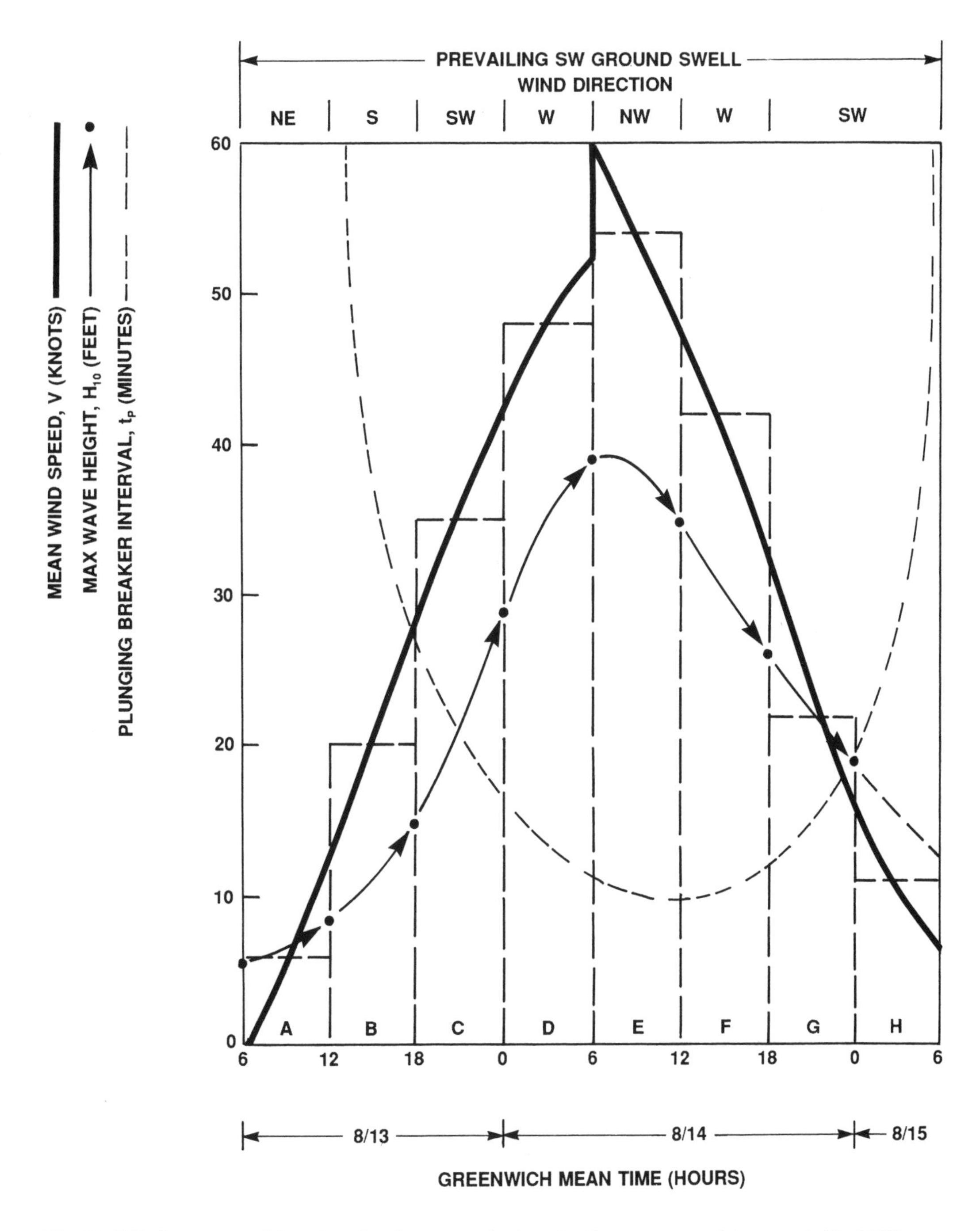

Fig. 84. Wind, wave, and breaker development during the Fastnet race of August 8–13, 1979.

Table 7. Fastnet sea state data

Fetch	Date (*GMT*)	V (*knots*)	Wind direction	t_w (*hours*)	$\sqrt{E}$ (*feet*)	t_s (*sec.*)	H_{10} (*feet*)	t_b (*min.*)	t_p (*min.*)
A	All	15	SW	FDS	1.5	6	5.5	∞	∞
B	13/1100	20	S	6	1.7	6	8.6	3.8	23.0
C	13/1700	35	SW	6	4.5	9	15	2.3	15.6
D	13/2300	48	W	6	7.4	11	29	1.8	11.1
E	14/0500	55	NW	6	9.4	12	39	1.7	9.6
F	14/1100	42	W	6	5.6	11	35	2.2	12.0
G	14/1700	22	SW	6	1.9	8	26	4.2	19.6
H	14/2300	11	SW	6	1.0	4	19	25.2	68.6

Maximum average breaker height H_{10} and plunging interval T_p are also shown in figure 84, from which we can now make the following general observations. Maximum wave height remains quite closely in phase with wind speed, rising to nearly 40 feet just after the wind veered northwest and peaked out at 60 knots. Under such extreme conditions, we expect wind gusting to as much as 80 knots, and there is a 5 percent expectancy of rogue waves as high as 64 feet (fig. 72) every half-hour or so, whereas the highest 33 percent of all waves in the vicinity would average only 30 feet, and would not be breaking. The expectancy interval for plunging breakers decreases from infinity as windspeed and wave height increase, reaching a minimum of 9.6 minutes at the peak of the storm, and then the interval increases again as wind and sea abate. The expectancy interval for encountering breaking

Plate 17. Breaking interactions between steep waves coming from several directions produce a chaotic sea state that is virtually unnavigable for small vessels.

turbulence is about six times shorter (table 7), and occurs as often as every 1.7 minutes at the peak of the storm.

But the most unusual and—from the point of seamanship—most difficult and dangerous aspect of this storm was the very confused and chaotic nature of the sea state, with high waves interacting from four different directions with no perceptible rhythm or regularity (plate 17). As shown in chapter 23 (fig. 135), for a period of about 24 hours, the Western Approaches were virtually unnavigable for boats smaller than 40 feet.

17

Breakers and Surf

Toward the end of Part IV, we showed how ideal periodic waves entering shallow water are variously refracted, diffracted, and reflected back into deep water or trapped against the coastline. In the real ocean, these processes are modified to some extent by the irregularity of incoming waves, and by breaking, in which most wave energy is dissipated through turbulent mixing, and very little is reflected. Although coastal engineers are now applying statistical methods to near-shore wave behavior, most of what we know about this complicated regime is still based upon a patchwork of oversimplified theory, laboratory experiments with periodic waves, and a few field experiments. The surf zone is a difficult place to work, and results are slow in coming. The best we can do here is to describe the present state of progress, and try to extract some concepts useful to our subsequent discussions of anchoring and boat handling in surf.

MECHANICS OF WAVE BREAKING

It may seem surprising that so familiar a phenomenon as wave breaking still lacks a satisfactory mathematical explanation. Physically, it is generally conceded that breaking is an instability process related only to wave steepness and beach slope. But, while there are several theories that closely predict the observed profiles and speeds of steepening waves, none of them correctly deduce the particle velocities and accelerations of steep waves, and hence they cannot properly assess the terminal instability and inception of breaking. These theories assume—for mathematical tractability—that waves are symmetrical fore and aft of their crests, and that the bottom is flat—or "very gently sloping"; neither of these assumptions is valid near the breaking point. Moreover, the theories treat waves as individual entities, ignoring backwash from preceding waves as well as the general buildup of water in the surf zone, which is related to the cumulative flux of water from groups of incoming waves.

But all of these factors have been examined in many laboratory and field studies, and the gap between theory and experiment is rapidly closing. It may not be too presumptuous to expect that another facet of Nature's mystery may soon be exposed to view. For the present, we can give only a phenomenological picture, without being able directly to relate cause and effect.

In the following, we shall first treat the behavior of steady periodic waves that are normally approaching a straight, uniformly sloping beach. Later we will modify our discussion to include such factors as wave groups, variable periods, irregular topography, waves approaching at arbitrary angles, and the dynamic effects of tides, currents, and bars.

Wave Shoaling: In Part IV it was noted that, as deep-water waves pass into shoaling water near shore, their phase velocity loses its dependence on the wavelength, and becomes a function only of the water depth: $\sqrt{gL/2\pi} \rightarrow C \rightarrow \sqrt{gh}$ ft./sec. Thus, C continuously diminishes as the depth h becomes shoaler. But the wave period T and the energy flux $\rho gH^2Cg \rightarrow \rho gH^2C$ remain constant throughout the transition. Hence the wavelength ($L = CT$) correspondingly decreases, and the wave height H and steepness H/L both increase, as the depth h shoals shoreward (fig. 85).*

More precise analysis shows that the shallow-water wave velocity is also a function of wave height: $C \doteqdot \sqrt{g(0.75H + h)}$ ft./sec. The effect of increasing wave height is opposite to that of decreasing depth, and acts to procrastinate the decrease in phase velocity when the height amounts to a significant fraction of the depth.

Wave Breaking: The current state of the art seems to consist of rather unsatisfactory—and often contradictory—theoretical attempts to explain what is experimentally observed. I will confine remarks to the latter, as summarized below:

1. Periodic waves will break if their deep-water steepness H_o/L_o exceeds a critical ratio to beach slope S given by: $H_o/L_o > S^2/2\pi$. Because wave period is easier to measure, this relation is more commonly given as $H_o/T^2 > gS^2/4\pi^2$. If this limit is not exceeded, the waves will simply surge up the slope and reflect seaward.
2. Waves destined to break, as above, will do so when their local height-to-depth ratio has increased to somewhere within the range $0.78 < H/h < 1.2$, depending again on H_o/L_o and S.
3. The offshore distance to the breaking point is also related to the absolute wave height: higher waves of the same steepness will break farther offshore. In general this point can be calculated only by considering the entire growth history from deep water.
4. Except for plunging breakers, the crest angle at the breaking point approaches 120°, and remains at this value after breaking.
5. The horizontal particle velocity U at the crest of a breaking wave is presumed to be close to the limit phase velocity $U \rightarrow C = \sqrt{g(0.75H + h)}$ ft./sec., although good experimental proof is still lacking.

Breaker Classifications: Although breaking is a go–no go phenomenon, there are all degrees of breaking intensity. Over very gradual slopes ($S < 1{:}50$), breaking commences when the height-to-depth ratio $H/h \doteqdot 0.78$ as a gentle *spilling* action that just removes enough energy to keep the above ratio constant as the wave moves shoreward. In such situations, several waves may be breaking simultaneously, one behind the other. If viewed from shore by an observer, kneeling so that his or her eye is just at sea level, the tops of all breaking waves will fall along a line that slopes

*In deep water $Cg = C/2$; in shoaling water $Cg \rightarrow C \rightarrow \sqrt{gh}$ as $h \rightarrow L/2$. But, during this transition, Cg goes through an intervening maximum. Since ρgH^2Cg is constant, H actually decreases about 10 percent in this transition before commencing its eventual increase to the breaking point.

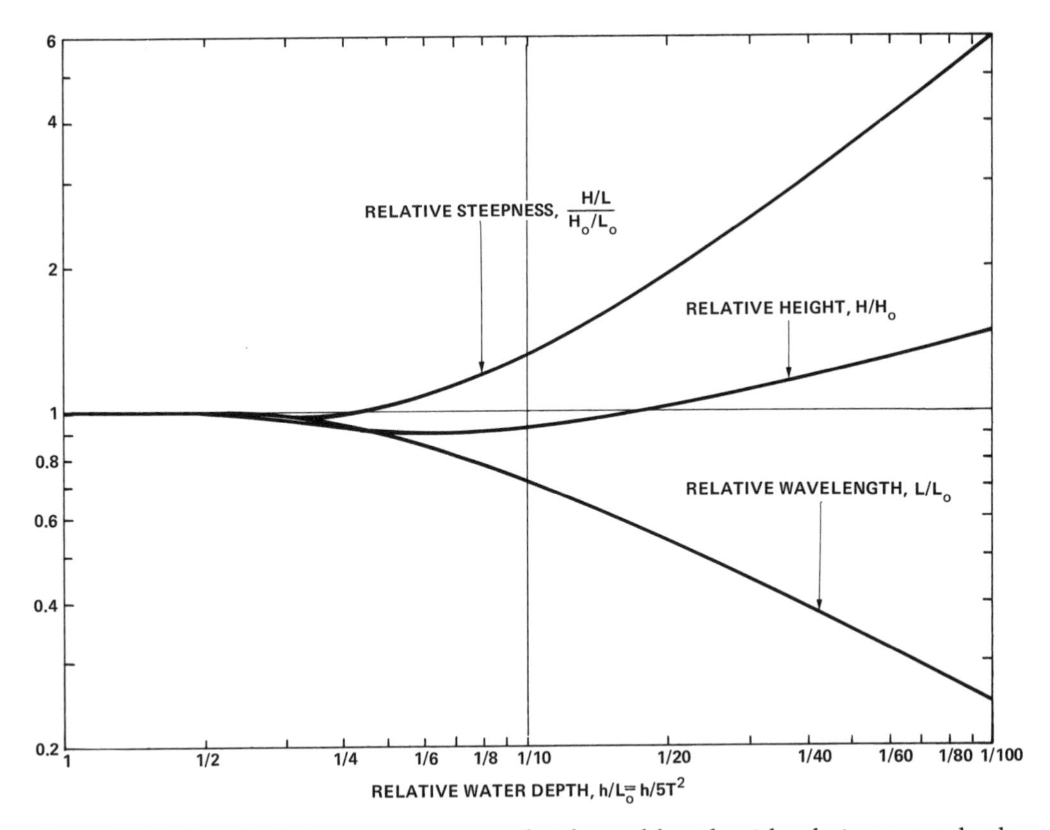

Fig. 85. Variation of relative wave steepness, height, and length with relative water depth. Subscript *(o)* refers to deep water. Note that H/H_o, passes through a minimum as the depth shoals, before increasing above its deep-water value.

shoreward at an angle α = arctan $0.78S$ (fig. 86). Thus, if α can be measured by a sextant or thumb rule held at arm's length, the (uniform) bottom slope can be determined directly.

Over steeper slopes, the critical depth ratio increases ($0.78 < H/h < 1.2$), breaking is progressively more vigorous, and fewer waves are breaking simultaneously. With increasing intensity, breaking action is successively described as spilling, curling, and plunging (plate 18). The latter marks the limit of breaking right at the shoreline, and all similar waves surge up steeper slopes without breaking. With plunging breakers, only one wave breaks at a time, and the intensity of each wave is greatly augmented by backwash from its predecessor. This situation is particularly dangerous for swimmers: not only does the plunging cascade fall in very shallow water—or on bare sand—but the strong backwash precludes easy exit from the breaker zone. One can enter the water easily by riding the backwash seaward and diving beneath the next incoming crest. Coming out, it is best to wait until a wave group is ending, and hope to make it to shore before the next group arrives.

NONUNIFORM BREAKING

Several independent factors may modify the somewhat idealized picture presented above. We will discuss them individually, although they often combine to produce more complicated situations.

Wave Variability: The range of wave periods present at a given shoreline location varies considerably, depending upon the distances of the contributing storm centers, and their respective intensities and durations. The more remote a storm, the more regular its swell; the largest and longest waves will arrive suddenly—often within 2 to 3 hours—and average wave heights and periods will decrease slowly over several days. The nearer the storm, the less regular its swell, and the broader the range of periods. Similarly, the regular swell from a distant storm is well subdivided into groups of 6–8 waves, while the wave groups from a nearby storm are irregular, contain fewer waves, and may be difficult to distinguish from a random sea.

These differences also are manifested within the breaker zone. There is an orderly migration of successive breakers from a long swell: the first low wave in each group breaks nearest shore, and subsequent, larger ones break farther out. After the largest wave has broken, the process is reversed, ending up again with near-shore breaking, followed by a relatively long period of calm between wave groups. Contrariwise, the short, random swell from a nearby storm breaks continuously and irregularly at many places, with no intervening calms.

Nonuniform Bottom: Irregular bottom topography can drastically alter the pattern of breaking from that observed along a straight shoreline. If the scale of the irregularities is larger than one or two wavelengths, refractive effects (see Part IV) enhance the local wave height and induce premature breaking over shoals and ridges, while reducing and postponing them over hollows and troughs.

Abrupt discontinuities, such as the reef platforms surrounding coral atolls, precipitate violent, plunging breakers, while a deep hole may prevent them altogether. In laying instrument cables out across the reef at Johnston Island, we were able to work comfortably in a deep reef fissure only 60 feet across, while 30-foot breakers pounded the reef on both sides.

Variable-Approach Angle: In general, waves approach the shore at arbitrary and somewhat variable angles. In all cases, refraction in shoaling water tends to swing the crests more into alignment with the shore—and hence with their neighbors. As a result, very long-crested breakers may appear to materialize suddenly out of a somewhat irregular, short-crested swell. This effect is made more prominent by the aforesaid temporary reduction of wave height as the group velocity passes through

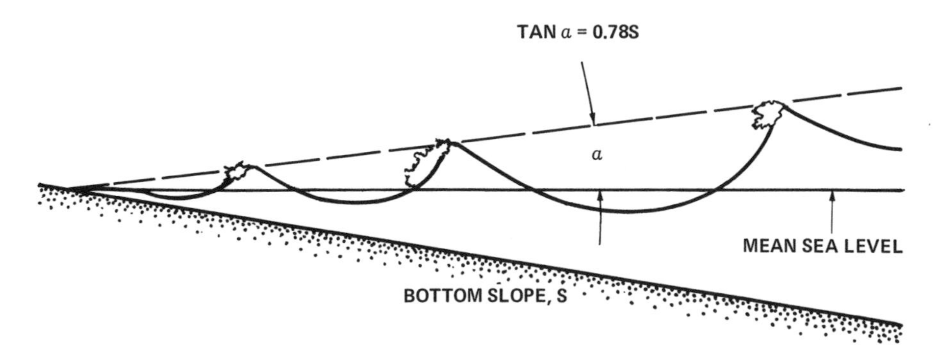

Fig. 86. The height of spilling breakers decays shoreward in fixed proportion to the water depth.

Plate 18. Breaker classifications: (*top left*) spilling breakers result from waves of low steepness over gentle slopes; (*top right*) plunging breakers emanate from steeper waves over moderate slopes, or over a discontinuity; (*bottom*) surging breakers occur where the beach slope exceeds wave steepness.

its transition from deep to shallow water, after which it increases rapidly to the breaking point. Thus, a quasi-random swell offshore sometimes appears to be bounded shoreward by an invisible barrier of quiet water, inside of which long, regular waves rise up out of nowhere to plunge and die.

DYNAMIC EFFECTS

Heretofore, we have treated the coastline as static, but there are also dynamic effects that can temporarily alter the character of an established pattern of breakers.

Tides and Currents: The general effect of tidal action is to move the entire surf zone onshore or offshore accordingly as the tide rises and falls. Over a uniform offshore slope, there is otherwise little change in the breaking pattern. However, such slopes are found only off of straight, mature coastlines that are plentifully supplied with sand, in equilibrium with local wave action. More commonly, the offshore slope is gentle and increases sharply toward the equilibrium beach slope, which may be as steep as 1:4 if composed of pebbles or cobbles. In such cases the surf zone at high tide may be very narrow, with a single line of plunging breakers, whereas low water exposes a wider expanse of gentler slope to breaking, and the surf zone may then rapidly increase to considerably greater width and include several lines of breakers. Here and there, isolated offshore reefs may shoal sufficiently so that local wave stability is exceeded, and sudden, heavy breaking may occur well outside the normal surf zone.

The upper reaches of the Gulf of California represent an extreme example of this migration. Here, the combination of 20-foot tides and a very broad low-tide shelf results in the transformation from a single shoreline break to a zone 3 miles wide in which breaking may occur anywhere at various tide stages. Similar effects are found in the English Channel, off the coast of Newfoundland, and in the Gulf of Alaska.

A current flowing contrary to the direction of wave travel acts to reduce wave speed and hence to increase wave steepness, by shortening the lengths and increasing the heights of all waves present (fig. 87). This effect may be sufficient to precipitate breaking where it would not otherwise occur, or to greatly increase the incidence and intensity of waves already breaking. Such conditions frequently occur even in deep water where strong winds from the northeast quadrant generate breaking seas that are opposed by the swiftly flowing Gulf Stream or Kuroshio. Similar conditions, known locally as *overfalls,* are produced when tidal currents in restricted waters, such as the English Channel, are opposed by southwesterly winds and waves. Very large, steep breakers of this type are often found off the mouths of large rivers along the Oregon and Washington coasts during southwesterly storms. Here, frequent fogs, strong tidal currents, and shifting offshore sandbars make the few tortuous inlets of this inhospitable region difficult to manage in calm weather. Even large ships stand off under storm conditions, when enormous waves, breaking clear across the bars, preclude all passage. The U.S. *Coast Pilot* gives detailed instructions for navigation in these waters, which should be closely adhered to.

The presence of a current running with the sea has just the opposite effect, lessening wave steepness. Thus, faced with the unavoidable necessity of entering an unfamiliar inlet on a lee shore under storm conditions, it is most provident to make your passage in daylight, just before flood tide, when all circumstances are in your favor.

Surf Zone Dynamics: So far, we have treated waves as individual entities, each of which grows in shallow water to the breaking point and then degrades shoreward. However, the breaking process not only dissipates energy, but also spreads wave momentum over the entire breaker zone. While energy is converted into heat, momentum is conserved in the form of a pulse of surface water moving shoreward *behind* the breaker. This pulse should not be confused with the more rapid orbital particle motion under the wave crest, but is distributed among—and contributed to by—all succeeding breakers in a wave group. The result of this cumulative action is a slow, shoreward, net transport of surface water, superimposed on the back-and-forth oscillations within individual waves, such that a barely buoyant piece of kelp can be observed to make large onshore-offshore excursions with each passing wave, but ends up each cycle a little closer to the beach.

This shoreward transport is smallest on steep beaches, where only one wave breaks at a time, and largest over gentle slopes comprising many lines of breakers (fig. 88a). In the latter case, the cumulative transport from many waves piles up a wedge of water against the shore, and sets up a slow subsurface return flow in one or

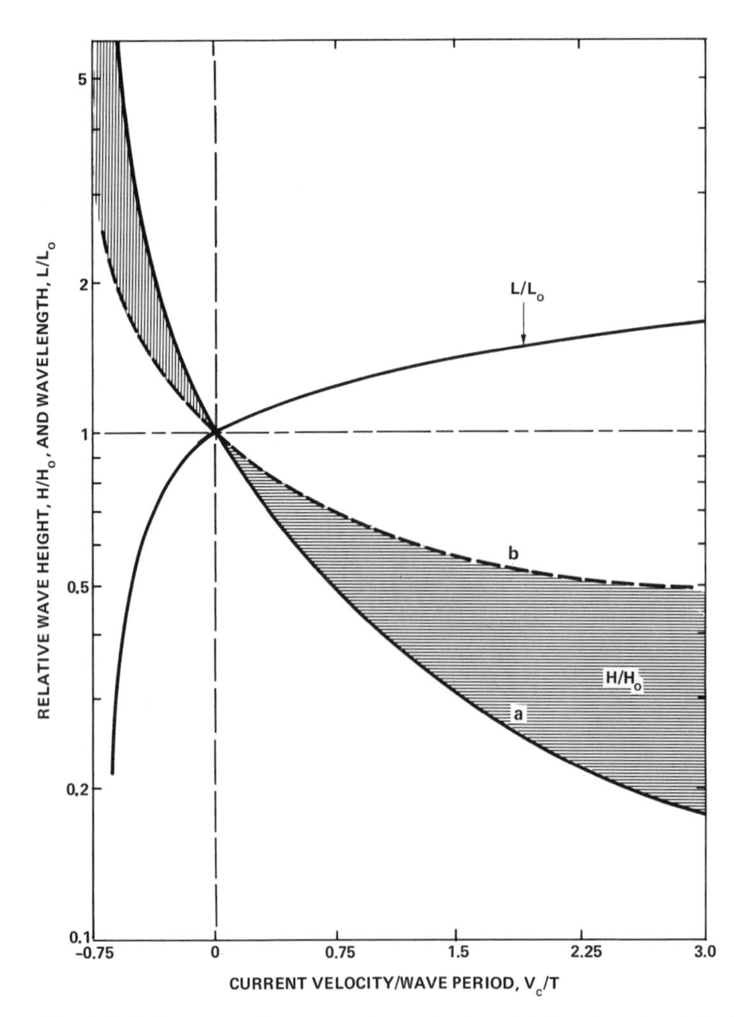

Fig. 87. Influence of contrary (negative) and following (positive) currents of velocity V_C on relative wave height and wavelength, for waves of period T seconds. The shaded band gives the range of variation from pure, periodic swell *(a)* to a random sea *(b)*. No swell can propagate against a current $V_C > 0.75T$ knots.

more vertical cells. No water is transported beyond the breaker zone. The piled-up wedge is not static, but oscillates up and down by about one-fifth of the breaker height, a phenomenon known as *surf beat.* The oscillation period is related to the wave-group interval, and to the ratio of wave steepness to beach slope—but usually falls within the range of 1 to 5 minutes.

In addition to the above vertical circulation, there is also a horizontal circulation that is linked to the predominate period of the breakers themselves. Conjunctively with the shoreward progression of breakers, a stationary pattern of coupled standing waves of the same period is established just outside the breaker zone. The nodes and antinodes of the standing waves are perpendicular to the shore, and interact with the breakers to produce a systematic longshore fluctuation of breaker height—maximal at alternate antinodes and minimal between them (fig. 88b). As a result, the greater breaker transport in the region of higher waves divides within the breaker zone into alternately directed longshore currents, which merge and flow

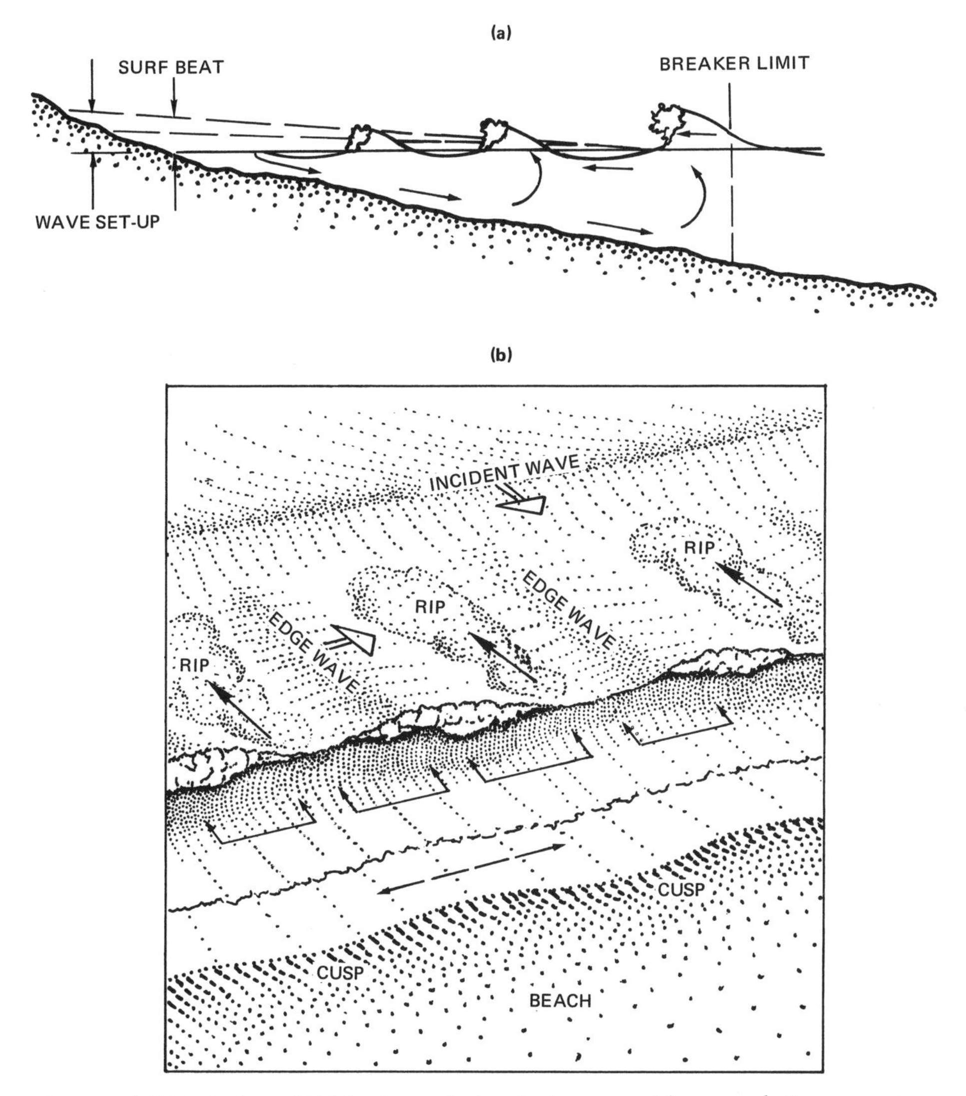

Fig. 88. *(a)* Vertical, and *(b)* horizontal circulation caused by cumulative wave transport in the surf zone. Cellular longshore circulation transports sand, deforming beach into cusps.

seaward as a system of swift, narrow rip currents along the antinodes where breakers are lower.

This characteristic pattern is rapidly established for a particular wave condition, and the combination of wave turbulence, longshore transport, and rip currents sculptures the beach into regular scallops, or cusps (plate 19), centered on narrow scour channels. Momentum carries the rip currents through the breaker zone, where they dissipate in mushroom-shaped eddies that are made visible after a rainstorm by their muddy clouds of fine sediment (plate 20).

If, as is usually the case, the waves are approaching at an angle, a general longshore drift will be superimposed on the above pattern, and there will be a slow migration of sand along the beach opposite to the direction of wave approach. The implications regarding operations in the surf zone are the following:

1. Rips are narrow zones of low waves and a seaward flow of up to 5 knots; they are the best places to launch boats through surf. Surfers often make use of rips to obtain free rides back out to sea.
2. Waves are highest and shoreward transport is strongest midway between rips, and landings should always be attempted here.
3. It is usually impossible to swim—or even wade—against a strong rip current. If swimming, and caught in a rip, look shoreward to detect any longshore drift, and then swim in that direction parallel to shore until you are in a region of high breakers, whose transport will assist you back to the beach.

These remarks pertain only to sandy beaches. Over coral reef platforms, the rip currents still exist, but are confined to permanent reef channels. Elsewhere, sharp coral and high breakers make operations difficult. Here the safest small-boat landings are effected by threading the edge of a large reef channel, thus striking a balance between the outward rip and the adjacent zone of high breakers. Reef channels vary greatly in width, and it is usually possible to make a provident selection with a little search for zones of lower-than-average breakers.

Maximum Breaker Heights: There is no absolute limit to breaker heights; they run in rough proportion to the incident waves offshore, which are limited only by storm size and intensity. Figure 89 shows some experimental results of the ratio of breaker height H_b to unrefracted offshore wave height H_o as a function of the offshore wave steepness parameter H_o/T^2 and beach slope S. In general, the ratio lies between $1.0 < H_b/H_o < 2.0$, and increases with decreasing steepness and decreasing

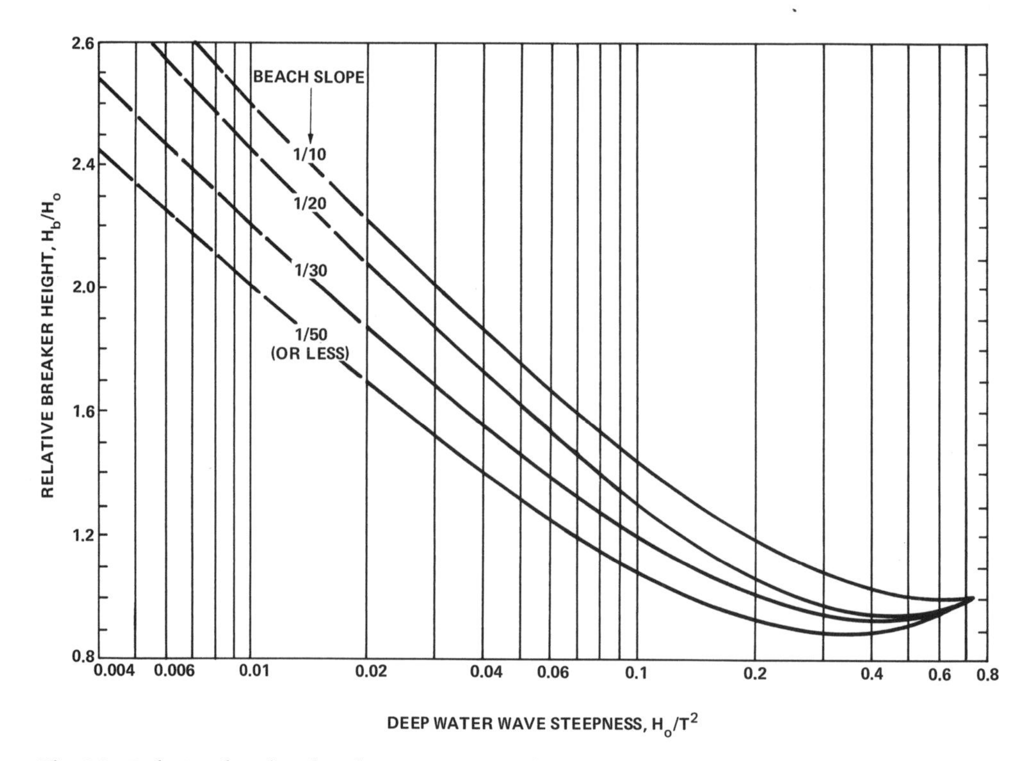

Fig. 89. Relative breaker height increases with increasing beach slope, and decreases with increasing offshore wave steepness.

Plate 19. Beach cusps result from cellular longshore circulation that transports sand away from zones of high breakers.

Plate 20. Pattern of rip currents is plainly visible because of silt from recent rain.

Plate 21. Solitary surfer is dwarfed by 32-foot breaker at Sunset Beach, Hawaii.

slope. Thus, in a storm-generating area, where the average waves are steepest, the breaker height on a lee shore will be of the same order as the prevailing significant sea state. Conversely, long swell from a distant storm, breaking on a gradual slope, may average twice the swell height in deep water. The highest official swell-breaker heights I know of (50–60 feet) were measured along the north coast of Oahu, Hawaii, by scientists from the Hawaii Institute of Geophysics during the storm of December 1–5, 1969 (see page 192). The largest I have personally witnessed were cresting the horizon at 55 feet above sea level, as sighted from the radar bridge of the Scripps Institution vessel *R/V Argo*, at anchor off Manihiki Atoll, Cook Islands, on June 2, 1967. This swell had a period of 21 seconds and an estimated deep-water height of 20–25 feet, and was approaching from due south.

Experienced surfers become expert judges of relative breaker height, but "professional" modesty shades their absolute estimates by about a factor of two. By simple scaling, the comber shown in plate 21 is cresting over 32 feet high, but the "surfer's estimate" here would probably run from 15 to 18 feet, depending on his or her audience, respectively.

More precise estimates are easily made by anyone from shore, by planting a pole in the sand, and marking the point on it that coincides with your line of sight across breaker crest to horizon. The elevation of this point above mean sea level can then be determined between groups of waves, when the sea is temporarily calm. Add 25 percent for trough depression to obtain the crest height. I have done this many times in the presence of witnesses—but surfers are hard to convince.

18

Tsunamis

CAUSE AND NATURE

Few natural phenomena have popular conceptions more surrounded with an aura of mystery than *tsunamis,* as exemplified even by their more-common misnomer: "tidal waves." At irregular intervals of from 5–15 years—and often without prior warning—the sea begins to heave and churn, sometimes receding to bare its floor, or suddenly rising far beyond the normal range of wave and tide, flooding over breakwaters, tearing ships from their moorings, and leaving widespread destruction to shoreline habitations and facilities. Although tsunamis are as old as history, they have only recently yielded to concerted scientific study as simply another manifestation of crustal readjustments, whereby the earth accommodates itself to internal radioactive heating.

General Phenomenology: Tsunamis comprise a special class of wave systems that originate within localized regions as the result of short, impulsive disturbances. Most large tsunamis are generated by vertical sea-floor dislocations associated with large, shallow-focus earthquakes, although only about 10 percent of the large quakes in a given region seem to produce them. This is thought to be due to relative differences in the type and extent of a sea-floor motion, which involves not only earthquake intensity, but also the underlying crustal structure and the focal depth of the quake. Historically, an average of 10 large and 100 minor tsunamis per century have occurred in the Pacific, which accounts for 85 percent of the total. Thirteen significant events have occurred within 250 years in the Atlantic, and 50 or so in the Indian Ocean and eastern Mediterranean Sea.

More rarely, tsunamis have resulted from explosive volcanic eruptions, such as that of Krakatoa, in the East Indies (1883). This 500-megaton explosion produced waves that inundated several nearby islands, drowning an estimated 36,000 people; the waves were recorded as far away as San Francisco.* Smaller tsunamis have been attributed to massive submarine sediment slides, but no large tsunami is known to have been generated in this manner.

Because even the deepest ocean depths are small compared with the breadth of the sea-floor dislocation required to produce a large tsunami, the resulting deformation of the water surface radiates in all directions as a system of shallow water waves.

*Anomalous waves recorded on tide gauges in the English Channel have been linked to the air-shock pulse from the explosion.

Much like the familiar pattern produced by dropping a pebble into a shallow pond, rings of waves spread out, with the longest at the front—or leading edge—of the system, and progressively shorter waves following behind. The velocity of the leading wave is given by the limiting speed $C = \sqrt{gh}$ ft./sec. for progressive waves in shallow water (450 knots in the Pacific), and all subsequent waves travel more slowly. In contrast to wind waves, all wave crests permanently maintain their identities, and their average heights diminish with distance because of geometric spreading and dispersion. As mentioned in Part IV, spreading reverses a little over halfway to the antipole of the source, and the waves somewhat increase in height after traveling more than 6,700 miles across the sea.

As the waves pass into shallower water, they build up again over the continental margins, where the previously symmetrical pattern is so altered by refraction and reflection processes that individual waves can no longer be identified. In the coastal regime, a tsunami is manifested by a long succession of periodic surges at intervals of from 10–20 minutes to 1–2 hours. With large tsunamis, these surges may produce swift, anomalous currents in harbors and estuaries, overrun low-lying land areas, and splash up as high as 60–100 feet on steep shoreline cliffs. Conversely, between surges, the sea retreats, exposing broad expanses of sea floor normally covered by 10–15 feet of water.

Most of the destructive effects from large tsunamis occur within a few hours after the first waves arrive, following which the surge intensity gradually diminishes. Decaying oscillations can be detected on tide records for several days (fig. 90).

Tsunami Generation: Because earthquakes are the primary source of tsunamis, we will consider only this mode of generation. As described in Part I, the ocean floors are composed of slowly moving crustal plates that plunge under the continental plates, forming the great, peripheral trench systems and adjacent mountain ranges in the process. Pressure between adjacent plates accumulates as elastic strain at the plate edges, and is discontinuously relieved in the form of earthquakes. In some places, as along the coast of lower California, the strains are oriented horizontally (fig. 91a), and the resulting horizontal displacements during earthquakes do not significantly perturb the sea surface. But, where the plates meet head on (Peru, Chile, and the Aleutian Island Arc), vertical strains result in earthquake dislocations that uplift or depress vast segments of the sea floor. The sea surface is correspondingly deformed, and a tsunami is the end product of its attempt to reestablish equilibrium (fig. 91b). The major tsunami source regions of the Pacific and their dates of most recent activity are shown in figure 92.

In large tsunamis, the volume of water displaced defies imagination: the great earthquake of March 28, 1964, in southeastern Alaska involved a dislocation averaging 6 feet vertically over 100,000 square miles—thrice the size of Florida (fig. 93)! About half this area was on land, and subsided; the other half, which included the entire 100-mile-wide shelf bordering the Gulf of Alaska, was bulged upward—in some places as much as 50 feet. The resulting mound of water poured off the shelf for two hours, and fanned out over the Pacific, creating widespread damage as far south as Crescent City, California, some 1,200 miles distant, where half of the business district was swept away.

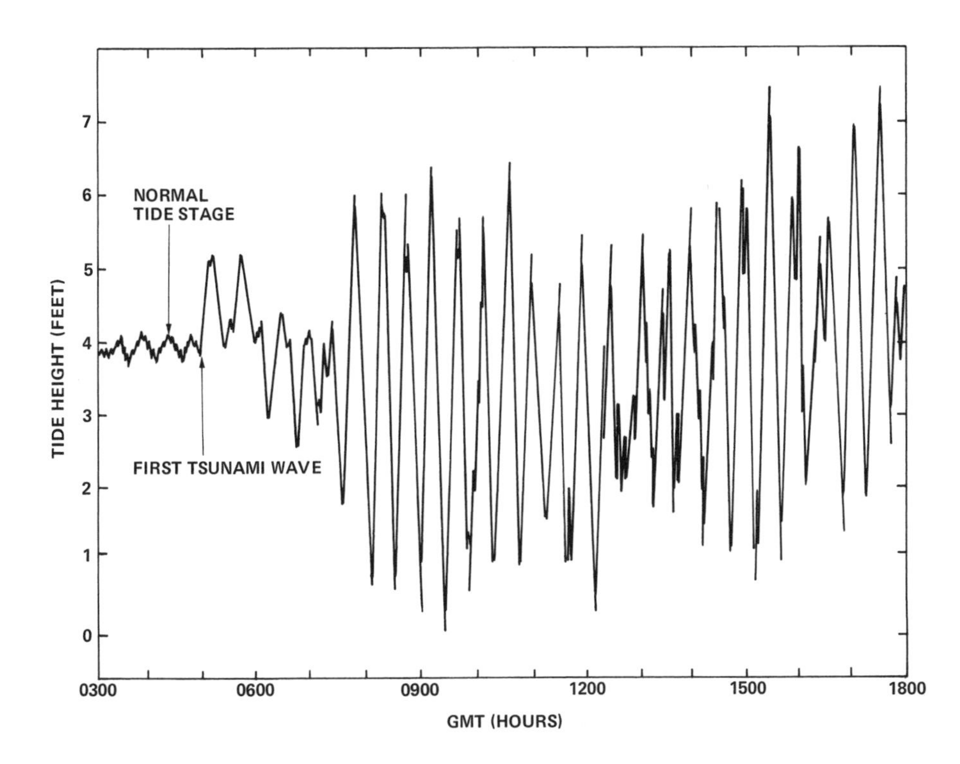

Fig. 90. Tide gauge record of the great tsunami of May 22, 1960, in southern Chile, as observed at Acapulco, Mexico.

Although this was by no means the most destructive tsunami in historical times,* five of Alaska's seven largest communities were devastated by the combination of earthquake and wave damage. Its fishing industry and most seaport facilities were virtually destroyed, and only massive federal aid and years of effort have sufficed to restore its crippled economy.

As a long-time student of tsunamis, I flew to Alaska within twenty-four hours of the quake, and made a ten-day aerial tour of the entire region affected, traveling by helicopter, navy aircraft, and chartered bush plane. Among accounts of calamities too numerous to mention is one of particular interest to mariners. Most of the Alaskan crab fleet was at anchor in Kodiak harbor when they received a belated warning of "50-foot waves" passing Cape Chiniak, 20 miles to the south. Getting rapidly under way, they were well out into Chiniak Bay when they encountered the first incoming crest, estimated to be 30 feet high and breaking. The next instant they were making 16 knots sternway, and were carried over a mile back into the harbor, over a protective mole, and on up into the center of the waterfront business district. The crab boats, being rather stoutly constructed, were not appreciably damaged by this excursion into unfamiliar territory, but a number of buildings were knocked down during their gyrations. The lesson here is that the fleet was within the epicentral area of the earthquake, which had been plainly felt by everyone. The first waves did not arrive until some forty-five minutes later. Had the fleet put to sea

*Recurrent tsunamis in Japan have taken over 100,000 lives and have caused untold property damages.

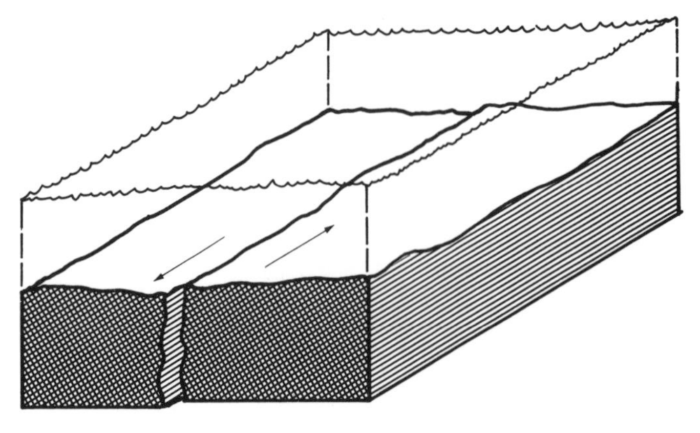

(a) HORIZONTAL (STRIKE) SLIP-NO TSUNAMI

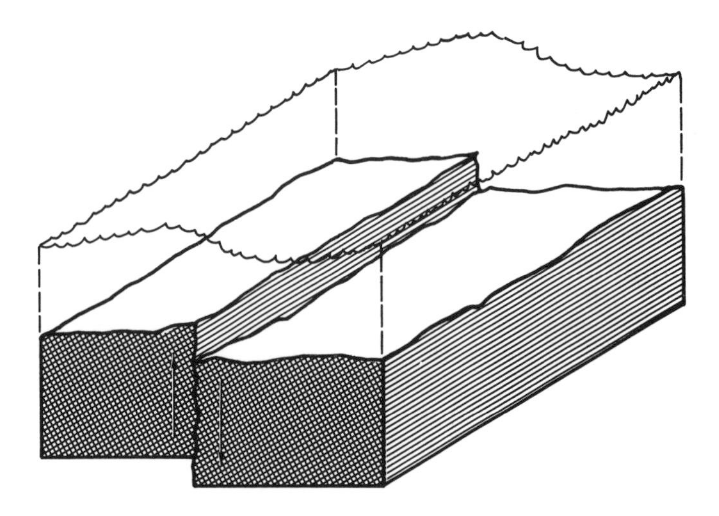

(b) VERTICAL SLIP-TSUNAMI

Fig. 91. Tsunamis are produced by vertical sea-floor dislocations that similarly deform the water surface. The vertical scale in this figure is enormously exaggerated.

directly after the earthquake instead of waiting for further advisement, the catastrophe might have been avoided.

Open-Sea Character: Because of their low heights (1–2 feet) and long wavelengths (50–250 miles) outside the source region, even large tsunamis pose no hazards to vessels at sea. Indeed, one would be hard put to measure them with even the most sensitive instruments from shipboard. However, shortly before Hawaii was devastated by the great Aleutian tsunami of April 1, 1946, Admiral Lloyd Mustin saw something on the cruiser *Baltimore's* radar screen that he interpreted as rapidly advancing waves from the north. This is, I believe, the only known report of a tsunami sighting by radar.

Coastal Effects: The severity of local wave effects at any place remote from the source of a large tsunami depends in a complicated way upon the source size and orientation, its distance, and the local offshore topography. Coastal regions protected by a wide, shallow offshore shelf experience a much lower intensity of

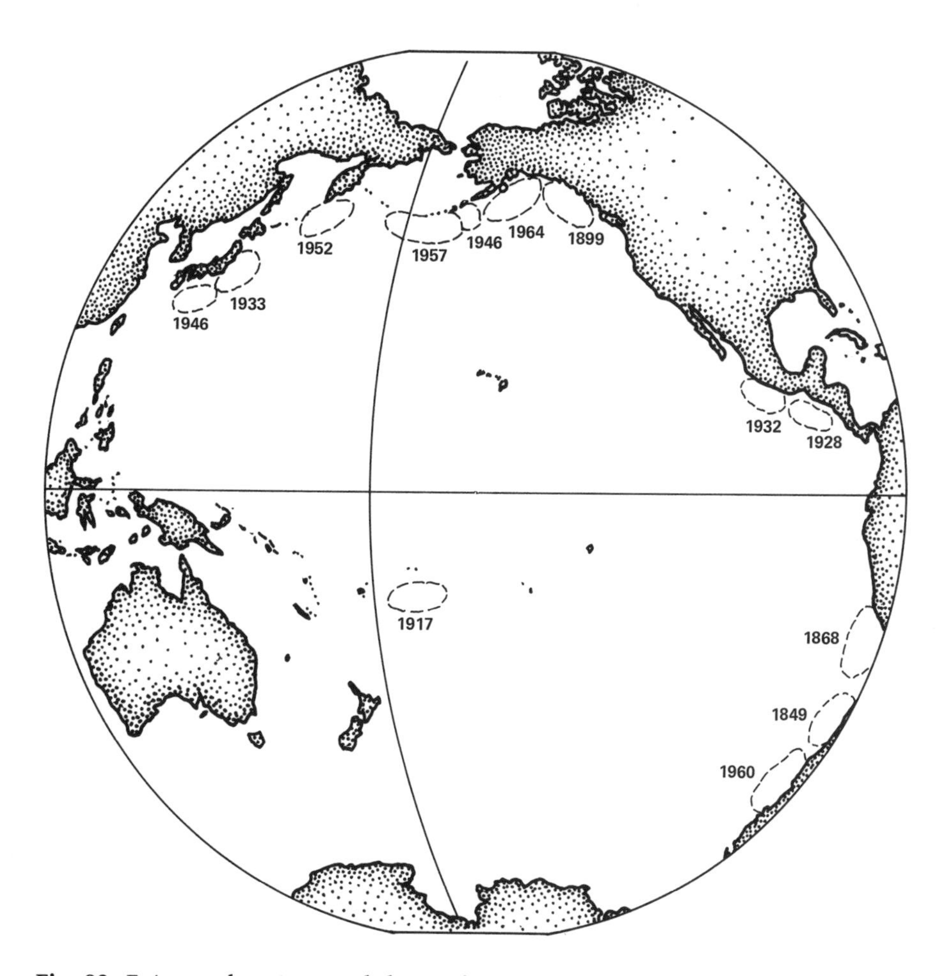

Fig. 92. Epicentral regions and dates of occurrence for recent destructive tsunamis in the Pacific Ocean.

shoreline wave action than those that slope rather steeply to great depths. Thus, southern California is relatively immune to direct wave damage, while Hawaii and Japan are particularly vulnerable. Aside from Hawaii, whose islands comprise the peaks of an undersea mountain chain, most small, isolated Pacific atolls offer little impediment to the advancing wave system, and effects are usually negligible.

Even in susceptible regions, the local inundation heights vary by a factor of ten from point to point along the coastline. As might be expected, most damage occurs in populated areas, where surging action may overtop—or carry away—protective breakwaters and seawalls, demolish waterfront structures, and tear vessels from their moorings (plate 22). Protected as they are, the San Diego and Los Angeles harbors both sustained substantial damage—principally to smaller craft—during the Chilean tsunami of May 22, 1960. Strong, oscillating currents were generated in response to the exterior rise and fall of water, and large sections of older, rotten, wooden wharfage came adrift, together with dozens of boat slips. These unwieldy assemblages collided with adjacent facilities, producing further damage. The San Diego–Coronado Ferry was forced to suspend operations for sixteen hours when its ships were swept 2 miles up and down the bay. All this, while the maximum sea-level excursion recorded at the Scripps Institution tide gauge on the open coast never exceeded 4 feet!

TSUNAMI WARNINGS

Through international cooperation after the devastating tsunami of April 1, 1946, an oceanwide Tsunami Warning System (TWS) was established among countries bordering the Pacific Ocean. The TWS headquarters in Honolulu contains a special seismograph, equipped with an alarm that rings whenever an earthquake of magnitude 7.0 (considered to represent the threshold for quakes capable of generating a dangerous tsunami) is detected. Immediate queries are then teletyped to seismic stations in Japan, Alaska, and California, whose responses confirm the quake magnitude and establish its epicenter—usually within minutes. Key tide stations nearest the epicenter are next alerted by radio or telegraph to report any indication of abnormal wave activity, and tsunami advisory messages are sent to interested military and civil agencies throughout the Pacific. Wave arrival times at principal ports are determined from travel-time charts and taped for transmission. If no waves are reported from the key stations, the alert is cancelled; but if the reports portend a dangerous tsunami, warnings and arrival times are broadcast, followed by additional reports, as received, until the danger is past. Local warnings are disseminated by civil agencies in most areas likely to be affected.*

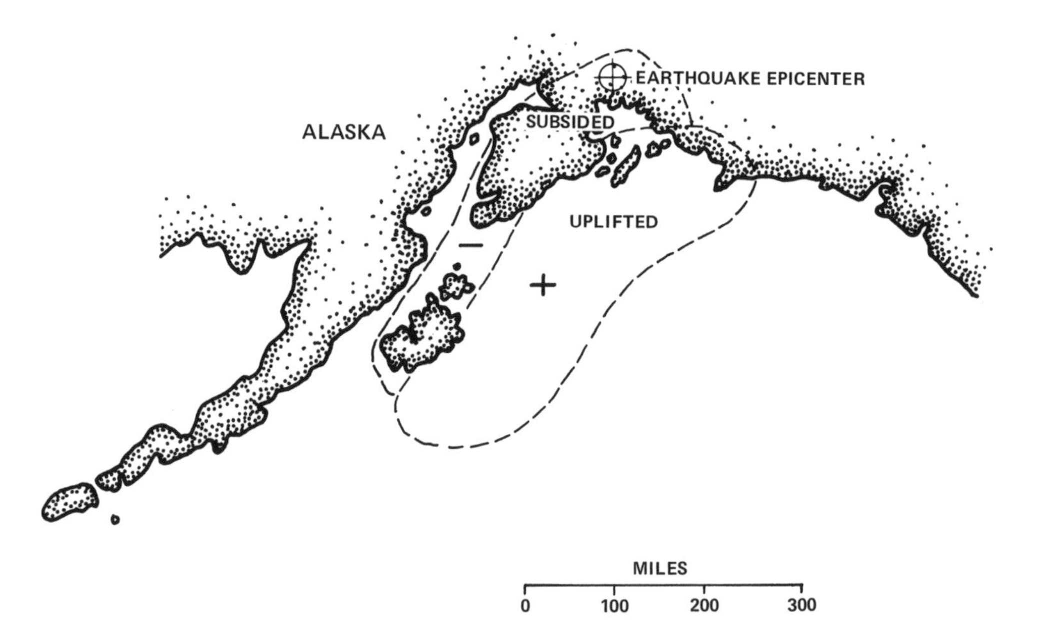

Fig. 93. The source area of the 1964 tsunami in Alaska was about 100,000 square miles. Uplifted portion extended over most of the shallow shelf bordering the Gulf of Alaska.

*In 1986, NOAA scientists tested for the first time a prototype of an early-warning system, whereby automatic seismic and tsunami sensors detected and transmitted signals via a Geostationary Operational Environmental Satellite (GOES) to a remote warning station, thus opening up a new era of rapid warnings for a variety of geophysical hazards.

Plate 22. Devastation at Hilo, Hawaii, during tsunami of May 22, 1960. More than a square mile of waterfront was swept by waves from 12 to 16 feet high.

PROTECTIVE MEASURES

Except in Japan, most harbor and port facilities frequented by oceangoing vessels are remote from known tsunami epicentral regions, and several hours' notice of an impending tsunami is usually provided where warning facilities are available. Such warnings are generally broadcast on military and commercial radio frequencies and, in especially susceptible areas, by local public address facilities at moorings and yacht clubs, or by portable civil emergency units.

In case of tsunami alert, only two courses of action are indicated: head for the open sea for a mile or two beyond the limits of the 10-fathom contour, or double up your slip lines and ride it out. This choice is predicated on the time available, the degree of protection afforded by the harbor—relative to previous experience—and the integrity of local mooring facilities. It should be emphasized that the TWS cannot reliably forecast tsunami intensity; the first choice is always to get to sea, if possible.

If circumstances require that a small vessel remain in port, there is an advantage in tying up to a fixed dock or wharf, in 20–30 feet of water, and well away from larger vessels, where there is a possibility of mooring failure and subsequent collision. The latter are particularly vulnerable to surge action and, even where securely moored, may impose severe stresses on wharves or docks, to the point where failure of the entire structure may occur. Mooring lines should be manned to maintain fender contact, and so that necessary slack can be veered to compensate for the rise and fall of water. Auxiliary fenders are useful to prevent impact during such motions.

Anchorages or buoyed moorings of any sort cannot be trusted, since the oscillating currents can swing compass-round within a few minutes. Collision with floating debris of all kinds is a common hazard during tsunamis.

While the severest local activity is usually over within 4–6 hours, some tsunamis exhibit renewals of intensity or adverse effects that depend upon the tide stage, and it may be necessary to maintain vigilance for 24 hours or more in extreme cases. The best guide is judgment, since local authorities usually cannot predict the duration of hazardous conditions. If several hours of diminishing activity follow high water, it is usually safe to consider that the worst is over.

19

Harbor Oscillations

NATURE AND CAUSES

Many harbors of the world are intermittently beset by long-period oscillations of magnitudes sufficiently serious as to interfere with normal operations. Concordant heaving and/or surging motions induced in vessels at dockside may be large enough to part mooring lines, to abrade rub rails and fenders, and—in some cases—to cause damaging collisions between adjacent ships. At best, such motions interfere with cargo transfer, sometimes forcing suspension of activities for several days. Vessels anchored or moored in open roadsteads are less vulnerable, but may swing to and fro in an anomalous fashion, obstruct ship traffic, or drag their anchors in poor holding ground. Large harbors, notorious in this regard, include Table Bay, Cape Town, South Africa; Hilo Harbor, Hawaii; and even the port of Los Angeles, California.

Harbor oscillations, known to oceanographers as *seiches,* are nothing more than the standing waves described in Part IV, and comprise the natural modes of oscillation of water in semienclosed basins (fig. 94). They can be initiated by a variety of external disturbing forces, such as gusty winds, heavy ground swell, or—more rarely—a remote tsunami. But in all cases cause and effect are similar: the pattern of wave motion is governed by the harbor shape and orientation; its severity is a balance between the intensity and frequency of the exciting forces and the rate at which energy is removed from the system by internal dissipation and radiation from the harbor entrance.

In principle, there are an infinite number of possible modes that can be excited in any basin. But in practice, only the fundamental (longest-period) longitudinal (*a*) or transverse (*b*) modes contain significant energy (fig. 94). These modes have periods equal to the time required for long waves ($C = \sqrt{gh}$) to travel the length (or breadth) of the harbor and back again. Sometimes these two modes are out of phase, and the result is a rotary, progressive mode that travels around the harbor (*c*).

Because strong oscillations represent accumulations of stored energy, rectangular harbors with steep—or vertical—boundaries are the worst offenders, since multiple reflections can occur with little loss of energy. Geographically, a harbor's orientation, mean climate, and degree of exposure to wind and waves are factors influencing the relative severity of conditions. Thus, there is often a distinct seasonal preference between suitable alternate ports within a limited region—a circumstance seldom noted in nautical advisories.

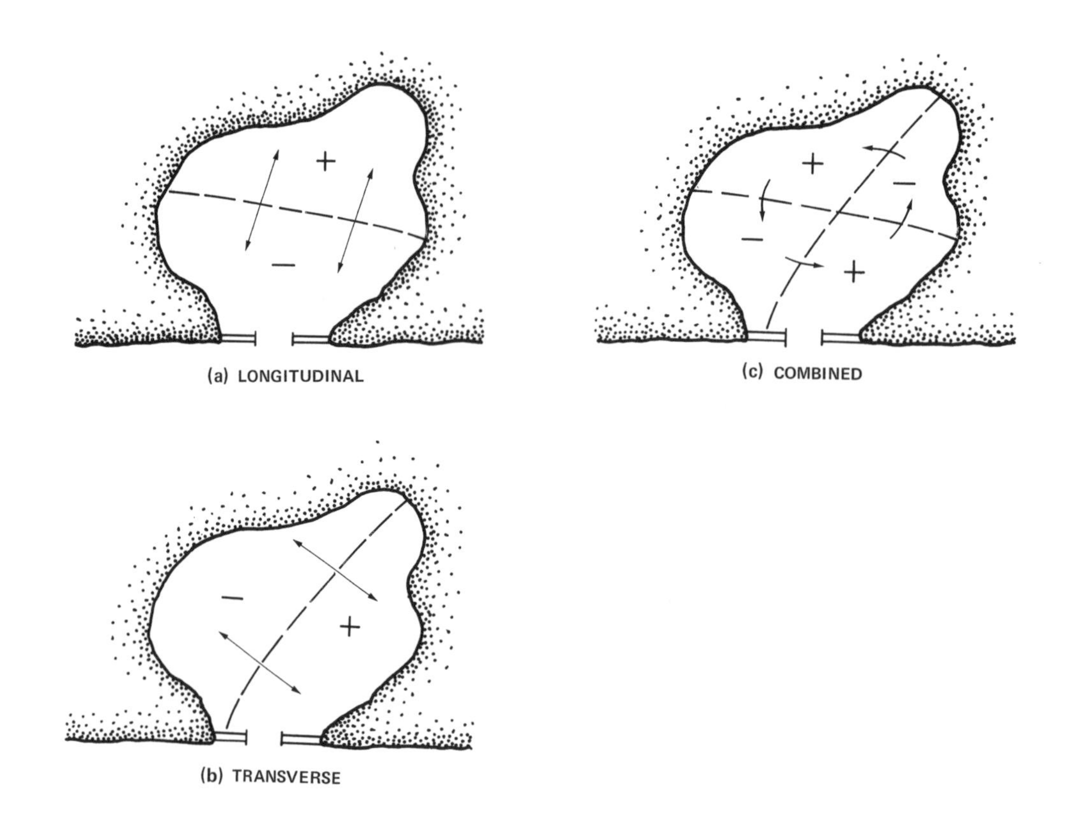

Fig. 94. Common modes of harbor oscillation. Plus and minus signs indicate water-level anomalies that reverse sign every half-period of oscillation. Currents (arrows) similarly reverse.

PROTECTIVE MEASURES

Most of the adverse effects on vessels in port, and suitable protective measures, are similar to those described above for tsunamis. If the oscillations are wind-generated, additional consideration should be given to direct wind effects when anchoring or mooring, as these may be more severe than those of water motion.

By reference to *Coast Pilot* or *Sailing Directions,* one can usually identify harbors where seiching is a problem. Port captains ordinarily direct boats to anchorages, but their advice is often motivated by convenience, rather than by concern for safety, and should be considered advisedly. If you are lacking other information, observe how local boats are disposed and moored, and inquire about conditions from their owners. Part VIII includes some additional comments on anchors and anchoring.

PART VI

Ship Dynamics

20

A Little Background

INTRODUCTION

Having wet our feet, so to speak, in preceding parts on the nature and characteristics of the sea and her various moods and mysteries, the remainder of this book is devoted to practical applications of this information, most of which fall under the general category of seamanship.

In the broadest sense, *seamanship* covers a host of things that would be inappropriate in a book dealing with the sea as an operating environment, and the behavior of vessels thereon. Pilotage, navigation, and the care, maintenance, and normal operation of boats and ships are treated in many other works—and one would be foolish to venture out without a basic grounding in these subjects. Accordingly, we here take the narrower view that most books on seamanship suffer from an inadequate representation of the real ocean, and that while experienced sailors may well know what to do in most circumstances, they may not know why it works. It is a curious dichotomy that "why it works" can be found only in books on naval architecture, and seems not to have filtered down to the level of instruction in shiphandling.*

Having throughout this book adopted the philosophy that why it works (or happens) is as much of interest as what works, in the ensuing sections we shall borrow a few secrets from the naval architect's bag of tricks, with the objective of aiding and abetting seamanship—particularly in heavy weather—through a better understanding of the principles underlying ship design and performance in a seaway. In this, we should not suppose that all of the problems are solved; there are still substantial gaps between theory and experiment. But we address ourselves here to the widest gap—that between designer and skipper.

THE SEARCH FOR SPEED

Naval architecture, which is concerned with the design and seaworthiness of ships of all classes, has advanced within the last 150 years from a nearly occult art to a highly technical science.** According to *Lloyd's Register of Shipping,* which keeps

*For those interested in the broader aspects of seamanship, the more important works are included in the References.

**The famous yacht designer L. Francis Herreshoff apparently does not share my views, especially as regards yachts. In his celebrated work *The Common Sense of Yacht Design* (now, sadly, out of print and almost unobtainable), he declaims against naval architects, mathematical analysis, marine toilets, overlapping jibs, and balloon spinnakers, among other things now commonly accepted.

tabs on all seagoing merchant ships of 100 tons and upwards, average annual ship losses from all causes during this period have diminished tenfold, roughly from 1,500 to 150 ships per year. This reduction is attributable to many factors. The transition from sail to screw propulsion has increased speed and maneuverability and has largely eliminated dependence upon wind and tide, and the development of steel hulls has reduced maintenance and has made possible the construction of much larger and stronger ships. The modern seaman has the advantages of charts, navigation aids, communications, and worldwide weather facilities; the invention of radar has greatly mitigated collision risk. Lastly, facilities for prompt rescue and assistance to ships in distress in well-traveled coastal waters have steadily reduced loss of life from the disasters that still occur.

It is primarily the constant economic pressure for increased size and speed of commercial vessels* that has elevated naval architecture to its present high technical level. Apart from the functional requirements to be satisfied in designing a particular vessel, optimum performance usually implies the maintenance of highest sustained speed consistent with safety on a given heading. This objective holds not only for cargo and military vessels, but also for power cruisers, and for racing sailboats and hydroplanes as well. Even in cruising sailboats, speed is considered a desirable factor, since there is an innate satisfaction in making a fast passage.

Yet, with increasing sea state, sustained speed is limited, not by the additional power required to drive a hull in the face of wind and wave resistance, but by the necessity for voluntarily reducing power in order to moderate increasingly violent ship motions. Figure 95, derived from review of typical cargo ship logs in the north Atlantic, shows the dramatic difference between possible and tolerable ship speeds as a function of wave height. Although this figure is only qualitative, voluntary power reductions commence when the significant wave height has risen to as little as 5 feet; in 25-foot waves, most vessels studied can barely maintain steerageway.

The adverse effects of ship motion are roughly divisible into two categories: those related to the *amplitude* of motion, such as shipping water, slamming, propeller racing, and steering difficulty; and those produced by severe *accelerations*, including cargo shifting, dislodging fixtures and equipment, and the discomfort of passengers and crew. Perhaps surprisingly, there seems to be very little information on acceptable limits for these effects, which run the gamut from those of a sedate passenger liner, where comfort is of paramount importance, to those of the racing sailboat or hydroplane, where speed is routinely pressed to the threshold of disaster.

The end result of this search for speed is that the naval architect must ever more closely tread the path between those factors controlling propulsive effort, maneuverability, strength, seaworthiness, and sea kindliness, on the one hand, and ever-increasing dynamic forces on the other. For large ships, these challenges have been met by progressively larger and more complex model-testing facilities, in which sea conditions can be generated in almost any desired scale, and by concomitant advances in oceanography, hydrodynamics, engineering, and structural theory, abetted by the use of large computers. At this writing, I have on my desk a brochure from the Institute for Marine Services in Bilbao, Spain, that describes a womb-to-tomb computer program, into which one feeds the specifications for a desired ship and its anticipated operating environment, and receives back a com-

*Military vessels pose their own particular exigencies.

plete hull design, even including specific shapes to which the hull plates should be cut for optimum fabrication! But in the United States, at least, tank testing is still employed extensively—if only to keep the computers honest.

In a more restricted sense, the great advantages of computer-aided design (CAD) are making inroads in the burgeoning small-craft industry, whose total tonnage now compares with that of all U.S. commercial shipping. Through interactions between naval architects and CAD specialists, techniques that could formerly be applied only to America's Cup contenders are now being applied to the specialized design of many classes of smaller commercial vessels, and to private yachts, both sail and power.

Nevertheless, yacht design today remains as much art as science; the rash of structural failures associated with the massive transition to fiberglass construction, and with the use of exotic composites such as carbon-fiber masts and rudderposts, suggests that there is still a great deal to be learned if we are to take optimum advantage of the great strength and versatility of these new materials. As noted below, the failure problem seems also compounded by sailing yacht rating rules that result in designs favoring speed over structural safety and seaworthiness.

IRON SHIPS AND WOODEN MEN?

While ship design has taken a great leap forward, the pleasure boat explosion seems not to have improved the average quality of seamanship. Thousands of amateur yachtsmen are now venturing beyond sight of land with little or no knowledge of the hazards involved. The U.S. Coast Guard currently spends 80 percent of its time

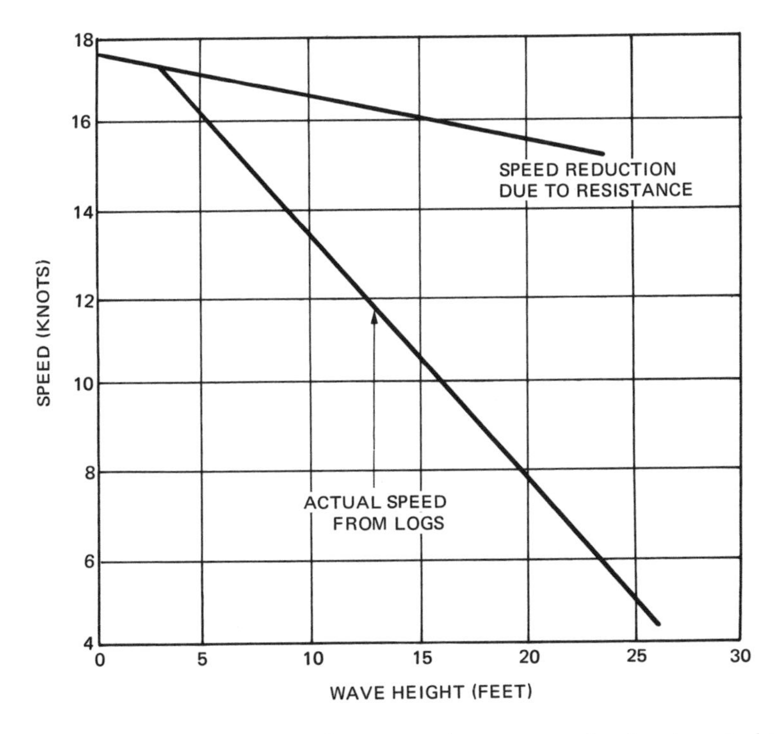

Fig. 95. Resistance-imposed versus voluntary speed reductions in high seas, as determined from logs of cargo ships in the north Atlantic.

rendering assistance to private yachts, as compared with about 19 percent for small documented and commercial vessels, and only 1 percent for registered vessels. Although only about 4 percent of all U.S. marine casualties occur in offshore coastal waters, the average number of lives lost in small-boat accidents substantially exceeds the worldwide total for registered vessels.* Of these, the Coast Guard attributes the majority to bad seamanship, leading to capsize as a result of overloading, ignoring weather warnings, or venturing into sea conditions that exceed the skipper's training or experience; secondary and tertiary causes are crew overboard and swamping, respectively. Most injuries result from collision of small, high-speed power craft, of which the primary cause is failure to maintain adequate forward lookout. Manifestly, many of these casualties could be avoided if more boat owners were to take advantage of free courses in basic seamanship and boat handling offered by the U.S. Coast Guard Auxiliary and the U.S. Power Squadrons at more than 350 locations throughout this country.

From the foregoing discussion, one is led to wonder whether we are currently undergoing a transition from a long tradition of "wooden ships and iron men," where every seaman knew how to hand, reef, and steer, to the reverse. Ships and yachts are now larger, stronger, and faster, but there seems to be room for improved seamanship.** To this end, in this part, we attempt to understand how the naval architect goes about determining the behavior of a vessel in a seaway, and in the next, how these results can be applied to the problems of ship handling.

*Coast Guard statistics for 1990 alone list 865 fatalities and 3,822 injuries among 6,411 accidents involving 8,591 vessels. My almanac lists an average of 426 lives lost per year for all major marine disasters between 1980 and 1990.

**Even the U.S. Navy might stand a little practice. Some years ago, as commander of a scientific task group aboard a light carrier, I witnessed a series of ludicrous attempts to launch a 12-man motor whaleboat, with the object of recovering a piece of scientific gear that had been dropped alongside by helicopter. The ship was hove-to to weather of the drop point in 6-8-foot seas and a 20-knot wind, and was rolling through about 30° of arc. As the first boat descended from beneath the flight deck some 50 feet above the water, she commenced to pendulate, and her swings increased to remarkable amplitude as she came into synchronism with the ship's roll. On the third or fourth swing, she struck the ship with such violence that her entire side was stove in, following which she was hastily retrieved. The second boat, lowered faster, reached the surface and bounced heavily on a wave crest. Her crew, considerably encumbered in bulky cork life-vests, managed to cast off the stern fall, but the bow remained secured, with the result that she was upended and lifted clear of the water on the ship's next roll, dumping all occupants into the sea. A third boat was then launched successfully to rescue the survivors, but, this accomplished, great difficulty was experienced in hooking up the falls simultaneously. When they finally got clear, it was discovered that the bow falls had tumbled, such that there was a twist in each part, and the boat could not be hoisted high enough to disembark her crew. After considerable discussion, the carrier was headed upsea, and the boat was warped into a side loading port, where she remained for the duration of the exercise. (N.B.: the gear was never recovered.)

21

Stability

HYDROSTATIC EQUILIBRIUM

That a vessel floats upright and on her marks in calm water is ordinarily taken for granted. Yet the interplay of forces that maintain her in this configuration is basic to ship dynamics and deserves some mention. According to Archimedes' Principle, any body partially or totally immersed in a fluid is buoyed up by a force equal to the weight of the displaced fluid. For a floating body, the proof of this statement is very simple: when the body is in equilibrium, it has no tendency to move and its entire weight is supported by the fluid. If somehow we could remove the body and instantly fill the resulting cavity with fluid identical to that surrounding it, no motion would take place: the body weight would exactly equal that of the displaced fluid (fig. 96a).

Archimedes' Principle gives not only the magnitude of the upward force of buoyancy, but also its line of action. Since the fluid filling the displacement cavity is in equilibrium under the action of all pressure forces acting perpendicular to the submerged cavity surface, it follows that the resultant of all horizontal components of these forces must be zero, and that of the vertical components must exactly equal the weight of the displaced fluid. Now suppose that we freeze this displaced fluid without change of volume, remove it from the cavity, and hang it from a thread in precisely its original attitude. We have, in effect, replaced all of the upward pressure forces with downward gravity forces of equal magnitude, and their sum—the total weight of (frozen) fluid—just equals the thread tension (fig. 96b). Because the suspended fluid is in horizontal equilibrium, the thread axis must pass through its *center of gravity* (CG).* Thus we can conclude that the resultant of the buoyancy forces must similarly act upward through the CG of the displaced fluid which, by convention, is called the *center of buoyancy* (CB).

The equilibrium attitude of a buoyant body floating in calm water is determined solely by interaction between the weight of the body, acting downward through its CG, and the resultant of the buoyant forces, which is equal in magnitude to the weight of the body and acts upward through the CB of the displaced water. If these two forces do not pass through the same vertical axis, the body is not in equilibrium, and will rotate so as to bring them into vertical alignment (fig. 97a). The body is then said to be in static equilibrium.

*In this and succeeding chapters, I have introduced a number of symbolic definitions that may not conform to "standard practice." There seems to be no universal standard terminology that applies to all maritime agencies, and the symbols used are generally phonetic and internally consistent.

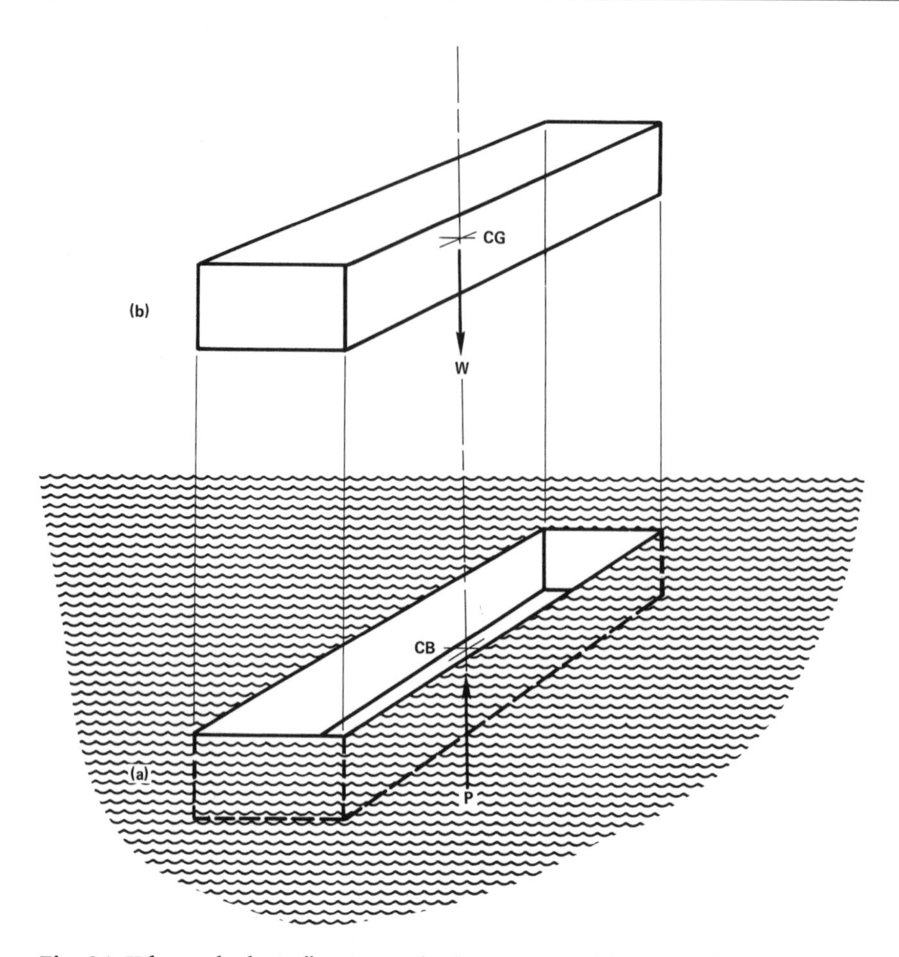

Fig. 96. When a body is floating in hydrostatic equilibrium, all horizontal pressures cancel. The resultant of all vertical pressures (P) is equal to the weight of displaced water (W), and passes vertically upward through its center of buoyancy (CB). The CB is also the center of gravity (CG) of the displaced fluid.

The measure of *static stability* is the *moment* (twist) that must be applied to rotate a body through any given angle away from its equilibrium attitude. *Dynamic stability* is the work required (moment × angle) to effect such a rotation.

With two exceptions,* there is at least one stable equilibrium attitude for any floating body—although there may be several alternate attitudes for a body floating at the surface, depending upon its shape, displacement, and internal mass distribution.** For homogeneous solid bodies of simple shape, stability is easily calculated. For example, if the density of a square wooden timber is less than 21 percent or greater than 79 percent of the density of seawater (13.6 and 51.4 lb./ft.^{3}, respectively), it will always float with one of its sides upward; if its density lies between these values, it will float corner-up (fig. 97b). This interesting duality in the stable attitudes of a square block seems not to be well known. While most common woods

*A homogeneous sphere is neutrally stable, since it will float in any attitude; a circular cylinder is neutrally stable about its axis of symmetry.

**A submerged body has only one stable position: it will always rotate so as to bring its CG directly beneath its CB.

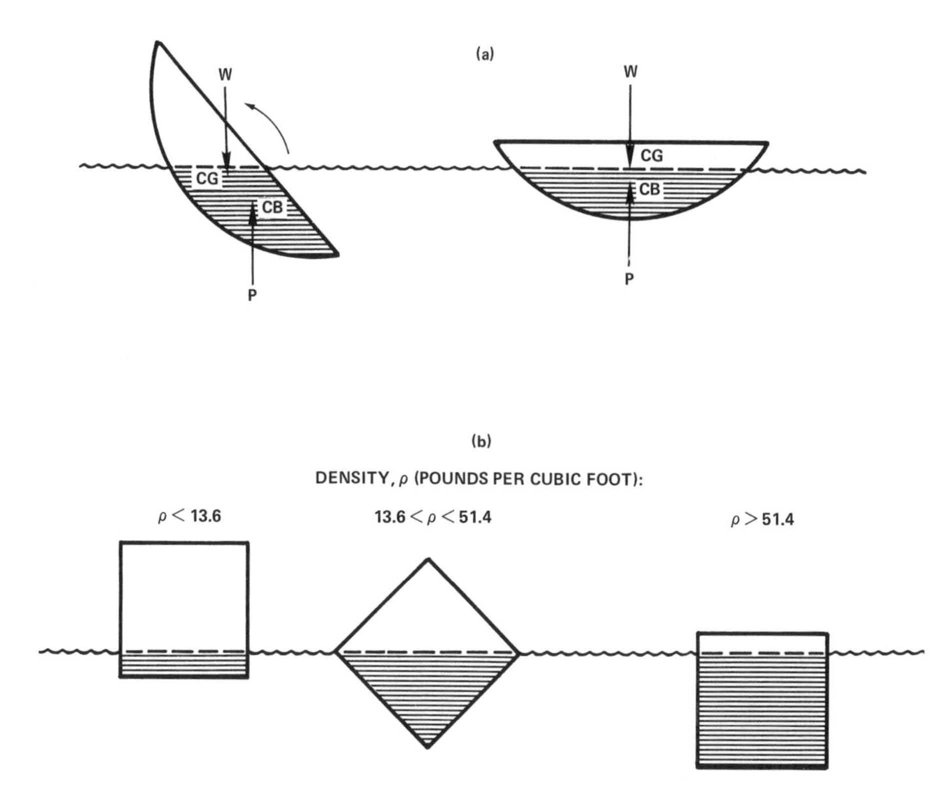

Fig. 97. *(a)* In calm water, a floating body will always rotate so as to bring its CG and CB into vertical alignment. *(b)* Depending on its density, a square, homogeneous solid block has two stable floating attitudes.

fall in the corner-up category, you can amuse your friends by floating small blocks of balsa, Douglas fir, and apitong, and asking them why only the fir block floats corner-up.

Unfortunately, no simple formulae suffice for calculating the displacement, equilibrium attitudes, and stability of an inhomogeneous body of irregular shape, such as a ship hull. Yet the designer must determine these quantities with great care, so that a vessel will not only float upright at her design waterline(s), but will also maintain an adequate safety margin of static and dynamic stability under any intended operating condition. The latter involves consideration of all anticipated dynamic forces acting on the vessel, as well as inadvertent changes of static forces that might occur as a result of cargo shift, partial flooding, or accidental grounding.

In practice, the above quantities are ordinarily calculated by a rather tedious procedure, which can be boiled down to the following.

Center of Gravity: The entire hull is first subdivided into discrete elements whose individual weights (and their CGs) can be accurately estimated. These are conventionally grouped into classes, such as hull structure, machinery, and equipment. The individual weights are then tabulated, together with their distances from each of three mutually perpendicular reference planes, often taken as the vertical *centerline plane* (CP), a vertical plane through the intersection of the stem and the waterline, called the *forward perpendicular* (FP), and a horizontal plane tangent to the lowest point of the underbody *molded baseline* (MB; see figure 98). Multiplying these distances by the individual weights gives three sets of moments—one for each reference plane—that are separately summed, and then divided by the entire weight of the vessel to obtain three coordinate distances *(moment arms),* which locate the CG with respect to the reference planes. If the designer has done the work right, the CG will fall fairly close to the point intended, although some juggling of weight may be necessary to refine its position. The designer may also compute alternate CGs for different loading configurations.

Once located, the CG fixes that point within the vessel at which its entire weight can be considered to be concentrated, and through which the force of gravity always acts vertically downward, irrespective of the vessel's attitude. In most vessels, the CG lies in the midship plane and slightly abaft the midship section. Its vertical position may vary considerably, depending upon the particular hull shape and condition of loading.

Center of Buoyancy and Load Displacement: The location of the upright center of buoyancy (CB) is determined by a somewhat similar procedure. In effect, the vessel's underbody beneath her *design waterline* (DWL) is subdivided into elemental volumes by imaginary transverse and horizontal planes. These volumes are converted to equivalent weights of displaced water; then moments are taken about the FP and MB planes, separately summed, and divided by the vessel weight to obtain the coordinates of *longitudinal* (LCB) and *vertical centers of buoyancy* (VCB), respectively. (When upright, a vessel's transverse CB lies in her midship plane.) At

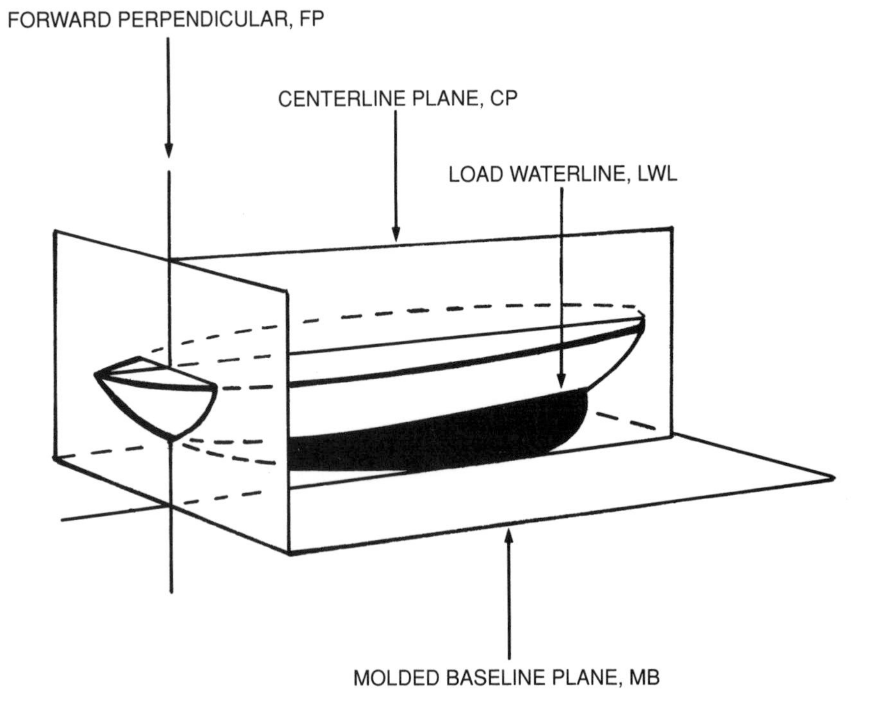

Fig. 98. Coordinate system for CG and CB calculations.

the same time, the weights of displaced water between consecutive horizontal planes are tabulated, and a curve is plotted, giving displaced weight against depth of immersion. Again, if the designer is lucky, the CB will fall vertically beneath the CG, and the immersion curve will intersect the DWL at a displaced weight equal to the vessel weight. Other displacement points on this curve define alternative load lines (Plimsoll marks*), from which a vessel's load displacement can be determined afloat by inspection.

All of this is not so much guesswork as might be implied; initial positions of these centers are estimated from experience with similar designs, and are continuously revised as the design progresses. It is rare that a vessel is launched badly out of trim or off her marks, although it is reputed that prior to World War II the Japanese constructed several destroyers from (somehow procured) American plans, the first of which capsized and sank on launching, owing to various modifications and additions, and to the lack of adequate stability analysis.

TRANSVERSE STABILITY

Having located the positions of the CG and the upright CB of a vessel, one can now investigate her transverse (lateral) stability. This is done without regard to external forces, merely by considering the hull to be inclined through several angles and calculating the respective moments exerted by the vertically opposing forces of gravity and buoyancy. These moments are generated by horizontal displacements** of the CB, relative to the CG, as a vessel is inclined (fig. 99), such that these forces are no longer colinear, but are separated by same distance d, which is a function of the angle of inclination, θ. The magnitude of both forces remains always the same,

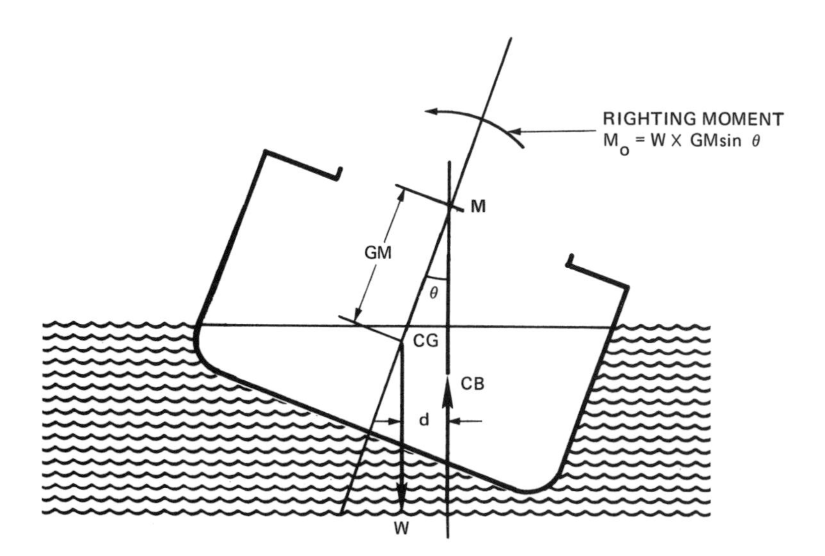

Fig. 99. Forces and moments in transverse stability.

*After Samuel Plimsoll, a British member of Parliament, who was active in promoting regulations against vessel overloading (circa 1875).

**Stability calculations are usually referenced to a coordinate system that rotates with the vessel, wherein vertical displacements must also be determined.

and equal to the vessel weight W, but their moment ($W \times d$) is similarly a function of θ. If the moment of the buoyancy (or any other) force acts to rotate the hull about its CG opposite to the direction of inclination (as shown), it is called a *righting* moment; if in the same direction, a *heeling* moment.

In figure 99, the line of action of the buoyancy force through the CB intersects the hull centerline at point M, called the *metacenter,* and its distance from the CG is the *metacentric height* (usually written GM). From the geometry of the figure, it is evident that the moment arm $d = \text{GM} \sin \theta$, and that the moment ($M$) of the force W through CB, and about CG, is: $M = W \times \text{GM} \sin \theta$,. Thus the determination of stability reduces to finding GM as a function of θ, for all desired displacements W. In essence, this involves the calculation of new positions of the CB and waterlines for each angle of inclination and displacement by the method outlined for the upright CB, except that, for transverse inclination, the moments of all component weights about the midship plane are also required.

The Stability Curve: The end result of all this mumbo-jumbo is a set of static stability curves, an example of which is shown in figure 100 for transverse inclinations of a typical cargo vessel. In this figure, the height (ordinate) of curve O–A–B–C at any angle θ, gives the righting moment exerted by the vertical displacement and buoyancy forces (shown schematically in the ship sections above) or, alternatively,

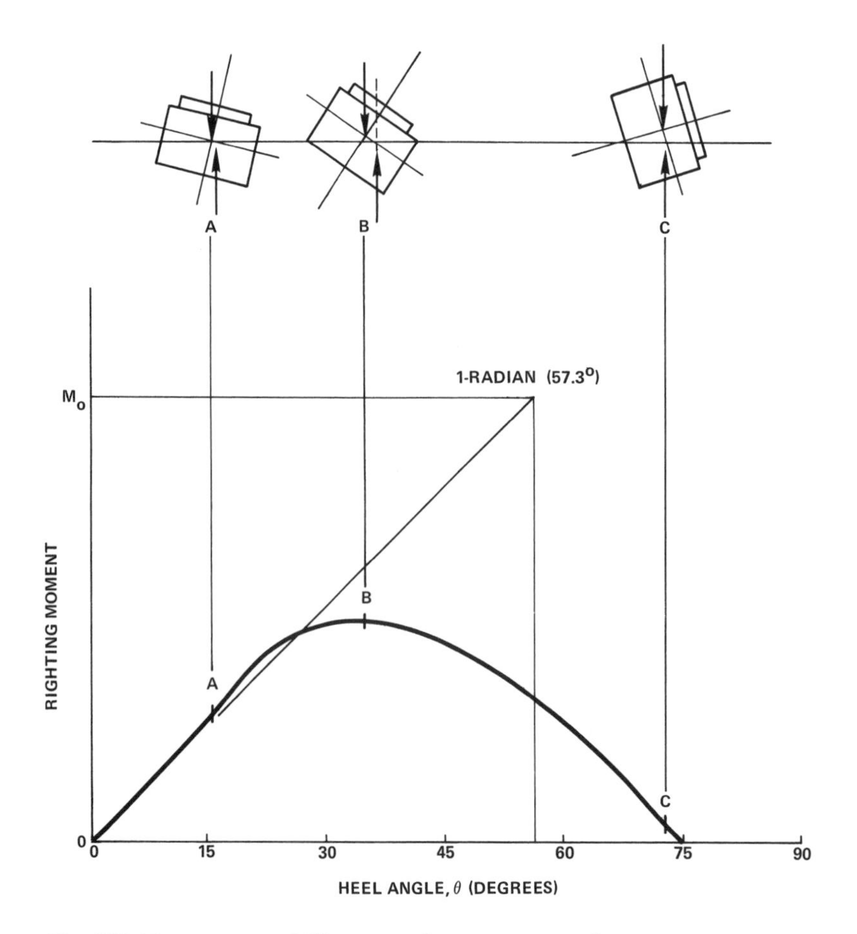

Fig. 100. Transverse stability curve for a cargo vessel.

the external moment that would be required to incline the vessel to that angle. The area under the curve to the left of any angle is a measure of dynamic stability and gives the total work (moment × angle in radians) required to effect this inclination.

At small inclination angles ($\theta < 10°$), the slope (M/θ of the stability curve is nearly uniform and independent of θ.* Because the sine of a small angle is nearly equal to the angle itself, expressed in radians, our righting moment equation can be written $M/\theta = W \times GM$. Thus, within this range, a vessel's metacentric height can be determined by extrapolating the straight portion *(O–A)* of its stability curve to its intersection with a vertical at 57.3° (one radian), projecting a horizontal line to the moment axis, and dividing this moment by the vessel's displacement: $GM = M_O/W$. For small inclinations, GM can be found experimentally in a completed vessel by moving a known weight transversely amidships, and measuring the resulting inclinations with an accurate pendulum *(inclining experiment)*. This method is often used as a final check on computed stability.

At angles greater than 10°, the heeled wedge moments are no longer equal—and GM is no longer constant. The shape of a stability curve now depends in a complicated way upon the vessel's displacement, section profiles, and CG location. In general, the CB continues to shift outboard, and the righting moment to increase, until the vessel is heeled to some point *B* where her rail dips under. This is the angle of maximum (and safe) positive stability, beyond which it becomes increasingly difficult to recover.

At still greater angles *(B–C)* progressively smaller righting moments are provided by the restoring forces, whose lines of action are now coming closer together (shorter moment arms), until finally (point *C*) the curve plunges below the 0-moment axis. The vessel is then in neutral equilibrium: any righting force will cause her to roll upright; any heeling force will capsize her. While most ballast-keel sailing yachts can survive being forced down to neutral equilibrium, this attitude is untenable for a cargo vessel, which would already be in jeopardy if listed beyond point *B*, owing to the probability of taking water topsides. In fact, *dynamic* rolling even to this angle would be dangerous, because progressively smaller transient forces (wind and waves) would roll her farther, not to mention the possibility of cargo shifting, etc. We will return to this question later, when discussing ship dynamics.

While stability curves are generated for many trial shapes and loading configurations when designing a commercial vessel,** the same principles apply to smaller boats of all classes, for which such detailed analysis might be too expensive. Therefore, we consider here only some qualitative features that have a bearing on seaworthiness, seamanship, or purchase selection.

Effects of Displacement Changes: Any significant alteration of a vessel's design load displacement (such as by the addition or removal of cargo, fuel, water, etc.) affects her inclined stability in two ways: first, the magnitude of the displacement

*This is because for small inclinations the emerged and immersed wedges have roughly the same shape, and thus exert nearly equal moments.

**For a detailed discussion of ship stability and most other aspects of ship design, interested readers are referred to *Principles of Naval Architecture*, published by the Society of Naval Architects and Marine Engineers.

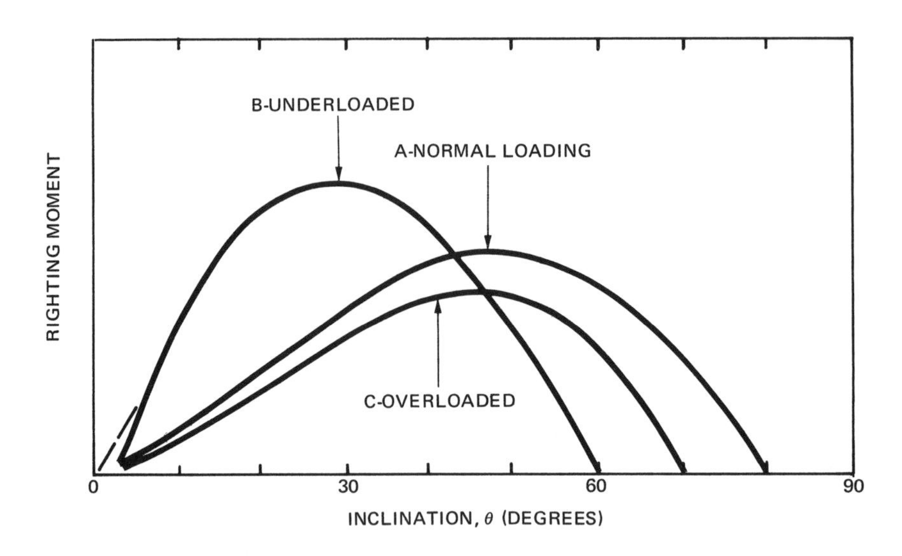

Fig. 101. Influence of loading on stability. Dashed portion of curve *B* shows stability improvement obtained by adding water ballast to an underloaded vessel.

and buoyant forces will be similarly altered; and, second, the positions of her CG and CB will both shift, such that their horizontal separation (moment arm) will be different at all angles of inclination off the vertical. As a result, both righting moment *M* and metacentric height GM will vary with inclination in a different manner than for the previous displacement, as reflected by a new stability curve.

Because a vessel's upright equilibrium derives principally from the symmetrical shape of her underbody, she may be capable of undergoing great extremes of load displacement and still float upright in calm water, although her inclined stability and range of safe dynamic motion in a seaway will be marginally restricted. This is true for both extreme underloading and overloading. Figure 101 shows the relative effect of large changes in load displacement for our hypothetical cargo ship, although these curves might apply as well to a small power cruiser of boxlike section. Curve *A* is similar to that of figure 100, and represents the vessel's inclined stability at normal design load displacement; her righting moment reaches a maximum at about 45° inclination, and diminishes to zero when heeled to 80°.

In an extreme light-load configuration (curve *B*), the vessel's peak righting moment is about 30 percent higher than that for her design displacement, but it occurs at a much smaller inclination, and her range of positive stability is similarly curtailed to about 60°. Tankers, and some older cargo vessels, have such an extremely large ratio of heavy- to light-load displacements that they may be slightly unstable when empty, and must take on water, or other ballast, to avoid lolling in calm water (dashed branch of curve *B*).

Curve *C* illustrates the effect of gross overloading; the combination of increased displacement and reduced freeboard results in the lowest righting moment at all heel angles, and the angles of maximum and neutral stability are substantially smaller than for normal displacement.

Recalling that the area beneath a stability curve is a measure of the work required to incline a vessel, it can be seen that the effect of increasing displacement is to progressively reduce both her static and her dynamic stability, and thus the vessel's margin of safe stability in roll is greatest at normal displacement. As we shall see in the section on "Ship Motions" (p. 296), a vessel should not normally be operated under conditions whereby the rolling angle exceeds that corresponding to the maximum of her stability curve, and this angle is substantially diminished by either under- or overloading. Lastly, both extremes are innocuous; a vessel may appear perfectly stable at dockside, only to capsize prematurely in rough weather.

Influence of Hull Sections: Irrespective of displacement, the shapes of a vessel's transverse sections have a marked influence on her transverse stability, which can be summarized as follows:

1. Increased beam confers increased righting moment at small heel angles, thus providing greater initial stiffness in roll.
2. Increased freeboard raises the CG, thus reducing righting moment at small heel angles and increasing it at large angles. Hence, the maximum (rail under) heel occurs at a larger angle.
3. Softening bilges increases maximum righting moment, and conversely.
4. Flare increases maximum righting moment and tumble home reduces it.
5. Any of these changes can significantly alter the neutral equilibrium angle.

From the above, it should be apparent that upright stability is conferred mainly by increasing a vessel's beam,* whereas a low CG and ample freeboard are most effective at large inclinations. While many small, light-displacement power craft possess good upright stability and adequate freeboard, their CGs are often located *above* the waterline, and thus they may be inherently more stable inverted than upright. This is particularly true for boats not fully decked over, which may be partly flooded upon overturning. The safe stability limits for small craft are further compromised statically by overloading (which raises the CG), and dynamically when running at high speed (which raises the entire hull). Under such conditions, even a sharp turn, or a transverse wave impact, may tip the balance and induce capsize—as eloquently testified to by Coast Guard statistics. Once the craft is bottom up, only a crane can effect recovery.

Stability of a Sailing Yacht: Somewhat dearer to the heart of the blue-water sailor is the stability curve for a 60-ton ketch (fig. 102).** Because of her 48,000-pound external ballast keel, she is capable of being heeled to 122° before reaching neutral equilibrium. However, if heeled farther, she develops *inverted* stability, as shown by hull profiles for 150°. The line of action of the buoyancy force now lies to the left of that for the gravity force, and the resulting heeling moment will continue to roll her over to a fully inverted attitude, where these forces are again in alignment. The curve shows that a righting moment of more than 100,000 ft. lbs. would then be required to roll her upright again. Interestingly, this might be accomplished by partial flooding (allowing air to escape through her sea cocks). This would progressively reduce her buoyant heeling moment, relative to the righting moment of her keel. At some point—hopefully short of foundering—the latter would dominate and

*Excessive beam, however, makes for a "stiff" boat, i.e., one that attempts always to respond quickly to inclinations of the water surface, leading to jerky and uncomfortable motion in a seaway.

**The positive stability portion of this curve was obtained from Uffa Fox's excellent book *Sail and Power* (p. 42); I have deduced the dashed inverted-stability portion by somewhat approximate methods.

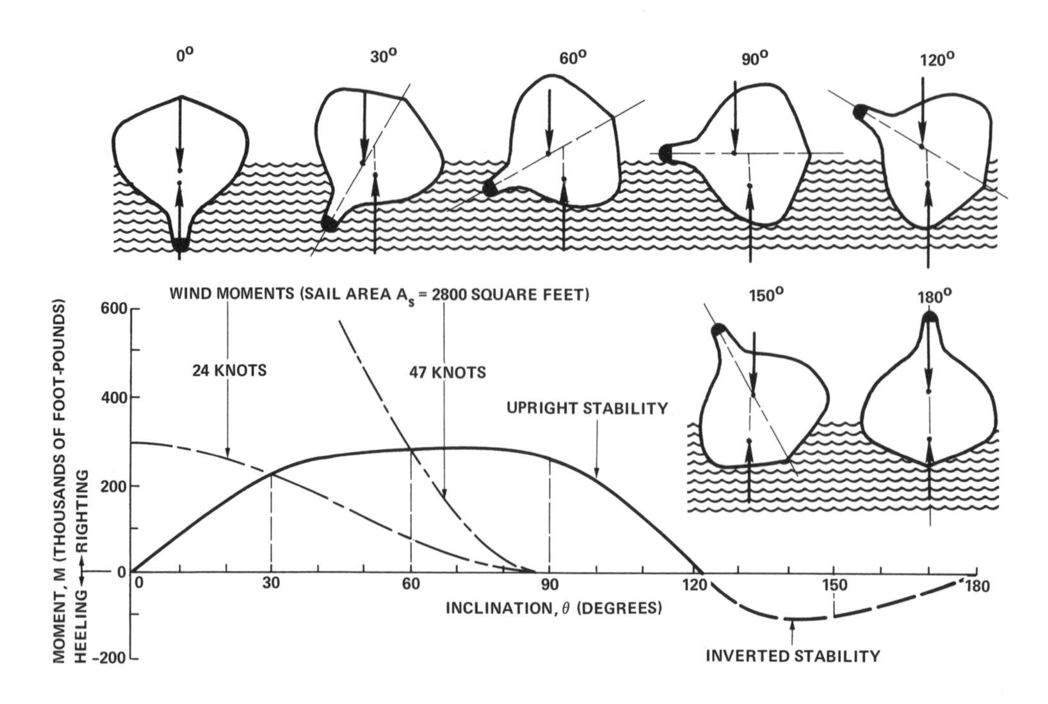

Fig. 102. Stability and wind moment curves for a 60-ton ketch. At heel angles greater than 122°, she develops inverted stability, and will continue to roll upside down.

roll her back. Before readers attempt this with their own boats, I should point out that the possibility of success can be precisely determined by advance calculation (see "Flooded Stability").

Relative to inverted stability, it is of interest that the modern trend to aluminum (vs. wooden) masts for sailing yachts may make the difference in whether recovery from a capsize is successful. Once under water, a wooden mast exerts a righting moment, whereas an aluminum mast continues to heel a yacht. This tendency could be more than compensated by plugging the latter to entrap air. For the 60-ton ketch discussed above, I have calculated that hollow alluminum masts would add about 100,000 ft. lbs. of righting moment at 122° heel, and would give her positive stability at all angles.

Figure 102 also shows the heeling moments produced by steady winds of 24 and 47 knots, assumed to act athwartships on the ketch's working sail area, A_S = 2,800 sq. ft. These moments are proportional to $\cos^2 \theta$ and tend to zero at 90° heel. Their intersections with the stability curve indicate the equilibrium heel angles (30° and 60°, respectively) for these wind speeds. Thus, close-hauled under full working sail in a 24-knot wind, the yacht would heel 30°, and would have her rail awash—a maximum speed configuration. If she encountered a sustained 47-knot squall gust (quadruple the wind force), and if nothing carried away, she would be forced down to only 60°. This is because heeling moment decreases very rapidly at large heel angles, whereas the ketch's stability curve is still rising, and she still retains about half her dynamic stability. By calculating other wind moment intersections with the stability curve, we would find that heel angle is roughly proportional to wind speed between 10 and 60 knots—a useful racing kink in variable winds. This proportionality holds for most ballast-keel yachts. If your stability curve is known, you can apply it regardless of sail area, provided that the heeling moments stay within the same relative range.

Influence of Rating Rules on Stability: Nowhere is the search for speed more evident than in yacht racing, where winning times in one-design classes often differ from second-place times by a fraction of a percent. To enable yachts of different sizes and types to compete against one another, various rating rules have been devised, and (repeatedly) revised. But, as Peter Johnson* puts it, ". . . once the rule is established yachts will be designed to its formulae, . . . and the management of rating rules in effect becomes a running battle between the rule authority, which tries to maintain a mathematical formula that reflects the racing potential of the yacht, and the designer's attempt to construct the fastest boat for the value of the rating, naturally exploiting any weaknesses that are detected in the formulae."

Between 1922 and 1969, this procedure worked quite well, evolving into the Cruising Club of America (CCA) Rule, which favored comfortable, beamy, shallow-draft (centerboard) boats, and the Royal Ocean Racing Club (RORC) Rule in England, which tended toward deep, narrow (wineglass) hulls. Neither of these rules posed any real threat to safety. But, largely to promote international competition between boats of dissimilar characteristics, the CCA and RORC Rules were combined in the current (1970) International Offshore Rating (IOR) Rule. According to Olin Stephens, who chaired the International Technical Committee appointed to work out the new rule, "The main effort was to eliminate old problems, especially the difficulty of measuring length and displacement under the CCA Rule and loopholes surrounding hull weight and scantlings under the RORC Rule."

In effect, the IOR Rule combined RORC hull measurements with CCA sail measurements, and attempted to solve the stability and scantling problem by adopting a center-of-gravity factor (CGF) that favored inside ballast over deep keel weight. The end result of intense IOR competition was a new breed of fast, light-displacement, beamy, wedge-shaped boats, with skeg-mounted or cantilevered rudders separated from a narrow fin keel, and more sail than could reasonably be carried without a large crew to act as movable ballast. As eloquently argued by C. A. Marchaj, only one of many who investigated the causes responsible for the Fastnet catastrophe, many elements of structural integrity and stability were sacrificed for speed.

As an example of this disparity, consider the difference between the dynamic stability curves for *Assent*, a traditional RORC one-design racer-cruiser,** and *Grimalkin*, a custom IOR racer, both of which were entered in Class V of the 1979 Fastnet race (fig. 103). From the following specifications, it can be seen that *Grimalkin* was slightly longer, beamier, and lighter than *Assent*—all hydrodynamic qualities that emphasize speed. She had a lower ballast/displacement ratio, and carried one-third of her ballast internally. As a consequence, her dynamic stability curve lies everywhere much lower than that for *Assent*, and becomes negative at all heel angles greater than 115°, while that for *Assent* becomes neutral at 165°, and scarcely drops below zero. Thus, from purely static considerations, *Assent* could be rolled 40° farther and still recover. Dynamically, *Grimalkin*'s area of negative

**Ocean Racing and Offshore Yachts.*

**Any yacht can race under any rule, but a boat not designed for that rule may incur measurement penalties that diminish its chances of winning.

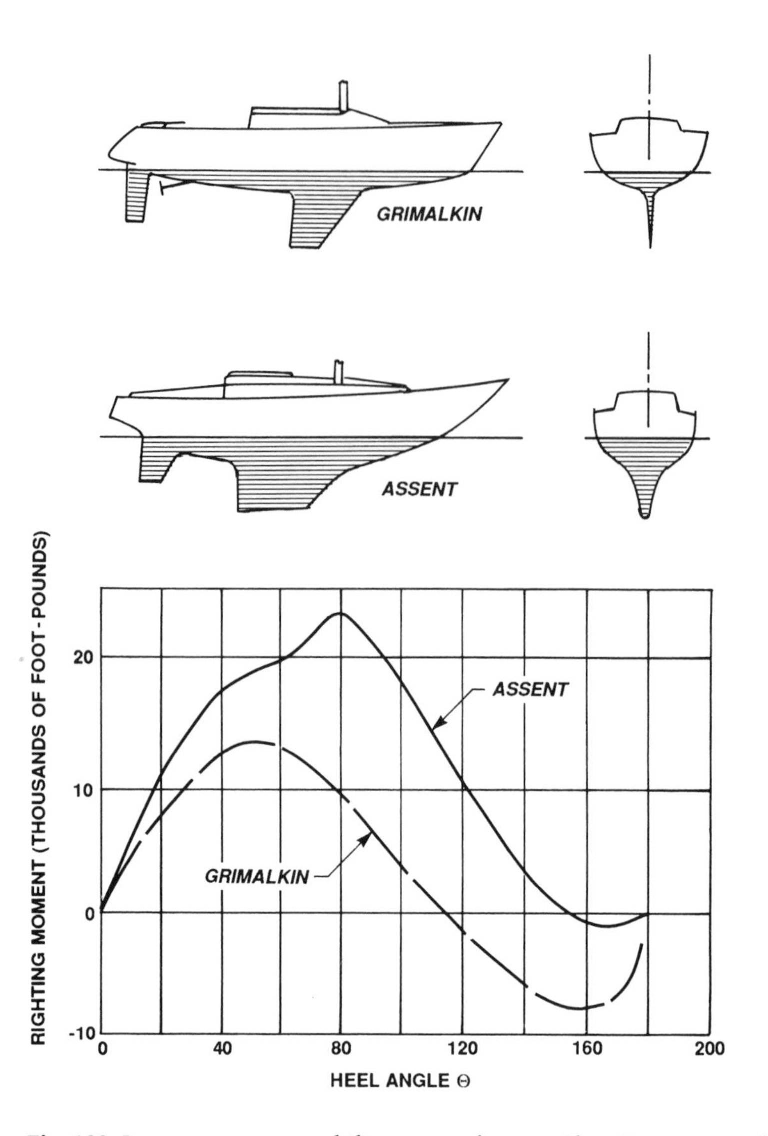

Fig. 103. Intact transverse stability curves for two Class V entries in the 1979 Fastnet race.

stability is half as great as her area of positive stability; i.e., once capsized it takes almost half as much work to roll her back upright as it does to roll her down. Significantly, in the 1979 Fastnet storm, *Assent* was the only Class V boat to finish; *Grimalkin* was knocked down six times and rolled over twice, remaining upside down for several minutes and drowning two of her crew.

Specification	*Assent*	*Grimalkin*
Displacement (lbs.)	10,112	8,320
Waterline length (ft.)	24.0	24.8
Beam (ft.)	7.2	8.0
Draft (ft.)	5.0	5.75
Ballast ratio	44%	34%

The Fastnet imbroglio brought about two important improvements to sailing safety. First, an international volunteer research and development effort was sponsored by the ad hoc USYRU/SNAME (U.S. Yacht Racing Union / Society of Naval Architects and Marine Engineers) Joint Committee on Safety from Capsize. Among the committee's findings was the recognition that capsizing is not merely a stability problem, but one that intimately involves the dynamic interaction of boats with (breaking) waves. The committee therefore developed a method of ranking boats in order of capsize resistance, as well as a method for estimating how long a capsized boat is likely to remain inverted (see chapter 23).

The second improvement was the development of a new rating rule, the International Measurement System (IMS), which is intended to coexist with the IOR Rule as a forum for racer-cruiser competition. The IMS is considered to be virtually foolproof against designer manipulation. Instead of a formula length based on a sparse set of hull and sail measurements, the IMS converts lofted or measured lines into a dense data set, from which sailing speed can be estimated for various wind speeds and directions by computerized fluid flow equations. Provisional handicaps for each race are based on predicted speed, and are later revised and incorporated into the data set to improve future estimates. A byproduct of the IMS is a transverse stability analysis out to the angle of neutral equilibrium, and boats found to have neutral angles less than 120° are cautioned against offshore racing.

The IMS is designed to make safer, more traditional designs competitive—and in this it appears to have succeeded. But its detractors have objected to what they regard as inconsequential auxiliary provisions such as interior arrangements (bunk, galley, and head design), and, more particularly, to provisional handicaps, often revised after the fact, which tend to dull the keen competitive edge of racing, since one is never sure which boat one will actually be competing against.

Thus IOR racing persists as a technical and atavistic challenge; like thoroughbred horses or Formula 1 race cars, IOR-type boats are recognizably fragile, difficult to handle, and exhausting to sail in rough weather, but provide maximum speed and exhilarating performance. IMS boats give the casual racer a chance to be competitive and still enjoy comfortable cruising. But both must still eschew sea conditions where they cannot prevail.

LONGITUDINAL STABILITY

Longitudinal—or *trim*—stability is calculated by exactly the same procedure as that for transverse stability, but this time taking vertical moments about the forward perpendicular (FP) plane. Most vessels have much higher longitudinal than transverse stability, and these calculations are ordinarily undertaken only for larger vessels at small angles of inclination, in order to investigate their equilibrium trim attitudes under various loading and flooding configurations, as well as to determine dynamic response under design operating conditions. In small craft, the designer adjusts sections and weights so that the longitudinal position of the CG coincides with that of the CB—or, in sailboats, with a position slightly aft (stern trim) of the CG, to compensate for the bow-down moment produced by the forward component of wind force under way. However, the best trim in small craft is somewhat arbitrary, because the CB will shift about a good deal in various performance attitudes, and some ballast adjustments may be required to fix the optimal CG location.

Whether for trim or for lateral stability, the shifting of (live) ballast is standard practice in sailboat racing—even for boats as large as the America's Cup defenders. Dead ballast shifting probably reached its practicable maximum in the Australian Sandbagger, a racing class whose only restriction was the overall length of hull, and in which several tons of sandbags were frantically heaved thwartships between tacks to balance an overwhelming spread of sail.

But by far the most impressive display of delicate and precarious equilibrium I have personally witnessed was that of single-handed racing of Gilbertese sailing canoes at Canton Island. These very light, and seemingly fragile, double-ended racing machines are lovingly crafted from small bits of driftwood, laced together with pandanus fiber, and then pitched, sanded smooth, and painted to a mirror finish. Their deep, narrow V-section hulls are held upright at rest by an outrigger consisting of a flexible "rose trellis" of slender wooden strips, terminating in a light pandanus log. There is a forked branch lashed to the leeward gunwale* at each end, through which a paddle is thrust for steering. When not under way, the mast lies on the outrigger trellis, bundled together with its permanently attached lateen sail. Two permanent stays run slightly fore and aft from outrigger to mast top (plate 23).

When ready to sail—as at a race start—the skipper maneuvers his canoe so that its outrigger is to windward, stands erect with one foot on the trellis and the other on the weather rail, grasps the sheet in one hand, and, with the other, slides his paddle through the appropriate fork. At the start signal, he thrusts the heel of the mast into a socket in the weather rail and flips it up so as to catch the wind. In the wink of an eye, he is pounding along through the choppy lagoon seas at up to 20 knots with his outrigger wildly gyrating through 30°–40° of arc, as he is raked by gouts of spray from his plunging bow. I had ridden a surfboard for nearly thirty years in waves up to 20 feet and higher, but the Gilbertese single-hander handed me! After two months of sporadic and frustrating attempts to get up and away without capsizing within the first five seconds, I decided that there are still some things a reasonably agile and versatile sailor might take a year to master.

DIRECTIONAL STABILITY AND CONTROL

Directional stability is conventionally defined as the property that constrains a vessel to a relatively straight course despite random perturbations by wind and waves. But what it really means is that, once trimmed and balanced on a steady course, she essentially steers herself, requiring minimal course corrections.

Proper trim and balance are most easily achieved in a vessel whose hull is long compared with the size of pertubing forces (waves and wind gusts), and whose buttock lines are straight enough to minimize pitching and yawing—such as a cargo ship or passenger liner. These qualities are most difficult to realize in sailing yachts, which are rocker-bottomed, and which must usually sail crabbed to windward by a small angle (α) so that their keels can develop the lift necessary to counteract the lateral component of wind force. This requires that the sails' *center of effort* (CE) be

*The canoe sails always with the outrigger to weather. In coming about, the boom and paddle are reversed end-for-end.

Plate 23. Two hands struggle to keep a Gilbertese single-hander upright in a stiff breeze.

trimmed slightly abaft the hull's *center of lateral resistance* (CLR),* so as to turn the bow into the wind. This tendency is resisted by a light "weather helm" (β) on the rudder (fig. 104).** Thus, the rudder has two functions: to provide the minimum effort (and minimum drag) necessary to maintain a steady course; and to similarly disturb this equilibrium when altering course. But these two functions are in themselves somewhat contradictory; it appears impossible to simultaneously achieve good directional stability and quick rudder response—particularly in high seas. Accordingly, every rudder-keel combination is a compromise, and the designer must proportion them according to performance priority.

*The CLR is an imaginary point through which passes the lateral component (F_h) of all hydrodynamic forces acting on the hull. Because the CLR moves about, depending upon a yacht's attitude and point of sailing, designers traditionally use instead the *center of lateral plane* (CLP), defined as the longitudinal balance point of the underbody profile. The CLP, unfortunately for realists, lies *abaft* the CE by a fraction of the waterline length known as the "lead" (fig. 104), which varies from 0–5 percent for split-rig yachts to 5–18 percent for sloops and cutters.

**Owing to asymmetry of her heeled waterplane, a well-trimmed yacht may sail herself on upwind headings. Heeling causes the center of buoyancy to shift aft, thus lifting the stern and canting the mast (and CE) forward. The boat then rounds up, which relieves the wind pressure and reverses the heeling effect, so that the bow falls off. When the boat is well trimmed and the tiller is locked, these motions are almost imperceptible. Correspondingly, on a small boat, and in light winds, one can steer by moving forward and shifting one's weight from side to side.

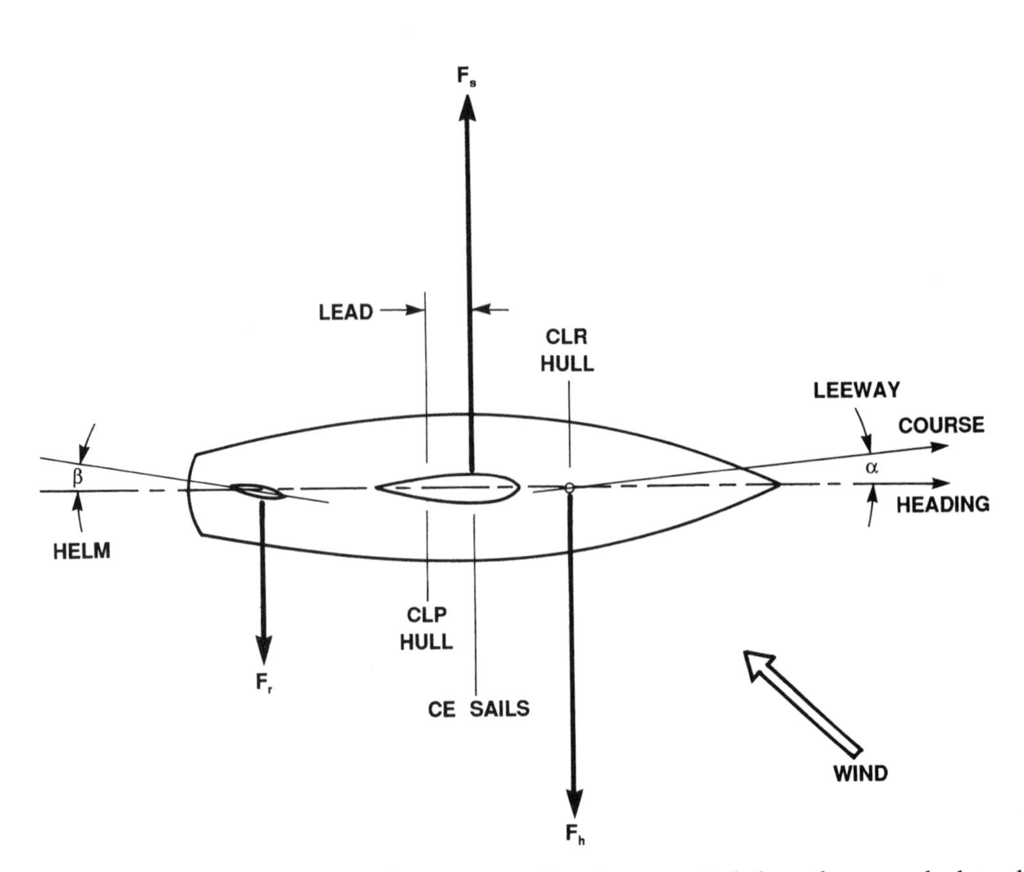

Fig. 104. Directional trim in a sailboat requires that the torque imbalance between the lateral sail force F_s and hull forces F_h be balanced by rudder force F_r.

On large ships, there is usually no external keel member; instead, the after quarters are pinched in to form a skeg or horn upon which the rudder is hung—usually in the wake of the propeller slipstream, which provides positive control even when there is no way on (fig. 105). The rudder itself is usually "square" (aspect ratio, AR=1), and may be fully balanced, partially balanced, or unbalanced, as shown, depending upon whether the pivot axis is located at about 25 percent, 15 percent, or the leading edge of its chord, respectively.

Rudder area, usually dimensioned by experience, varies between 1.7 and 2.5 percent of the ship's *lateral plane*, or LP (the shadow image of the ship's underbody on the *midship plane*). Although skeg-mounted, unbalanced rudders prevailed worldwide until after World War II, the present trend is toward horn-mounted, semi-balanced units, which greatly reduce torque loads on the steering engine, and are easier to dismount for service or repair.

In addition to rudders, large vessels are increasingly equipped with transverse bow (and stern) thrusters, which greatly aid maneuvering in confined spaces. I was recently a passenger on a new, ultra-modern, 600-foot cruise ship that was equipped with cycloidal, variable-pitch propellers and a 7-foot bow tunnel-thruster. Her captain would drive her straight toward a crowded dock, spin her on a dime, and warp her sideways into a berth not 10 feet over her length, to the consternation of the local docking crew.

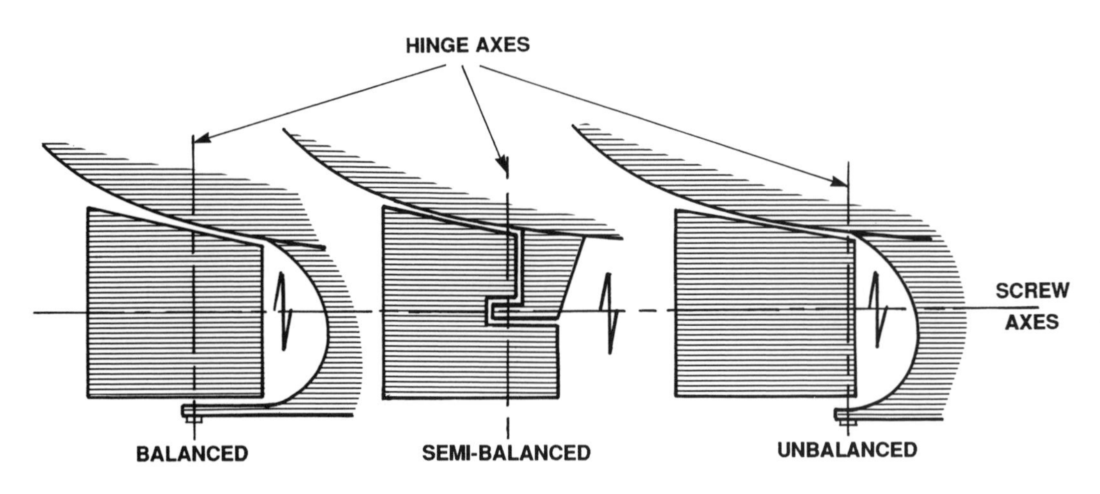

Fig. 105. Typical ship-rudder profiles of AR=1.

There are no hard and fixed rules for sizing powerboat rudders. Some designers use 2–5 percent of LP, depending upon cruise speed, and others use curves and tables from books on yacht design. I have found that Skene's method can be fit rather closely by the simple expression

$$AR = 4\sqrt{(W/\sqrt{L_W}\ )}\ \text{sq. ft.}$$

where W is displacement (metric tons), and L_W is waterline length (ft.). Thus, rudder area increases with displacement and decreases with speed (see chapter 22).

Planing power craft with fixed screws should have a rudder area about equal to the developed screw area, have an aspect ratio of 2, and be balanced about 10 percent aft of the leading edge. Most racing boats have steerable drives with only vestigial fins ahead of the screw to direct inflow.

Sailboat rudders are generally sized to about 8–10 percent of LP, but their disposition and attachment are matters of continuing controversy. As mentioned above, the rudder acts in concert with a sailboat keel to provide directional stability. Until about 1965, rudders were traditionally hung from the after end of a relatively long keel, which provided high stability and smooth—albeit sluggish—directional control. Although several custom designs had appeared by then, featuring skeg-mounted rudders set close behind the keel, the first true, mass-produced, fiberglass cruiser-racer is attributed to C. W. Lapworth, who, in the interests of simplicity and economy of mold design, omitted the skeg, and placed a spade rudder farther aft of a "square" fin keel. Because of its light displacement, and because there was no "wetted surface" penalty in the newly formulated IOR rule, the Cal-40 had a low rating and became a highly successful racer.

Since then, the trend has been toward ever wider, shallower hulls, with narrow, high–aspect-ratio fin keels and spade rudders set as far aft as possible (with rudder stock still below the waterline). In this position, it is still possible to trim the yacht so as to have "finger-tip" balance in light and moderate seas. But the crew must be kept aft and hanging over the weather rail to keep the rudder in the water when running under spinnaker. Moreover, the rudder is very liable to ventilation, and to loss of steering control and broaching in steep waves.

Cantilevered spade rudders carry many other penalties; it is astonishing to what extreme designers will go to gain the last micron of speed. These rudders catch kelp and weeds (almost every racer has a weed pole). If the rudder is wheel-steered, the wheel must be connected to the rudder quadrant by cables, which invariably stretch and allow flutter in rough seas. And, most seriously, spade rudders are subject to high bending stresses, a frequent cause of structural failure in heavy weather. In the 1979 Fastnet race alone, six boats lost their spade rudders completely! Most of these problems could be avoided by reintroducing a faired support skeg, which would also increase rudder efficiency.

FLOODED STABILITY

No one needs to be told that flooding imperils the safety of a vessel at sea. Not only do her motions become sluggish and erratic, but performance and control are adversely affected. In rough seas, these effects are easily recognized by a vessel's behavior. But, as with overloading, a progressive rise of water within the hull may not appear dangerous, under moderate conditions, until GM (metacentric height) and righting moment vanish. Then, like an iceberg, the vessel may suddenly capsize, or she may heel excessively, take more water topsides, and sink.

Commercial vessels of upward of 150 tons are ordinarily designed with a view to the possibility of accidental flooding. Their hulls are compartmented by a plurality of decks and watertight bulkheads, so that damaged sections can be closed off and upright equilibrium reestablished by compensatory flooding elsewhere. Each ship has a *margin line* slightly below her upper bulkhead deck, the immersion of any part of which indicates a critical stability condition requiring abandonment.

However, smaller vessels are more often uncompartmented, so that flood water has access to all parts of the hull. The presence of free water in a vessel's bilge has an additional effect on stability over that of an equivalent weight of fixed ballast or cargo. Although either produces the same draft increase and loss of upright freeboard, the water is free to run in the direction of inclination as the vessel pitches or rolls, thus shifting her inclined CB and altering the righting moment. There is also a dynamic effect, owing to the inertia of water sloshing back and forth, that makes the vessel's motions jerky and irregular. Although dynamic effects are difficult to appraise, the magnitude of the static CB shift for a given inclination can be calculated by assuming the flood water to be always in equilibrium within the hull, and by using the method of moments, as before, to determine the CB of the void space between the flooded surface inside and the new waterline outside. While the *volume* of this buoyant void space remains constant and equal to the intact (unflooded) displacement, its CB depends upon the extent of flooding, the shape of the vessel's sections, and the angle of inclination. The CB may shift either toward or away from the direction of inclination, and no simple generalizations can be made.

We can, however, distinguish between two extreme examples of hull shape in which these tendencies are just opposite. Figure 106 compares the normal (unflooded) transverse stability curves for an open-cockpit power cruiser of rectangular section and a keel yacht of wineglass section, with those where each is flooded to her original waterline. The intact *(d)* and flooded *(d')* moment arms for each vessel are shown for several inclinations in the profiles above.

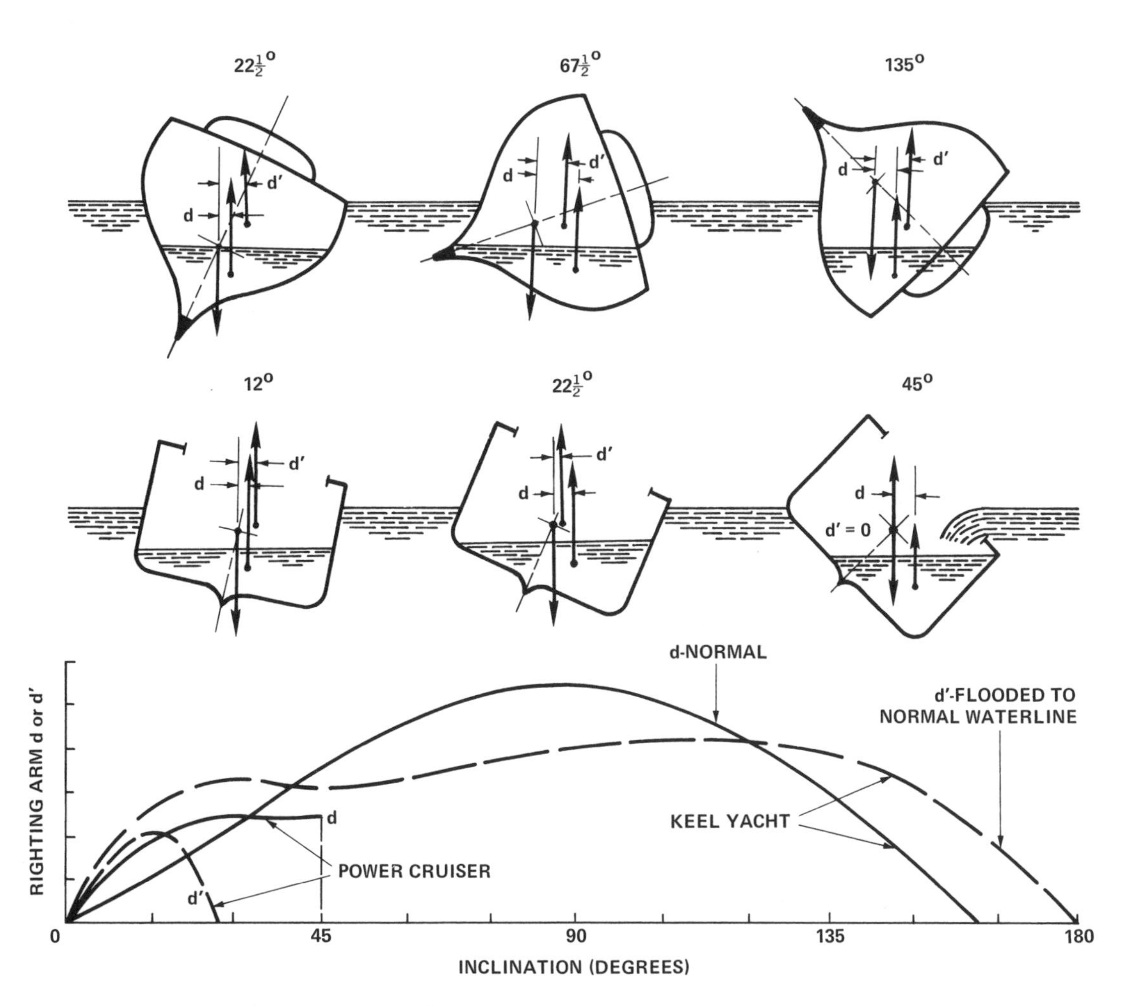

Fig. 106. Variation of intact *(d)* and flooded *(d')* righting arms for a keel yacht and a power cruiser at various heel angles. Both vessels are considered to be flooded to their respective intact waterlines *(d')*.

Consider first the power cruiser. Her intact righting moment increases continuously as she is inclined to a maximum angle of 45°, beyond which she would take water into the cockpit and abruptly lose stability. In the flooded configuration, her stability is a little higher up to about 15°, but then rapidly decreases to neutral equilibrium at 28°, as her moment arm vanishes; she would be unsafe in a seaway if rolling more than 15°. Thus flooding greatly reduces her safe heeling angle.

Conversely, the keel yacht has much higher intact righting moments than the cruiser at all angles of inclination above 30°. She reaches neutral equilibrium at about 170°, and has negative (inverted) stability at larger angles. When flooded, her moment curve is lowered and broadened. While it never rises as high as the intact moment curve, the yacht now has positive stability at *all* heel angles,* and will always right herself if rolled under (hence our earlier reference to righting a capsized keel yacht by partial flooding).

*It is assumed that the vessel has watertight integrity when inverted, and that flooding is somehow arrested at the indicated level.

Lastly, dynamic stability (area beneath the moment curve) is drastically reduced by flooding the power cruiser, while it remains about the same for the sailing yacht, whose stability is conferred mainly by her lead keel. Although margin lines are not ordinarily computed for small vessels, it is not difficult to determine just how far one's own boat can be flooded before it loses overall buoyancy and sinks. To the extent that the sides of most boats are vertical at the deck line, the safe margin line can be approximated as the lower surface of the buoyant volume below the minimum-freeboard deck level that will just support the boat's displacement. For practical purposes, the vertical height H_m of the margin line above the waterline is

$$H_m = F_m - 45W/(B \times \text{LOA} - A_c) \text{ ft.}$$

where W = displacement (tons); F_m, B, and LOA are the vessel's minimum freeboard, beam, and overall length, respectively; and A_c is the area of any cockpit space below deck level. By painting such a line inside your boat, you will know when to stop pumping and start jumping!

On a smaller scale, the free surface of liquids in tanks, or of water in an open cockpit, has much the same effect on stability as free water in the bilge; both comprise movable weights that reduce the righting moment as a vessel is inclined. For a given liquid volume, this reduction is proportional to the height of its centroid of volume above the CG, and to the *cube* of its free surface dimension in the direction of inclination. For this reason, it is advisable to place tankage low in a vessel, and to compartment—or otherwise restrict—its transverse dimensions, since this effect is much more serious in heel than in trim. Similarly, open cockpits should be restricted to the smallest practical dimensions and preferably be located amidships. A large stern cockpit is more susceptible to pooping and, if flooded, can cause considerable stern trim (squat) in addition to its influence on dynamic stability.

22

Resistance and Propulsion

SMOOTH-WATER RESISTANCE

A vessel moving at uniform speed in calm water is in equilibrium between the propulsive force of her screw(s)—or sails—and the resistance produced by the passage of her hull through the water. The total resistance R_t is conventionally divided into two parts: frictional resistance R_f due to water flow past the hull, and wavemaking resistance R_w caused by a fore-and-aft pressure imbalance associated with the familiar pattern of surface waves generated as the hull moves through the water.

Except at very low speeds, frictional resistance can be approximately represented by the relation $R_f = C_f S V_s^2$ (lbs.), where V_s is vessel speed (knots), S is the wetted surface area (sq. ft.) of the underbody, and C_f is an experimental coefficient that decreases with the logarithm of $V_s \times L_W$ (L_W = waterline length). (See figure 107.) These "smooth" values should be doubled for a year of temperate fouling, and trebled for a year in tropical waters.

Wave resistance, on the other hand, results from very complicated wave interactions with the hull, and cannot be simply formulated. However, for a particular vessel, it can be determined indirectly by towing a model (or a full-scale vessel) and measuring R_t, from which R_w is obtained by subtracting R_f.* Although ship waves have been measured very accurately, and rather elaborate theoretical models have been constructed to explain them, there are still some discrepancies between computed and measured resistance. We shall content ourselves here with a general description of induced waves and their influence on vessel behavior.

The characteristic deep-water ship-wave pattern (fig. 108a) consists of two similar wave systems roughly originating at the fore and aft ends of a vessel's L_W. Each wave system comprises a train of diverging waves (conventionally called *bow* and *stem* waves, respectively) and a secondary train of *transverse* waves—one for every bow or stern wave—propagating in the same direction as the vessel. The consecutive intersections of the diverging waves with the transverse waves are marked by a V-shaped array of prominent cusps, making a half-angle of 19.5° with the vessel's center line. Because the wave pattern moves with the vessel, the transverse waves travel at the same speed ($C_t = V_s$) and will have deep-water wavelengths:

$$L_t \text{ (ft.)} = 0.20V_s^2 \ (V_s \text{ in ft./sec.}) = 0.56V_s^2 \ (V_s \text{ in knots}).$$

*Various corrections are needed to account for model scale effects.

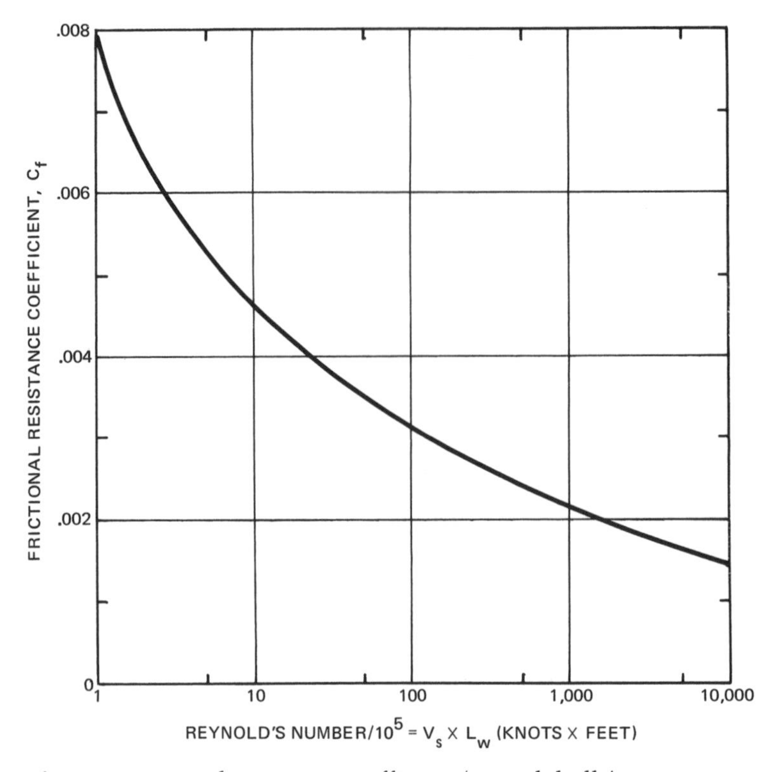

Fig. 107. Frictional resistance coefficient (smooth hulls).

This result can be obtained directly by eliminating the wave period T from the deep-water wave relations $L = CT$ ft. and $C = 5T$ ft./sec. (Part IV). If you wonder why waves traveling at vessel speed are left behind in the wake, it is because the energy of deep-water waves travels at group velocity Cg, equal to half the wave speed. Similarly, the divergent wave velocity (C_d) is equal to the component of vessel speed in the direction of wave propagation, which is about 30° off the vessel heading:

$$C_d = V_s \cos 30° = 0.87 V_s.$$

The divergent wavelengths are thus about three-quarters those of the transverse waves.

Because the wavelength L_t of the transverse waves is a function only of vessel speed, there may be several waves from the bow system along her waterline at low speeds, or only a fraction of a wave at high speeds. If we let $N = L_w/L_t$ be the number of such waves, we can rewrite the transverse wavelength formula as:

$$V_s/\sqrt{L_w} = 1.34/\sqrt{N}.$$

The parameter $V_s/\sqrt{L_w}$ is known as the *speed/length ratio,* and is a standard performance index; that is, a model ship in a model sea will have wave resistance characteristics similar to those of a full-size vessel if their speed/length ratios are the same.* The number of crests along the waterline in the latter system and their corresponding speed/length ratios are shown in the following table:

*A more common speed index in naval architecture is the Froude number, $F_n = V_s/\sqrt{gL_w}$ (V_s in ft./sec.), related to speed/length ratio by $F_n = 0.3V_s/\sqrt{L_w}$. "Speed/length ratio" is actually an archaic and cumbersome misnomer, but is so deeply rooted in design literature that one is obliged to retain it.

N	6	5	4	3	2	1	3/4	1/2	1/4
$V_s/\sqrt{L_w}$	0.55	0.60	0.67	0.77	0.95	1.34	1.55	1.89	2.68

Both divergent and transverse bow wave systems commence with a crest rising from undisturbed water at the forefoot (fig. 108b). The stern transverse wave system is similar to the bow system, but upside-down; the leading wave is a trough, commencing where the vessel's underbody begins to taper aft. Thus, for $V_S/\sqrt{L_W} > 0.95$, there will always be less than one complete wave under the stern.

While all of the above are free waves, in the sense that they are generated by the pistonlike action of a moving vessel, there is also a forced *displacement wave* of constant length that always accompanies the vessel and leaves no wake behind. This wave commences with a crest under the bow, rising from some distance ahead of the forefoot; it has a trough amidships, has another crest under the stern, and dies away a short distance astern.* If a vessel were exactly symmetrical fore and aft, the displacement wave would produce no net drag, but in most cases waterlines are fine forward and fuller aft, and consequent flow separation at the stern produces *induced* drag. Additional drag also results from flow interference by propellers, rudder, struts, etc., all of which are lumped with total wave resistance under the heading of *residuary* drag.

Although the heights of these various waves differ considerably, depending upon a vessel's hull form, for a given vessel the bow waves are generally higher than the stern waves, and all waves increase in height as the square of vessel speed. However, total wave resistance increases roughly in proportion to the square of wave height; as the vessel moves faster, her bow tends to rise higher on its growing bow crest, and her stern tends to squat in its deeper trough. In addition to altered trim, the vessel's hull sinks bodily into the trough of its displacement wave, and the potential energy of this sinkage is abstracted from her thrust potential. Thus, in effect, a vessel creates her own wave-resistance trap; like a car stuck in sand, the harder she pushes, the deeper she sinks; and only light displacement and high power will get her out.

At speed/length ratios below $V_S/\sqrt{L_W} = 1.5$, this resistance increase fluctuates systematically with speed (particularly for vessels having a long, parallel middle body) as the lengthening bow and stern transverse waves constructively or destructively interfere, causing the stern to rise or fall, respectively, about its average trim.** These fluctuations are not significant for small vessels when $V_S/\sqrt{L_W} < 0.95$, since there are then two or more waves along the underbody that provide more or less uniform support. But above this speed, the first bow wave trough begins to

*The displacement wave is a consequence of Bernoulli's principle for steady, free surface flow of a liquid around an obstacle, which states, in effect, that the sum of the local surface elevation and the square of local flow velocity must be constant. That is, a vessel moving in calm water behaves exactly as if moored in a moving stream. At the bow, the stream is diverted and slowed down, resulting in a local rise in elevation (the bow wave). Amidships, the flow is faster than in the free stream some distance abeam, and the level falls below normal (midships trough). Near the stern, the flow converges and is again slowed down, and the elevation rises (stern wave).

A corollary to this principle is that two vessels moored parallel and a beam's width or so apart in a moving stream will be driven together, because the average velocity between them will be higher—and the average elevation lower—than that to either side. The same effect occurs between two vessels moving in calm water on parallel courses, as in fueling operations at sea. Here, careful steering is necessary to prevent collision.

**Recommended cruise speeds for large ships are often intended to coincide with the hollows of their speed resistance curves, where a little extra speed can be maintained without much power increase.

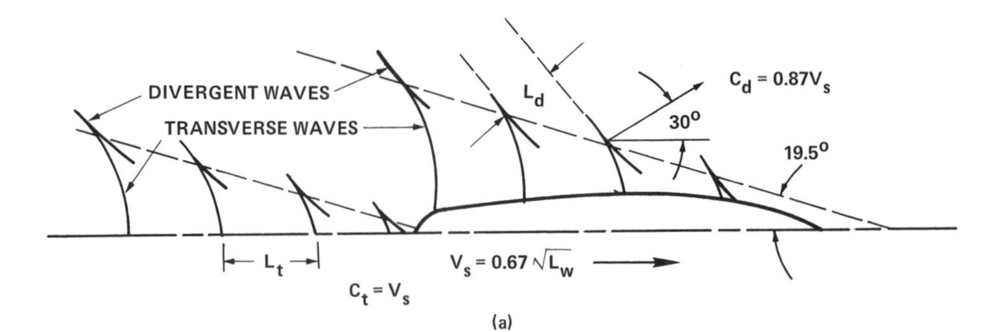

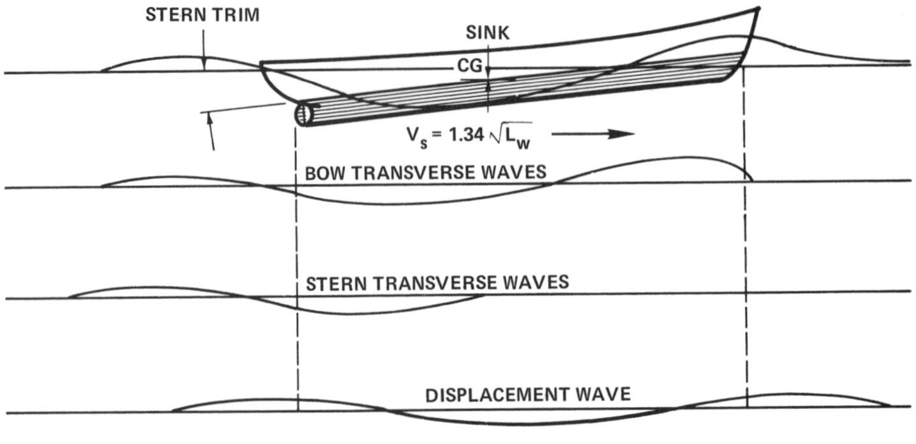

Fig. 108. *(a)* Plan view of ship-wave pattern at $V = 0.67\sqrt{L_W}$. *(b)* Total wave pattern along hull *(top)* consists of superimposition of transverse bow and stern waves, and a permanent displacement wave. Pattern shown is for speed $V_s/\sqrt{L_W} = 1.34$, where troughs of bow and stern waves coincide, producing maximum sink and stern trim.

encroach upon the leading stern wave trough, and resistance rises very rapidly, reaching a maximum rate at about $V_s/\sqrt{L_W} = 1.5$, where these troughs coincide. At higher speeds, the influence of the bow crest extends far enough aft to partly compensate for the stern trough, and no further perturbations occur. Here, principal wave resistance results from the diverging bow waves, which have now become very large.

For a given displacement, otherwise arrived at, both frictional and wave resistance are very sensitive to hull form, clear up to the sheer line. Subject to many other design constraints, speed will be maximized and power minimized by reducing resistance as much as possible. However, the above constraints, which depend largely on a vessel's intended service, often substantially affect her hull form and hence her resistance and speed range with available power. Thus, hull design is always a compromise between conflicting objectives; there is no optimum design for all speeds. Fast hulls have excessive resistance at low speeds, and conversely. The ultimate test of a vessel is that she perform with minimum power at her smooth-water design speed, and have an adequate power reserve in rough seas.

In order to simply represent the disparate effects on resistance of hull form, length, weight, and displacement, designers commonly resort to several performance indices, the most important of which are the following:

1. **Prismatic coefficient:** $C_p = D/(A_m \times L_w)$,
   where D = displacement volume (cu. ft.), and A_m = area of largest section (sq. ft.).
2. **Displacement/length ratio:** $\delta = W/(L_w/100)^3$,
   where W = load displacement (tons).
3. **Entrance angle:** EA = angle between tangents to the waterline at the bow.

Because they all involve vessel dimensions, these indices are interrelated, but each affects a vessel's resistance vs. speed differently. A vessel's C_p is a measure of her relative fullness: if her displaced volume is concentrated amidships (keel yacht), she will have a low C_p; if it is distributed more or less uniformly along her length (tanker), she will have a high C_p. Her δ is a measure of average density (weight per unit length), and her EA largely determines bow-wave resistance.

Table 7 lists representative values for these indices for four widely different hull types, and their normal variance with speed/length ratio is shown in figure 109. Here, the solid curves indicate optimum values that minimize total resistance (principally wave-making) for small vessels, while their dashed alternates show practical maxima, limited by available power. At speeds below $V_s/\sqrt{L_W}$ = 1.0, wave resistance is small, and wide variations of hull form are possible without much penalty in power. Minimum frictional resistance is obtained with a low C_p and low δ, because this combination results in minimum wetted surface area. However, a low C_p implies a short, pointy hull with a larger EA than would be tolerable at higher speeds.

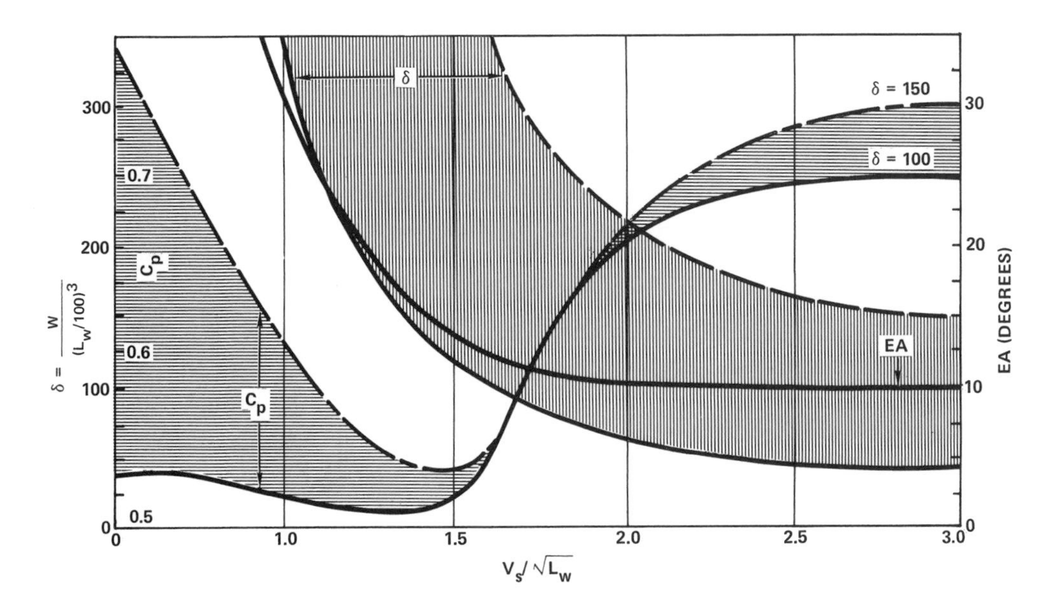

Fig. 109. Normal ranges of displacement/length ratio *(δ)*, prismatic coefficient *(C_p)*, and entrance angle *(EA)*, as functions of speed/length ratio, $V_s/\sqrt{L_W}$.

Table 7.

Vessel type	L_W (ft.)	W (tons)	δ	$V_s/\sqrt{L_W}$	EA (deg.)	C_p
Keel yacht	32	12	340	1.4	48	0.52
Express cruiser	50	20	160	2.6	40	0.70
Destroyer	350	4000	64	1.7	8–10	0.63
Int. sailing canoe	17	0.2	41	4.0	34	0.70

At the mid-speed range ($1.0 < V_s/\sqrt{L_W} < 1.5$), as previously described, the bow and stern wave troughs overlap, and wave resistance takes a sudden jump. Here, all three indices have intermediate values—optimum perhaps only for low-performance powerboats, since few other vessels are specifically designed to operate within this range.

Speeds above $V_s/\sqrt{L_W} = 1.5$ are the province of high-performance vessels and small racing craft. Great power is needed to overcome wave resistance, which can only be minimized by a high C_p, and low δ. Little latitude is possible without a large power penalty, and vessels having displacement/length ratios above 150 are ordinarily too heavy to be driven at such speeds.

These features are further illustrated by the resistance curves shown in figure 110 for the vessels listed in Table 7. The solid curves give total resistance R_t in lbs./ton (2,240 lbs.) displacement, as a function of $V_s/\sqrt{L_W}$. The dashed curves refer to frictional resistance R_f. Except as noted below, the curve lengths indicate approximate operating ranges of the respective vessels.

In order of decreasing displacement/length ratio, consider first the resistance curves for a 45-foot LOA, 12-ton keel yacht, of the racer-cruiser type (*A*). With her smooth, symmetrical underbody, she has the lowest R_f at all speeds of the four types shown, but her broad entry angle and low C_p are very conducive to wave-making at high speed; below $V_s/\sqrt{L_W} = 1.0$, her resistance is mainly frictional, but at $V_s/\sqrt{L_W} = 1.1$, $R_W \doteqdot R_f$, and at $V_s/\sqrt{L_W} = 1.4$, R_W is about ten times larger! On her best point of sailing and with all the sail she can carry, she cannot much exceed this speed.* The hyperbolic shape of a sailboat's resistance curve explains, in part, why one-design racing is such an art: at low speeds, a slight sail change can cause a large speed variation, but above $V_s/\sqrt{L_W} = 1.3$, everything must be pressed to the limit to attain the last fraction of a knot. It also explains why boat-for-boat handicapping is so difficult, since very slight changes in hull form can significantly alter the shape of the entire resistance curve.

The rectangular sections and square transom of the 55-foot LOA, 20-ton express cruiser *(B)* give her a slightly higher wetted surface per ton and higher R_f than those of the keel yacht at low speed/length ratios. But her finer entry and bluff transom greatly suppress wave-making at high speeds; not only are her transverse bow waves relatively lower, but, since she has little hull taper aft, the leading stern transverse wave is generated well abaft her transom, where its interaction with the bow waves does not affect her trim. Thus, there is only a moderate rise in R_t near $V_s/\sqrt{L_W} = 1.3$; her twin 250-brake horsepower engines can drive her smoothly up to about 19 knots. The slight downturn of her R_t curve above $V_s/\sqrt{L_W} = 2.0$ indicates

*Cup defenders can attain about $V_s/\sqrt{L_W} = 1.45$ under optimum conditions, but in triangle racing the average speed is more like 1.0–1.2. Heavy-displacement cruising sailboats average 0.8–1.0 over long distances.

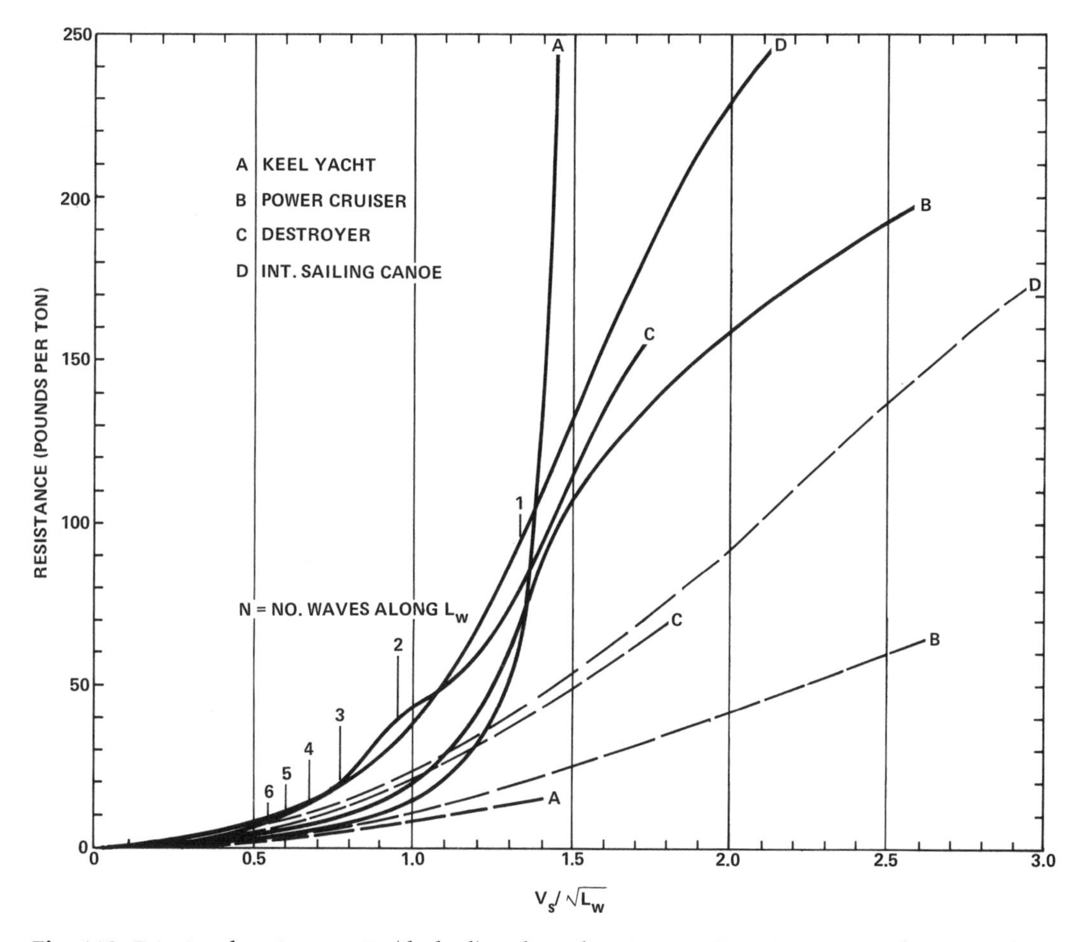

Fig. 110. Frictional resistance R_f (dashed) and total resistance $R_t = R_f + R_w$, as function of speed/length ratio $V_s/\sqrt{L_W}$, for four different hull types.

that she here develops enough dynamic lift on her flat underbody to begin to plane, and could probably achieve speeds above $V_s/\sqrt{L_W} = 3.0$ with larger engines.

Because of her very fine entry and low displacement/length ratio, the 400-foot, 4,000-ton destroyer *(C)* has a smaller wave-making potential than the express cruiser, although her relative wetted resistance R_t is somewhat higher. The wiggly nature of the R_t curve reflects transverse wave interference along her slender middle body; the depression near $V_s/\sqrt{L_W} = 1.1$ suggests an optimum cruise speed of 22 knots, but she must ring up full power (70,000 shaft horsepower) to get over the resistance hump at $V_s/\sqrt{L_W} = 1.5$ (30 knots) and achieve flank speed (34 knots).

Lastly, we consider the 17-foot, 435-pound International Sailing Canoe *(D)*, whose long, slim, flat underbody and very light displacement are designed for planing. Her R_f curve is slightly higher than that for the destroyer, and her bluff bow and high C_p also produce higher wave-making resistance at low speed/length ratios. But, with a 180-pound man hanging from a trapeze to weather, her 108-sq. ft. sail can be balanced in a 20-knot breeze, providing over 300 lbs./ton of drive—enough to lift her out of the displacement trap and boost her up to a planing speed of about 16 knots ($V_s/\sqrt{L_W} = 4$), which is completely off our diagram.

In summary, wave-making resistance presents a virtual speed barrier to heavy-displacement hulls that can only be transcended by thrust approaching 5 percent of displacement. This barrier is less formidable for long, slender hulls, or for those designed to plane, but the power requirements still remain high, and such hulls have increased resistance at low speeds. Thus, most commercial vessels, and sailboats having displacement/length ratios over 150, are forever committed to low speed, and have hulls designed to minimize frictional resistance. In racing sailboats, considerable improvement in performance has been achieved by very light construction and cutting away deadwood aft (both of which reduce wetted surface), and by rounding the after buttock lines so as to control stern transverse wave interference.

While hull form is not easily altered post facto, changes in several other factors may substantially improve your boat's smooth-water performance:

1. **Frictional Resistance**
   In general, the smoother the hull finish the less the frictional resistance at all speeds. However, this requirement is incompatible with good antifouling paint, which depends upon gradual solution for effectiveness. Despite manufacturer's claims, there is no happy intermediate. If you have to leave your boat in the water, weekly scrubbing will keep an enameled or plastic surface in good condition.

   Smoothness implies fairness—the elimination of all bumps, hollows, and discontinuities, however small. There should be no declivity at the stem rabbet, and all through-hull fittings should be faired flush. Manufacturers have yet to come up with a flush, popper-type sea cock, but some modern racing boats have eliminated all openings below the waterline.

2. **Residuary Resistance**
   This category includes all drag-producing hull appendages. Troubles and remedies include:

   (a) Sharp cutwaters are not hydrodynamically efficient, except under the very restrictive conditions of going dead ahead in smooth water. Rounding the cutwater to a parabolic section will reduce resistance under most conditions.
   (b) Propellers are a perennial problem for sailing craft. Free-running props produce more drag than if stopped; offset props more than in line; and fixed props more than folding ones. The most efficient propeller and location often produce the highest drag. The most effective combination of high thrust and low drag seems to consist of a large, slow-turning, full-feathering prop set in line ahead of the rudder. Such props are available in Europe (including reverse-pitch), but are not popular in America, owing to the extra cost of propeller and reduction gears.
   (c) Too small a rudder is a common source of excess drag; turned sharply, it brakes the boat, rather than cambering the water flow. This problem is aggravated in unbalanced sailboats, which require a heavy helm to hold course on some headings. It is relatively easy to increase the size of a rudder, and there is little penalty in wetted surface.

3. Wave-making Resistance

The three controllable factors most affecting R_w are weight, trim, and power, all of which interact to some extent:

(a) Lightening a vessel reduces δ and can significantly improve performance at all speed/length ratios, with relatively less effect near the resistance hump ($V_s/\sqrt{L_w} = 1.34$). Stability will also be improved if the weight is removed from above the vessel's CG.

(b) Optimum trim is vital for maximum speed. Because squat increases with speed, increasing forward trim is desirable, and a bow trim of as little as $0.001 \times L_w$ can make a great difference. This is equivalent to moving a weight equal to 1 percent of a vessel's displacement forward by 10 percent of her waterline length. In small boats, bow trim can be obtained tactically by sending crew or passengers forward, or by any other means that increases the bow-down moment of sail or propeller thrust. Strategically, this may imply moving the engine, mast, or ballast keel forward.

Lateral trim is of equal importance to sailboats. For a given thrust, almost all hulls sail faster upright than heeled, and any measure that confers uprightness will improve both comfort and performance.

(c) It is almost redundant to say that power begets speed. But, short of planing, the steepness of most resistance curves invokes a progressively diminishing return in speed vs. power. Possessed of an adequate power reserve for rough weather, optimum power may imply optimum avoidance of power waste—that is, the best matching of engine and propeller, a reinvestigation of location and shaft angle, and (quite often) a little more care and feeding of the engine itself (see "Propulsion," p. 279).

As a guide to testing the effect of trim or of hull alterations on an existing sailboat, or as a simple means of evaluating bottom roughness in racing competition, anyone can determine his or her own resistance curve, aided only by an accurate speedometer and a stopwatch.* While readers are referred to the cited article for details, the method, in brief, is as follows:

1. In relatively calm water with the true wind slightly abaft the beam, run the boat under bare poles up to full power and, having obtained uniform speed, shift to neutral. Record speedometer reading every few seconds by stopwatch until boat comes to rest. Repeat twice and take the average for better accuracy.
2. Plot speed against elapsed time on graph paper, and draw a smooth, descending curve through the points (fig. 111,a). At any point, the slope of this curve is a measure of the boat's deceleration at that instant. According to Newton's Law $F = M_a$, any slope, multiplied by the boat's effective mass, gives the force (resistance) at that instant.
3. The boat's effective mass is given by $M = (W/g)(1 + k_m) = 0.031\ W(1 + k_m)$, where W is displacement in lbs. (including crew), and k_m accounts for the

*See *Testing Hull Resistance* by John R. Stanton, *Sail* magazine, December 1971, pp. 48–50.

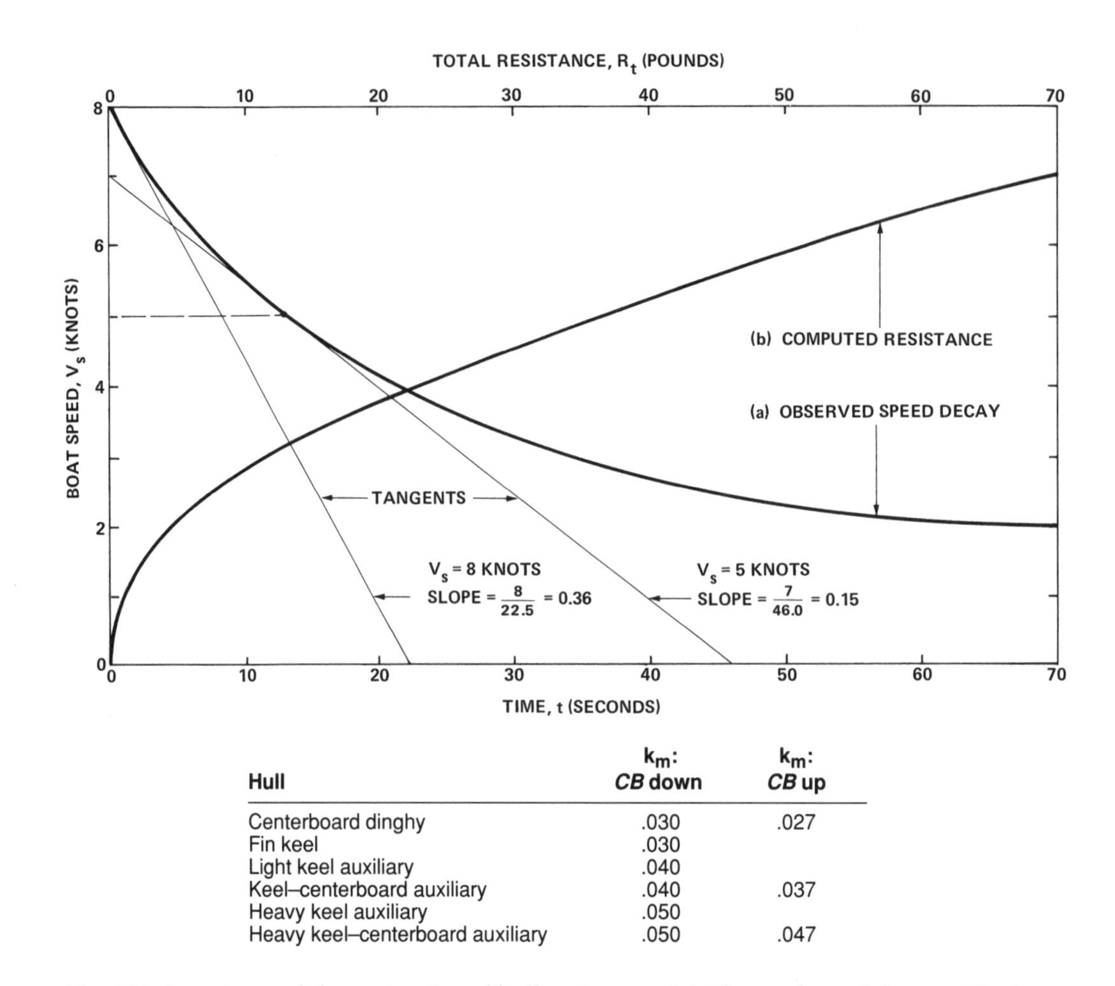

Hull	k_m: *CB* down	k_m: *CB* up
Centerboard dinghy	.030	.027
Fin keel	.030	
Light keel auxiliary	.040	
Keel–centerboard auxiliary	.040	.037
Heavy keel auxiliary	.050	
Heavy keel–centerboard auxiliary	.050	.047

Fig. 111. Experimental determination of hull resistance. *(a)* Observed speed decay with time after cutting power. Tangents to this curve define instantaneous deceleration. *(b)* Resistance curve obtained by multiplying tangent slopes by vessel mass. Data are for 4,500-pound, 20-foot L_W sloop.

added mass of water entrained in her wake. Values of k_m appropriate for various sailboat hulls are given in the figure.

4. To obtain the final resistance curve, draw a number of tangents to the velocity curve, calculate their slopes, multiply them by $1.69M$, and plot them against their corresponding velocities (fig. 111,b).
5. By repeating this test with an inclining weight hung from an outboard boom, one can also test a boat's heeled resistance. This will usually be found to increase by a percentage roughly proportional to heel angle.* Do not forget to add the extra weight in calculating M.

SHALLOW-WATER RESISTANCE

A vessel moving under constant power at moderate speed in calm water will experience a substantial speed reduction when passing from deeper water to depths

*Despite increased frictional resistance, a boat with long overhangs may sail faster heeled to some optimum angle, owing to an effective increase in L_W, and consequent reduction of wave-making resistance.

several times her average draft. This reduction is insidious with gradual shoaling, but is dramatic enough to feel as if brakes were being applied when the depth change is abrupt.

The cause of shoaling resistance is twofold. First, proximity of the bottom acts to accelerate relative flow past the hull, increasing the amplitude of the displacement wave, and aggravating both sinkage and frictional resistance.* Second, the free waves generated by the moving hull undergo a transition from deep- to shallow-water waves (Part IV). Like swells entering shallow water, their heights increase, their wavelengths shorten, and their phase speeds lose their dependence upon wavelength. At the limit, all waves approach a common velocity dependent only upon the depth:

$$C_S = \sqrt{gh} = 5.67\sqrt{h} \text{ ft./sec.} = 3.36\sqrt{h} \text{ knots.}$$

These changes are visibly manifested by a broadening of the divergent bow and stern wave patterns from a deep-water half-angle of 19.5° to a shallow-water limit of 90°, such that they coincide with (and augment) the transverse waves. The pattern then consists of a single transverse crest under the bow, and a single stern trough, both of which die away several widths abeam. The net result is that, as the water shoals, the vessel is caught in an ever-increasing resistance trap; her bow continues to rise, and her stern to squat. Unless power is greatly increased, she cannot exceed the critical wave speed ($V_S = C_S$). This situation has a close analogy to the "sound barrier" in aerodynamics, wherein an aircraft encounters a very sharp increase in resistance at sound speed (620 knots at sea level). In either case, given enough power and a high-strength, low-drag shape, the barrier can be overcome, and at higher speeds the excess resistance disappears.

The added resistance commences when the depth has decreased to about half a vessel's waterline length, and increases very rapidly with further shoaling, until there is danger of grounding by the stern due to excessive squat and sinking. These effects have been experimentally determined in some detail by measuring the forces necessary to tow model vessels at constant speeds in water of various depths. While I have been unable to dig up such data for sail or power yachts, figure 112 shows several shallow-water resistance curves for a model destroyer, which illustrate most of the interesting features. The heavy curve *(A)* refers to normal deep-water resistance, which is to be compared with the broken curves *(B–E)* in order of decreasing depth/length ratios (h/L_W). Moving upward along any $V_S/\sqrt{L_W}$ line gives the change in resistance with water depth at constant speed; moving horizontally to the left until resistance equals propulsive effort gives the speed reduction in decreasing depth.

Note, first, that the general effect of shallow water is to abnormally increase resistance at low speeds and to slightly decrease it at very high speeds. This is because, if a vessel has enough power to drive her over the crest of her own bow wave, she will then trim slightly by the bow, and encounter less resistance than in deep water.** Second, the shallower the water, the more rapidly resistance rises, and the lower the speed at which this rise occurs. While a destroyer has adequate power

*Frictional resistance is further increased by the presence of a lateral boundary (say, the bank of a narrow channel), wherein the corollary to sinkage is that the vessel is increasingly attracted to the boundary as the vessel's distance from it diminishes, and will collide with it unless diverted by strong rudder action.

**European barge horses discovered this interesting hydrodynamic fact centuries ago. A horse soon learns that, if he pulls a canal barge past a critical speed ($V_S = 3.36\sqrt{h}$ knots), it will coast with relatively little effort on the crest of its own bow wave.

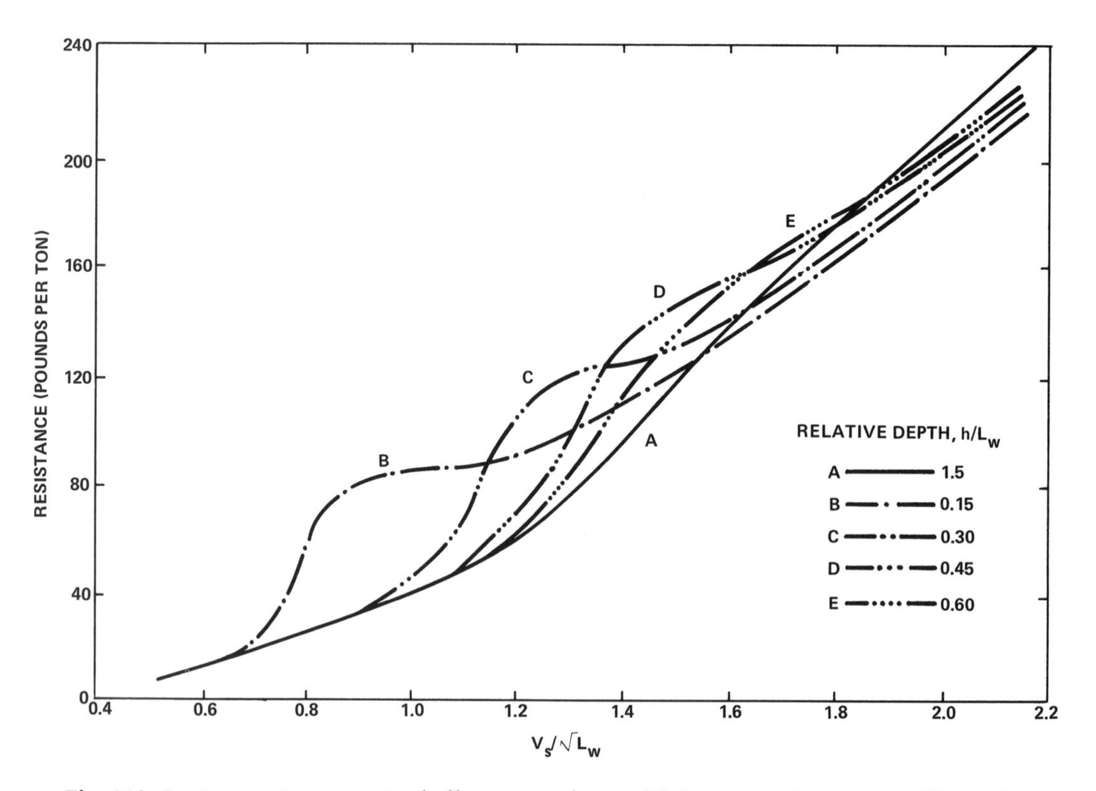

Fig. 112. Resistance increases in shallow water for model destroyer. Curve *A* is effectively in deep water. For $V_S/\sqrt{L_W} > 1.8$, the effect reverses, and resistance is less than in deep water at all relative depths, $h/L_W < 1.0$.

to surmount this resistance barrier at all practicable operating depths, commercial vessels—and particularly sailing boats—will be stalled at the first hump, and can lose as much as 40 percent of their deep-water speed before running aground. Third, at low speeds ($V_s/\sqrt{L_W} < 0.8$), where wave-making is unimportant, there is no sharp resistance rise, although a significant speed reduction may still occur in very shallow water.

Shallow-water resistance also increases with increased relative beam and draft, and the increased draft implies greater risk of grounding over an irregular bottom. In racing, these disadvantages must be weighed against the possible speed increase afforded by slower tidal currents and smaller wave resistance off a weather coast (see below). If you elect to sail in shallow water where you cannot clearly see the bottom, keep a sharp watch on your fathometer or your bow wave angle. If there is a sea running, with the highest waves longer than your hull length, you will also notice a significant height increase over dangerous shoals—even without evidence of breaking.

ROUGH-WATER RESISTANCE

Except when surfing downwind, the presence of wind waves imposes an additional resistance to vessel motion over and above the normal wave-making resistance encountered in smooth water. Because its magnitude depends not only upon the statistics of random seas but also upon a vessel's erratic motions as it encounters them (see "Response to Random Seas," p. 296), wave resistance cannot be simply described in terms of hull shape and speed.

Most of what we know about wave resistance comes from model towing tests or from propulsion measurements with full-sized ships, in which resistance is observed to increase roughly in proportion to *E*, the sea state index, and to the square of relative speed. Resistance is maximum when heading up-sea, much smaller down-sea, and minimum—but not zero—when running in the troughs. However—as earlier noted—upwind speed is more likely to be limited by adverse motions than by increasing resistance in high seas.

As a result of recent developments in computer technology, it is now possible to evaluate the influence of small changes of hull form and weight distribution on wave resistance at arbitrary sea states, speeds, and headings. This can be done much more cheaply than by repetitive model tests, although the latter are still necessary for confirmation of final designs. In a popular article,* scientists of the Netherlands' Delft Shipbuilding Laboratory described the results of such calculations aimed at evaluating the *upwind* performance of three displacement variations of an Admiral's Cup racer, all having the same length, beam, and draft. Displacement changes were obtained by increasing the depth of the immersed fairbody, at the expense of reducing the fin–keel aspect ratio. Calculations were repeated for four different sea states, and for three longitudinal weight distributions (*gyradii*). In brief, they concluded:

1. Irrespective of displacement, distributing weight toward the ends of the boat increased wave resistance with respect to that where weight was concentrated amidships. The percentage increase was greatest for the lightest displacement. Apparently, concentrating weight reduces a vessel's pitch period, shifting her maximum response toward the short-period end of the wave spectrum, where exciting forces are smaller.**
2. Irrespective of gyradius, when sailing close-hauled to windward, the lightest design encountered the least wave resistance. This is attributed to the reduced leeway afforded by a higher-aspect keel.
3. At all true wind speeds and sea states considered, the lightest displacement design made the best speed to windward. The speed advantage was greatest with low winds and high waves, and became insignificant in either low waves or high winds.

Most of these effects are well recognized—at least qualitatively—by experienced yacht designers and racing skippers. But it is comforting to think that modern science is catching up with long experience, and should pave the way for increasing refinements in this competitive field.

In a short *following* sea, wave resistance is much lower, because the wave crest particle motions are favorably directed. In choppy seas, resistance can be minimized by running at wave speed and adjusting one's position relative to the larger crests so that the vessel's transverse waves are partially canceled out, and a considerable power reduction is often possible without loss of speed.

*"Seakeeping Performance," by J. Gerritsma and D. Moeyes, *Sail*, April 1973.

**Dynamically, one can also argue that reducing a vessel's pitching inertia allows her to follow steep wave profiles more easily, instead of boring through them. The advantage of lighter displacement accrues from a greater reduction of momentum lost to wave impact.

Plate 24. Rare shot of a 45-foot yawl surfing at high speed on a big swell.

The ultimate extension of this technique is the art of planing, or surfing, on the forward face of relatively long, steep swells. With sufficient initial speed, the forward component of wave orbital velocity, along with the bow trim that occurs as the vessel's stern is lifted by an overtaking crest and the downslope component of gravity, combine to accelerate her to wave speed. By suitable power and/or heading adjustments, the vessel may be maintained at a *constant elevation* below the crest, and, under optimum conditions, may fly along at 15–20 knots, until the wave melts away to be replaced by another (plate 24).

Although this technique has long been practiced by small powerboats and fast cruisers, it is not clear whether its adoption by racing sailboats awaited the transition of surfboard enthusiasts to sailing yachtsmen, or is the natural consequence of light displacement designs. In any case, it is now standard practice, and mandatory if one hopes to place in long, downwind ocean races.* Surfing is now recognized as an appropriate survival tactic to avoid pooping in storm waves (see Part VII).

*Actual planing can only be achieved by relatively light-displacement sailing boats ($\delta < 150$), but even heavier boats can enjoy a considerable boost in a semiplaning configuration for short intervals. Surfboards have displacement/length ratios between 50 and 140, but with no power assist they can only plane in relatively steep, near-breaking waves.

WIND RESISTANCE

In addition to its role in producing waves and currents, the wind also exerts drag forces roughly proportional to the square of wind speed on all exposed surfaces of a vessel. If under way at an angle θ_t from the true wind V_t (fig. 113a), she also counters an effective head wind V_s equal to her speed (corrected for local currents, if any). The relative (or apparent) wind V_r will be the vector resultant of V_t and V_s and will act on the vessel at an angle θ_r, which is always equal to (up- and downwind headings) or less than θ_t.

The geometrical relationship between these five quantities can easily be constructed for any practical purpose. At sea, V_r and θ_r are obtained from the ship's anemometer, and V_s from her log or DR plot. These are laid out to scale, as shown, and the triangle is completed to give V_t and θ_t. Use of such plots is, of course, very basic to sailing and pilotage, but is described in so many nautical books that there is no point in attempting to elaborate here.

Knowing V_r and θ_r, the calculation of wind resistance would be almost as simple, were it not for the fact that the resultant wind force vector acts, not from the relative wind direction (θ_r), but at some other angle that depends upon the shape and attitude (angle of incidence) of any obstacle disrupting the air stream. It is principally this change in angle between the resultant wind and the force vectors that enables a sailboat to work to windward (fig. 113b). Here, the change $(\theta_f - \theta_r)$ is large enough that the resultant force F_t actually has a forward component F_a that drives the yacht ahead, while its larger, athwartship component F_h, working against high lateral water resistance, produces only a small leeward drift.

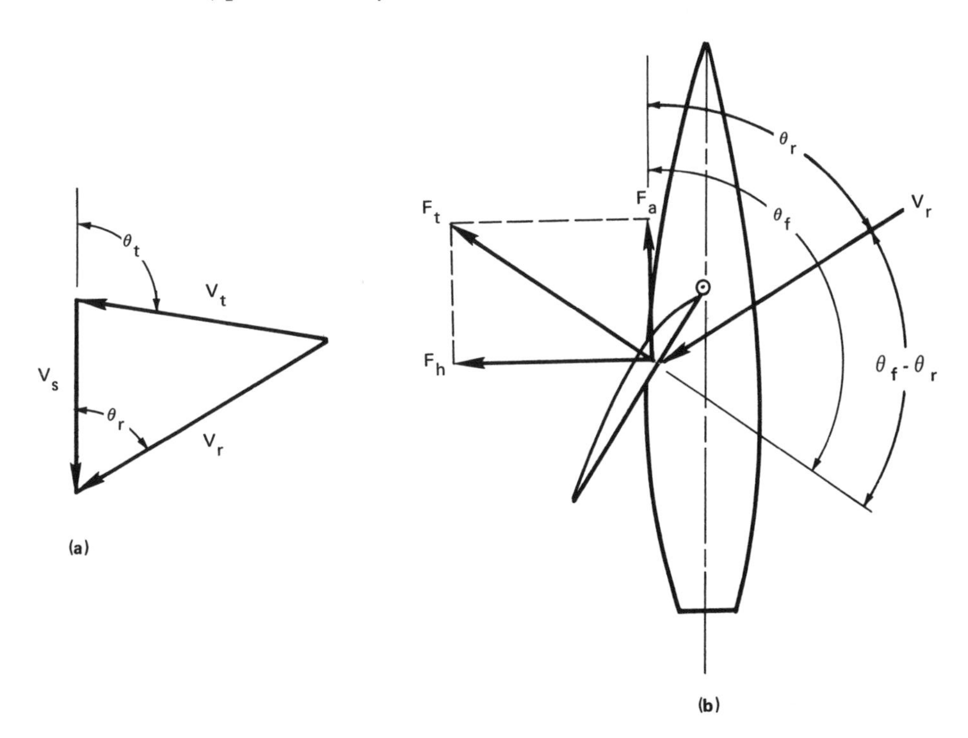

Fig. 113. *(a)* Construction for determining true wind V_t, knowing ship speed V_s, relative wind speed V_r, and its direction θ_r. *(b)* Sailboat can move against apparent wind V_r because its resultant force F_t on sail has a net forward component F_a.

Sails excluded, the net wind resistance encountered by a vessel comprises the resultant of all of the individual forces acting on various elements (hull, superstructure, spars, rigging, etc.) in disparate directions, most of which also vary with the relative wind direction θ_r. Because of the work entailed in summing up all these bits and pieces, designers ordinarily resort to shortcuts which—further abbreviated—amount to the following. To a reasonable approximation, the total wind resistance acting on a vessel can be represented by the formula

$$F_t = 0.004V_r^2 \, (A_T \cos^2 \theta_r + A_L \sin^2 \theta_r)/\cos \, (\theta_f - \theta_r) \text{ lbs.}$$

where V_r is in knots, and A_T and A_L are the totals of all elemental areas (in sq. ft.) exposed to windage, projected on a vessel's athwartship and longitudinal planes, respectively. Because of hull fineness, only 30 percent of the fore- and aft-hull projections are used in calculating A_T. In using this formula for a particular vessel, one need merely take her bow, stern, and lateral profiles, measure the areas—including spars and rigging, if any—deduct 70 percent of the bow and stern hull areas, and plug in the remainder.

This works fine for the designer, but for anyone interested in estimating wind drag on his or her boat, I have worked out a more general shortcut, by statistically averaging the component areas for a number of representative sailboats (bare poles only) and powerboats, and reducing them separately to a common basis, according to the square root of displacement. This involved considerable work and resulted in a few interesting comparisons:

1. The spars and rigging of a sailboat average about 77 percent of her fore- and 70 percent of her aft-projected areas, but only about 28 percent of her lateral area.
2. The frontal areas of sail- and powerboats of the same displacement are about the same, but the latter have about 20 percent greater lateral areas, owing to higher freeboard and greater length of deckhouses.
3. The effective after areas of powerboats are about 35 percent larger than fore areas due to bluff transoms.

The relative wind force distributions for these two vessel types are plotted as functions of θ_r in the polar wind-force diagram (fig. 114), where the radial distance to either curve, in arbitrary units, is a measure of the wind force acting in the direction of the relative wind; that is, the force vectors have been corrected for change of incidence ($\cos(\theta_f - \theta_r)$). The projection of any force vector (such as F_t) on the vertical (0°–180°) axis is similarly the ahead (or stern, for $\theta_r > 90°$) component of resistance for that relative wind direction, indicated by F_a (or F_s). The latter are useful in estimating mooring forces, power required to combat head winds, or the driving component of following winds.

From this diagram (fig. 114), one can see at a glance that wind force rises from its smallest value with the relative wind ahead, to a maximum 4.5–5.5 times larger with the wind abeam, and then falls to slightly larger minima with the wind astern. Forces are generally larger for powerboats, but in both cases, they attain about two-thirds of their maximum (beam) values when the relative wind has shifted from ahead to only 30°–35° off the bow or stern. These angles also correspond to maximum values of F_a and F_s, respectively. Thus, a vessel being towed upwind or lying at

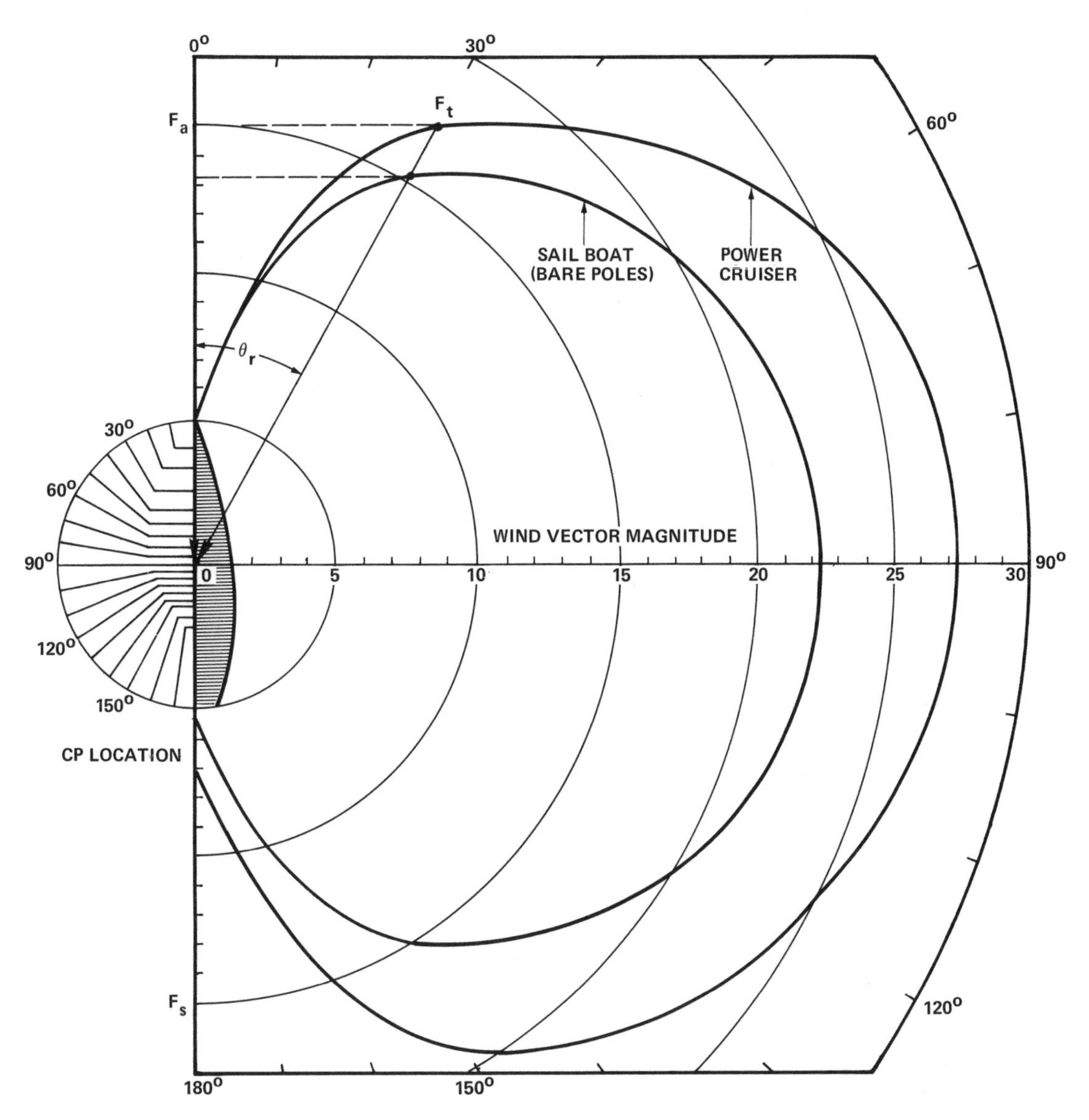

Fig. 114. Wind-force diagram shows direction and magnitude *(right)* and point of application *(left)* of relative wind force. The ahead (or astern) component is the projection of F_t on the vertical axis.

anchor will experience 2–3 times her head-on resistance if allowed to veer (or swing) as much as 30°.

To calculate the actual force magnitude, we need figure 115, which includes the effects of relative wind speed and vessel displacement. Entering this figure at the bottom with displacement in tons (2,240 lbs.), project a vertical upward to any wind speed, and then move horizontally left to obtain the wind-force factor. This factor, multiplied by the appropriate vector magnitude F_t from the wind-force diagram, gives the total wind force (lbs.) acting on the vessel from the relative wind direction θ_r; alternatively, multiplying by F_a or F_s gives the ahead or stern resistance, respectively. For example, the vector F_t (shown at $\theta_r = 30°$ in figure 114) intercepts the sailboat resistance curve at a magnitude of about 15.2. From figure 115, the wind-force factor for a 15-ton (L_W = 38-ft.) sailboat in a 40-knot relative wind is 95, and the

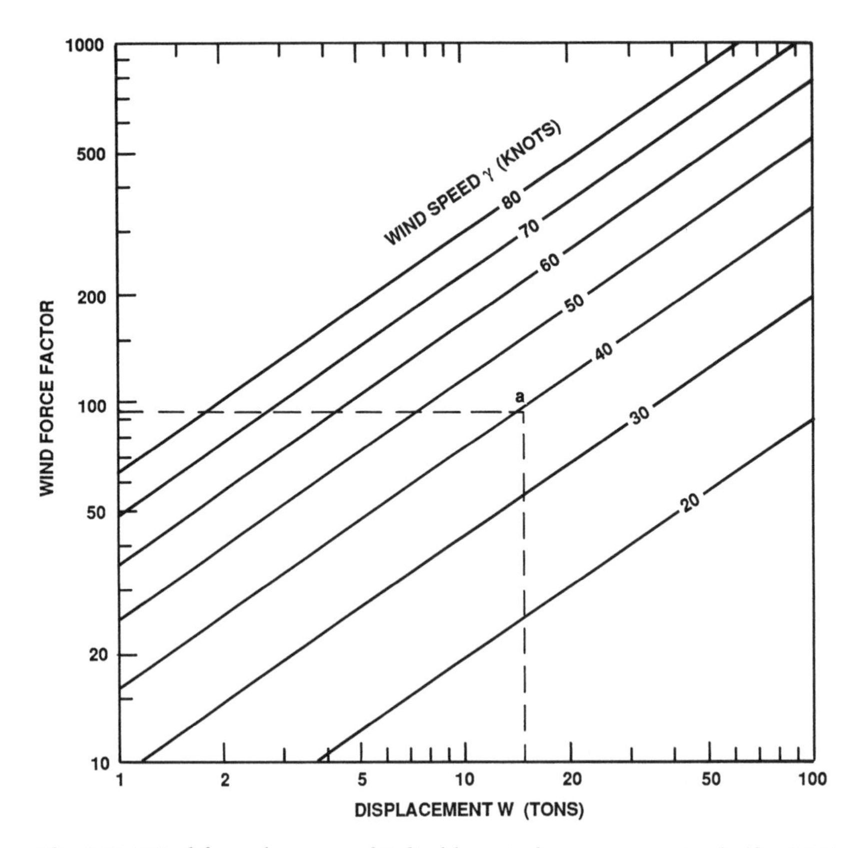

Fig. 115. Wind-force factor, multiplied by wind vector magnitude (fig. 114), gives total wind resistance in pounds.

total wind resistance on such a boat would be F_t = 15.2 × 95 = 1,444 pounds. Similarly, her ahead resistance would be about 13.3 × 95 = 1,234 pounds. The same vessel, anchored *(without swinging)* in a 40-knot head wind would exert a pull on her anchor of only 5 × 95 = 475 pounds. She could conceivably ride to such a wind on a 25-pound anchor (in good holding ground) and a rode of 1/8-in. (600-pound) nylon parachute shroud, but a relative wind shift of 30° would part the rode. However, in Part VIII we will show that, under storm conditions, emergency anchoring can actually be accomplished safely on ordinary tackle, with a few special modifications.

The vector diagram to the left of the polar wind force curves (fig. 114) shows the positions of the effective centers of pressure (CPs) along a vessel's hull for winds from various directions; in general, the closer the relative wind is to the bow, the farther forward the CP, and conversely. Hence, for all quartering winds, there is always a force component acting to turn a vessel broadside that must be compensated for by rudder control to hold a straight course. In strong winds, this moment may be so large as to prevent a low-powered vessel from turning either into or off the wind. Turning off generally will be easier, because the CP is nearer amidships and its moment smaller.

Quite similar diagrams could be drawn for wave and drift forces acting on a vessel's underbody, and a drifting vessel will always orient herself so that the resultant of these combined forces is in equilibrium. Because the resultant of wind and wave forces is usually slightly forward of amidships, and that for drift forces slightly aft, most vessels will lie-to with the wind somewhat abaft the beam, and drift to leeward at 5–8 percent of the wind speed, depending upon the ratio of these forces. Because lateral drift speeds are much lower than when running off, heaving to may be preferable if sea room is a problem and the seas are not breaking heavily. A vessel's precise attitude can to some extent be controlled for maximum safety and comfort by power or rudder adjustment, or by setting a scrap of sail (see Part VII).

PROPULSION

Whether by engine or by sail, *propulsion* is a very complicated subject, with a vast popular and technical literature that we could hardly improve on here. Therefore, we shall content ourselves with outlining the important points and adding a few suggestions that may aid in evaluating or improving a vessel's performance.

At all speeds, of course, a vessel is in equilibrium between the forward thrust delivered by her engine(s) or sails, and the ahead component of all of the above resistances prevailing; thus, resistance is the basis for most power requirements.* While wind resistance can be calculated by the methods outlined above, the wave and wave-making resistances of large ships are conventionally measured directly by model—or full-scale—towing tests under representative sea conditions, or are inferred from previous tests of similar hull shapes.

Because of the expense of towing tests, unless a design is very unusual, small-craft propulsion systems are almost exclusively arrived at by more approximate methods, matching available engine and propeller specifications against resistance tables for typical hull parameters.** Such methods are not always completely reliable: quite often a boat ends up with a different propeller—if not a different engine—from that originally supplied; a sailboat may be completely rerigged to alter her trim, balance, or total sail area.

Engine Propulsion: Irrespective of size, a vessel's power requirements are determined by much the same procedure, differing mainly in that propulsion systems for large ships are specially designed and constructed, whereas those for boats under 100 feet or so are assembled from off-the-shelf components. Before getting involved in design, however, we need to know a little about the force and power balance of a vessel, and how the latter is distributed among the various propulsion components according to their respective efficiencies.

Figure 116 illustrates the elements of a typical system and the principal force and velocity components acting on the vessel. When under way at uniform speed V_s, she is in equilibrium between her total ahead resistance R_a and the forward thrust of her screw propeller (*a*). This thrust, of course, is obtained from acceleration of water through the screw, and is imparted to the vessel through a thrust bearing that is set

*In special-purpose vessels, such as tugs and trawlers, high static thrust is required to overcome inertial resistance and to provide steerage control.

**Much the same can be said for yacht design itself; in a foreword to Henry and Miller's *Sailing Yacht Design*, Olin Stephens remarks, ". . . a new boat frequently will develop from the one before. Local or specialized types can grow from one to another, a little like Topsy, shaped by intuition and improved by experience until at sea under sail the results may turn out to be better than the best products of sophisticated procedures. And yet we should remember still that it helps to know both the intuitive and analytic approach. . . ."

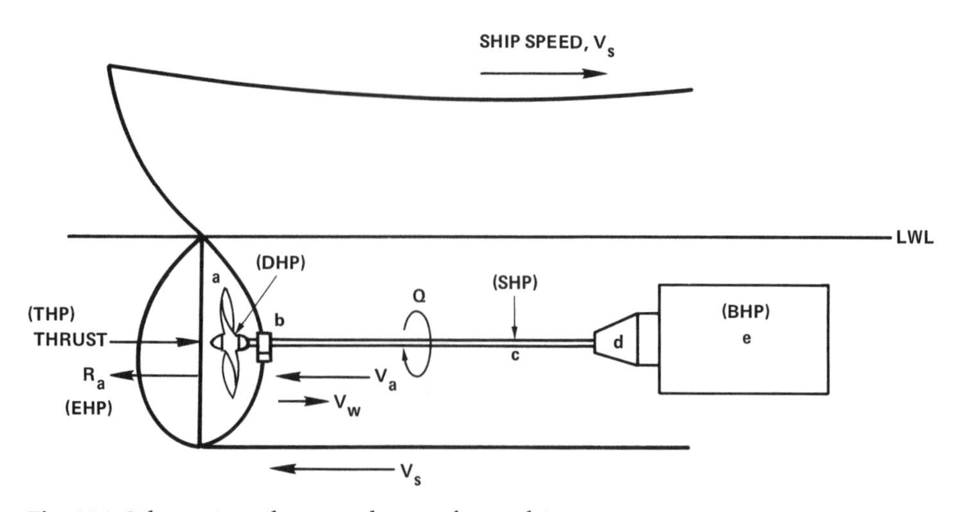

Fig. 116. Schematic and nomenclature of propulsion-system components.

in the *thrust block* (*b*). From here forward, the power system is rotary; the torque Q generated by the engine (*e*) is transmitted to the screw by way of a gear train (*d*) and propeller shaft *c*) running through the thrust bearing. There may be other guide bearings to prevent whipping if the shaft is longer than about fifty times its diameter.

The arrows shown beneath the propeller shaft indicate the direction and magnitude of fluid velocities important to propeller action, as measured relative to the vessel's hull. At some distance from the hull, the relative velocity is directed aft and is equal to the ship speed V_S. Close to the hull there is a smaller relative forward flow V_W, called the *wake stream*, arising from water frictionally entrained by the hull, and augmented by the orbital motion of the stern waves. Their difference $V_a = V_S - V_W$), the actual velocity encountered by the advancing propeller, is called the *speed of advance*. Owing to convergence of flow under the counter, V_W is strongest for in-line screws, and much weaker for screws set outboard.

Because V_W decreases with distance from the hull, and also because of other perturbations produced by struts, bosses, or deadwood ahead of the screw, inflow to the blades is unequally distributed, and at certain critical speeds can result in strong vibrations of the propeller and shafting. Such vibrations are often synchronized with the blade frequency, and are apt to be most severe for large, slow-turning, two-bladed propellers. Again, these effects are minimized for outboard screws.

A vessel's power balance can be written schematically as (see figure 116)

engine HP	>	shaft HP	>	delivered HP	>	thrust HP	<	effective HP
(BHP)		(SHP)		(DHP)		(THP)		(EHP)

where the inequalities represent various efficiencies, or loss coefficients, defined by the following relationships:

(a) BHP $\times \eta_e$ = SHP (η_e = engine and gear efficiency);
(b) SHP $\times \eta_s$ = DHP (η_s = shaft and bearing efficiency);
(c) DHP $\times \eta_p$ = THP (η_p = overall propeller efficiency);
(d) THP $\times \eta_h$ = EHP (η_h = hull efficiency).

The reversed inequality sign (<) between THP and EHP requires some explanation. By conventional definition, EHP = $R_a V_s/326$ HP* and THP = $TV_a/326$ HP. Since a vessel's thrust balance is given by $T \doteq R_a$, and because $V_s = V_a + V_w$, it follows that $\eta_h > 1$; e.g., hull efficiency is *greater* than unity. Put more simply, the propeller gets a little boost from the wake stream over what it would have in open water. For most small vessels, this effect is usually negligible, but it can amount to as much as 25 percent for large, single-screw tankers.

All of these efficiencies change with vessel speed and the revolution rates of engine, shaft, and propeller, and one of the principal objects of propulsion design is to maximize their product. With a fixed-pitch screw, this can be accomplished only over a narrow range of hull speeds—usually that corresponding to 80–90 percent of maximum rated engine power. Table 8 lists the normal ranges of achievable efficiencies and their products (overall propulsion efficiency $\eta_t = \eta_e \times \eta_s \times \eta_p \times \eta_h$ = BHP/EHP), whose considerable variation reflects the degree of compromise inherent in designing power systems. While propulsion efficiency has a relatively high priority in commercial vessels, sailing auxiliaries are relatively inefficient, and fuel consumption is a minor consideration in powered racing craft.

Table 8.

Engine	BHP range	Efficiencies				
		η_e	η_s	η_p	η_h	η_t
Steam turbine	3,000+	.90–.95	.97	.70–.80	1.10–1.25	.67–.92
Diesel engine	20–3,000	.60–.80	.98	.60–.70	1.00–1.10	.35–.60
Gas engine	1–1,500	.40–.60	.99	.50–.60	1.00–1.05	.20–.37

Whereas the functions and characteristics of engines, gearing, and shafting are probably familiar to the average reader, those of propellers are not so obvious and merit some discussion. In addition to producing the required thrust at a vessel's design speed, her propeller must be effectively matched to the other propulsion components. In this, there are many options, and the basic design problem is to select the most efficient combination, subject to a number of limitations, the most important of which are:

1. **Torque.** Ideally, the propeller should be capable of absorbing all the power the engine can deliver without running away. This means that the propeller's restraining torque must be high enough to limit engine speed, but low enough to avoid overloading the engine at full power.
2. **Thrust loading.** This should be low enough to prevent cavitation at all power settings.**
3. **Diameter.** While efficiency increases with screw size, the latter is often limited by vessel dimensions, depth of immersion, or the desirability of minimizing propeller drag on sailing auxiliaries.

*For similarly proportioned vessels moving at hull speed ($V_s/\sqrt{L_W}$ = 1.34), R_a and L_W increase roughly as W and $W^{1/3}$, respectively. Thus the total power at wave-making speeds increases approximately as $W \times \sqrt{W^{1/3}} = W^{7/6}$.

**Some high-speed (*supercavitating*) propellers are designed to operate at unusually high-thrust loadings.

Contrary to popular opinion, a propeller does not screw itself through the water, but rather flies through it. Like an airplane wing, the propeller blades are shaped hydrofoils, and are designed to have a maximum lift-(thrust-)to-drag ratio when operated at the proper angle of incidence θ to the relative flow. This angle, together with other important propeller dimensions and flow parameters, is shown in figure 117, to which the following definitions apply:

- D = *diameter* (ft.)
- P = *pitch* (ft.), the distance a propeller that delivers no thrust would advance in one revolution through an unresisting solid, such as butter
- P/D = *pitch ratio,* propeller advance per revolution in diameters
- N = *rotation speed* (rpm)
- ϕ = *pitch angle,* the angle between the blade face and the plane of rotation

When rotating in a real fluid, owing to flow acceleration in developing thrust, the propeller does not advance by P ft./revolution, relative to its inflow velocity (V_a), but by some smaller distance JD, where $J = 101.33V_a/ND$ = the *advance ratio* (physically, J is the actual advance in diameters/revolution). The distance $S = P - JD$ (ft./revolution), by which a working propeller lags behind the advance it would have if delivering no thrust, is called the *slip.**

The advances P and JD can be converted into relative velocities, if multiplied by $N/60$ (ft./sec.), or $N/101.33$ (knots). The relations between these velocities and V_s, V_w, and V_a, previously defined, are shown below the propeller-advance diagram in figure 117, from which it can be seen that the propeller inflow velocity V_a, is just the propeller's speed of advance. Furthermore, for those trigonometrically minded, the working propeller's effective incidence angle is

$$\theta = \phi - \arctan J/\pi = \phi - \arctan (101.33\, V_a/\pi DN) = \phi - \arctan 101.33\, V_a/\omega$$

where $\omega = \pi DN$ is the propeller's angular velocity (radians/sec.). Put in plain English, angle of incidence = pitch angle – arctan (inflow velocity/angular velocity).

With this introduction, the optimum propeller efficiency, size, and pitch ratio can be determined directly from a standard propeller chart** after calculating a single parameter (B_p), which contains only known—or measurable—quantities:

$$B_p = N\sqrt{\text{DHP}}/V_a^{2.5} = \textit{propeller coefficient}$$

An example of a chart for three-bladed propellers having a 0.5 blade/area ratio is shown in figure 118. Its horizontal axis is in units of B_p, and the vertical axis in units of P/D; and it comprises two sets of intersecting contours of propeller efficiency (η_p) and a parameter called the *propeller constant:* $ND/V_a = 101.33/J$. In addition to being well adapted to design, such charts provide a good perspective of the interrelation-

*In many design references, slip is used as an index of efficiency. But it actually has no physical significance and, as outlined below, efficiency can be more intuitively related to real physical variables.

**The best-known standard propeller charts were compiled by L. Troost from review of hundreds of open-water propeller tests. Separate charts are available for propellers having from two to seven blades, and for a number of blade/area ratios. Because the charts represent averages from many tests, extreme accuracy may require more sophisticated analysis. However, they are ordinarily quite adequate for stock propeller selection.

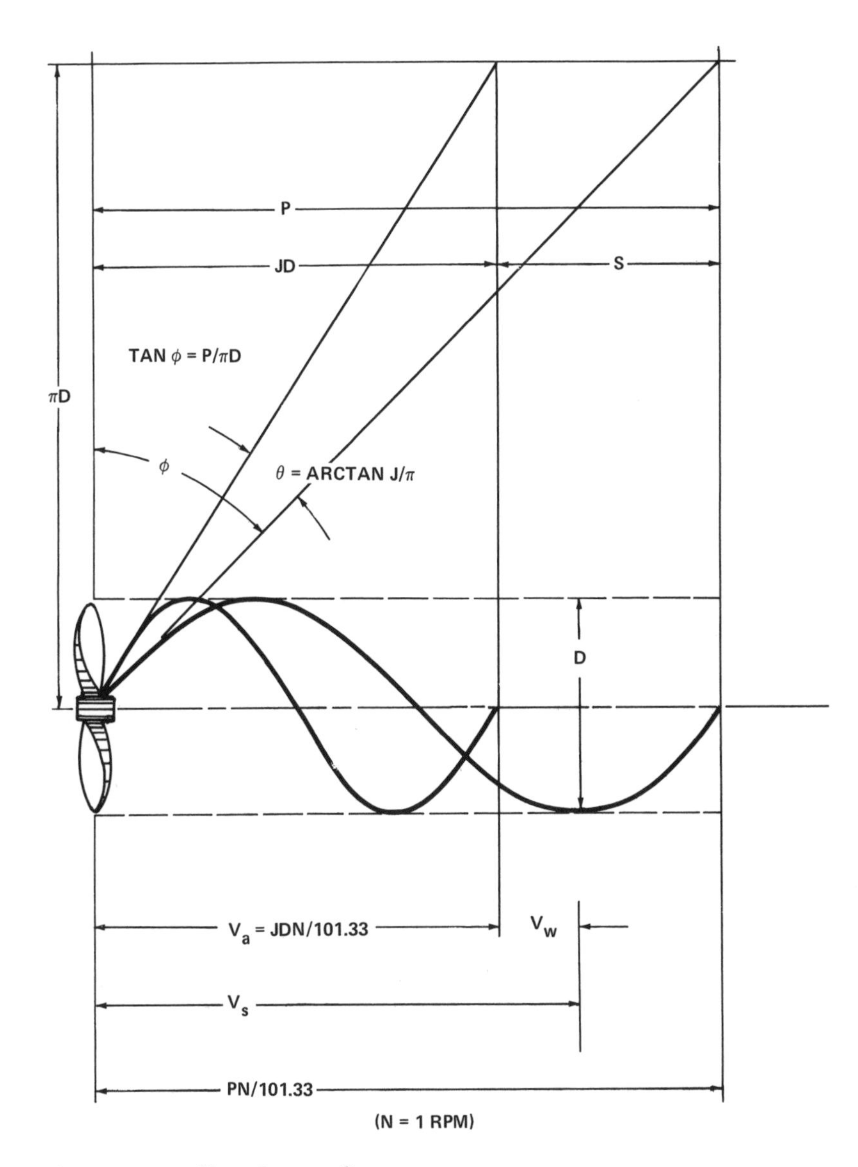

Fig. 117. Propeller-advance diagram.

ships between all propeller variables, and how changes in these variables affect efficiency.

The chart is entered at a computed value of B_p by projecting a vertical upward to its intersection with the dashed maximum-efficiency line. From this point, the optimum efficiency as well as ND/V_a and P/D are determined by interpolation. Knowing N, V_a, and ND/V_a, D is fixed; knowing P/D and D, P is also fixed. In searching for a stock wheel, one would select the nearest available values of P and D, and (assuming no change in V_a or N) would then calculate P/D and ND/V_a, whose intersection defines a new operating point on the chart. If this point falls between curves *a* and *b*, the choice is satisfactory. If it falls much below curve *a*, the diameter may be too large and tend to overload the engine; if much above curve *b*, it may be too small and the engine may run too fast. The first condition can be improved by reducing N without altering DHP (viz, a slower engine or reduction gearing) or by reducing the blade/area ratio or the number of blades; the second condition, by just the opposite means.

From the general layout of this chart, it is easy to see that low values of B_p correspond to high efficiency, large, slow-turning screws, and low thrust loading, and conversely. Thus, low-speed vessels ($V_s/\sqrt{L_W} < 1.34$) will usually have propeller coefficients in the range from 1 to 20; high-performance vessels from 21 to 100. Planing boats operating at $V_s/\sqrt{L_W} > 4$ generally use special super-cavitating propellers that have altogether different hydrodynamic characteristics, to which this discussion does not apply.

The dotted curve *c* in figure 118 defines the region above which cavitation may occur as a result of high thrust loading. This region is only approximate, since cavitation also depends upon temperature and the depth of propeller immersion.*

Returning, now, to the question of power system design, it can best be described as a somewhat circular trial-and-error process that ordinarily can be accomplished in less time than it takes to describe it. The procedure may differ a bit, according to individual preference, but usually involves the following steps:

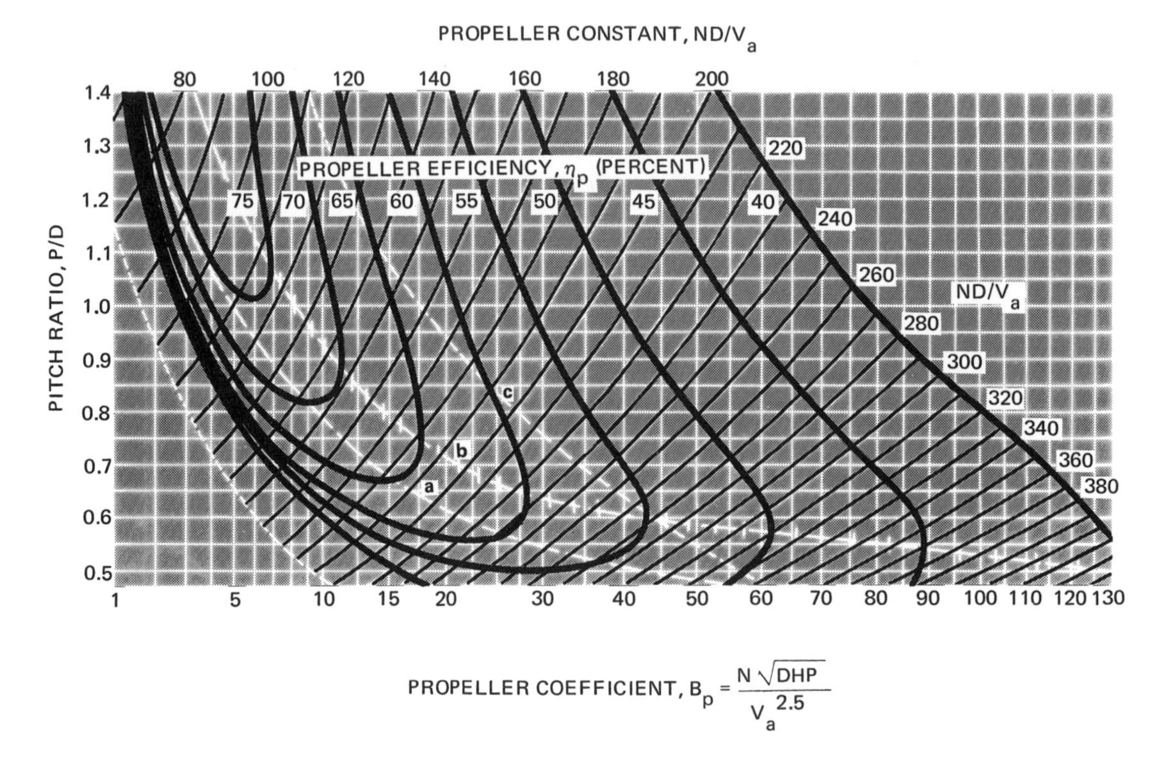

Fig. 118. Troost standard series propeller design chart for three-bladed screws of 0.50 blade/area ratio. Dashed curve *(a)* is optimum design line; *(b)* is maximum efficiency curve. Region above curve *(c)* is liable to cavitation.

*Physically, cavitation is local boiling, caused when the pressure in the vicinity of any blade element falls below the ambient vapor pressure at the prevailing water temperature. Collapse of small steam bubbles can result in extensive local erosive pitting, and is accompanied by a hissing or crackling noise and loss of thrust. It is most apt to occur in high-pitch propellers that are operated at thrust loads in excess of about 15 lbs./sq. in., under which conditions the blades are *stalled*, in an aerodynamic sense, and abrupt flow separation occurs behind the blade tips.

1. Estimate total vessel resistance R_a at maximum design speed in smooth water V_s and compute effective horsepower. $\text{EHP} = \frac{R_a V_s}{326}$

2. Estimate propulsion efficiency η_t and compute trial brake horsepower (BHP). Add 10–25 percent reserve depending upon intended service. $\text{BHP} = \text{EHP}/\eta_t + 10\%\text{–}25\%$

3. Select engine for this BHP, note maximum power rpm (N), and corresponding shaft speeds N_1, N_2, etc., if there are gear-reduction options.

4. Estimate engine efficiency η_e (including gears), and compute shaft horsepower(s). $\text{SHP} = \text{BHP} \times \eta_e$

5. Estimate shaft efficiency η_s and compute delivered horsepower. $\text{DHP} = \text{SHP} \times \eta_s$

6. Estimate speed of advance $V_a = V_s - V_w$,* and compute propeller coefficient(s) B_p. $B_p = \frac{N\sqrt{\text{DHP}}}{V_a^{2.5}}$

7. Enter a suitable propeller chart (figure 118 is a good start) and select stock propeller(s), as outlined above. This takes a bit of juggling; the object is to find a wheel of the right diameter whose efficiency agrees with that assumed in step 2. If this can be done, well and good. If not, try a different number of blades, a new blade/area ratio, or (last resort) a different engine.

8. With final propeller efficiency η_p, compute thrust horsepower and thrust T. $\text{THP} = \text{DHP} \times \eta_p$; $T = \frac{326\ \text{THP}}{V_a}$

9. Determine projected area of propeller blades $A_p = 0.79D^2 \times$ blade/area ratio (0.5 for this chart), and compute thrust loading. For displacement-type hulls, this should fall in the range 3–6 lbs./sq. in.; for high-performance boats, 6–15 lbs./sq. in. The latter should always be checked for cavitation if tip speed, πDN, exceeds 9,000 ft./min.

The problem of propeller location in sailing craft is a notorious can of worms, of which no less an authority than Captain H. E. Saunders has remarked, "There is no good place. In almost any position the propeller is a nuisance for sailing and a misfit for propulsion." While this is certainly true for fixed-pitch screws, it is much less so for full-feathering, controllable-pitch propellers, whose open efficiency at optimum pitch approaches that of an equivalent fixed-pitch wheel, but with as little as

*For practical purposes, $V_w = 0$ for twin-screw vessels; for single, center-line screws, $V_w = V_s(0.7C_b - 0.21)$, where C_b is the vessel's *block coefficient:* the ratio of her displacement volume to that of the rectangular box which would just contain her hull (fin keel excluded) up to the waterline.

5 percent of the drag in the feathered position. A controllable propeller is preferably mounted in line and ahead of the rudder, where its thrust will provide steering control at zero speed. The deadwood should be smoothly streamlined to a small radius to minimize turbulence and increase the inflow velocity.

Considering their many advantages, it is really quite surprising that controllable propellers have not made greater inroads on the American small-craft market. The causes seem to be mainly lack of education, inertia, and expense, although the additional cost is somewhat offset by the elimination of reverse gearing. The propeller can absorb maximum power over a wide speed range, a feature particularly useful for tugs, trawlers, and draggers, which require high thrust at both low and high speeds. With the addition of automatic pitch control, the engine can operate at constant speed, and at the most efficient power setting, independent of load.

Precisely the same advantages accrue to propeller-driven aircraft, which—except for the least-expensive models—have almost completely converted to constant-speed propellers.

Sail Propulsion: If adequately rigged (see Part VII), a boat's power to carry sail is limited only by its transverse stability. Other things being equal, sail thrust is proportional to sail area (A_S), and, in many books on yacht design, you will find attempts to relate A_S and other stability parameters to various powers of displacement and waterline length. But such approximations are unsatisfactory because different rigs and hull shapes require separate sets of curves, and these are rarely simple powers of W.

Most good designers work directly with stability, proportioning sail area and the height of its *center of effort* (H_{ce}) so as to obtain a desired relation between heel angle (θ) and L_W for a given wind pressure (P), or conversely, to obtain the desired wind pressure at a specified angle. Both methods are variations of a stability relation called the *power to carry sail* (P_s),

$$P_s\,(L_W) = \frac{2240W \times \text{GM}}{P \times A_s \times H_{ce}}$$

where (L_W) indicates that P_s is some function of L_W. Actually, P_s has nothing to do with power, but is a nondimensional number expressing the ratio of righting/heeling moments for a particular design.

Without getting involved in mechanics, it is instructive to see how these quantities vary among actual sailboats, and how they are related to power and speed. Figure 119 presents average curves of A_s, H_{ce}, GM, and L_W against W for some fifty sailboats of all categories selected at random from CCA handicap listings. Irrespective of rig, and whether internally or externally ballasted, 95 percent of the data (shown by the shaded bands) fall within 10 percent of the median lines. This deviation diminishes to about 3 percent for heavier yachts, reflecting increasing attention to physical principles rather than rough dimensional rules, which would argue that all of these lines be straight. Any boat whose dimensions fall much outside these shaded bands can be considered abnormal. This does not necessarily mean she won't sail well (America's Cup defenders have markedly higher A_s/W ratios), but most mutations behave poorly in one manner or another.

In this figure, I have also shown a curve of P_s (assuming P = 1.2 lbs./sq. ft.), whose upward trend is a measure of the increasing stability of larger boats. A boat whose P_s falls 10 percent above this line will be too stiff; 10 percent below, too tender. Again, P_s is often taken as a straight line having a slope of 1:6 in this representation, but one cannot draw a line that everywhere fits this curve within anything like 10 percent; ergo, rough rules give rough boats.

From the above data, I have compiled the five curves of parameters relating to power, speed, and performance, as shown in figure 120. At the top, you will find our old friend the displacement/length ratio $\delta = W/(.01L_W)^3$, which, as you may recall, is a measure of a vessel's relative density per unit length. It has a maximum for boats of about 10 tons of displacement, and diminishes rapidly to either side. Its smallest value (δ = 175) indicates that even boats as light as 1 ton are too heavy to plane in smooth water.*

To calculate the actual power a sailboat can produce, we need to know her total hull resistance R_t at maximum design speed $V_s = 1.4\sqrt{L_W}$, which is about all a displacement-type hull can muster. The latter curve is easily obtained from L_W (fig. 119). The R_t curve was obtained from Taylor's standard series resistance curves (see *Skene's Elements of Yacht Design*), and is proportional to δ for $\delta > 180$. Again, we see that 10-ton, 31-ft. boats have the highest resistance per ton within the range considered. Since such boats have relatively high stability (low sail area per unit displacement), they would do poorly in boat-for-boat racing, were it not for favorable handicapping.

The relative effective horsepower can now be obtained from the relation EHP/ton = $R_t/W \times V_s/326$, and is shown as the lowest curve in figure 120. To obtain total EHP, this curve must be multiplied by W; it increases from a minimum of about 1.5 EHP for a 1-ton, 18-foot boat to 410 EHP for a 100-ton, 72-footer. Considering their relatively low propulsion efficiencies, gas—or even diesel—engines of much greater horsepower would be required to provide the same thrust. However, it would be very uneconomical to operate an auxiliary at $V_s/\sqrt{L_W} = 1.4$. Near the peak of the resistance hump, speed varies only as the fourth root of power, and thus 38 percent of the indicated EHP would drive any boat at comfortable cruising speed ($V_s/\sqrt{L_W}$ = 1.1). From table 8 (p. 281), 38 percent is about the maximum overall efficiency for gas engine propulsion, from which we can conclude that the EHP curve shown could also be considered as a reasonable estimate of the minimum BHP per ton required to provide cruise power. Normally, one would add about 20 percent reserve. For diesel engines, the curve could be used as is.**

What wind speeds are required to generate this much power? This gets down to the question of sail efficiency, which so far we have left open. But we can make some estimate, using data from Marchaj's book, *Sailing Theory and Practice*.

Most boats (spinnakers excluded) sail fastest with the true wind V_t on—or slightly abaft—the beam, and with sheets trimmed so that the sails make an angle of incidence $\alpha \div 27°$ to the relative wind V_r (fig. 121a), which is blowing from some

*Thrust approaching 10 percent of displacement is also required to plane.

**Somewhat greater power is required at low displacement, owing to the lower efficiency of small, high-speed screws (dashed branch of EHP curve).

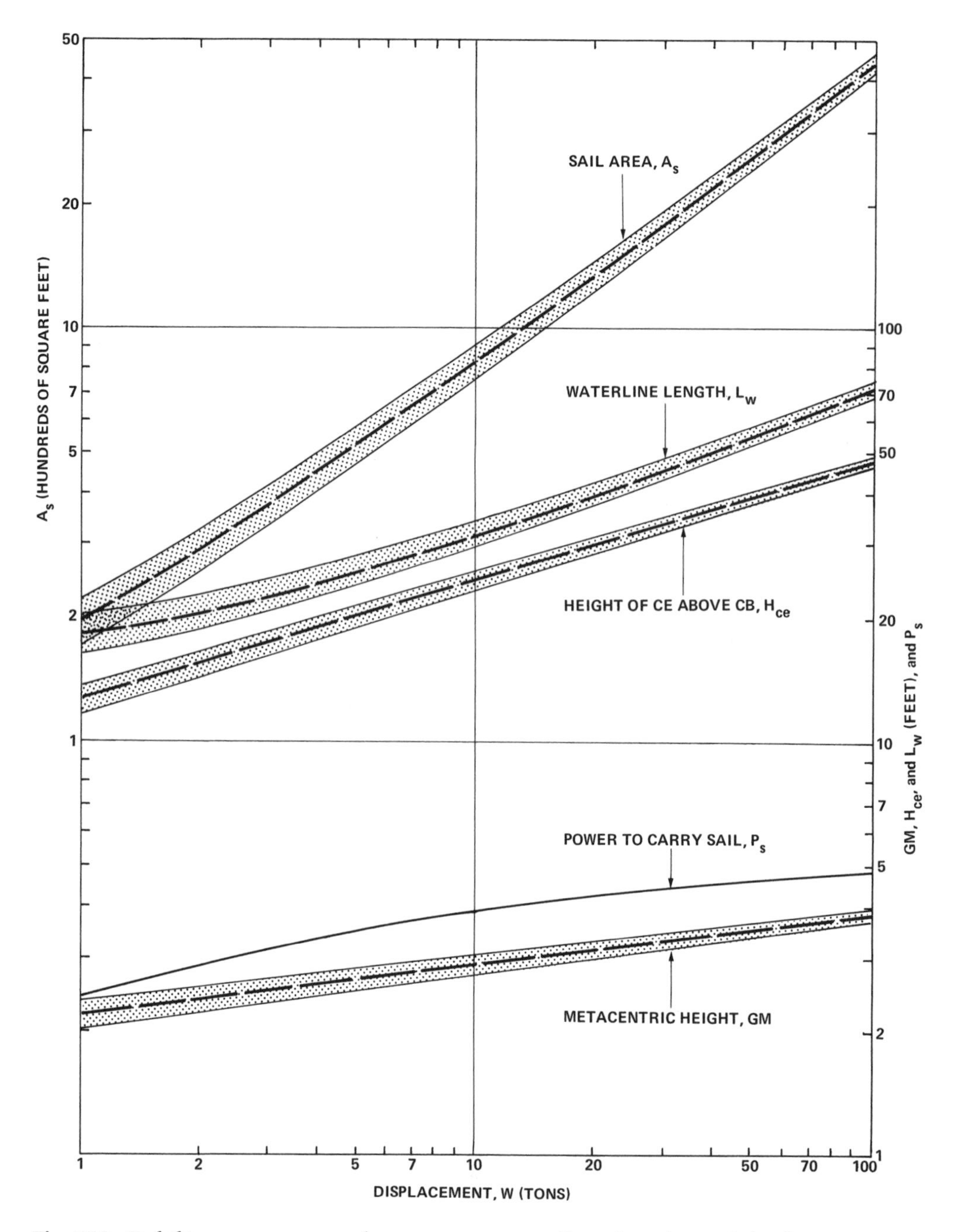

Fig. 119. Stability parameters and power to carry sail, as functions of displacement. P_s is computed for a wind pressure of 1.2 lbs./sq. ft.

unknown angle φ off the bow. The resultant sail force is then approximately*

$$F_t = \tfrac{1}{2}\rho C_L A_s V_r^2 = 0.0034 \times 1.38 A_s V_r^2 \text{ lbs. } (V_r \text{ in knots})$$

acting at an angle $\theta = 107° - \phi$ off the opposite bow. Its ahead component F_a is in equilibrium with the boat's total resistance R_t at speed $V_s = 1.4\sqrt{L_W}$. From the

*Sail twist, leeway, and angle of heel are neglected.

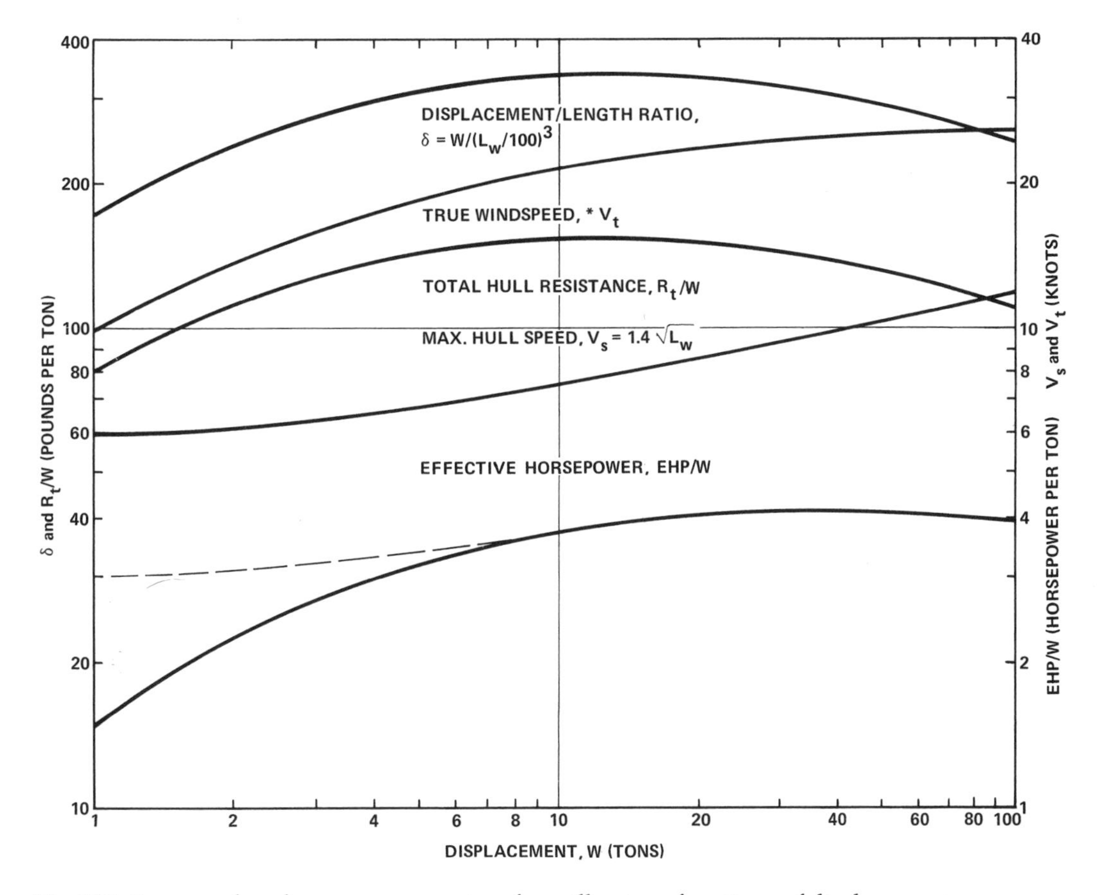

Fig. 120. Power and performance parameters for sailboats as functions of displacement.

geometry of the figure, we can write the following relations:

$$F_a = F_t \cos(107° - \phi) = R_t$$
$$V_r = V_s / \cos\phi$$
$$V_t = V_r \sin\phi$$

Knowing A_s, R_t, and V_s, these equations can be solved explicitly for ϕ, V_r, and V_t. To avoid cluttering up figure 120, I have shown only the curve of V_t, which can be seen to be almost parallel to that for EHP. This surprising result implies that, irrespective of displacement within the range considered here, any boat sailing at maximum speed $(V_s/\sqrt{L_W} = 1.4)$ on a beam reach develops at best about 0.18 EHP/ton per knot of true wind.

On any other heading, speed, resistance, and EHP will be lower—but not much so. Figure 121b, taken from Marchaj's book, shows a polar speed diagram obtained from laboratory data for a 6-meter (L_W = 23.5 ft.) sloop. With V_t = 15 knots, her maximum speed is about 6.3 knots with the wind abeam, but varies by only 6.5 percent at all headings greater than about 40° off the wind. Although the relative wind speed (and power) varies by a factor of three over this range of headings, the small change in V_s is principally due to the steepness of the sloop's resistance curve near hull speed. At lower speeds, where the resistance curve is less steep, her speed variation will be greater on downwind headings.

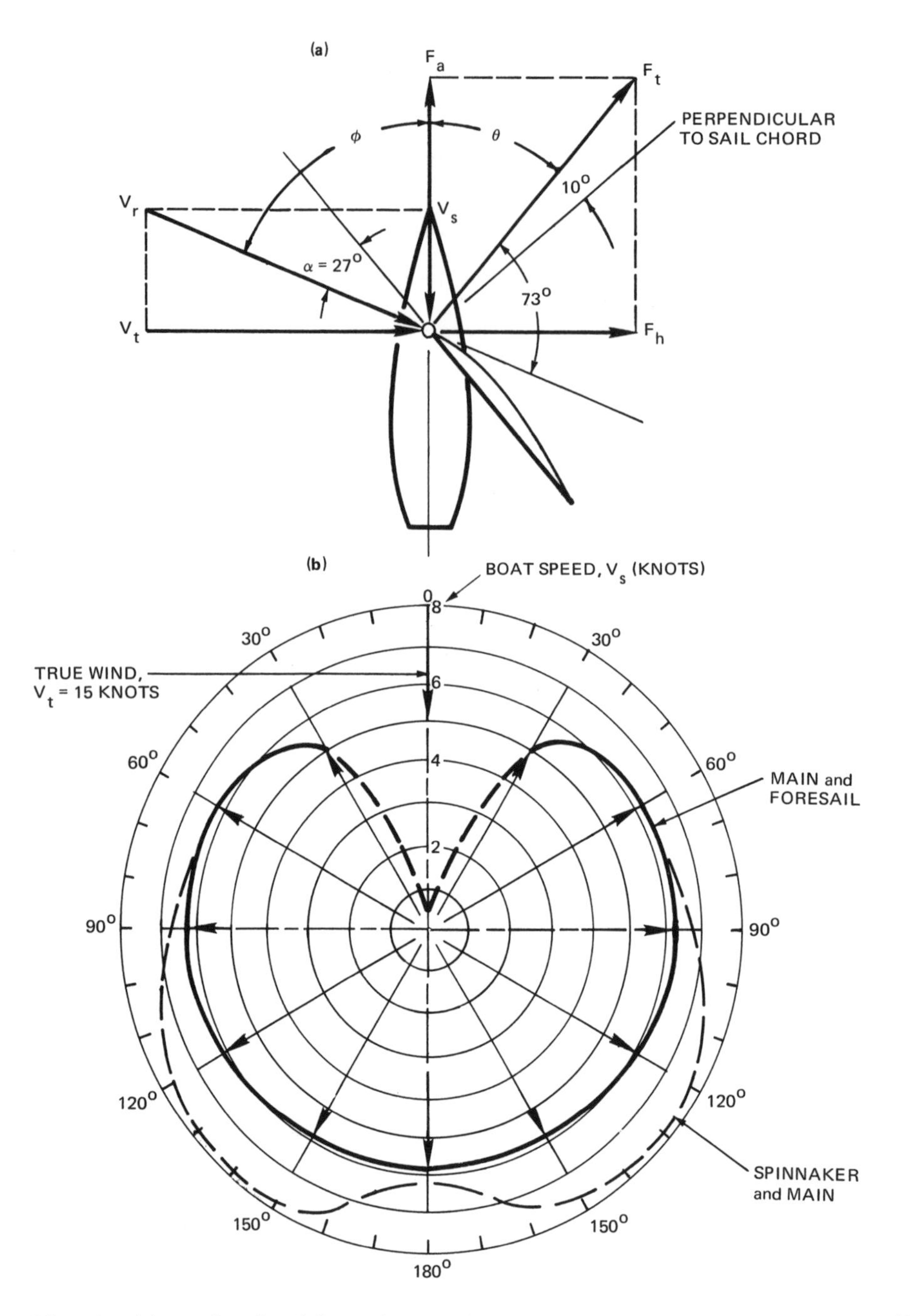

Fig. 121. *(a)* Wind and sail force diagram for vessel sailing at maximum speed $V_s/\sqrt{L_W} = 1.4$ with beam wind V_t. *(b)* Polar speed diagram for 6-meter sloop in 15-knot wind.

The dashed curve shows the improved performance in reaching and running obtained by setting a spinnaker. The sloop's maximum speed is now 8 knots, and is attained at about 135° off the wind. But her dead-off performance is not much better, owing to blanketing by the mainsail, a lower V_r, and poorer effective angle of incidence.

How do the modern IOR-type racers stack up against the CCA data employed in the above analyses? Taking design specifications from USYRU (United States Yacht Racing Union) measurement listings for ten contemporary Admiral's Cup competitors having displacements ranging from 4 to 16 tons, I have found that all parameters vary with displacement in much the same way as in figures 119 and 120, but are displaced by fixed factors, as shown in the following schedule:

Parameter	GM	A_s	H_{ce}	P_s	L_W	V_s	δ
IOR/CCA ratio	0.84	1.55	1.12	0.48	1.18	1.09	0.6

From these ratios it is evident that, for the same displacement, IOR racers have, on the average, 16 percent smaller metacentric heights, 55 percent greater sail area, and 12 percent higher centers of effort, which together confer about 52 percent less stability. But with 18 percent greater waterline lengths, these racers can sail 9 percent faster on the same wind. Moreover, with 40 percent lower displacement/length ratios, they can plane in following seas. The measurement data also include the angles of neutral stability for the IOR boats, of which only two attain the 120° considered safe to compete in ocean races.

We could play many more games with these curves, but our main object in this discussion is to explain how performance is related to design; the details are much better covered in the references cited.

PLANING POWERBOATS

Planing is hard to define, but easy to recognize. Whether one is in a light-displacement sailboat as it is lifted by a following sea, on a sailboard when the wind gets much above 15 knots, or in a high-performance powerboat when the throttles are shoved forward, the hull climbs up out of its displacement-wave trap and skims over the surface, now more like ice than water. Physically, like a skimming stone, planing is the partial support of a vessel's weight by dynamic water pressure against her underbody. Dynamic pressure increases as the square of speed; it is quite negligible at low speed, but, given enough power, can lift a hull of the proper shape right out of the water and cause her to become temporarily airborne (plate 25).

Figure 122 is an expanded version of figure 110, beneath which I have added a graph showing the vertical positions of a vessel's bow, stern, and center of gravity (CG), relative to still water, and a third scale of boat attitude, all as functions of the speed/length ratio $V_s/\sqrt{L_W}$. The resistance scale on the left is equivalent to the force per unit weight required to drive the vessel to the indicated speed. Of the three curves shown, the keel yacht and express cruiser are carried over from figure 110; the third curve epitomizes performance of the modern open ocean racer described below.

With regard to planing performance, the speed/length axis is divided into three regions. Within the displacement region ($V_s/\sqrt{L_W} < 2$), the hull maintains a bow-up attitude, with her weight almost entirely supported by buoyant displaced water. As speed increases, the bow, stern, and CG all sink below their still-water levels, and the bow-up angle is in the range 6°–10°. With her hull unsuited to and underpowered for planing, the keel yacht is limited to a speed/length ratio of about 1.4.

Plate 25. A 46-ft., 1,200-HP ocean racer becomes airborne in rough water, with actor-pilot Don Johnson flying.

Within the mixed region ($2 < V_S/\sqrt{L_W} < 4$), the hull becomes partly supported by dynamic water pressure *(P)*; the stern continues to sink, but the bow and CG rise, and the bow-up angle first increases and then decreases to about 4° as power is increased to lift the hull out of her displacement-wave trap. This is the semiplaning performance region of most power cruisers; the onset of planing, in which about 30 percent of displacement is provided by dynamic support, requires about 200 lbs./ton of thrust, equivalent to about 40 HP/ton, depending upon engine and propeller efficiency.

Full planing ($V_S/\sqrt{L_W} > 4$) is only achieved with a power/displacement ratio of about 60 HP/ton, in which configuration the bow-up angle diminishes to 2°, and dynamic pressure contributes 60-80 percent of total hull support. Because the center of dynamic pressure moves progressively aft as speed increases, it is necessary to locate the engine(s) as far aft as possible so that the CG will remain over the resultant of dynamic and buoyant lift; small trim adjustments are made by hydraulically activated tabs, usually transom-mounted and flush with the underbody.

At full planing attitude, with the forward third of its hull out of water, the ocean racer is in precarious equilibrium in both transverse and longitudinal stability—even on smooth water. Transverse stability is maintained by building the underbody with deep V sections, averaging about 20° deadrise, so that when tipped (say) by wave impact, the immersed half-plane produces more dynamic lift than the emer-

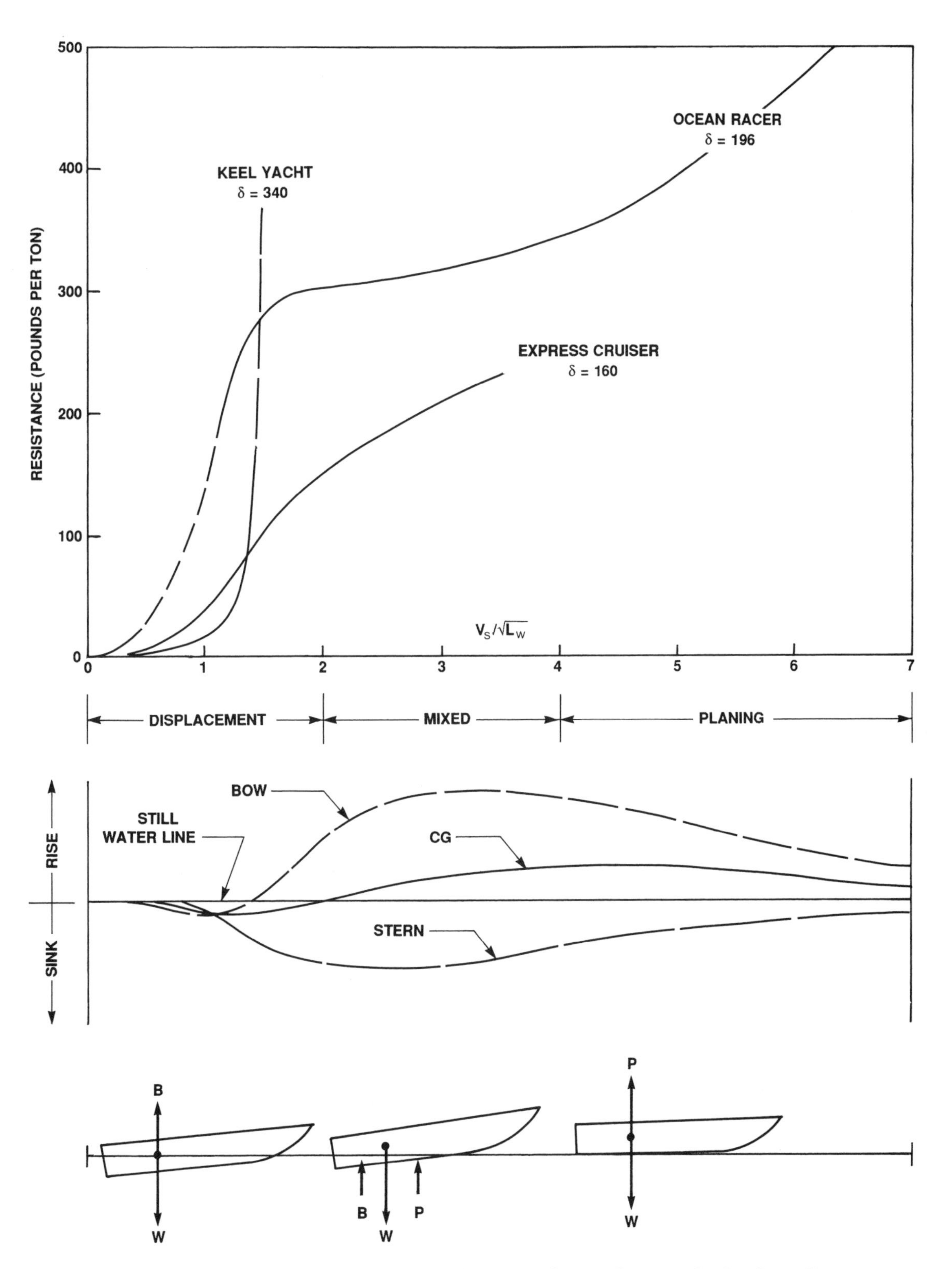

Fig. 122. Resistance of planing hulls is characterized by a rapid rise at low speed, a leveling off as dynamic lift reduces wave resistance, and a final increase of frictional resistance with the square of speed in full-planing attitude.

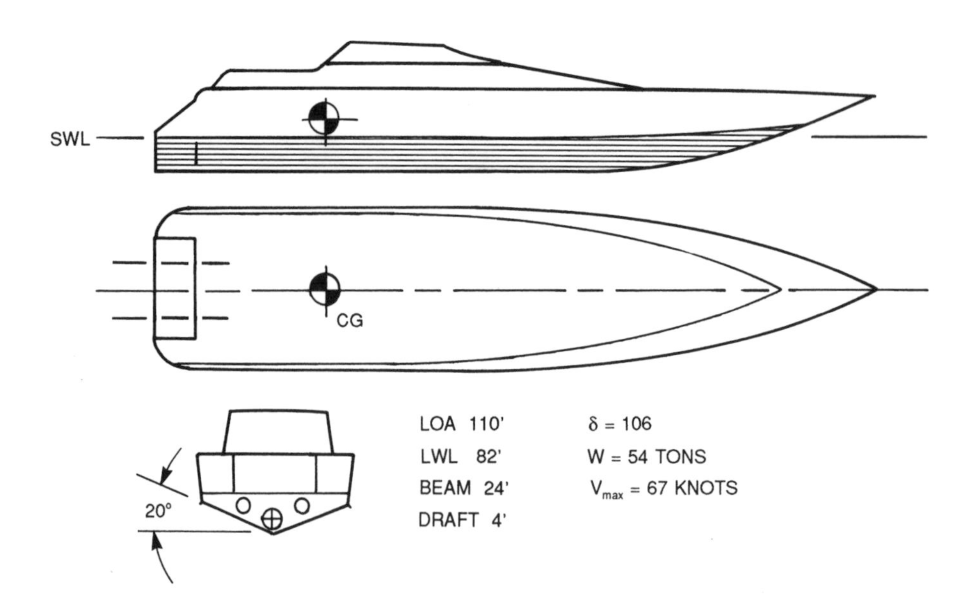

Fig. 123. *Gentry Eagle,* holder of the transatlantic record, is powered by two 3,000-HP dieseljet pumps, and a 4,000-HP gas turbine driving an Arnesen five-bladed surface-piercing propeller.

gent plane, which acts to restore balance. Steady lateral pressure, such as from transverse winds, is compensated by individual trim-tab adjustments. Longitudinal instability is manifested by a tendency of the long bow overhang to *pendulate* (porpoise); this is compensated by throttle control, which alters the bow-up moment produced by the screw(s). But, since a high-powered ocean racer can produce enough propeller thrust to balance her weight when standing vertically on her screws—like a rocket on lift-off—almost any attitude is temporarily possible, and it is up to the driver to keep things within tolerable limits.

Although many ocean racing boats can attain speeds of 90–100 knots on smooth water, speed is progressively reduced as sea state increases in order to moderate impact loads and accelerations. Even so, 50–60 knots can be maintained in 4- to 6-foot seas, and a new transatlantic speed record (2.6 days at 47 knots) was set in July 1989, by the megaracer *Gentry Eagle* in 6- to 8-foot seas (fig. 123).*

Special engineering and construction is necessary to survive such punishment, and a highly specialized industry has sprung up within the past twenty years to support this rapidly growing sport, which now boasts greater audience participation than any aquatic event except the America's Cup races.

Open ocean racing in the United States is under the ad hoc jurisdiction of the American and Pacific Ocean Power Boat Racing Associations. Boats are divided into seven classes, according to length and horsepower, with the unlimited class constrained to $35 < LOA < 50$ feet and 1,200 HP. Races are scheduled and sanctioned by

*The oldest prevailing (Blue Riband) trophy race among powered vessels across the Atlantic was first won by the steamship *Great Western* in 1838 (12.3 days at 18.2 knots), and was most recently held by the *USS United States* in 1952 (3.8 days at 35 knots).

the associations, but, by general acclaim, the guru of powerboat racing is 79-year-old Robert A. Nordskog, who "sets the races and wins most of them."*

Irrespective of race duration (up to 32 hours), Nordskog drives standing up in a padded armpit enclosure, with one foot on the throttle, one hand for the wheel, and the other for trim-tab control. His strategy is "all-out," his technique "seat-of-the-pants," decelerating just before becoming airborne, and accelerating the "instant the prop grabs [water]." The two great hazards are porpoising and engine runaway, if the props should emerge under power. At 150 ft./sec., this technique requires exquisite timing.

*Of three roomfuls of trophies, Nordskog's favorite is a lifetime award of a case of Hennessy cognac, delivered within twenty-four hours at any major city. He earned this award during a 1972 race from Long Island to Bermuda. An hour out, while encountering heavy weather, his engineer suffered spinal compression. Nordskog took him back and started out again alone. Twenty-eight miles from Bermuda he sank from massive hull leakage, all others having long since abandoned the race. After being picked up in his rubber dingy by a Dutch freighter the next morning, they stood off Hamilton, unable to receive the pilot launch because of heavy seas. Nordkog swam ashore to greet the massed media, who had long presumed him lost. "Weren't you afraid?" they gasped. "Hell, no," he replied, producing a flask from his britches, "not as long as I had my Hennessy along!"

23

Ship Motions in a Seaway

Up to this point, we have been talking only about smooth-water performance, wherein a vessel's dynamics are not much affected by natural waves. Outside the harbor, when the sea gets up, we have a whole new ball game. The rules are still the same, but, whereas before we have considered only steady motions, we now must deal with the vessel's interactions with these waves.

Although a vessel's hull and propulsion system are normally designed to maintain some specified service speed in moderate seas, they must also be able to tolerate much more severe wave conditions at reduced speed. The range between moderate and tolerable obviously varies widely, depending upon the class of vessel. At the low end, waves are principally an impediment to speed, in that they induce adverse motions that increase resistance. Somewhere in midrange, more violent motions may cause cargo shifting and discomfort or injury to passengers and crew—any of which can seriously interfere with normal operation. The acceptable upper limit of this range is usually governed by the onset of impact forces large enough to cause structural damage to hull or machinery. In this, all things are relative: the smaller the vessel, the worse pounding she will take in rough seas—and the more she can withstand.

While, as mentioned earlier, small-yacht designs often spring one from another without much consideration for physical principles,* with a new one being beefed up here and there to overcome weaknesses discovered in its predecessor, the naval architect goes to great efforts to estimate wave forces and ship motions when designing a large ship. Because of their complexity, the necessary calculations are often performed on computers, but the underlying principles apply equally to vessels of all sizes, and there is much to be learned from ship-motion studies that is of direct benefit to seamanship. As before, we shall first look at things as a naval architect sees them, and then try to interpret them from the standpoint of shiphandling.

DYNAMIC EQUILIBRIUM

Whether or not a vessel is under way, its motions *(response)* in a seaway can, at any instant, be considered as the *dynamic* equilibrium between the resultant of her *inertial* resistance and that of all external forces acting on her; that is, just as in hydrostatics, the sum of all forces and their moments about her CG must be zero. The vessel, of course, will not remain stationary, but will be accelerated in the

*Sturdy Chinese junks have been built "one-off" for thousands of years without any drawings whatsoever.

direction of the external force resultant. If the line of action of this resultant passes through her CG, she will move bodily *(translation)*; if not, she will also be rotated. An instant later, she will have moved to some new position (and/or attitude), as will have the waves, and she will again be in equilibrium with a new set of forces.

Nevertheless, if a vessel's initial attitude and state of motion were known, and all external forces (wind, waves, flow resistance, rudder and propeller thrust, etc.) could be specified as functions of time, her subsequent motions could (in principle) be calculated step by step, just like a rocket's trajectory in space. The catch here is that the rocket's thrust and fuel consumption (instantaneous mass) are well known at all times, whereas the wave forces are not. As you may recall (Part V), ocean waves are best described as a random pattern of bumps and hollows that never repeats itself.

Fortunately, there is a way out of this dilemma. A vessel's response *can* be calculated for small-amplitude, single-frequency, periodic waves, since such waves can be mathematically expressed as continuous functions of time. Thus, to the extent that any sea state can be represented as a spectrum of many such periodic waves, the vessel's *spectral response* can be assumed to be equal to the sum of her responses to the component waves.* Thus, if supplied with a vessel's characteristics and a set of sea state spectra, the computer can readily provide the data needed for design. However, because a wave spectrum is only a statistical description of how wave energy is distributed among the component wave periods, what one gets from the computer is not a time history of a ship's motion, but a set of curves indicating the relative probabilities of experiencing certain accelerations, amplitudes, and displacements, and how these quantities will vary with speed and heading. Interestingly, the speed and heading enter the problem only implicitly, insofar as they affect the frequency and direction of wave encounter, for the computerized vessel is assumed to remain nearly stationary, and merely to gyrate about in response to the spectral exciting forces.

But statistical data are not readily generalized for the purposes of seamanship without first understanding a vessel's responses to ordinary periodic waves. For this, we must digress, temporarily, to the more traditional, precomputer concepts outlined below.

RESPONSE TO PERIODIC WAVES

In practice, a ship's motion is conventionally broken down into six component motions *(degrees of freedom)*: three translations, *surge, heave,* and *sway,* along longitudinal, vertical, and transverse axes, respectively, intersecting in the waterline plane directly above her CG; and three rotations, *roll, yaw,* and *pitch,* about these respective axes (fig. 124). Of these six degrees of freedom, three (heave, pitch, and roll) are called *oscillatory* modes, because they involve vertical changes in a vessel's equilibrium displacement that invoke restoring forces. These, coupled with her inertia,** cause her to oscillate back and forth (or up and down) for a few cycles,

*While this assumption has not been proven theoretically, numerous tank and sea trials have shown it to be fairly reliable, up to the point where wave-breaking commences, or where other assumptions regarding extreme motions are invalidated.

**A vessel's inertial resistance is given by $(W + w)/g$, where w/g is the "virtual" mass of water accelerated by her hull.

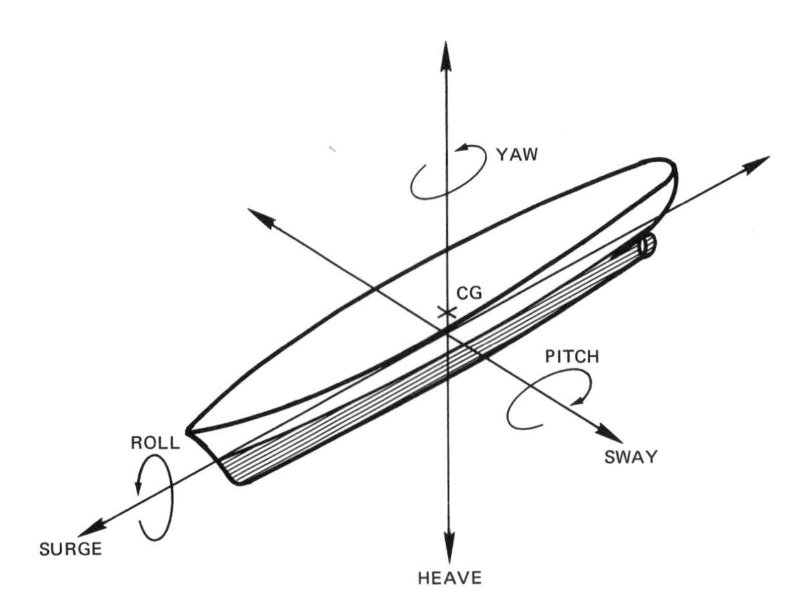

Fig. 124. Coordinate axes for ship motions intersect in waterline plane beneath CG.

until the motions are damped out by frictional resistance in the water, or by radiation as surface waves. Conversely, the *nonoscillatory* modes (surge, sway, and yaw) do not involve displacement changes or restoring forces;* the vessel is simply translated (or rotated) until brought to rest by dissipative forces.

For small-amplitude motions in calm water, the oscillatory modes have relatively constant periods. Thus, if we depress the rail of a floating yacht by a small amount and then release it, her angular rolling motions plotted against time will be a damped cosine curve (fig. 125), whose consecutive maxima define her free roll period T_r. Similar curves—but of different periods—for pitching and heaving could be obtained by depressing her bow, or her entire hull, respectively. These free (or natural) periods of oscillation are a property of a specific vessel, and depend primarily upon her shape, load displacement, and internal weight distribution. Additionally, the free roll period and its rate of damping are both increased by factors that increase rolling resistance, such as sails, hull roughness, and underwater appendages.

While the free periods are not easily calculated, owing to uncertainties regarding virtual mass and resistance, they can be determined experimentally (as above) for smaller vessels. Although seldom calculated directly for large ships, they appear implicitly in the computerized dynamical solutions, and can be approximated by the relations:

$$\text{Roll:} \quad T_r = C_r \times \sqrt{\frac{\text{rotational inertia about longitudinal axis}}{\text{static stability about longitudinal axis}}}$$

$$\text{Pitch:} \quad T_p = C_p \times \sqrt{\frac{\text{rotational inertia about transverse axis}}{\text{static stability about transverse axis}}}$$

$$\text{Heave:} \quad T_h = C_h \times \sqrt{\frac{\text{load–displacement volume}}{\text{load–waterplane area}}}$$

*When a vessel is elastically constrained to small horizontal motions, as by mooring lines, free oscillations can occur in surge, sway, and yaw.

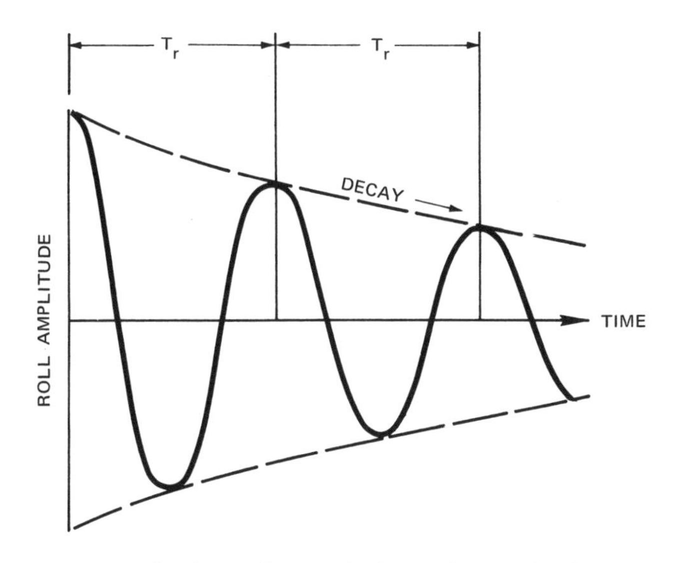

Fig. 125. The free-roll period of a yacht can be determined by timing her damped rolling motions in a calm sea.

where the coefficients C_r, C_p, and C_h are taken from experimental curves for similar vessels. In general, all free periods increase roughly as the sixth root of load displacement, and their approximate ranges for various vessels are shown in figure 126. Although there are substantial variations between vessels having different hull forms, roll periods tend to exceed pitch periods, with heave periods falling somewhere between. In larger vessels, there is a considerable overlap between heave and pitch, and the two may be nearly the same, as indicated by the dashed portion of their line of division.

The oscillatory modes strongly influence the dynamic response of a vessel through the phenomenon of resonance, a condition whereby synchronism between any free period and a properly directed, periodic exciting force results in strong energy coupling and abnormally large oscillation amplitudes. In roll and pitch, this effect is analogous to the motions of a child's swing; each impulse adds energy to the system, which is only partially removed by dissipation during the next cycle. Thus, the system energy continues to increase until limited by increasing dissipation, or by catastrophe; the swing loops the loop; the vessel capsizes or is pitchpoled.

Roll Response: Figure 127a shows the dynamic rolling response of several typical vessels to periodic waves approaching abeam. The vertical axis gives response in terms of the ratio of maximum vessel inclination to effective* wave slope (θ_r/θ_W); the horizontal axis is in units of the ratio of the wave period to the free roll period (T/T_r). The several curves in this figure refer to different degrees of damping, in units of the percentage reduction in rolling angle per cycle in calm water. Except for the effect of damping, these curves are generally similar and have the following common characteristics:

*The effective slope depends upon the vessel, but usually lies between the maximum wave slope and a line from wave crest to trough.

1. For very short-period waves ($T/T_r < 1$), the vessel's response is always less than unity, and she rolls less than the effective slope.
2. For very long-period waves ($T/T_r > 1$), the response is nearly unity; that is, her maximum rolling angle is equal to the effective wave slope.
3. For waves near synchronism with the free-roll period ($0.4 < T/T_r < 1.4$), the response is always greater than unity (unless the damping is so great as to extinguish the roll in less than one cycle) and increases rapidly as damping decreases.

Although a vessel theoretically would have infinite response at synchronism for zero damping (no viscosity or added mass), roll-damping coefficients for most vessels are usually so small that very large oscillations can occur near resonance. For example, table 9 gives representative resonant periods (T_r) and damping coefficients for the vessels in figure 127, together with the hypothetical heights of nonbreaking waves (at those respective periods), that would produce roll amplitudes of 50°. The table is ordered downward in terms of increased damping, but the maximum heights are not similarly ordered: the passenger liner, having the longest roll period, can tolerate the highest waves; the battleship, with a very large GM (for gun steadiness) and moderate damping, does nearly as well, despite her short roll period; the keel yacht has high roll damping and dynamic stability, and thus does much better than the relatively undamped cargo vessel and the notoriously uncomfortable destroyer.

A significant feature of the roll response curves is that, for small damping, the high response values occur over a very narrow band of wave periods; somewhat like a sharply tuned radio, slightly "off station" the response is very low. Secondly, many large ships, such as the passenger liner, have free roll periods T_r longer than 20 seconds, which is roughly the longest natural wave period normally found at sea. However, as will be shown, for a vessel under way, the *effective* wave period encountered is a function of speed and heading, and may be much longer than 20 seconds. Thus, certain combinations of speed and direction can induce heavy rolling in otherwise stable vessels.

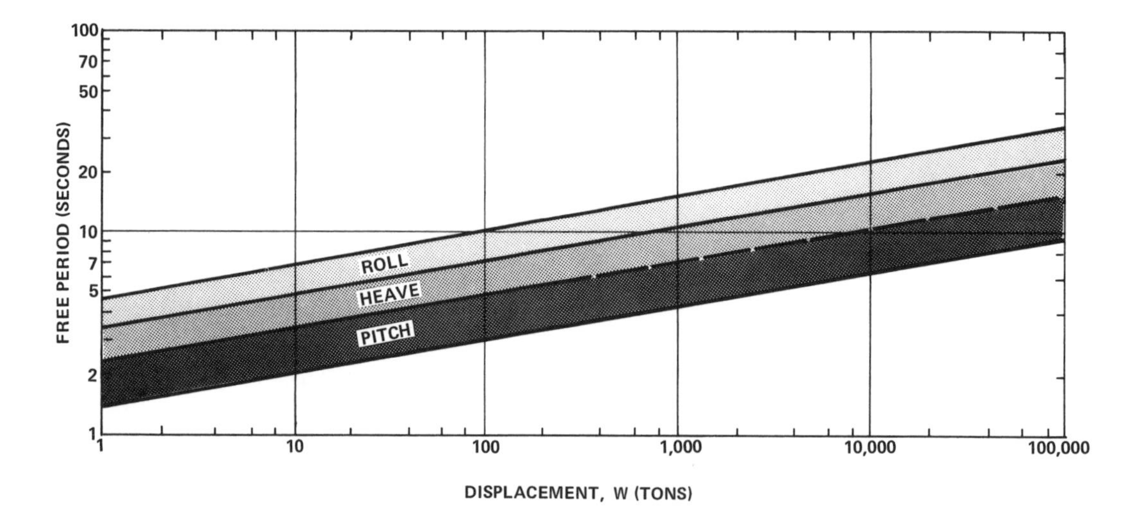

Fig. 126. Free periods of ship oscillation are roughly proportional to the sixth root of displacement.

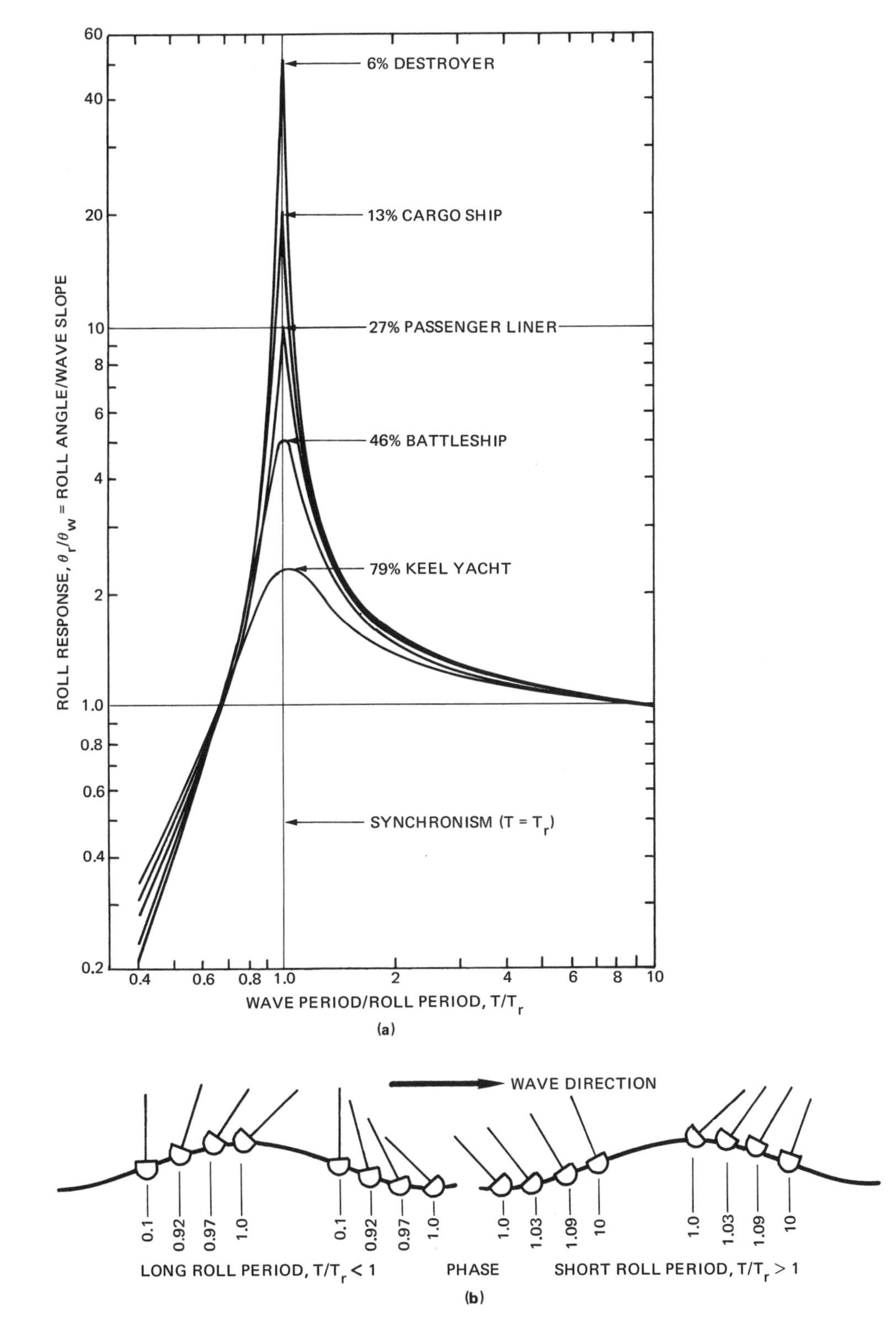

Fig. 127. *(a)* Roll response to beam sea, as function of degree of damping (percent per cycle) and ratio T/T_r of incident wave period to free-roll period. *(b)* Direction of inclination relative to an advancing or receding wave crest (phase angle) for a vessel with 27 percent roll damping per cycle.

Relative to an oncoming or receding crest, the position on a wave (phase angle) at which a vessel achieves maximum inclination is also a function of the period ratio T/T_r and the degree of damping. Figure 127b shows the direction of maximum inclination and corresponding value of T/T_r for 30° increments of phase angle* and a damping decrement of 27 percent per cycle (passenger liner, table 9). For higher damping, the intermediary values of T/T_r would be farther off resonance, and conversely, but the important phase relations can be summarized as follows:

1. For short-period waves ($T/T_r < 1$), the vessel rolls very little, but its maximum inclination occurs midway between crests and troughs.
2. For long-period waves ($T/T_r > 1$), the maximum inclination equals the wave slope, and also occurs midway between crests and troughs.
3. At resonance ($T/T_r = 1$), maximum inclination is much larger than the wave slope, and occurs at the crests and troughs.
4. For all $T < T_r$, the vessel heels towards an advancing or receding crest, and conversely.

Table 9.

Vessel Type	Roll Damping (%/Cycle)	Free Period T_r (sec.)	Roll Enhancement (50°/θ)	Maximum Wave Height for 50° Roll $H_{max.ft.}$
Destroyer	6	14	1.0	6
Cargo ship	13	16	2.5	18
Passenger liner (with bilge keels)	27	22	5.0	67
Battleship	46	15	20.0	62
Keel yacht	79	8	25.0	45

Roll Stabilization: Because uncontrolled rolling is so hazardous to ship operation, many devices have been invented to mitigate synchronous rolling before it becomes dangerous. Of these, only two have proved so successful as to be generally adopted.

Passenger ships, in which even moderate rolling is an economic "discomfort" factor, have almost universally adopted antirolling fins that protrude perpendicularly from the ship's sides, like stubby aircraft wings; their chord axes can be inclined oppositely to the water flow past the hull so as to produce a torque that counteracts the roll moment. Fin incidence to the flow is controlled by gyro-activated servomotors, which actually sense the inception of rolling and produce just enough torque to damp it. When not required, the fins retract into slots in the ship's side. Performance measurements show that such fins can reduce rolling to less than 10 percent of its uncontrolled magnitude.

Fin stabilization is too expensive for all but the larger power yachts and fishing vessels. Smaller, slow-speed (12-knot) power vessels, particularly on the West Coast, have resorted to a cumbersome but effective paravane device known to the trade as the "flopper/stopper" (F/S). The F/S is a horizontal steel triangle, equipped with a vertical steel rudder and a cylindrical nose weight. Dependent outboard from a

*The point from which the phase angle is reckoned is somewhat arbitrary. But a complete wave cycle comprises 360°; thus the phase difference between a crest and mean sea level is 90°, or 180° between a crest and its succeeding trough.

moving vessel on a guyed pole (fig. 128,a), it is towed along with very little drag. If the vessel rolls toward the F/S, it dives like a bomb, but then flattens out, and strongly resists the counter-roll that lifts it. The F/S can be towed singly to weather on a motor sailer, or in a pair from a fishing boat or power cruiser, and can be retrieved aboard and its pole stowed when not required (fig. 128,b). Despite the stalwart rig required, and the inconvenience of launching and retrieving it, the F/S enjoys a steadily expanding market among long-range vessels under 50 tons.*

Pitch and Heave Response: Analogous maximum response curves for the pitch and heave modes of a vessel encountering periodic head seas are shown in figure 129; the response in heave is given in terms of the ratio of vertical heave amplitude to

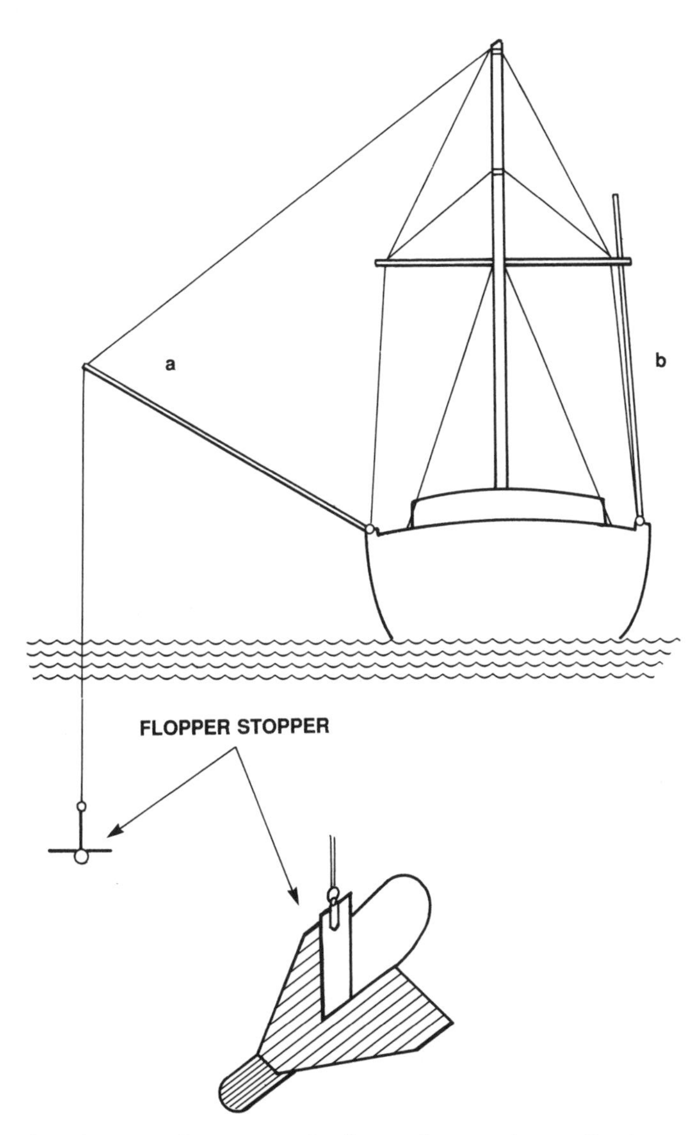

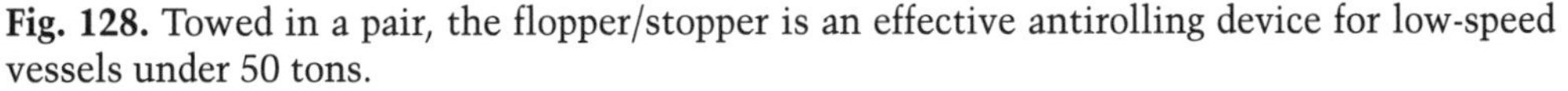

Fig. 128. Towed in a pair, the flopper/stopper is an effective antirolling device for low-speed vessels under 50 tons.

*For design details, see Robert T. Beebe, *Voyaging under Power*, South Seas Press, NY, 1975.

wave amplitude (A_h/A). These curves are generally similar to those for roll, but differ in several important ways. First, the relatively high trim (pitch) stability and inertial and damping forces limit the maximum response ratios in both heave and pitch to very small values (0–1.5), as compared with those for roll (50). Moreover, both forces increase rapidly with vessel speed, whereas they are nearly independent of speed in roll. As a result, we cannot simply relate pitch or heave response to the degree of damping, but instead to the speed/length ratio $V_S/\sqrt{L_W}$. Second, while a vessel's beam is usually small compared with the wavelengths of important waves at advanced sea states, her L_W may be of the same order as—or even longer than—the wavelength. As a result of straddling a wavy surface, L_W becomes a factor in determining pitch and heave response, acting to suppress the high-peak response characteristic of roll, and to broaden it to cover a much wider range of effective periods.

To illustrate better the geometric influence of ship length, the response curves in figure 129 have horizontal axes in units of the ratio L/L_W. The responses at synchronism are indicated by the intersections of the response curves for different speed/length ratios with the dashed curves labeled $T = T_p$ and $T = T_h$ for pitch and heave, respectively. Despite these differences, response in pitch and heave can be interpreted similarly to that for roll:

1. For waves shorter than about three-quarters the waterline length ($L < 0.75\,L_w$), response is small at all speeds; the vessel slides smoothly through head seas.
2. In very long waves ($L > 2.0\,L_w$), response approaches unity at all speeds; the vessel generally follows the wave contour.
3. Response is greater than unity only for $V_s/\sqrt{L_W} > 0.5$; maximum response occurs near synchronism and is associated with waves between 1.0 and 1.5 times a vessel's L_W.

Unlike roll, excessive pitch and heave motions occur at high speed near synchronism, running the gamut from wet decks in moderate seas to plunging and slamming in higher waves. As every experienced shipmaster knows, synchronous pitching is to be avoided at all costs, and best progress can be achieved by running at high speed in short waves, and at progressively slower speeds in increasingly longer waves.

Period of Encounter: For a vessel lying dead in the water, synchronism with periodic waves can occur only when the wave period happens to coincide with the free period of any oscillatory mode. However, when under way in any direction other than parallel to the wave crests, the effective period of wave encounter *(T_e)* will change, according to the speed and heading, becoming shorter when headed into the seas, and conversely, and the possibility of synchronism will similarly be altered. For a vessel traveling at speed V_S (knots) whose heading subtends an angle α with the direction of wave *approach,* the effective period is given by:

$$T_e = \frac{3T}{3 + (V_s / T)\cos\alpha}$$

Figure 130 is a resonance diagram, comprising a set of graphical solutions of this equation, from which the possibility of synchronous excitation of any mode can be determined by inspection, given a vessel's free periods, her speed and heading, and the prevailing wave period. One enters the diagram by locating the intersection of

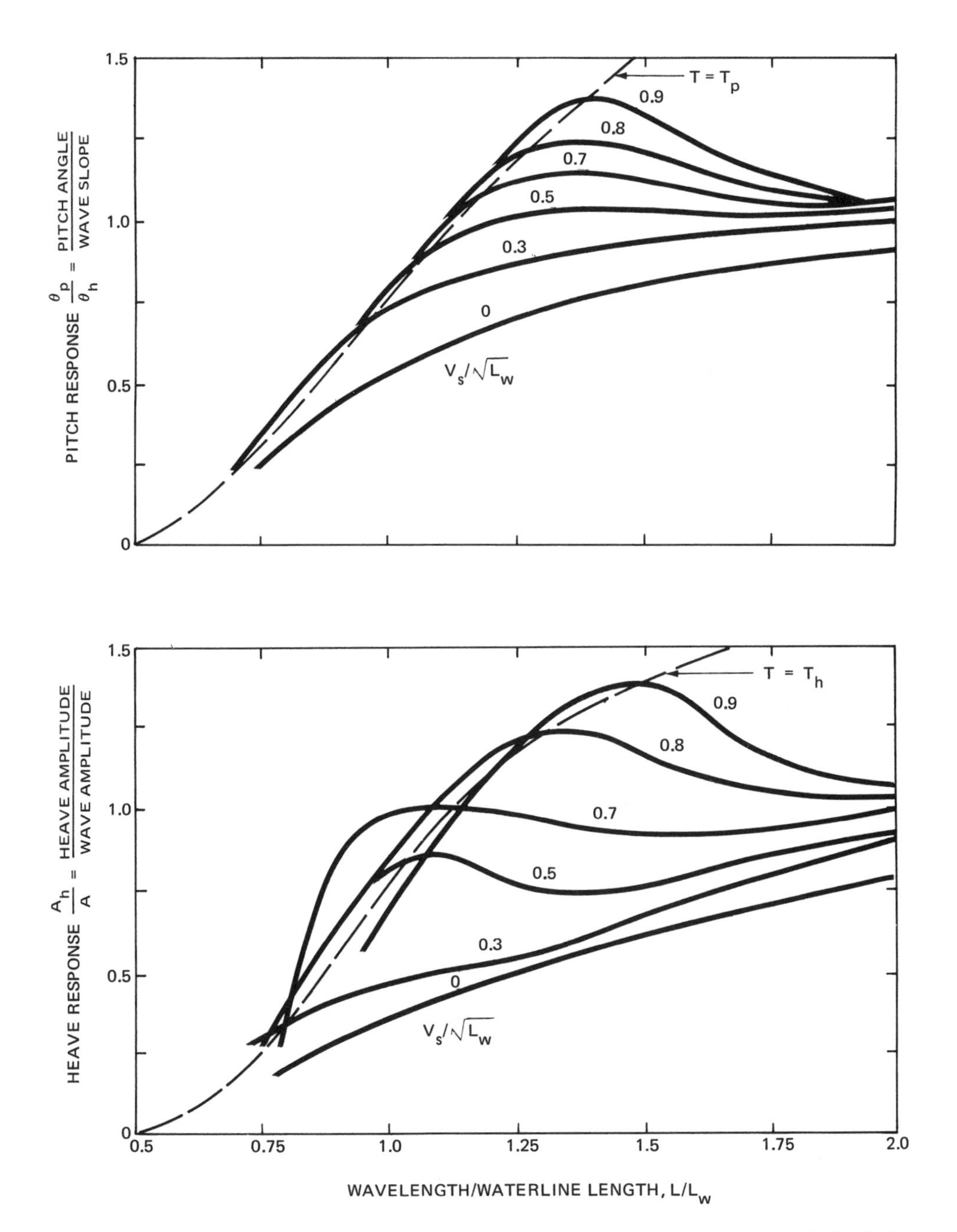

Fig. 129. Pitch and heave responses for a cargo vessel in head seas. Dashed line in both figures indicates synchronism between wave period and pitch or heave period, respectively.

any free period with the prevailing wave period. Then, projecting a horizontal line to the left from this intersection across the speed-direction rose, one obtains all combinations of speed and heading that will result in a synchronous period of encounter.*

*The range of the diagram is doubled by using numbers in parentheses for small values of period and velocity, which are appropriate for vessels up to about 100 tons displacement, and using unbracketed numbers for larger vessels.

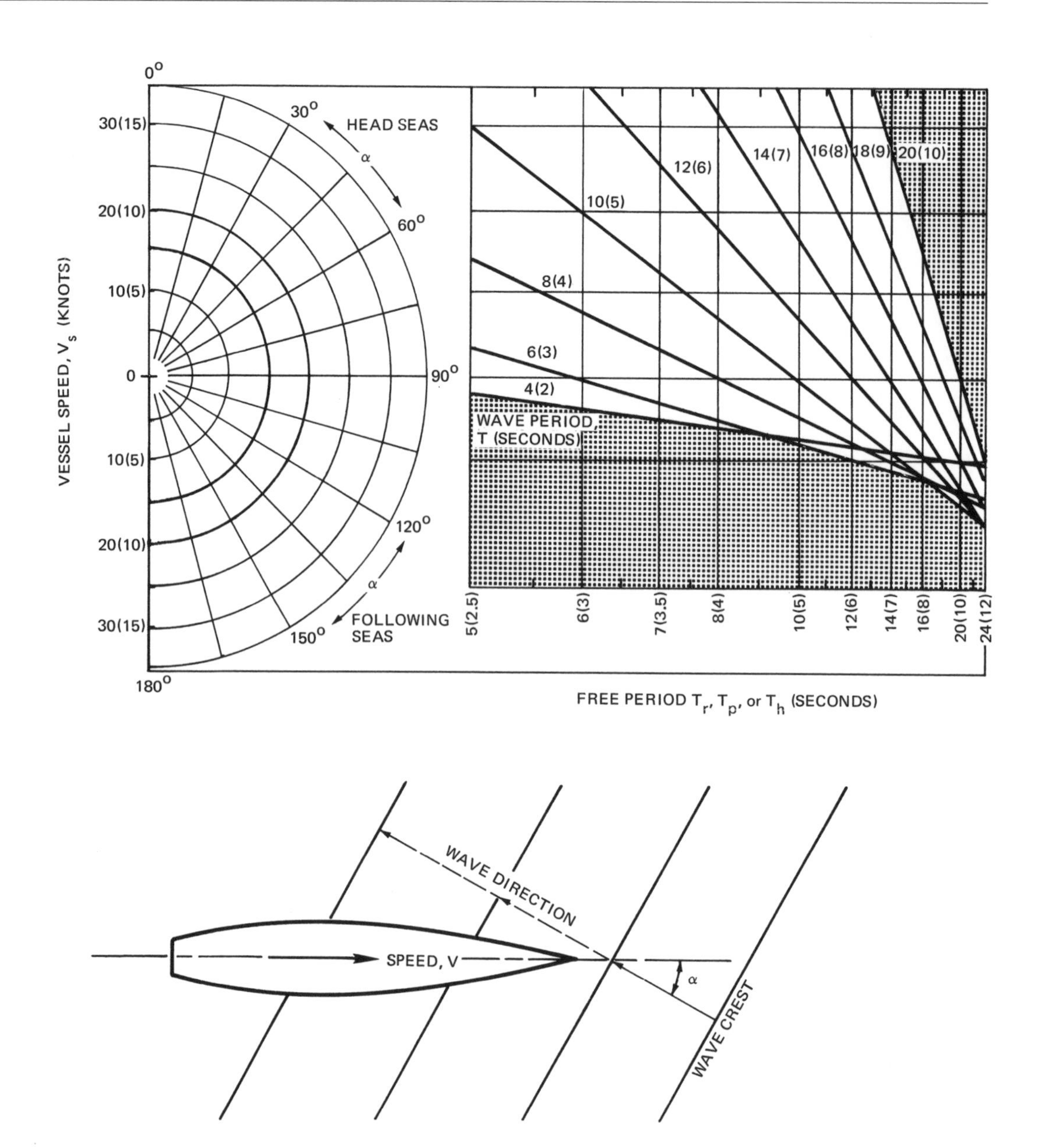

Fig. 130. Resonance diagram, giving speed and heading combinations that will result in synchronism between true wave period *T* and any free period of a vessel.

Like the CSS diagram, the resonance diagram condenses a great deal of information into a few lines. From the concentration of wave period lines in the upper half of the diagram, it is clear that there are a great many possibilities for synchronism in head—as opposed to following—seas, and no chance at all within the shaded sectors outside the range of significant waves in nature (4–20 sec.). These circumstances are not altered—but are, instead, compounded—by the fact that advanced sea states comprise a broad spectrum of wave periods, which largely overlaps the range of oscillatory free periods of most vessels. Thus, we have immediate, quantitative confirmation of the common experience that most vessels experience much more violent motions in head than in following seas, not only because of synchronism, but also because impact forces increase as the square of the relative velocity between ship and wave: beating into the wind is just that!

However, running off may not always be the optimum course of action. As figure 126 showed, vessels upwards of 10,000 tons may have roll periods in excess of 20 seconds. Places on this diagram where any of a vessel's free periods coincides with the *intersection* of two wave period lines are especially to be avoided, since this situation brings the vessel simultaneously into resonance with *two* different wave components whose energies are additive, and motions will correspondingly be more severe. All such intersections occur at low speeds in *following* seas, and result from the fact that the vessel is overtaking the waves of shorter period and is being overtaken by waves of longer period; in each case, the period of encounter is the same—and always equal to the *sum* of the intersecting periods. Thus, a vessel having a free-roll period of 20 seconds will be in resonance with waves of 4 and 16 seconds on any speed-heading combination along a horizontal line passing through the intersection of these two periods. The general convergence of the wave period lines toward the lower right sector of the resonance diagram indicates that vessels operating within this sector may have difficulty avoiding roll synchronism with any period between 4 and 20 seconds, and should reduce speed to below, say, 5 knots and head into the seas to avoid this situation.

RESPONSE TO RANDOM SEAS

In the preceding sections, we have treated the somewhat idealized case of small-amplitude motions in periodic waves. In a real seaway, several factors act to modify a vessel's behavior.

First, the free periods of the oscillatory modes are only relatively constant for small departures from stable equilibrium (even in periodic waves), and all tend to increase as motions grow larger. This is particularly true in roll; for heel angles above about 10°, the free roll period increases progressively with inclination, and becomes infinite when heeled to an attitude of neutral equilibrium. The pitch period may be little changed up to the point where a vessel's bow or stern emerges or buries, beyond which it increases sharply. Heave periods are nearly constant in slab-sided vessels—short of taking water aboard amidships—but increase with amplitude in vessels having flaring sections or long overhangs. All of these period changes act to reduce a vessel's maximum resonance response, since she automatically detunes herself if motions become too large; conversely, her motions in longer waves off resonance may be larger than indicated by her small-amplitude response curves. Thus, the effect of higher waves is to flatten and broaden the response curves in the direction of longer periods, relative to those for lower waves.

Second, we have ignored the effects of *coupled motions,* such as heave and pitch, in which energy is considered to be exchanged between two (or more) degrees of freedom by virtue of the rigidity and inertia of the vessel. In this context, the concept of separate degrees of freedom itself is only a convenient artifice: a wave crest whose resultant force does not pass through the vessel's CG will exert a pitching moment; if this resultant has a vertical component, heaving will be excited. But, in fact, both modes are simply components of the common motion. Similarly, it is practically impossible to induce rolling or pitching without heaving; yawing induces rolling and swaying, etc.

While we cannot simply represent coupled motions in the form of response curves, spectra computed from records of ship motion (acceleration), in the same manner as described for sea state (Part V), usually exhibit energy peaks that can be associated with a vessel's free periods, showing that she *does* respond by rather complicated combinations of simple motions.

Lastly, we recall that a real storm sea is nonperiodic, but instead comprises a random pattern of bumps and hollows that never repeats itself. This stumbling block virtually arrested progress in ship design for many years; naval architects could find no way to apply periodic ship-response functions to a random sea. Only within the past thirty years has a breakthrough occurred, as a result of three corollary developments: (1) the recognition that sea states could be treated as statistical combinations of component periodic waves; (2) the assumption—now justified—that ship response to random seas could be treated as the sum of individual responses to these components; and (3) the availability of large computers to perform the necessary arithmetic.

Returning, now, to the real world, modern ship-response calculations proceed somewhat as follows. The computer is supplied with a sea state spectrum, a set of experimentally determined response functions (for a particular vessel), and various speeds and headings. The spectrum is broken down into a number of directional spectra, and the energy is computed for all significant* component wave periods in each direction. These periods are then changed to effective periods of encounter (T_e), as appropriate for each heading and speed. The component response for each effective period is obtained by multiplying the component wave energies by their corresponding *response factors.* Lastly, the component responses are then summed (over all wave directions) to give the total response energy—as a function of effective period—for each degree of freedom. In short, the computer output is a set of *response-energy spectra* for each speed and heading, showing how the energy in the various modes of motion is subdivided according to (effective) wave period.

For example, figure 131a shows such a set of pitch-response spectra for a hypothetical vessel encountering irregular head seas at various speeds; figure 131b illustrates the effect of heading changes under the same conditions, but at constant speed. The vertical scales give relative pitch response, as a function of T_e, the effective wave period. The area under any curve is proportional to the square root of the total pitching energy E_p at these respective speeds and headings. Given the appropriate numerical coefficients, the vertical scales can be interpreted directly in terms of pitch angle or pitch displacement at any point within the hull. The dashed lines in these figures illustrate hypothetical limiting motions which, if exceeded, would result in (say) slamming or plunging; the vessel would have to reduce speed to $V_s/\sqrt{L_w} \approx 0.33$, or alter course by 22.5°, to operate safely under these conditions. The Frontispiece of this book shows a graphic example of a destroyer, committed to uncontrolled pitching in heavy seas by refueling constraints, leaping half her length out of water.

The pitch energy (E_p)—and its corresponding equivalents for other modes of motion**—is directly related to our old friend, E, the sea state energy. Thus all response spectra will grow with increasing sea state in much the same manner as the

*Those having significant energy at anticipated periods of encounter.

**Including surge, sway, and yaw.

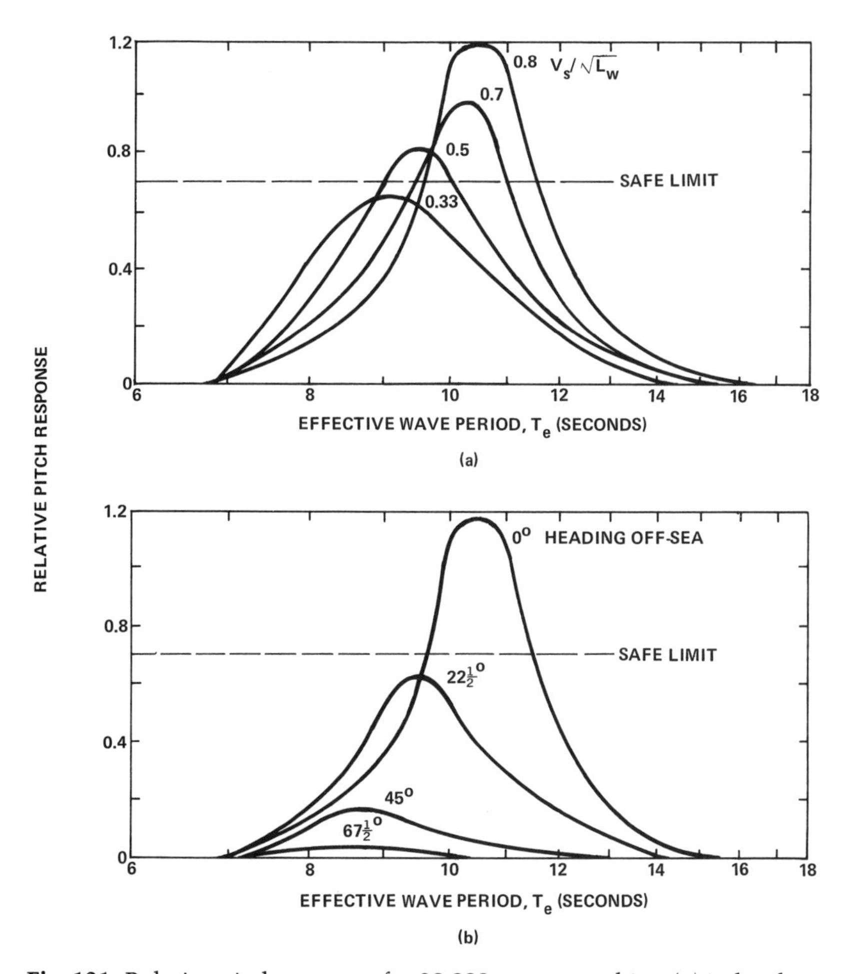

Fig. 131. Relative pitch response for 20,000-ton cargo ship: *(a)* in head seas, as a function of speed; *(b)* at $V_s/\sqrt{L_W} = 0.8$, as a function of off-sea heading. Sea state is assumed to be fully developed under 40-knot winds.

wave spectrum does. Moreover, probability curves for average or extreme motions, analogous to figures 72 and 73 (pp. 175 and 177), can be forecast for a given sea state, once a vessel's response spectra have been calculated. By comparing corresponding spectra for a suitable range of speeds, headings, and sea states, the naval architect can evaluate all aspects of a vessel's behavior in the early design stage. Like an organist composing a melody, by punching the keys of a computer the designer can alter her shape at will, seeking some happy compromise between performance and the forces of nature.

But we are concerned here with seamanship, and shipmasters have fewer options. Their keys are rudder and speed controls, whose effects can, to some extent, be modified by shifting ballast, trimming sails, trailing lines astern, etc. Their computers are their brains, within whose memory banks are stored response curves in the form of their feel for their own ships. These we can supplement here with a general description of ship response in confused seas, and attempt to explain why what works for one vessel in a particular situation may be completely wrong for another.

WHY IT WORKS

The amplitude of any mode of ship motion is related to the *product* of the prevailing exciting forces (sea state) × speed-heading factor (period of encounter) × ship response factor (a property of the vessel) × coupling factor (interaction between modes). Before making generalizations about ship motion, we shall consider these factors separately.

Exciting Forces: As noted in Part V, in addition to other waves that may be propagating through a region, any wind-generated sea contains wave energy at all periods and wavelengths up to some limit related to wind speed and its fetch or duration. This energy is distributed among the spectral periods in such a way that average wave height increases in concert with wave period (fig. 74, p. 179). Thus there will be exciting forces at all periods present in the spectrum, and the largest forces will tend to be associated with the highest and longest waves, particularly if they are breaking.

Steep, near-breaking waves are markedly unsymmetrical; the upward-directed particle velocities and accelerations are substantially higher in the forward face of a wave than in their downward counterparts behind the crest. If the wave is breaking, these upward forces will be immediately succeeded by even larger forward impact forces. Thus, to the extent that the wave spectrum is directional—as it will be if the wind has persisted for, say, 12 hours or longer—the largest exciting forces will come from windward.

Lastly, in a storm sea, the exciting forces are as irregular as the waves; there may be no rhythmic pattern to which a vessel can adjust, although the intensity and frequency of ship motions can often by mitigated by some of the techniques discussed below.

Speed and Heading: Commensurate with a skipper's objective in getting somewhere, speed and heading control should be aimed at achieving the greatest possible mismatch between prevailing exciting forces and the vessel's natural responses, assuming these are known by experiment or calculation. In heavy swell outside a storm area, the principal wave period can be observed directly by running slow, timing the passage of ten to twenty waves, and then determining an optimum course and speed from the resonance diagram. Here we note that the diagram has bilateral symmetry, and, except for courses dead into or dead off the swell direction, there are two alternate courses for any speed that will produce identical responses. Thus there is usually a choice that optimizes speed made good (V_{mg}), or that avoids a possible storm encounter.

In a growing storm sea, the largest waves usually come from some general sector—roughly centered on the wind direction. If a storm forecast is available, the average period of the largest waves can be estimated from the CSS diagram (fig. 79, p. 189), and the resonance diagram will indicate course and speed combinations for minimizing the highest exciting forces. These may be quite different for vessels of different types or sizes, but in general certain sectors of the resonance diagram will be proscribed for lack of adequate power or directional control, or because of increasingly violent motions.

The forbidden area of the speed-heading rose grows—and the available operating area shrinks—roughly in proportion to $\sqrt{E}$, the sea state index, as a result of an inexorable sequence of events. When wind and sea have risen past the point where a course objective can reasonably be maintained, progressive speed reductions are required to moderate ship motions; reduced speed brings about progressive diminution of rudder control with which to oppose the increasing tendency of wind and sea to swing a vessel broadside and expose her to uncontrolled rolling. At some stage, a skipper will be forced to one of three elections:*

1. Heading directly upwind at minimum control speed, and attempting to counteract wave yawing forces by minor course adjustments
2. Running in some optimum direction off the wind, at a speed just sufficient to prevent being pooped by overtaking seas
3. Heaving-to and trying to minimize adverse motions by some combination of power, rudder, sail, and drag device.

Which choice and when to make it depend upon the type and condition of the vessel, the storm prognosis, and the proximity to other hazards, such as a lee shore or other vessels. These factors, and various operating techniques in the above situations, are further discussed in Part VII.

Ship Response: If a vessel were but a cork bobbing at the surface in a random seaway, the statistics of her motions would much resemble those of the sea itself. But, because of her physical size, inertial resistance, and maneuvering capability, and also because most storm seas are quasi-directional, these statistics can be favorably altered to moderate her motions until the seas become so large that these properties dwindle by comparison, and she indeed becomes a cork at the mercy of wind and wave.

Because the exciting forces are random, ship response in the nonoscillatory modes (surge, sway, and yaw) will similarly be random, and maximum response will occur for components of the exciting forces that act fore and aft, broadside, and quartering, respectively. Being nonperiodic, these modes are independent of the period of wave encounter, but the intensity of motion (acceleration) will be proportional to the square of the relative velocity between ship and wave, and thus can be moderated by reducing speed on upwind headings. Normally, surge, sway, and yaw are not limiting to ship operations, except insofar as they couple energy into the oscillatory modes. However, excessive yawing makes for hard steering, and when coupled with surging on a following sea, can lead to broaching.

Motions in the oscillatory modes are quite different; a vessel will attempt to respond *periodically* to any exciting force component in roll, heave, and pitch, and thus will undergo a continuous succession of damped, transient oscillations that show up as energy peaks in spectra computed from harmonic analysis of her motions. Since these forced motions *are* periodic, it is important that a vessel avoid periods of encounter that bring her into synchronism with large, exciting forces.

Skippers should familiarize themselves with their vessels' natural periods, and how they vary with different load conditions, so that they can make best use of the CSS and resonance diagrams to avoid dangerous speed-heading combinations. They can do this by practicing in moderate and heavy—but not extreme—weather, since

*The *Admiralty Manual of Seamanship* regards all three as variations of heaving-to.

motions will be generally similar to (but less intense than) those at more elevated sea states. It is this characteristic of accumulating energy in the oscillatory modes that renders a vessel liable to extreme motions, and skippers who understand their vessels' responses and their safe limits can employ meteorological information and the laws of wave statistics to determine an optimum course of action in any navigable sea state.

Of the oscillatory modes, roll and pitch are the most dangerous, with the accent on roll. Their response factors differ in two important respects. First, irrespective of her period of encounter, *a vessel will usually roll at her natural period,* whereas she will pitch heavily in any succession of waves whose "wavelength" is between three-fourths and two and one-half times the projection of her L_W in the direction of wave advance. Second, a vessel can usually alter course or speed to moderate excessive pitching before it becomes dangerous, whereas she need roll only once beyond her safe stability limit to founder. The onset of dangerous pitching conditions is usually betokened by progressively more frequent slamming or bow immersion, but it takes only two or three waves at the proper period of encounter to roll a vessel under. Again, the cardinal rules of heavy weather seamanship are to avoid synchronous rolling at all costs and to reduce speed or alter course before dangerous pitching can build up.

Coupling Factor: Energy coupling essentially bridges the gap between the intuitively simpler concept of individual, uncoupled motions in six degrees of freedom and the actual erratic behavior of a vessel in a random sea. A practical compromise is usually taken in ship motion analysis by separately considering only certain combinations of strongly coupled motions, and ignoring weaker interactions with other modes. However, coupling as such does not ordinarily appear as a separate factor, but is hidden implicitly in the equations of motion solved by the computer.

Coupled motions comprise a skipper's feel of his or her vessel. Long experience tells the skipper how she will respond as she climbs obliquely over a steep crest and plunges into the next trough, although it may not occur to the skipper that she is jointly undergoing strongly coupled pitch, heave, and surge, and at the same time, more weakly coupled roll, yaw, and sway.

Some combinations of these modes are even less intuitive. Turning a vessel at speed in calm water simultaneously invokes roll, sway, and yaw. When the rudder is put hard over,* keel and rudder act as a cambered hydrofoil, forcing the stern outboard of the turn (fig. 132). At this initial stage, both the resultants of water resistance and the inertial forces lie above—and oppositely directed to—the horizontal rudder lift force, and the combination of these forces heels (rolls) the vessel into the turn. As the turning motion becomes established, the inertial (centrifugal) force reverses direction; the vessel rolls outboard to some equilibrium angle consistent with speed and turn radius.** In turning, the vessel pivots, not about its midsection, but about some point farther forward, so that her bow is always inside (and stern

*The optimum effective rudder angle is about 35°. At greater angles it acts more as a brake, and flow separation greatly reduces the thwartship lift component.

**In rolling from inboard to outboard at high speed, a vessel may overshoot to an alarming heel angle. An inexperienced helmsman may attempt an abrupt correction by applying reverse rudder. But this action would only produce a transient rolling torque in the *same* direction, further aggravating the condition. The proper action is to reduce power as rapidly as possible.

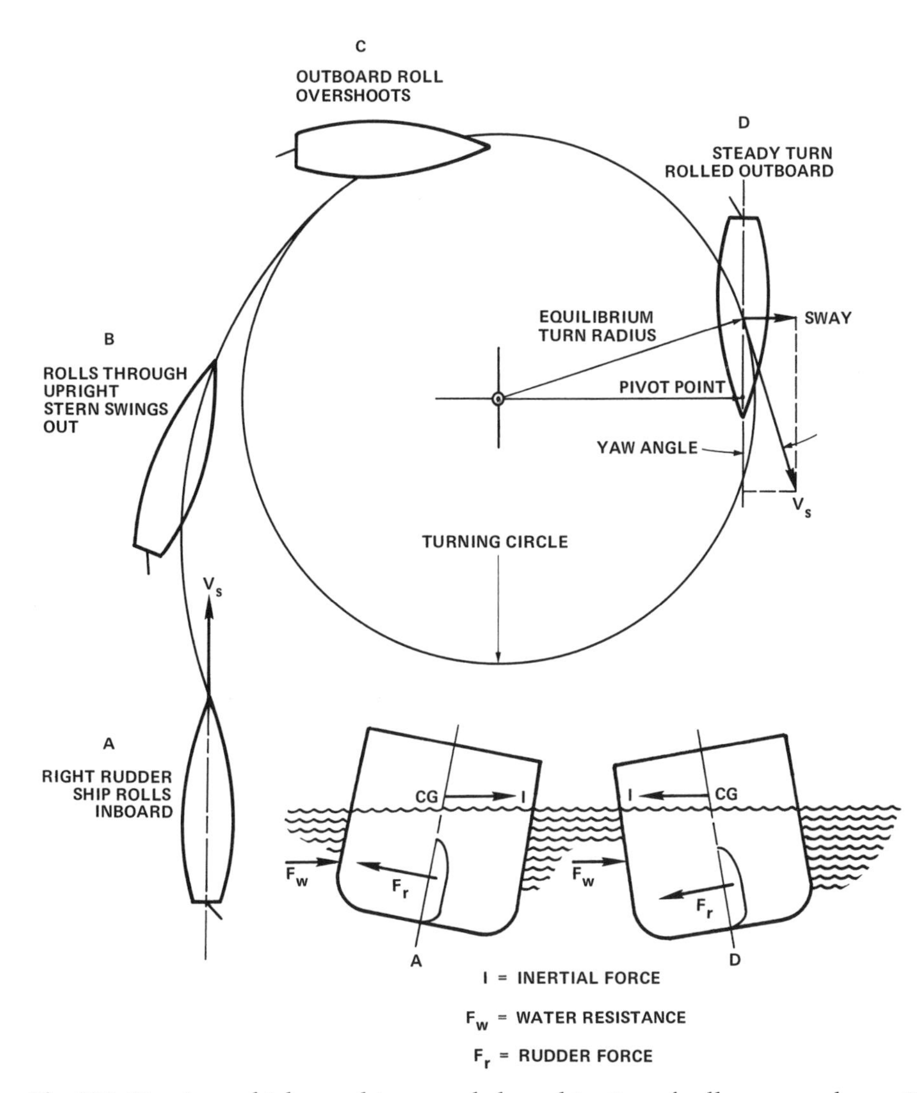

Fig. 132. Turning at high speed is a coupled combination of roll, sway, and yaw. The vessel initially rolls inboard when right rudder is applied *(A)*, then passes through upright equilibrium *(B)* as turning commences, overshoots in rolling outboard *(C)* as she enters her steady turning circle, and stabilizes at equilibrium outboard heel *(D)*.

outside) the equilibrium turning circle; i.e., she is yawed into the turn. At the same time, she is effectively swaying with respect to her instantaneous velocity vector, which is tangent to the turning circle.

Recalling that a vessel's total response will be the product of all of the above factors, we can now draw some general conclusions regarding her performance in an irregular seaway:

1. Because of the number of factors involved, it is not possible to predict vessel motions in an irregular seaway by means normally available at sea; but relative trends due to changes in these factors can be estimated.
2. On a given speed and heading, the average range and intensity of motions can be expected to increase in proportion to $\sqrt{E}$, the sea state index, up to the

point where a vessel's safe operating limits are exceeded. An exception to this statement is that, if the addition of longer and larger waves to the growing spectrum brings about a condition of resonance, this safe limit may be exceeded at a lower sea state.

3. At a given sea state, and on any heading, wave impact forces will increase roughly as the square of the relative velocity between ship and wave. At any speed, the relative velocity is maximum when headed into the seas, and varies as the cosine of the angle off the seas. Because there are many periods (and velocities) present, these forces are never zero. If the waves are breaking, the up-sea impact forces are nearly doubled.
4. At a given sea state, the intensity and type of ship motion are related to the product of sea state, speed-heading factor, and response factors; a vessel is most sensitive to speed-heading combinations that bring it into synchronism with normally acting exciting forces. Thus, synchronous pitching and heaving are most apt to be excited in head or following seas; rolling and heaving in beam seas. In quartering seas, any or all may be excited to some degree.
5. Because of low damping and high roll response, most vessels will tend to roll at their natural periods in random seas. Vessels under sail will roll much less, and more in concert with wind gusts. Conversely, because of high damping and low, broad response, pitch and heave motions will be most severe for "apparent" wavelengths of the same order as a vessel's L_W. Thus, smaller vessels will be first affected in a growing sea, and will subsequently always undergo more violent motions and accelerations than larger vessels; i.e., accelerations are inversely related to displacement for a given exciting force. Consequently, larger vessels are more sea-kindly at all sea states, and smaller vessels, by virtue of relatively more rigid construction, can take more punishment.
6. The effect of coupling is to redistribute energy between modes, such that quite unexpected motions may occur in a random sea.

In summary, randomness and irregularity in a seaway act to smear out a vessel's well-defined motions in periodic waves, so that they statistically resemble those of the sea itself. She will still respond most strongly to wave components in synchronism with her natural modes of oscillation, but may also undergo transient, random motions of greater magnitude. The resonance diagram can still be used to avoid speed-heading combinations that produce synchronism with the largest exciting forces, but the optimum combination is best found by experiment. This advice may seem redundant to the experienced ship handler, but the function of this chapter is to explain what works to those who may not know, and why it works to those who do.

EXTREME MOTIONS

Just as there are extreme waves in a random sea, so are there extreme vessel motions. Two questions of general interest arise: What are the probable magnitudes and frequencies of occurrence of extreme motions at a given sea state? How large a wave does it take to capsize or pitchpole a vessel?* The first question can, to some extent, be answered by considering response statistics as related to sea state, and the second, by considering some recent hydrodynamic model studies of vessels in breaking waves.

Influence on Ship Routing: When shipmasters think of heavy weather, they are less apt to be concerned about the elements than they are about the waves, and the effects of the waves upon their ships. At the least, the waves may slow her down or force her to detour, at some cost of fuel and delayed arrival. At most, the waves can strain her hull, wash cargo overboard, or even capsize her, for there is no ship so large but there is a wave that can sink her.

Although commercial shipping companies have been subscribing for more than three decades to meteorological services for the purpose of storm avoidance, it is only within the past ten years that it has become feasible to incorporate ship motion analysis into route planning. This improvement is due to three technological advances: long-range storm tracking through satellite imagery, shipboard computer communications via satellite link, and computer correlations between ship response and sea state.

While it is not practical at sea to calculate ship motions directly from sea state, one can, alternatively, "calibrate" individual vessels by installing instruments to record ship motion and sea state information for any heading, speed, and vessel configuration (loading, power, fuel consumption, etc.), until sufficient data have been accumulated to make ship-response estimates for all operating conditions of interest.**

With this accomplished, effective ship routing can be conducted as follows (fig. 133). Ship routes, constrained to fairly rigid time windows by departure and arrival loading schedules, are planned and executed in segments. Initial routing is predisposed to seasonal tracks, modified at each segment, if necessary, by local wind and sea state forecasts. Strategic decisions are made at a land-based routing station, but shipmasters make operational decisions based on computerized forecast displays and their own experience. A long-term goal, of course, is to transfer the data base to the ship itself, where decisions can then be made autonomously.

Computerized ship-motion prediction has proved so successful that it is spreading throughout the western shipping world: one more small triumph over natural adversity.

Catastrophic Motions: Capsizing and pitchpoling (flipping end for end) must be regarded as catastrophic—rather than extreme—motions, since few vessels, except the Eskimo kayak, are designed for such maneuvers. Capsizing is common enough

*These questions, considered in the first edition of this book (1974) to be almost academic, have since become the focus of worldwide attention: the first because it has sparked a revolution in cargo ship routing, where minimization of adverse motions is a factor in optimum route selection; and the second because the 1979 Fastnet catastrophe has provoked a storm of inquiry into the causes and prevention of capsize among racing yachts. I take comfort in the fact that the important suggestions appeared here long before the fact.

**The bulk of this work was pioneered by Ocean Systems, Inc., Oakland, California, with the cooperation of American President Lines.

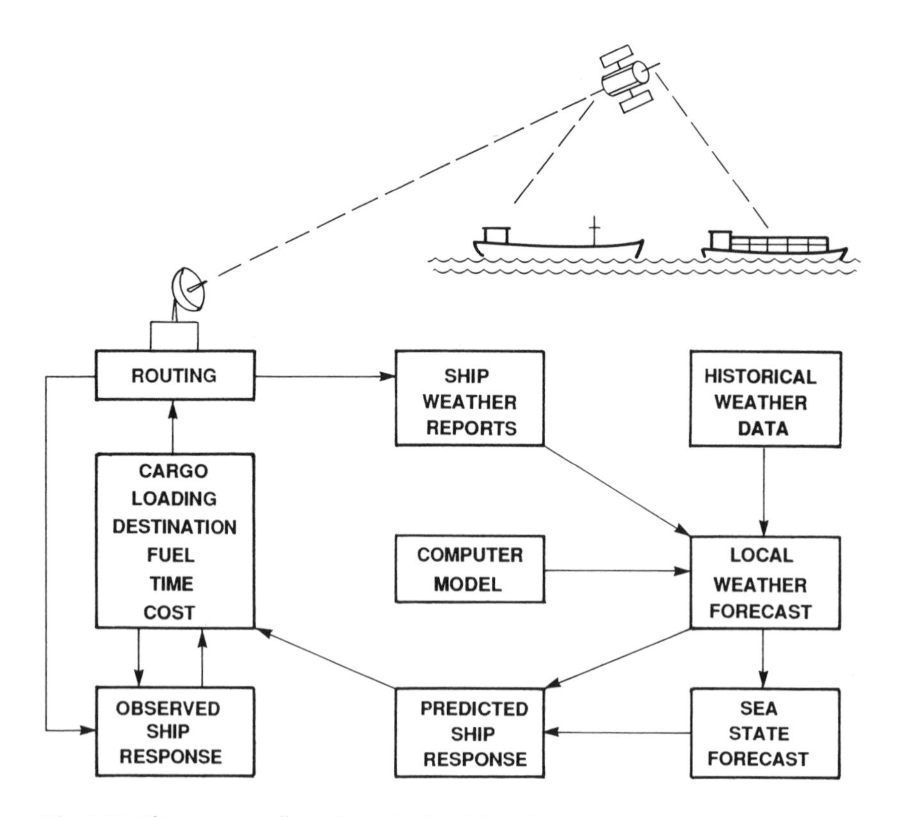

Fig. 133. Ship routing flow chart for land-based routing station.

among light sailing craft and powerboats overdriven in choppy water, but we refer here to relatively rare accidents to larger vessels in advanced sea states. Many a large ship has rolled over in the trough of large waves, as a result of loss of power or control, but I know of none that have been pitchpoled, the latter seeming to be confined to vessels of LOA less than about 50 feet.* Of these, the current record seems to be held jointly by the 45-foot ketch *Tzu Hang,* which was pitchpoled and rolled over in separate South Pacific storms, and by the 30-foot sloop *Grimalkin,* rolled over five times in the 1979 Fastnet storm. The latter storm set many catastrophic records: 5 boats sunk, 19 abandoned, 15 broken rudders, and 6 broken masts among 110 knockdowns and 78 rollovers.

Such circumstances, of course, are quite beyond the scope of ordinary ship motion analysis. But a review of the literature on naturally occurring events suggests a characteristic pattern: (1) a vessel is operating in a marginal sea state under minimal control, or hove-to; (2) catastrophe results from encounter with a single, extraordinarily large, steep, breaking wave; and (3) such events are rarely repetitive. In the first edition of this book (1974), I cited some results of my own laboratory experiments with ship models exposed to breaking and near-breaking deep-water waves, together with government tests with powered 1:100 scale models of navy

*Since writing this, I have been advised that trawlers as long as 90 feet have been pitchpoled off George's Bank, west of Cape Cod. But abnormal wave heights are to be expected in such shallow water (8 fathoms).

ships exposed to breakers as high as 600 feet (plate 26)*, with the following observations:

1. No vessels capsized or pitchpoled in single, nonbreaking waves.
2. Multiple *nonbreaking* waves caused some vessels to roll so heavily as to immerse their topsides, which could cause loss of power or foundering in a prototype vessel.
3. All vessels tested occasionally capsized or pitchpoled in single, *breaking* waves when the wave height approximated their beam or overall length, respectively. This observation is consistent with observed limits for pitchpoling of full-scale vessels, since the maximum probable height of natural breaking waves appears to be about 100 feet.
4. Under similar conditions, capsizing and/or pitchpoling was often averted by a tendency for the vessel to slew or yaw from its beam or head-on attitude, respectively, and to eventually recover—albeit after suffering heavy immersion.
5. In capsizing, a vessel was characteristically caught in the curl of the horizontal breaking vortex, becoming, as it were, part and parcel of the breaker. Pitchpoling was quite different; whether head- or stern-to, the vessel first pitched up to an angle of about 45°, and then slid down the face of the wave to bury its lower end up to about one-third the length, after which forward breaker impact completed the rotation.
6. Vessels under way at scale speeds of 20 knots into the waves were capable of surviving slightly larger breakers. Unfortunately, no tests were conducted with vessels running before the waves.

While these results may not be directly applicable to other vessels, such as sailing yachts under bare poles, there is some reason to suppose that hull characteristics make little difference under these conditions. When breaking waves reach a critical size, stability and integrity have little to do with a vessel's dynamic behavior, although they may determine whether or not she eventually recovers. Apparently, much higher breakers can be survived end on, and chances are still better if they are taken slightly on the quarter, and if one employs any tactic that resists backsliding or surfing to the point of bow or stern immersion. Means for accomplishing the latter are discussed in Part VII.

The Fastnet disaster provoked a host of laboratory studies, both here and abroad, with the object of quantifying sailboat capsize dynamics, hopefully leading to recommendations for safer yacht designs and racing rules. The final report of the Capsize Committee, which coordinated these rules, includes the statement: "The sea can, and does, produce waves big enough to roll any yacht. Therefore absolute roll-proofing is unattainable. There are, however, various yacht design characteristics that can render a boat better able to resist rolling, and quicker to re-right herself if rolled over."

*These tests were an evaluation of the "Van Dorn Effect," in which I had predicted analytically that large nuclear explosions in deep water off the continental slope could convert the entire continental shelf into a surf zone, inimical to most ship operations.

Plate 26. Experimental test of ship survivability in high waves. Model destroyer is being pitchpoled in breaking wave of height equal to its length; longer aircraft carrier has survived.

The committee's conclusions and recommendations are most simply reviewed in *Desirable and Undesirable Characteristics of Offshore Yachts* (see References). Among the most important are the following:

1. Capsizing is a dynamic process, in which the rotational influence of the wave vortex is resisted by the yacht's rotational inertia.
2. Half a yacht's rotational inertia is in her mast, without which she is twice as liable to capsize. Larger yachts have a proportionally greater rig inertia than smaller ones, and hence a relatively greater capsize resistance.
3. Any factor increasing a yacht's range of positive stability will increase her resistance to capsize and her ability to recover—such as more ballast, lower CG, higher freeboard, etc.

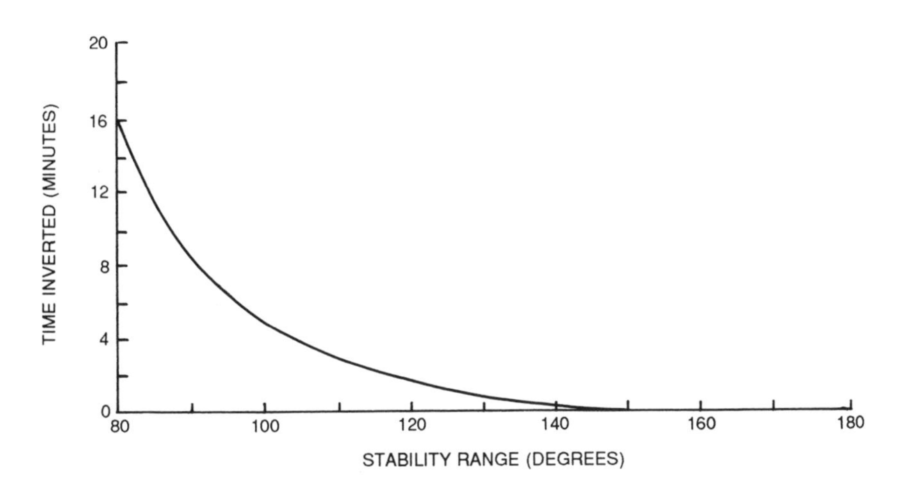

Fig. 134. Estimated recovery time for capsized sailing yachts as a function of the range of positive stability. Example shows recommended minimum of 120° for offshore sailing.

4. Plugged hollow masts and watertightness topside favor quick recovery.
5. The worst capsize attitude is beam-to; end-on attitudes greatly augment rotational inertia and resistance to capsize.
6. From wave spectral analysis and stability considerations, the committee generated a capsize recovery time chart (fig. 134). Assuming most passengers and crew can survive two minutes with the ship inverted, they recommend a 120° range of positive stability as the minimum required for offshore sailing.
7. Because stability calculations are made only under the IMS rule, a simple estimate of capsize vulnerability is the formula $64B/D < 2$, where B is maximum (feet), and D is displacement (pounds).

But all of the above factors only slightly modify the actual sea state that a yacht can survive. Ultimately, survival depends upon the probability of encountering extreme breaking waves in storm seas. To answer this question, we can make use of an extreme probability formula from which the curves of figure 73 (p. 177) were calculated. In terms of the total number *(N)* of waves passing a stationary observer, the most probable value for the maximum expected wave height is, approximately

$$H_m = 2\sqrt{E \log_e N} \text{ ft.}$$

This formula applies only to steady conditions, which restricts us to fully developed (FDS) sea states, for which we previously (p. 180) gave the equilibrium relationships: $\sqrt{E} = 0.0068\,V^2$ ft., $T_a^* = 0.29\,V$ sec. Lastly, recall that the time for N waves to go by is $t = NT_a^*/3{,}600$ hrs.

With a little juggling, we can combine the above to obtain the desired formula for the time expectancy of encountering a wave of height H_m as a function only of the FDS wind speed V (knots):

$$t = 0.0008(V/10)\exp\left(0.54\,H_m^2 / (V/10)^4\right) \text{ hrs.}^*$$

*Here, *exp ( )* means "*e*" raised to the power in brackets, where $e = 2.718$ is the base for natural logarithms.

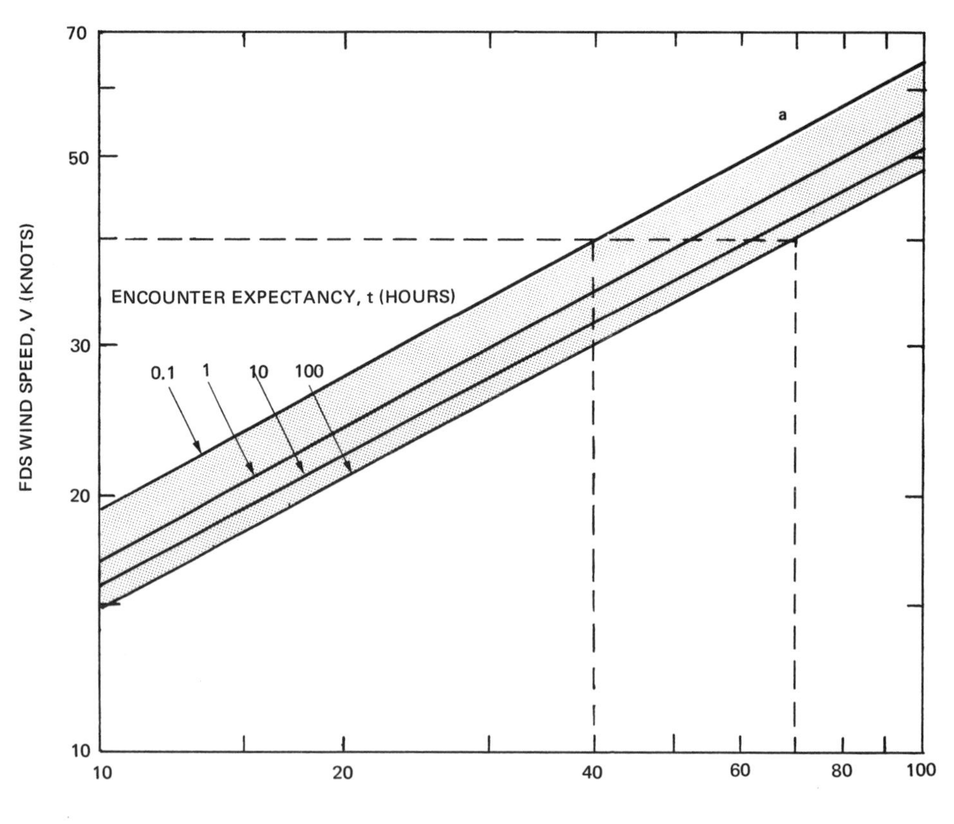

Fig. 135. The catastrophic probability (*CP*) diagram indicates time expectancy of encountering breaking waves high enough to capsize or pitchpole a vessel, assuming an FDS sea state at indicated wind speeds.

This expression is a little messy to calculate because of the large numbers, but the results are shown in figure 135, which I shall call a catastrophic probability (CP) diagram.

Assuming that the highest waves present in an FDS sea will be breaking, and that a vessel runs some risk of capsizing or pitchpoling if the wave height exceeds her beam (if broadside) or her LOA (if end-on), respectively, the CP diagram indicates the expectancy of encountering such a wave, as a function of wind speed. For instance, in a 40-knot FDS sea, the dashed lines show that a 40-foot breaking wave could be expected every 6 minutes or so, or a 70-foot wave every 100 hours. Additionally, the diagram shows that, for a constant expectancy interval, maximum wave height increases in proportion to the square root of wind speed. Conversely, for a fixed maximum wave height, the expectancy interval decreases very rapidly as the equilibrium wind speed rises. Thus, if the wind increases, and sea state *slowly* approaches a new FDS level, maximum height will increase a little, or the same height will occur much more frequently.

The CP diagram, published with little verification in the first edition of this book in 1974, can now be tested against the abundant casualty statistics from the 1979 Fastnet race, using our previous reconstruction of the Fastnet sea state. According to table 7 (p. 266), the maximum breaker height every 6 hours was H_{10} = 39 feet, corresponding to an FDS wind speed of about 40 knots on the CSS diagram (fig. 79). By coincidence, this happens to be the sample wind speed shown on the CP diagram (fig. 135), which makes specific encounter expectancies easy to estimate. Table 10 lists the number of entrants and finishers in the 1979 Fastnet race by class and IOR length, to which I have added the percent of all entries retiring (including those abandoned or sunk), and the number of breakers per hour having heights equal to the IOR lengths of boats in each class, assumed to be making little headway into the seas.

Table 10. Comparison of breaker expectancy with percent of boats failing to finish the 1979 Fastnet race.

Class	IOR Length (Ft.)	Starters	Finishers	% Retired	No. Breakers/Hr.
0	55–79	14	13	7	<1
I	45–59	56	36	35	1–5
II	39–43	53	23	57	5–12
III	34–38	64	6	91	12–26
IV	33	58	6	90	30
V	28–32	58	1	98	30–57

We note immediately that the percentage of boats that finished was directly related to breaker frequency, and that even the probability of 1–5 breakers per hour caused 35 percent of the starters to withdraw. The lesson seems clear: when the breakers get as big as the boats, the boats can no longer make headway, and become much more liable to damage. In the Fastnet, all the boats abandoned or sunk were in the last three classes, among which the wave frequencies exceeded 12 per hour, and less than 10 percent finished.

In closing Part VI, we cannot avoid commenting that the foregoing simple relationships between boat length, breaker size, and frequency, available since 1974, appear to have escaped scrutiny by the Capsize Committee. In retrospect, it would seem much simpler to base class entry in ocean races upon wave forecasts than upon small differences in the capsize potential of individual boats.

PART VII

Heavy Weather Seamanship

24

Ships, Shoes, and Sealing Wax

Having acquired some knowledge about how ships and boats behave in the real ocean, we turn now to the question of its practical application. While many a lucky novice has sailed great distances in an ill-found and poorly equipped boat and has arrived convinced that hazards at sea are vastly overrated, many an expert has set out under apparently benign circumstances and has disappeared without a trace.* Clearly, the important point is to prepare for the worst, even if it is rarely to be expected. Having just dealt with the probability of catastrophic conditions, we now discuss seamanship in the context of such conditions.

In view of its surpassing importance to survival, it is somewhat surprising to note how lightly most standard references treat the subject of seamanship in heavy weather: Knight's *Seamanship* devotes but five pages to the subject, and the *Admiralty Manual of Seamanship,* seven; even Crenshaw's *Naval Shiphandling* contains only seven pages of rather elementary advice. On the positive side, two new editions of older works, *Heavy Weather Sailing* and *Heavy Weather Guide,* summarize quite well the experiences and opinions of cruising yachtsmen and commercial and military skippers. And many of the questions discussed in the first edition of this book have since been re-examined independently by boating experts in the aftermath of the 1979 Fastnet catastrophe and of more recent races involving heavy damage and loss of life (see Recommended Reading).

Because most commercial and military vessels are designed to rigid specifications as to their strength, seaworthiness, and performance,** and their safe operation is the responsibility of licensed and (hopefully) qualified officers, there is little point in reviewing operating procedures that are largely prescribed by regulation. Accordingly, the following is directed to the much larger number of small-vessel operators (particularly cruising yachtsmen) who may become embroiled in circumstances outside of previous experience that are not covered in most books or manuals on boat handling. The interesting psychological point about calamity is that one tends to regard it in the third person—it always happens to someone else.

*To wit, Joshua Slocum, Richard Halliburton, and—more recently—Lee Quinn all disappeared. The latter had sailed 45,000 miles about the Pacific with female crews, and vanished in the fall of 1970, en route from Yokohama to Vancouver.

**The same cannot be said of emergency equipment. Most lifeboats, life vests, rafts, and locating devices are woefully behind recent developments in ship design and marine technology. A recent government survey attributes this inadequacy to the overlapping and often contradictory safety requirements of at least six federal agencies, which together make it exceedingly difficult to obtain approval of new devices.

In response, I can only say that anything of merit in these pages is much more easily absorbed in the comfort of one's living room—or while tied dockside—than when standing up to one's waist in cold water in a pitch-dark cabin, trying to read by flashlight while a howling wind and breaking seas are rolling the boat on its ear every few seconds.

25

Strategy

CONSTRUCTION

Seamanship, like charity, begins at home. Any vessel intended for rough-weather service should be designed for it from the keel up, not only as to structure and rigging, but also as to stowage. Prime examples of rugged design are the North Sea trawlers and the Kodiak crab boats, both of which operate routinely in some of the world's roughest waters, combating heavy icing as well as extreme wind and sea conditions.

Unfortunately, most pleasure craft do not meet these standards, as attested by mounting Coast Guard rescue statistics. Among racing sailboats, so many dismastings and structural failures have occurred in recent years that the IMS Technical Committee has instituted more stringent entrance regulations. Perhaps the most telling comment appears in John Rousmaniere's foreword to *Desirable and Undesirable Characteristics of Offshore Yachts:* "While the trend toward complicated, highly specialized, fragile, and unforgiving racing boats and gear is obvious and disturbing, what most frightens many observers is the influence that these boats are having on the design of cruising boats."

Mass-produced fiberglass boats are among the worst offenders. Despite the great strength and flexibility of glass-resin laminates, poor design and shoddy construction often inspire a false sense of confidence that will lead owners to expose themselves to what would ordinarily appear to be moderate weather, only to have their chain plates pull out, their spade rudders snap off, or their keels fracture clear around and fall to the bottom. Don't laugh—all of these failures have occurred many times.

But space here does not permit a complete chronicle of boating defects; we shall confine our discussion instead to a few points that bear on seamanship and survival, with emphasis on private yachts, whose owners or prospective purchasers have a considerable choice of designs and/or equipment. In this, one should keep in mind the worst conditions that might be encountered. At elevated sea states, green-water impact forces can exceed a ton per square foot on deck or cabin trunk—roughly equivalent to dropping a vessel bodily into calm water from a height of 32 feet, or to submerging her to an equal depth. In such conditions, water will squirt like a fire hose through every crack or pinhole. Sixty-knot wind gusts apply pressures of 12 lbs./sq. ft. to sails and spars, equivalent to about one ton per ton of displacement. Accelerations can exceed that of gravity, such that objects of any size will exert forces equal to their respective weights against fastenings or, if unsecured, sail gracefully in parabolic arcs to some new resting place.

Lastly, one should also ask, "Can my boat recover from a knockdown, or roll clear over and come up all standing, without losing watertight integrity, without topside gear coming adrift, and without emptying the contents of all drawers and lockers in a hopeless jumble on the cabin sole, or into the bilge, to slosh back and forth in a mixture of seawater, fuel oil, and battery acid?" If the answer is yes, you have passed the first test of seamanship, and good shiphandling will carry you through all but the worst that Nature can devise; if it is no, the following pointers will get you started in the right direction. Otherwise you will be forever committed to fair-weather sailing, or to courting disaster.

Currently, there are such a variety of construction methods that it is sometimes difficult to distinguish good from bad—even for a marine surveyor. Whereas, with traditional wood or steel construction, one can readily determine the size and integrity of scantlings and fastenings by inspection, the surveyor is now also confronted by plywood hulls and decking, which are subject to slow delamination; by glass-resin (fiberglass) laminates, whose elasticity and impact resistance are only conjectural; by aluminum hulls, whose greatest enemies are electrolytic corrosion and embrittlement of weldments; and, more recently, by ferro-cement boats, for which there are as yet no required specifications as to density of reinforcement, composition of cement mix, or plastering and curing techniques. But today, the overwhelming majority of private boats are constructed of fiberglass, in which evidence of structural incompetence is fairly easy to detect—if not to repair.

Problems with Fiberglass: Even with good design, there are problems with fiberglass boats. It is not commonly realized that resins almost never cure absolutely, but continue gradually to stiffen and embrittle over the years. This process is accelerated by sunlight (ultraviolet rays), and thus decks and cabin roofs are first affected, as evidenced by the appearance of crazing, or small cracks, at inside corners or at other points of stress concentration. Painting, decking, or otherwise covering the topsides greatly reduces sunlight damage. Most mass-produced hulls are so redundantly strong below the sheer line that fatigue fracture due to loss of elasticity is not a major problem, except around the root of deep, high-aspect keels when accidently grounded, or around separated rudderposts or skegs in heavy-weather areas.

More insidious and just as inevitable is the problem of gel-coat blistering below the waterline. A federally sponsored study has shown that blisters are caused by gradual water absorption into resin laminates, where it becomes concentrated in focal points of water-soluble nuclei. These then swell by osmotic pressure until the laminate separates. The number and size of blisters are directly proportional to the time a boat has been in the water; if allowed to continue, they can permeate the entire laminate and damage the hull beyond recovery. Sample surveys suggest that the process is accelerated by continuous immersion, but that temperature and salinity appear not to be factors. The rate and extent of hull damage can be minimized by early detection and repair, but no means has yet been found to prevent the slow water absorption that causes blistering.

Because of their stiffness, light weight, and good thermal insulation, composite sandwiches of an end-grain balsa or foam core, bonded between fiberglass lamina, have found increasing use in boat deck, cabin roof, and even hull construction. But their use brings a host of problems. Unless carefully supported by broad filets to distribute shear loads over a wide area, the panels tend to delaminate under flexural or impact loading, particularly where attached to some rigid auxiliary structure (as in decks attached to hull sides, coamings, toe-rails, etc.), or where penetrated by bolted attachments, such as stanchions or deck fittings. Even wave impact to flat hull sections can produce incipient delamination. Once water is admitted to a delaminated region, whether by seepage or by diffusion, the whole structure comes unglued.*

Most structural failures are preceded by local evidence of incompetence that a surveyor is trained to detect, and your best guarantee of seaworthiness is periodic survey—preferably during haulage. This not only exposes a vessel's bottom for examination, but the transfer of weight from hull to keel considerably wracks the former, such that inadequate joints become misaligned, seams may gape, and all structural elements are stressed as if working in a heavy sea. At the same time, watertight integrity can be tested by directing a high-pressure (50 psi) water jet against all possible points of leakage; these include plank and deck seams, through-hull fittings, mast boots, all exterior doors, ports, hatches, and ventilators, and all exposed screw or bolt connections—particularly those securing chain plates, stanchions, and deck winches. If your boat passes the hose test without taking a drop, you stand a good chance of keeping a dusty bilge in heavy weather.**

Cockpits: Open cockpits are often oversize—and drains undersize—relative to vessel displacement, and, if flooded, they can seriously jeopardize stability and control. Safe design recommendations are that cockpit volume should not exceed 4 cu. ft. per ton of displacement, and that the cross-sectional area of drain tubes should be at least 0.1 sq. in. per square foot of cockpit area, in order that it empty within 5 minutes. Drains should have through-hull cocks, should be straight, should cross each other, and should be located at the forward cockpit corners. Seat lockers should have positive latches and be limbered to drain rapidly if accidently flooded. It should be possible to fill the cockpit to overflow without taking water through the companionway, and the latter should be provided with wedge-shaped drop-boards equipped with slide bolts to prevent them from floating out if the boat is rolled over (fig. 136).

Needless to say, the large, open stern cockpits of many high-performance power runabouts and sportfishermen do not satisfy these requirements. Unless provided with rigid, waterproof decking above the waterline, and unless adequately scuppered for rapid drainage, such cockpits are grave threats to survival in rough seas. The limiting safe sea state for power craft with nondraining (not self-bailing) cockpits can be taken as that in which a boat immerses not more than 60 percent of her freeboard in twenty rolls, when drifting broadside with her engine idling.

*I know of a $3.5-million custom power cruiser of almost entirely composite design, built with indifferent engineering supervision at a major West Coast boatyard, which virtually disintegrated on its maiden voyage to Alaska, although exposed to only moderate seas. An insurance investigation disclosed that most of the damage was due to inadequate core-panel bonding. She is now undergoing a $2-million refit—but she will never fulfill her owner's expectations.

**Incipient leakage is a common plague at sea. Topside, in rainy weather, it is often remarkably difficult to track down. After their recent circum-Pacific cruise, Hal and Margaret Roth thoroughly renovated their 35-foot fiberglass ketch, *Whisper.* Only after much searching and prying did they discover the source of a distressingly persistent leak: water was penetrating and collecting inside a continuous void space behind the bolts that held the toe rail to the deck-hull sandwich. They were forced to remove the toe rail, plug 144 bolt holes, and bond the entire seam with glass and resin.

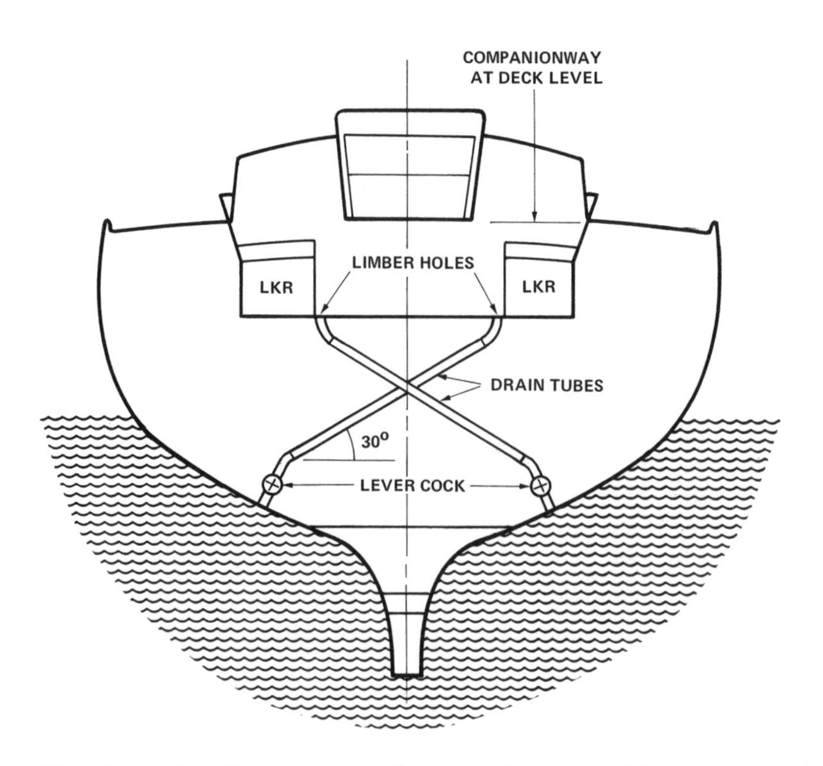

Fig. 136. Preferred arrangement of cockpit drains. Total drain area should be 0.1 square inch per square foot of cockpit area to void cockpit in 5 minutes.

Structure: All exterior cabin openings should have elastic seals and screw or dog latches, and should be strong enough to withstand a dynamic water pressure of one ton/sq. ft. It is very difficult to design a sliding companionway hatch to seal adequately, and dogged, double-lift hatches are much preferable. Ports and port lights should be made of safety glass having a minimum thickness of one-half inch per foot of least dimension. The lights should have exterior metal reinforcing rings screwed or bolted to the hull or deckhouse (rubber-sealed lights on fiberglass yachts have a disconcerting habit of starting, or popping out all in one piece). Ports are best fitted with soft plastic rain shields and, if recessed, should have drain holes, drip pans, or rain shelves that will permit leaving them ajar for ventilation without risk of wet bunks.

Decks, pilothouses, and cabin roofs are particularly vulnerable to wave action, especially when fabricated from foam or balsa-cored sandwich, which is apt to delaminate when it is dimpled by wave impact or locally crushed by deflection of bolted fixtures.

Because strengthening after the fact is difficult and expensive, the integrity of overall construction is of prime concern to a prospective purchaser. Although detailed specifications exist for all classes of construction, they are only recommendations, and the designer or builder is under no compulsion to observe them. An advertisement that a boat is "built to Lloyd's specifications" has no bearing without a Lloyd's inspection certificate, which applies to only the particular boat that carries it.

Although the recent trend to ultra-light-displacement, high-performance racing boats has led to new materials and techniques, most are not amenable to mass production, and copy-cat emulation has often led to dangerous compromises in strength and safety. While a reputable designer may meet this challenge with great skill and ingenuity, a profit-conscious competitor may cut every corner to reduce weight and improve performance. The end result has been a selective deterioration of quality and seaworthiness; important strength members such as floors, ribs, and deck beams have been reduced in size or eliminated altogether, fastenings are often inadequate or too widely spaced, and such "anachronisms" as mast partners, knees, diagonal strapping, and structural bulkheads have all but disappeared. One is thus often left with a Dixie Cup that pants, hogs, and twists in a moderate seaway, but which may cross the line first a few times before she breaks in two!

What can be done about all this? Before purchasing a boat, you should:

(a) Investigate the reputation of her designer.
(b) Examine earlier-produced examples of the same design for evidence of structural failure or incompetence.
(c) Make sure that your surveyor is familiar with recommended scantling rules and that they have been followed.
(d) Before haulage, measure the distance in the main cabin between keel and cabin roof; if it changes on the ways by more than 1 inch, you are probably headed for trouble.*
(e) Any point on the deck, hatches, or cabin roof that deflects more than one-eighth inch when jumped on by a 200-pound man is too weak to sustain breaking waves.
(f) Be especially suspicious of fiberglass construction if there is no evidence of ribs, stringers, deck beams, and (particularly) thwartship floor members that are through-bolted to the keel, and in way of masts and engine bed. Most reliable designers still employ these basic structural members.

RIGGING

Most sailing vessels are underrigged for heavy weather service. This is particularly true of racing boats, where every effort is made to reduce weight and drag aloft. Because rigging stresses can be determined fairly precisely, intentional underrigging is a calculated risk; when a vessel is hard-pressed, a sudden gust can cause loss of top-hamper and of the race as well. That this risk is too often underestimated is amply demonstrated by the current high incidence of rigging failure. There is now considerable pressure to require minimum specifications for one-design ocean racing classes.

There is no excuse for underrigging a cruising boat—particularly one intended for lengthy voyaging—although there is some disagreement as to reasonable maximum design loads. Most modern authorities design for a uniform normal wind pressure of 1–1.5 lbs./sq. ft. (corresponding to beam winds of 16–20 knots) against the working sail plan, suitably distributed among the various tension members

*I have a friend with a new 43-foot fiberglass racing sloop that loses 5 inches of headroom on the ways. He is about to lose his keel.

(shrouds and stays) by the method of moments. These loads are then multiplied by safety factors of 2–5, at the designer's discretion,* and the appropriate wire sizes are determined from tables. While this method of veriform analysis makes some allowance for inaptitude, it assumes a skipper will not be so foolish as to be caught close-hauled with his laundry out when a squall front moves through, and the wind suddenly veers 60 degrees and whomps up to 50 knots.

However, it is not difficult to show that a much more severe loading condition exists when a yacht is blown down on her beam ends and is simultaneously accelerated upward on a steep wave crest. Here, the weather shrouds are, in effect, supporting twice the keel weight, multiplied by the ratio of draft to half-beam. For keel yachts, this is roughly the same as lifting the entire hull clear of the water by its weather shrouds—a design load condition espoused by the old masters, N. G. Herreshoff and Uffa Fox. Very few modern yachts could meet this test, but the rigging specifications for the 67-foot yawl *Maruffa,* designed by Phil Rhodes, are given in Fox's book, *Sail and Power;* her combined lower shroud design tensions exceed her 70,500-pound displacement by 9,000 pounds.

The fore-and-aft components of the standing rigging are more properly proportioned to sail pressure, since the inertial loadings imposed by ship motion are always much smaller in these directions. Of these, by far the greatest tension is applied to the bobstay (if any), because it makes a small angle with the bowsprit and sustains the entire rig. It is also continually exposed to the highly corrosive action of salt spray concentrated by repeated dipping. Since there is no weight penalty, a bobstay should be redundantly strong, and chain or heavy rod is much preferable to even stainless wire rope. Second in strength come the head stays; the forestay is usually proportioned equal to—or larger than—the standing backstay, and the latter is at least as large as the main lower shrouds. Principal loads in these stays come from pretensioning to maintain a taut foresail luff.** Other stays are smaller in proportion to the square roots of their design loads, but every effort should be made to ensure that a mast is symmetrically supported to prevent bending, and consequent local crippling or buckling. Figure 137 shows the standing rigging dimensions averaged across some twenty yachts capable of meeting the "lift out" test. If your rigging falls much below these standards, you are taking a calculated risk in trading safety for performance—or your designer is simply inept.

Nowadays, racing yachts increasingly favor streamlined-rod standing rigging and internal mast halyards, both of which considerably reduce drag on upwind headings. However, the service life of rod rigging falls substantially below that of wire rope, owing to a high incidence of compression and fatigue failures, which seem to result from flogging of slack lee shrouds or headsail lulls. Moreover, if a single weak point develops, rod failure is abrupt, while wire rope is more apt to fail by progressive parting of individual strands, a circumstance that usually can be observed and corrected before complete failure. Internal halyards have the advantage of weather protection, but impose a second point of wear at the deck

*Here, "discretion" usually refers to the designer's record of survival or failure. After some experience, rigging analysis seemingly becomes almost a matter of rote; L. F. Herreshoff relates that his father, N. G. Herreshoff, who designed some 1,800 boats in his lifetime, could whack one out as fast as he could write, with no reference to tables, but with suitable concern for the vessel's intended service.

**Racing yachts often have adjustable backstay tensioners to further tighten the luff when on the wind.

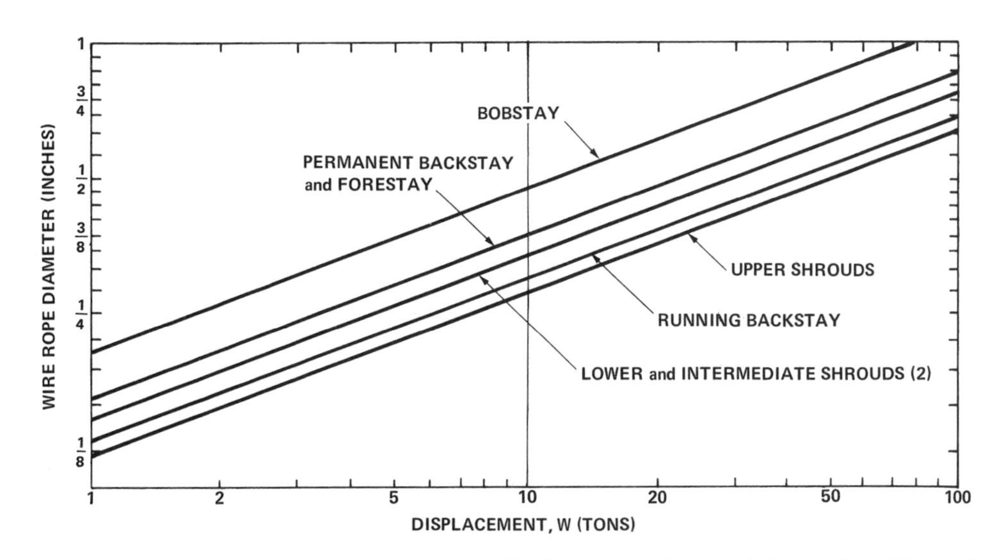

Fig. 137. Minimum standing rigging sizes for heavy-weather cruising yachts. Sizes refer to 7 × 7, stainless, preformed rope. For 1 × 19 construction, use an equivalent-strength table.

sheave; jamming upon overrunning is more difficult to correct, and replacement after failure requires a delicate fishing operation that may be difficult or impossible at sea. Both, therefore, are concessions to safety that seem unjustified for cruising yachts.

If the port or starboard shrouds should sustain a vessel's weight, her masts must be capable of equal compression loads, for a sailing yacht can be likened to a loaded crossbow, in which the combined stay tensions always act to shove the mast(s) through her bottom. Mast dimensions are conventionally determined from column-buckling diagrams, in which the critical factors are maximum compression loads and the lengths between support points (heel, deck, spreaders, and masthead). Buckling diagrams calculated for masts passing through the deck (or cabin roof) assume rigid lateral support against bending. In wood or metal construction, this support is provided by massive local reinforcement, such as angle braces *(partners)* or double-thickness plywood decking, designed to distribute the transverse shear forces throughout the entire deck width. Such support is difficult to come by in way of masts passing through a cabin roof and, even if passing through the deck the support is often omitted—or scrimped on—in fiberglass boats.

The mast step supports the entire mast compression (potentially equal to a yacht's weight), and should similarly be designed to distribute this load to the hull structure by way of her floors and ribs. Again, these features are often lacking in fiberglass construction, and mast compression is transmitted instead to the ballast keel. Since the chain plates are pulling upward with equal tension at deck level, the resulting *hoop* tensions in the planking or skin act to compress the hull laterally, squeezing it into an egg shape, and bulging the deck upwards (fig. 138). In the absence of strong deck beams, these forces can only be resisted by structural bulkheads in way of the mast—and then only if the bulkheads are edge-bolted all around. All too often such bulkheads are adequately resin-filleted to the hull, with the result that the deck comes unglued and lifts clear of the bulkhead by several inches. This tendency is greatly augmented by the practice of setting chain plates

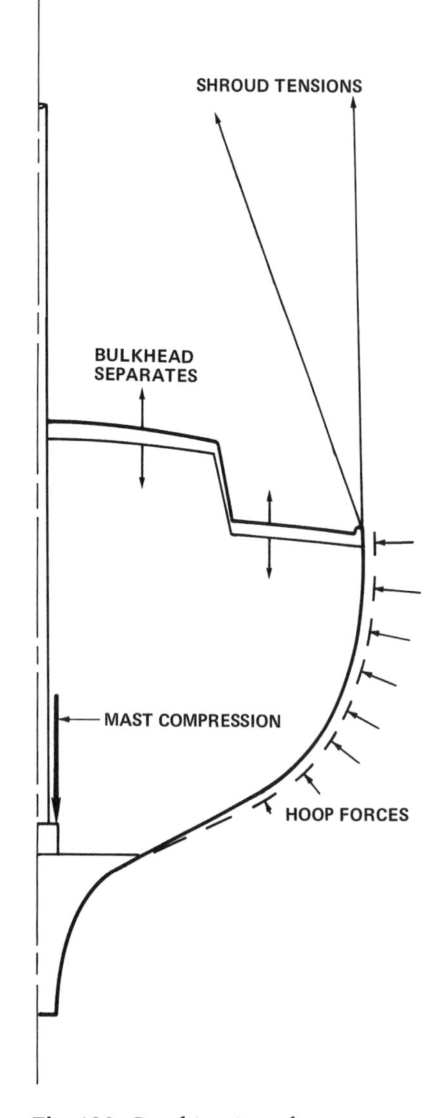

Fig. 138. Combination of mast compression and stay tensions produces hoop forces that tend to squeeze hull laterally. Unless bulkheads are securely bolted all around, they may separate from hull structure.

inboard of the rail, and attaching them by bolts to the deck itself.* Moreover, since mast-compression loads (and shroud tensions) are proportional to the draft/beam ratio, both are increased by locating the shrouds inboard, with a consequent weight penalty to achieve the same strength.

The practice of stepping masts on deck is acceptable in smaller yachts, provided that the compression loads are adequately distributed to the lower hull by a rigid pipe having the same sectional moment of inertia as the mast. But the same cannot be said for stepping on the cabin roof; there is no practical way to resist the

*I have recently seen a $40,000 fiberglass sloop where the inboard chain plates passed through the deck and were bolted to the sides of flimsy plywood hanging lockers. This was probably safe enough, considering that her shrouds were only strong enough to support one-tenth of her displacement.

horizontal loads of sails and of possible wave impact without interposing auxiliary lateral bracing that is more space-consuming than the mast itself.

Most modern sailing yachts employ extruded aluminum masts and spars because of their manifold advantages. These extrusions are necessarily of uniform section, which imposes a considerable weight penalty aloft, since a mast must be sized to withstand the much higher compression loads below the lower spreaders and is thus overstrength and overweight above them. Here, a considerable increase in performance without sacrifice of safety can be achieved by tapering the mast, or by going to sectional construction with progressively thinner sections aloft. Both methods entail greater expense, but this seems to be a secondary consideration in racing yachts, where the cost of sails and rigging often equals that of the bare hull.

In Part VI, we suggested the possibility of filling hollow aluminum masts with foam as a means of providing submerged buoyancy. Interior foam would also enhance stiffness and deaden the ringing produced by halyard slapping.

Running rigging has similarly undergone great changes in recent years, with many accessions to racing that would be better omitted from long-range cruising boats.* To my mind, if modern designs are underrigged, they are also overwinched. The great mechanical advantage of gear-driven winches makes it all too easy to overstress rigging and sails.

Except on very large boats, a "fire-hydrant" cluster of halyard winches is an unnecessary complication and expense; not only is it easy to overstress halyard and luff, but caged winches are prone to brake failure and to jamming when overrun.** While a single reel-winch is useful for large genoas (and almost a requirement for hoisting a bos'n chair to work aloft), sails up to 400 sq. ft. can be hoisted and lowered much more rapidly by one man on a rope-tailed wire halyard. Figure 139 shows a modification of a hoist arrangement suggested by L. F. Herreshot that I have found practical for sails up to 800 sq. ft. The wire halyard is eye-spliced at both ends and has a length equal to the hoist, less about 6 feet. The upper eye is shackled to the headboard in the usual fashion, and the lower to a double-becket snatch block. One end of the rope tail is spliced to the lower becket of this block, and the other, passed through the upper sail slide cringle and stopped with a figure-eight knot. The bight of the tail should just reach the deck.

After being hoisted freehand as far as possible, the tail is led around a cheek block set low on the mast, back over the snatch block, and under the halyard cleat above the cheek block, giving a three-part purchase that can be swigged up to a tension of 300–400 pounds without difficulty. The bitter end of the tail is now aloft, and serves as a downhaul. There is no rope surplus to dispose of if the bight of the downhaul is jammed under the last turn on the cleat, from which it can be released with a single jerk. Other advantages of this arrangement are that the wire halyard can be reversed to prolong its life, no rope-to-wire splice is required, and the endless loop is always under control and can never creep out of reach or be set flying in a strong wind. If extra lull tension is required, it can be obtained with a multipart gooseneck purchase.

*This includes the estimated 84 percent of one-design production racer-cruisers that are never raced competitively. In buying such a boat, you may be paying premium prices for a lot of unneeded hardware.

**While sailing solo around the world, Robin Knox-Johnston experienced repeated brake failures on both halyard winches, and ended up leaving their handles in place and lashing them to the mast.

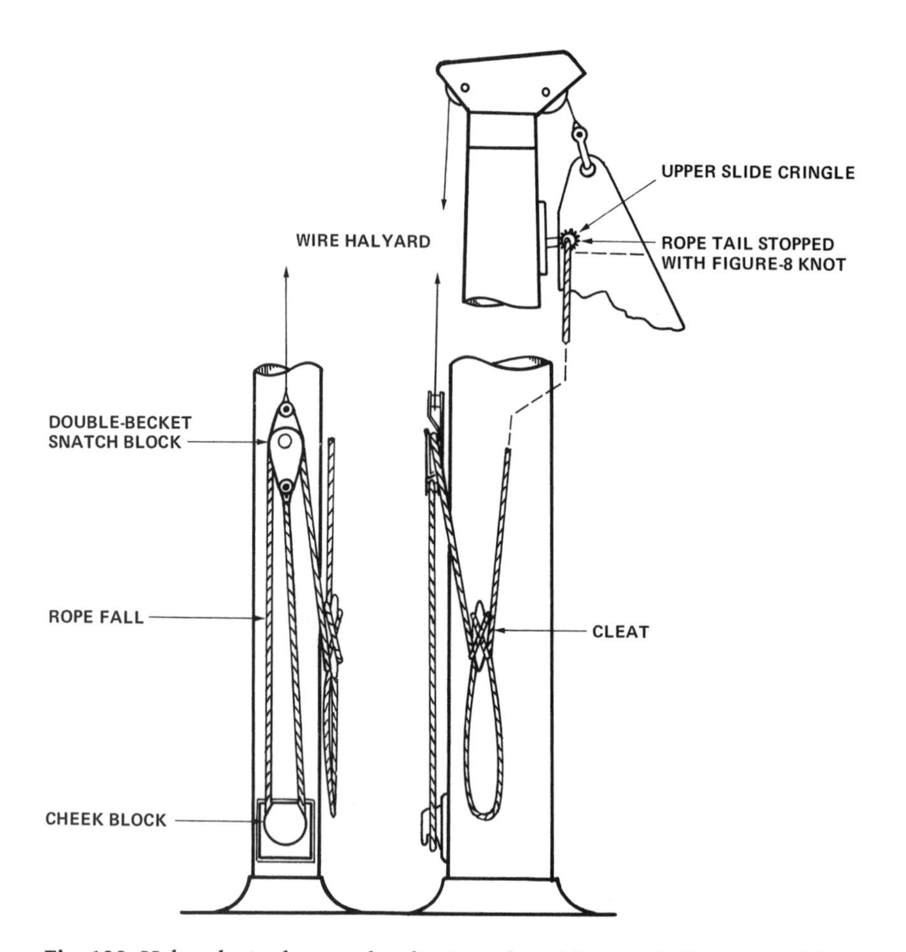

Fig. 139. Halyard winches can be eliminated, and large sails hoisted and lowered faster, with a simple multipart tackle.

Many large racing yachts are equipped with from four to eight pairs of sheet winches, suitably disposed to accommodate quick changes between a wide variety of headsails. But, in cruising yachts, two—or at most three—pairs should suffice. Irving Johnson's 50-foot ketch *Yankee* carried but one pair of Camper-Nicholson winches, where the crank and worm were permanently affixed inside the cockpit and could not be misplaced or lost overboard.

Aside from winches, the racing yacht carries a host of other expensive and fancy gadgetry that has no place on a cruising boat, where speed, perfect sail trim, and split-second timing are not of the essence. With fewer head sails to contend with, deck-mounted snatch blocks can be substituted for jib and genoa tracks. Even these can be eliminated when not in use if they are snapped into flush-deck pad eyes. Paired main and mizzen sheet blocks obviate the need for travelers, and, as Eric Hiscock points out, if these are mounted on a shoulder-height boom horse, the sheets will not chafe on guardrails when broad-reaching or running.

Club-footed jibs and staysails offer great convenience in tacking, and, on W. A. Robinson's *Verua,* sheets for the latter were altogether eliminated by bending their clews to special horses above head height in such a manner that they could slide back and forth on alternate tacks.

Thus, with a little research, thought, and ingenuity, running gear can be reduced to a comfortable minimum, decks can be kept almost clear of obstacles to catch the unwary toe, and, if the yacht is equipped with some means of self-steering, the deck watch can often confine his or her attention to lookout, and to the constant inspection and refurbishing of worn components necessary to ensure against failure in heavy weather.

OUTFIT

This category includes most permanent, nonstructural items, of which we will consider only those of principal importance to seamanship.

Stowage: Yachting advertisements frequently depict a cabin interior as a miniature version of a domestic apartment ashore; the dining table is neatly set, vases of flowers and other loose articles reside on every level area, shelves loaded with bric-a-brac adorn the walls, and there are colorful drapes over spacious "windows" and a heavy shag carpet on the cabin sole. At sea, nothing could be more illusory, and such a cabin would soon look as if a cyclone had struck, leaving everything in hopeless disarray.

Stowage should be viewed as if there were no such directions as up or down, and all objects had jumping legs and were capable of hurtling about like grasshoppers. Every item should have a place designed for it and not made do. While mass-produced yachts are notably deficient in this, in stowage—unlike in hull construction—improvements can be made post facto without much difficulty. All lockers and drawers should be compartmented to accommodate specific items; the former should preferably be equipped with bin fronts to avoid discharging their contents when the doors are open, and the latter with close-fitting overpanels to prevent objects from piling up and jamming the drawer or overflowing into void spaces. Both should have positive, noncorrodible spring latches that will sustain considerable wracking or jarring of the cabinetry without springing open,* and finger-hole pulls are preferable to knobs or handles. Because wood joinery tends always to swell or shrink, depending upon the humidity, the best drawers—and hatch lids—are made of hardwood, and have dovetailed corners, although these are rarely encountered nowadays because of the extra expense. Mechanical drawer slides are inferior to waxed hardwood mortise slides, and the latter can always be kept free-sliding if fitted with waxed filler strips (fig. 140) that can be planed down to the exact thickness for smooth action without side play if swelling occurs.

Objects that are preferably kept on shelves, such as books, sextant, chronometer, etc., should be provided with shock-cord retainers, or have their cases clamped down securely. Hanging objects, such as fire extinguishers and oil lamps, should similarly be secured, and the latter are better off screw-capped and stowed in rough seas.

Solid dinghies are best nested *upside down* in special cradles atop the cabin roof. Permanent eyebolts should be provided for multiple cross-lashing. Oars should

*In the 1970 Sydney-Hobart race, the 43-foot cutter *Rumrunner* was rolled some 140°. She came up all standing, to find the contents of several port lockers were now within the icebox and other starboard lockers, a circumstance which required that their respective doors flew open—and again shut—at precisely the right instant!

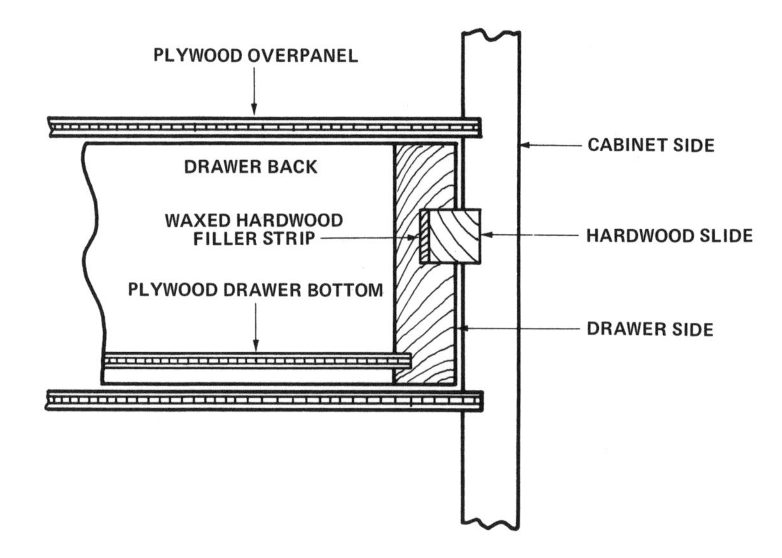

Fig. 140. Drawers will always slide easily if fitted with a mortise slide and a waxed hardwood filler strip. Close overpanels will prevent jamming or displacement of articles within.

have leathers and nonremovable ring rowlocks, and be lashed to the dinghy thwarts. From this position, launching and recovery are much facilitated by a rubber-covered roller set between two close guardrail stanchions. Inflatable dinghies are now rapidly replacing those of wood or fiberglass; on smaller boats, the former are best stowed in a cockpit locker for ready access.

All movable deck gear should have permanent lashings or attachment fixtures. Spinnaker booms are conventionally clipped to deck brackets, but an alternate arrangement is to lash them along the rail stanchions, where they provide extra protection against falling overboard. Similarly, a spare anchor is better off stowed on deck in special brackets made for this purpose (fig. 141).

In general, all interior joinery and furnishings should be looked at from the standpoints of convenience and hazard under adverse circumstances. Close your eyes and imagine yourself in a raging storm; ask yourself, "Where will I stand if heeled 30°? What will I hang on to? What might I hit and injure myself? Can I find any needed item blindfolded under these conditions?" Such questions will often indicate the desirability of rounding a sharp corner, providing an extra handgrip, or securing a flashlight in a suitable bracket.

Engine: To European eyes, a sailing yacht engine, like a telephone, is an object of great suspicion, to be used only when all else fails. But, for lack of adequate auxiliary power, many destinations are proscribed by inclement weather, dangerous approaches, or a strong tidal stream, and picking up a man lost overboard in heavy weather may be almost impossible. Lastly, an engine may be absolutely essential for maneuvering under emergency conditions, or for providing electric power for pumps, anchor winches, etc.

To be effective under storm conditions, an engine should have adequate power to maintain steerage control even where headway cannot be made upwind. This

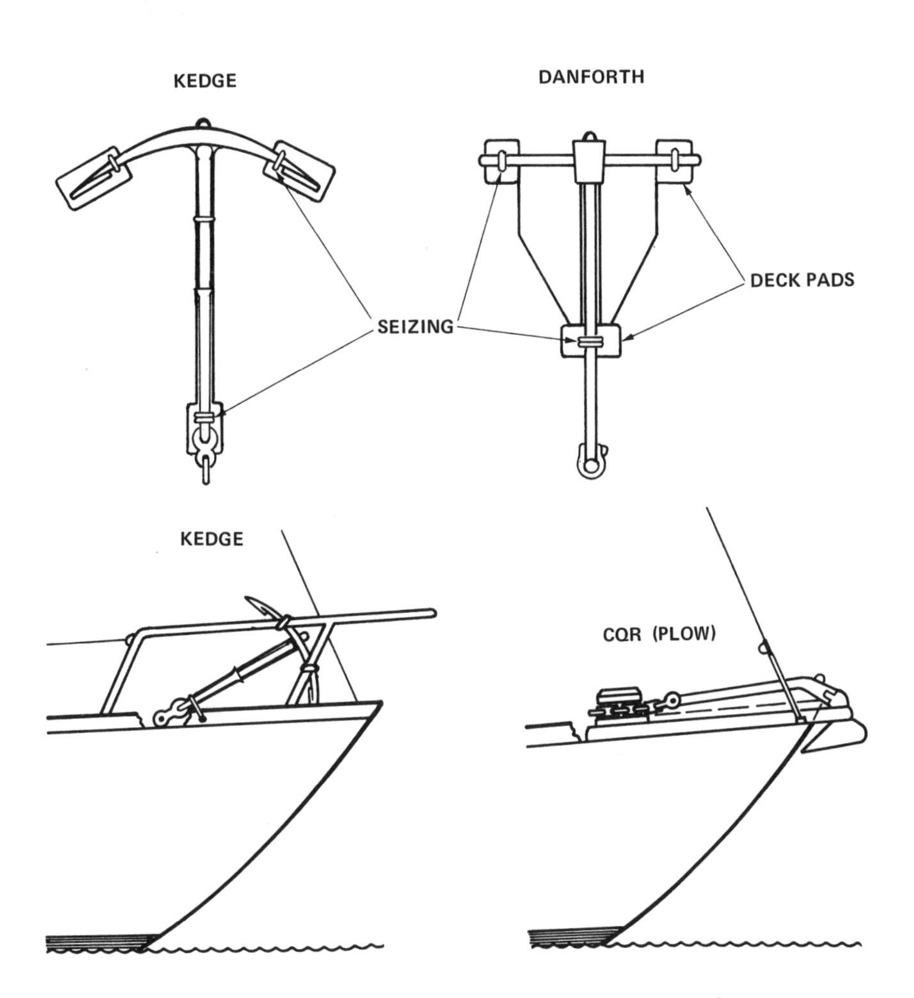

Fig. 141. Four methods of securing heavy anchors in storm conditions.

requirement will normally be satisfied by the power curve in figure 120, provided that the screw is large, slow-turning, and in line behind the rudder.

While there are many pros and cons as to engine installation, from the standpoint of heavy weather operation the following are requirements, in order of priority:

(a) good ventilation under all operating conditions;
(b) good accessibility for operation, maintenance, and removal;
(c) a location as high up in the hull as is consistent with stability, in order to minimize the possibility of loss of power in case of flooding;
(d) a removable sump pan to prevent fuel and oil seepage from running into the bilge.

Because engine weight increases less rapidly than displacement, these requirements are more easily satisfied in larger vessels. Where space permits, the engine is best accorded its own compartment, in which are also grouped such items as batteries, auxiliary power generator, refrigeration compressor, fire-control apparatus, and a small workbench provided with storage for tools and spare parts. The master fuel shutoff valve(s) should be located *outside* of this compartment, and should be readily accessible from the engine operating station.* In boats smaller than 25 feet long, the optimum arrangement may consist of an outboard, operating through a quarter-well in the cockpit, which can be stowed elsewhere when not needed, with the lower end of the well closed by a flush plate held in place with shock cord.

Sufficient spare parts should be carried to effect minor engine repairs, and operators should drill themselves in correcting common malfunctions, such as removing sediment, water, or air from fuel lines, removing moisture from the ignition system, cleaning and adjusting spark plugs, and replacing frayed or broken belts, etc., until they can perform these operations in total darkness, aided only by an emergency flashlight always kept handy for this specific purpose.

If a yacht has an auxiliary power generator, the same precautions apply, for this device may prove more essential in emergencies than the engine itself; a balky engine may refuse to start without protracted cranking,** which puts a heavy load on the batteries at a time when they may also be called upon to supply power for lighting and bilge pumps.

Bilge pumps: Too often taken for granted as to type, size, and functionability, bilge pumps should be regarded, not as a devices for ridding the bilge of the normal, slow accumulation of leakage from shaft glands, sea cocks, and the like, but rather as emergency means of removing massive amounts of water, such as might accumulate very suddenly in a storm by flooding through an open companionway, a burst port, or a hull perforation from the stump of a mast gone overboard and tangled in the rigging.

While opinions differ somewhat, most authorities would grant that a seagoing vessel should be equipped with a minimum of one high-capacity, engine- or auxiliary-driven rotary pump capable of voiding a full bilge in about 4 hours, and a second, manual pump of the largest capacity that can be operated by one man spelled at hourly intervals. For the powered pump, the required capacity figures out to roughly 1 gpm/ton displacement. Figure 142 shows the corresponding power requirements, as a function of discharge rate (or displacement), for a Jabsco model 10BBM pump with 1¼-inch ports, suitable for yachts of up to 30 tons of displacement. Although direct belt–driven pumps require less power, an integral electric pump is more reliable, since it can be driven by the engine, an auxiliary power plant, or batteries. Because of its high output pressure (46 psi), if fitted with an auxiliary outlet valve, hose, and nozzle, such a pump can also be used for fire control, for washdown, or for jetting loose sediment from under a stranded hull.

Detailed studies of protracted manual effort indicate that the greatest power with least fatigue can be delivered by a man standing erect and operating a vertical reciprocating handle, in which position he can produce about 0.2 HP for as long as 6 hours.*** Manual diaphragm pumps of this configuration, suitable for optimum cockpit mounting, can deliver as much as 1,500 gph at this work rate. Such pumps have intakes up to 4 inches in diameter, and will pass solid objects nearly this size

*The same applies to fuel shutoff valves for stoves and heaters; too often they are located behind these appliances, where they cannot be reached in case of fire.

**With automobiles, hand-cranking went out with the Model T, but it is still a useful provision in boats.

***A teeter-totter treadle pump is even less tiring, but one has not been developed for marine use.

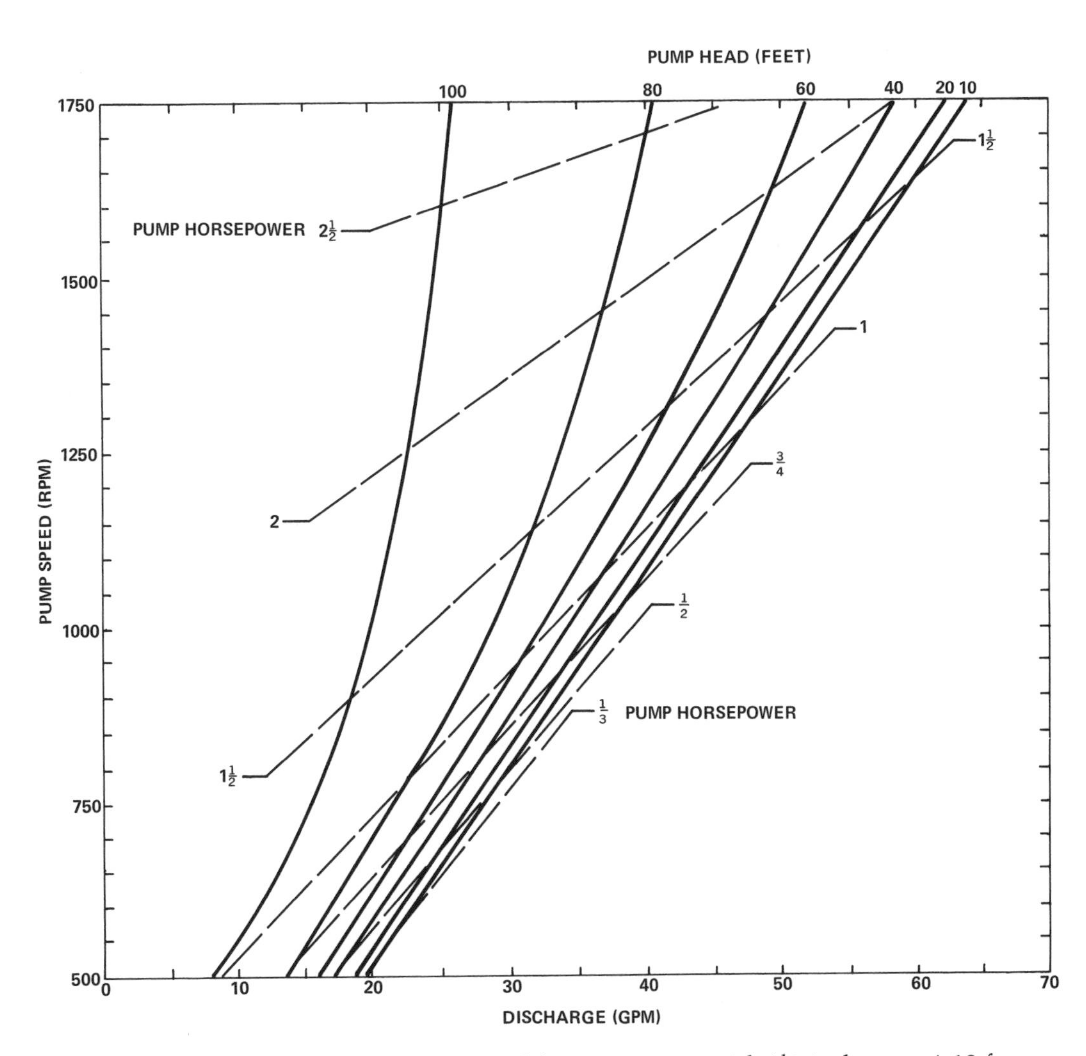

Fig. 142. Performance curves for a Jabsco model 10BBM pump, with 1¼-inch ports. A 10-foot head curve is sufficient for most small boats. Power should be adequate to maintain discharge of at least 1 gpm per ton displacement.

without clogging. Smaller, piston-type hand pumps are vastly inferior, but are useful for reaching odd corners. Lastly, a 5-gallon bailing bucket has saved many a boat where all else failed.*

Fixed-pump intakes should originate in a strum box at the lowest point of the bilge. The screening thereof should equal at least one hundred times the pump intake area, be stiff enough to resist the greatest pump suction when clogged, and have apertures slightly smaller than the maximum dimensions of objects that can be passed by the pump. Otherwise the screen will be susceptible to frequent clogging. Even so, the strum box should be readily accessible, in case a bit of waste or some article of clothing should get wrapped about it.

*In 1956, the 28-foot ketch *Marco Polo* was flooded to waist level, after being rolled over in a storm off Port Elizabeth, S.A. Her coskippers, Tig Loe and Tony Amit, first stuffed a mattress into a broken port, and then bucket-bailed for 24 hours without letup, until the weather eased. With mizzenmast gone, and the main dangling a single stay, they could not sail, but finally got their engine started and powered into the harbor, beached the boat, and collapsed from exhaustion on the sand.

While pumps are designed to scavenge the bilge, too often there is none to scavenge. The bilge is properly a void space where considerable water can collect without seriously affecting a vessel's operation. Modern planing hulls have precious little bilge to start with, and—particularly in smaller craft—the space beneath the cabin sole is often so crammed with miscellanea that a few gallons of water will flood above it. This common situation is rightly attributable to improper stowage; at least the lower parts of the bilge should be kept clear and made readily accessible through lift sections, not only to provide room for water to collect, but also for purposes of inspection and cleaning.

Although often neglected or insufficient, adequate limbering of keel floors in uncompartmented vessels is vital to proper drainage. Limber holes are more easily kept clear if a piece of chain is rove through them, such that it can be pulled back and forth to dislodge accumulations of debris. Above all, the bilge should be ventilated, for there is nothing so contributory to odors, corrosion, and rot as a wet and stagnant void, in which accumulated fuel vapor also poses a constant threat of fire or explosion.

Winches and Bitts: To my mind, the greatest laborsaving convenience aboard boats over 20 feet—particularly for weaker crew members—is a suitable anchor winch. This will be especially appreciated if you have been blown out of poor holding ground by shifting winds several times in a single night, as often happens in island waters. Secondly, a sturdy deck winch serves many purposes other than heaving up the anchor: it can be used to kedge off a grounded vessel, to careen her for bottom work, or to lift heavy objects—such as an engine from its bed.

For anchors up to 20 pounds, a ratchet-and-pawl snubbing winch will suffice. Releasing a pawl under tension is dangerous, and it should always be backed by a stopper (chain), or by belaying to a king post or anchor bitt (rope). Anchors from 20 to 40 pounds can be handled by a mechanical lever winch, but an electric winch is preferable for larger sizes. For combination rope-and-chain rodes, the winch should have a snubbing gypsy for hauling, and a lever-driven chain wildcat for breaking out and heaving the anchor. A high-pressure pump, hose, and nozzle are a great convenience in removing mud and slime as the anchor comes in.

In this connection, the disappearance of adequate chocks and bitts seems another sacrifice to modern design. A deep-throated roller chock is a must for chain; synthetic lines should never be led through metal chocks, unless protected by a split rubber hose, or by chafing gear; and a cleat is a poor substitute for a bitt, which, like the shrouds, should be strong enough to lift the boat out of the water. How many cleats have you seen that could pass this test?

Electrical: A vessel's electrical system should be completely moistureproof and capable of functioning even if submerged in salt water. Unfortunately, this is seldom the case in modern yachts, although the additional cost of waterproofing amounts to an insignificant fraction of total outlay. Ideally, electrical leads should be made up from round, nonhosing (solid-molded) neoprene or PVC marine cable having stranded conductors of twice the expected current capacity. All cables should pass through watertight packing glands into sealed junction boxes. Leads should be screw-fastened to insulating terminal strips, and should be color-coded or labeled for identification. The major electrical components (batteries, generators, voltage regulator, and main junction panel) should be grouped as closely together as possible, in order to minimize resistance losses in the main power cabling, although the master switch panel may be removed to a protected location within the main cabin, convenient to the helmsman. The latter panel should have a complete wiring diagram inside its watertight cover, and be provided with independent emergency illumination. All switches should be labeled, and the master switch should have an accidental-trip cover and a manual override.

Lightning is an ever-present risk in storm country and, aside from human hazard, can cause severe damage to a vessel that is not properly protected. Fortunately, the ground rules are fairly simple:

1. A vessel's seagoing (DC) power system should be left floating (ungrounded). Otherwise, a lightning bolt can jump right across open switches and melt a lot of costly electronic gear to puddles.
2. There should be a continuous, heavy-current conductive path from each masthead to the water. Aluminum masts fulfill this purpose, if jumpered with No. 8 stranded copper braid to an exposed metal keel, or to a square foot of copper ground plate set flush into the hull (a chain dangling overside is inadequate).
3. All other large metal components should be individually grounded to a central No. 8 copper bus that returns to the common ground. These include stays, chain plates, engines, generators, fuel tanks, pulpits, and steering column.
4. If steering with a metal wheel, wear rubber gloves. Otherwise keep clear of metal fixtures until the danger passes.

Even with these precautions, a direct strike may fuse your masthead light and wind vane, or weld your head sheave(s) and axle to the mast cap; of course, you will have spares in the bos'n's locker.

Battery requirements vary widely, depending upon appliance load, but, for engine-starting purposes alone, capacity should be at least one ampere-hour per horsepower of a gasoline auxiliary, and twice that for a diesel. No-load battery voltage is a good index of charge level, if corrected for temperature. Thus, if your battery compartment is equipped with a thermometer and a panel voltmeter, you can use figure 143 to calibrate the latter to indicate charge as a function of temperature. This will save you a great deal of messing with a hydrometer, although one should be used occasionally to check the calibration. In the figure, the slanting line shows the normal temperature (77°F) relation between specific gravity and no-load terminal voltage *per cell,* for heavy-duty, marine, lead-acid batteries. The changes in specific gravity for 30°(F)-increments of temperature and ½-inch increments of electrolyte level are indicated on the small correction scale, and should be added or subtracted, as indicated, to correct any reading to normal conditions. Except for extreme conditions, these corrections are small, and the lower scale in this figure will give you a good idea of average charge level.

Battery life can be considerably extended by observing three charging rules:

1. Keep the electrolyte within ¼ inch of its proper level. Never add water during charging—gas evolution may cause it to bubble out and run all over.
2. Never charge at an input voltage greater than 2.4 volts/cell.
3. Never allow battery temperature to exceed 125°F.

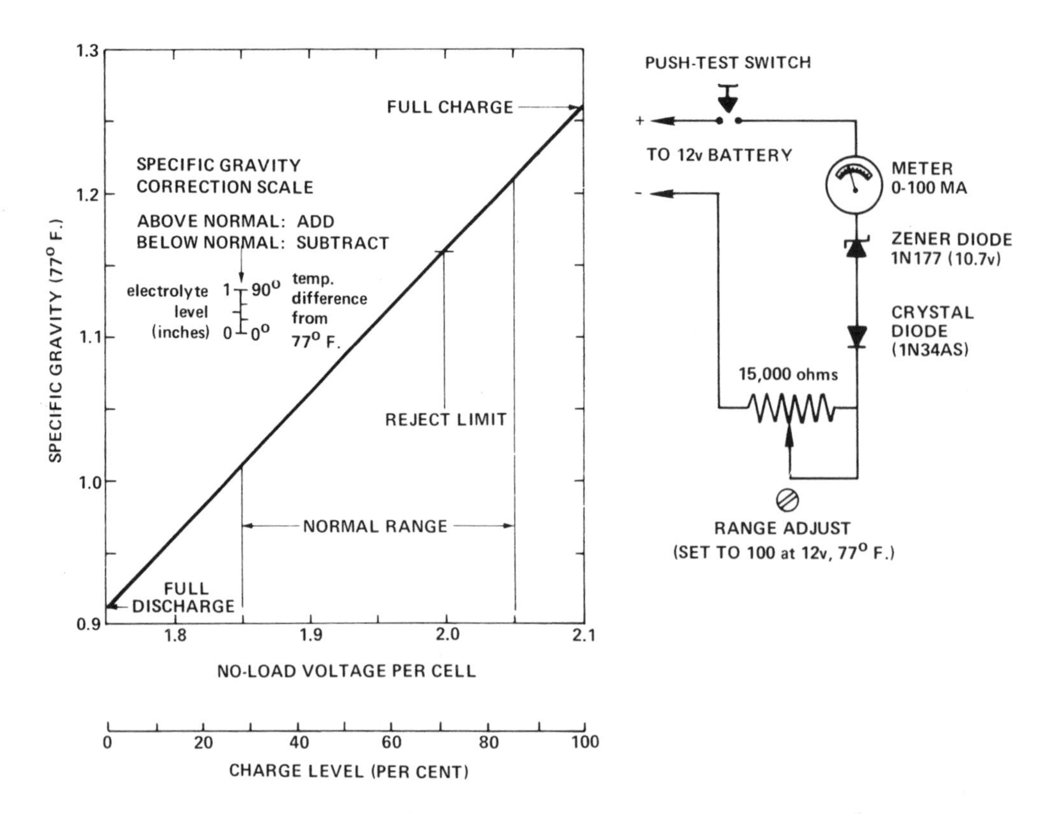

Fig. 143. Nomograph for interpreting no-load battery voltage (per cell) as an index of charge level (specific gravity). In circuit at right, 100-ma meter can be read directly as charge level (%) and then corrected for temperature and/or electrolyte level by scale on graph.

If the latter condition cannot be avoided, an ammeter should be installed, and the charge rate limited to 5 amperes.

All lead-acid batteries degrade with time—particularly if allowed to stand in the discharged state. The test of this is inability to accept a full charge, i.e., failure to reach full-charge voltage or hydrometer density when charged to rated capacity. When a battery cannot be charged to an open-circuit voltage of 2.0 volts per cell (specific gravity = 1.160 at 77°F), it should be replaced. Thus, it is advisable to carry a spare dry-charge battery when sailing in remote areas, if you do not have a standby to switch to.

Although it is not commonly realized, the internal resistance of seawater is quite high (about 400 ohms/in.), and, if provided with water-trap cell caps, the short-term performance of an ordinary storage battery is relatively unaffected by complete immersion, although it will discharge itself in two or three weeks under these conditions. Accordingly, such batteries can be placed low in the bilge to take advantage of their ballast weight without sacrifice of safety. However, they will discharge even faster with a damp salt incrustation between terminals; these should be covered to prevent accidental shorts, and coated with silicone grease.* Wherever

*Storage batteries are quite complicated electrochemical devices, and are too often abused or neglected—possibly owing to the difficulty of obtaining technical information on their operating characteristics. For a very good pamphlet, *The Storage Battery, Lead-Acid Types,* write to Exide Power Systems Division of ESB, Inc., P.O. Box 5723-T, Philadelphia, PA 19120.

located, storage batteries should be provided with separate hold-down clamps securely bolted to the hull structure. Allowing a 1-inch space between batteries will do a lot to prevent overheating during the charge cycle.

Safety and Comfort: One of the greatest hazards in rough weather is crew fatigue, which can only be combated by proper rest in a comfortable bunk. Despite repeated exhortations by experienced sailors, most modern yachts still have bunks in the form of plywood bins containing a foam mattress, sometimes supplemented by a rigid bunk board of just the right height to crack knees and elbows as one rolls violently about in a space much too wide. Most authorities agree that the most satisfactory bunk at sea consists of a 22-inch canvas or webbed pipe berth that is fitted with a 2-inch foam pad, and which (in sailboats) can be adjusted to tilt at any angle up to 20° from horizontal. The berth should have a similar overlay fastened along its inboard edge that can be secured to the overhead in rough weather, thus encapsulating the sleeper in a hammocklike cocoon (fig. 144).

Further protection against severe accelerations can be obtained by an aircraft-type safety belt at hip level, secured *beneath* the berth, and threaded between pipe and webbing. With this arrangement, a boat can be rolled clear over without throwing a sleeper across the cabin. Although the mattress pad should be contained in a zippered waterproof cover, the latter is best if fabric-coated on the outside to reduce perspiration in warm weather. Lastly, if fitted with elastic hook cords for securing berth, pad, and bedding together, the whole can be folded up out of the way when not in use, thus greatly increasing living space in smaller boats.

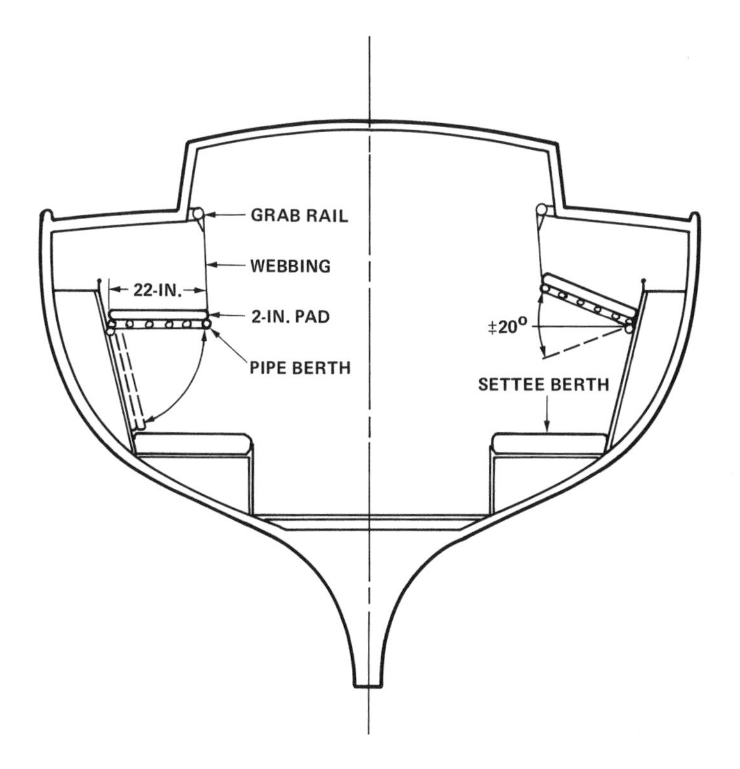

Fig. 144. Pipe berth installation. Although not necessarily replacements for settee berths, they are welcome alternates in rough weather.

Webbed body harnesses, capable of attachment to a suitable safety line or fixture to prevent falling or being swept overboard, are indispensable items at sea, and are now required in many ocean racing events. Unfortunately, some commercial models are poorly designed; they lack adjustable crotch straps and have the safety line secured at waist level, such that if one *does* fall overboard, he or she may be dragged along feet-first, or the belt will ride up to compress the diaphragm or chest and inhibit breathing. Parachute-type harnesses are constructed to avoid these difficulties, and can double as a bos'n chair for work aloft, or be clipped to suitable attachments at the wheel—or in the galley—to provide restraint against ship motion and greatly relieve fatigue. Because this type of harness requires individual adjustment, a separate harness should be provided for each person, and it should be worn even in calm weather when alone on deck; in rough seas, the harness is best supplemented with a flotation vest—just in case.

The guardrails or lifelines on many smaller yachts are inadequate to resist the impact of a human body hurtling against them at 30 ft./sec.—a likely situation in storm seas. Especially on mass-produced boats, rail stanchions are often simply screwed to the deck, whereas they should always be through-bolted to a doubler plate beneath. Welded tube rails are much preferable to those of wire rope because they are not only stronger, but also less apt to cause injury. The difference in total wind drag between 1-in. stainless steel tube and 3⁄8-inch plastic-covered wire rope is negligible. Considering that many yachtsmen will not hesitate to spend $1,500 for a titanium sheet winch,* the added cost of a welded tube rail seems trivial. Lastly, most life rails are too low—26 inches is a minimum, and a 30-inch rail will give you something to lean against, instead of trip over.

Although well-gimbaled stoves can be used under fairly rough sea conditions, there is no reason for not having hot meals no matter how bad the weather. Nowadays there are a variety of self-contained, thermoelectric heating (and cooling) devices that operate on either 12 or 120 volts, and any liquid or preprepared food can be heated in and consumed directly from its container with little effort and minimum risk of spillage. After a few days of such convenience, you may wonder why plates, cups, and saucers were ever necessary.

Foul-weather clothing is a matter of much controversy, and manufacturers keep coming up with new models of the same frowsy, baggy, and cumbersome garments, with the singular preoccupation that the patient be kept warm, if not dry. After thirty years of scuba diving, to me, there is no comparable equivalent for warmth, freedom, and comfort in cold, wet weather than a well-fitted wet suit. In a suit with zippered arms and legs, and complete with helmet, booties, and gloves—and even a face plate, if things are really bad—one can stand a four-hour stint at the wheel in near-freezing weather without undue discomfort. The suit affords considerable protection from bruising, and, if swept overboard, one could not be better equipped for survival. Wet suits are available in several thicknesses, depending upon exposure temperatures, and, if uncomfortably warm, one can shed all but the vest and still have adequate flotation (see Part VIII).

*In America today, the average cost of quantity-produced machine assemblies is about $1–$2/lb. This applies to automobiles, most household appliances, and even large structures, such as Ferris wheels and bridges. On this scale, most marine hardware is grossly overpriced. You can buy 500 Swiss watch movements or a 50-HP marine diesel engine for the price of a pair of titanium winches.

EMERGENCY EQUIPMENT

In addition to the usual list of fire, rescue, and signaling devices required by law, or recommended in most boating handbooks, a few extra precautions will add much to safety in heavy weather.

Lighting: Electric power failures are common at sea, and are most apt to occur in heavy going, when electricity is most needed. Therefore, it is advisable to carry two or three portable long-life electric lamps for general illumination in the cabin or on deck. Special, waterproof, 100-hour lamps, equipped with color visors for duty as running lights, are obtainable from most marine suppliers, together with separate mounting brackets that can be suitably disposed. A similar searchlight and signal lamp with a shorter duty life is a valuable asset for locating or identifying shoals, rocks, and buoys, and for rescue and towing operations. The waterproof variety can be lowered or carried underwater to assist in unfouling anchors, moorings, or propellers. Of course, ample spare lamp bulbs and batteries should be carried, the latter preferably of the alkaline type, which have longer lives than carbon-zinc cells, and do not deteriorate in warm weather.

Because of battery limitations, it is a matter of common experience that smaller vessels cannot maintain a continuously operating masthead light of sufficient brilliance to ensure detection at ranges great enough to avert nighttime collisions with large, high-speed ships. Although it is still an unresolved legal question, high-intensity, stroboscopic, flashing lamps do not seem to be proscribed by current international regulations. Such lamps draw only 1–2 watts of power on average, and can be seen with the naked eye at horizon distance (10–14 miles) from the bridge deck of a large ship. In the United States, small aircraft have almost universally adopted strobe lights as an optional collision-avoidance measure.*

Radar: Most large vessels maintain a radar watch, particularly at night or during poor daylight visibility. Because small boats present a low radar silhouette and, unless metal-hulled, have low signal reflectivity, collision risk is high. It may be considerably mitigated by carrying an adequate radar reflector at the highest practical point above deck level. Folding, metal reflectors are obtainable on the surplus market or from marine supply houses for only a few dollars, and, at low sea states, give a good return at ranges up to five miles, depending upon target elevation, which should be at least as high as the highest waves.

Communications: Aside from normal (and often abused) communication purposes, a vessel's radio may be its principal distress or emergency reporting device. Federal regulations require a 25-watt VHF transceiver as a primary communication link, but this is suitable only for coastal traffic, and can seldom be used at ranges

*Visibility notwithstanding, large ships' disregard of (or failure to note) distress signals seems to be a frequent occurrence. Nearing the end of his solo globe-girdling voyage, Robin Knox-Johnston was stricken with what he felt to be appendicitis. Although in the midst of a north Atlantic shipping lane, he was unable to attract the attention of a number of passing vessels. Some within several hundred yards failed to heed radio signals, red flares, loud hailing, or even rifle shots. Unless the neglect was intentional, it is apparent that these vessels were not maintaining adequate lookout, nor monitoring radio distress channels, both of which are required by law.

greater than 50 miles. At greater distances, and because of heavy traffic on VHF ship communication channels, a 100- to 150-watt radiotelephone is essential. Modern solid-state circuitry has greatly reduced the size and power requirements of these units, which can now be accommodated on any yacht large enough for offshore cruising.

Flash-lamp signaling is standard on military and commercial vessels, but has become almost a lost art among small-boat operators. It is a valuable asset, and well worth practicing on long voyages.

Survival Gear: All offshore racing yachts and most cruising sailors now carry emergency overboard survival gear, usually consisting of a marker pole, attached by a lanyard to a horseshoe buoy, which is in turn attached to a sea anchor and a self-actuated, flashing strobe light. About 14 percent of boats now also carry an overboard Emergency Position Indicating Radio Beacon (EPIRB). Often the light and the EPIRB are intended to be thrown over independently, so that they tend with time to drift apart, leading to some confusion in any search effort. I have written to manufacturers of these devices, suggesting that the light, beacon, and pole, at least, be incorporated in a common unit, with the antenna and light bulb atop the pole, and hence not occluded by waves from low-level sea searches. They have all replied negatively, stating that the cost would be too high to justify the purchase of such devices unless they were required equipment.

Meanwhile, I suggest that anyone can prepare a survival packet, to be attached to the buoy, consisting of a net bag containing:

(1) a 5 watt hand-held VHF transceiver in a waterproof baggy, with a rubber-band seal around the antenna, for communication with the master unit aboard your boat (you can see it much farther than it can see you);
(2) a survival suit, of which there are several commercial versions;
(3) a 1-qt. plastic container of fresh water;
(4) a 6-oz. packet of concentrated foodstuff in 1-oz. waterproof wrappers;
(5) a tube of opaque sun cream;
(6) a signal mirror;
(7) a packet of dye marker.

So equipped, a crew member overboard can survive in relative comfort for several days, giving ample time for search if the VHF unit fails.

Spares and Repairs: No rational person would think of putting to sea without a first aid kit to guard against sickness or injury. Similarly, every boat should have its own repair kit. As a minimum, the dictum at sea is to carry enough tools and spare parts to make jury repairs that will enable one to work the boat to port, where further repairs can be effected. The more distant the waters sailed, the more complete the inventory should be, and, in remote places, one should be equipped to make major repairs to most structural components. If an essential item cannot be repaired or easily replaced, two should be carried; two chronometers, two radios, and even two sextants are good insurance, although in a pinch one can take reasonable altitudes with a *three*-arm protractor.

Some really remarkable things have been accomplished by cruising sailors of the past, working under difficult circumstances and with limited resources.* But one needs a few essentials, among which are the following:

1. An assortment of hardwood timbers and planks for shoring up the interior of a stove hull, splinting a spar, replacing or battening a rudder, splicing ribs or deck beams, etc.
2. Enough wire rope of the most common size to rerig a mast, together with necessary thimbles, toggles, and a Nicopress tool and sleeves, or U-clamps.
3. Complete running gear replacements, a whole coil of small stuff, a dozen balls of tarred marlin, a Band-it tool, 500 ft. of ½-in. stainless steel, Band-it strapping, and a gross of clips.
4. A dozen tubes of silicone rubber for plugging holes and seams. It sticks to wet surfaces and will set up under water.
5. A dozen rolls of Scotch-brand No. 33 plastic tape. It can be applied under water and never slips, unless exposed to gasoline or oil.
6. A really husky vise that can be clamped to a cockpit coaming or set up ashore for heavy bending or pounding.
7. A half-dozen assorted spare anchors stowed as inside ballast; these are the best insurance you can buy.
8. About 400 feet of heavy nylon rope for trailing in heavy weather, although you will find a dozen other uses for it.
9. Replacement hose stock of every size employed aboard, and a box of suitable clamps. Replacement gaskets for portholes.
10. Precut plywood hurricane ports for all cabin openings. These should be predrilled and provided with drive screws (that can be driven with a hammer and removed with a screwdriver).
11. With some hardware cloth, galvanized staples, and a sack of Portland plastic cement, you can make a rigid concrete patch over really large holes in a wooden hull with very little effort (and sand is available most everywhere).
12. A small, heavy-duty, zig-zag sewing machine—even if hand operated—will save you hundreds of hours of hand stitching.
13. A *fothering blanket:* a piece of heavy canvas padded with soft foam, and fitted with corner grommets and pennants for hauling it into place over a hull leak.
14. Wooden drive plugs for every through-hull fitting.

We could continue this list *ad infinitum,* but, if you are contemplating a long cruise, take an inventory from someone who has just come back from one, and outfit yourself accordingly. Better yet, read the accounts of cruising yachtsmen. Many authors—particularly Hiscock—have taken the trouble to compile comprehensive

*In 1926, Alain Gerbault's 39-foot cutter *Firecrest* lost her 4-ton lead keel in Uvea Atoll, when she was pounding on the reef under gale conditions in a poor anchorage. While new keel bolts were provided by a French ship sent to his aid, the real test of ingenuity lay in lightering and manhandling the keel up to the stranded hull, and sliding it into place on coconut logs. All in all, some four months were required to refloat and rerig her for sea.

lists of equipment and supplies, and have given detailed accounts of their progress, adventures, and misadventures.

PREPARING FOR SEA

In addition to careful preparations, strategy in port also involves advance planning for a voyage—the longer, the more thorough. Planning includes the establishment of a tentative itinerary, and the laying-in of necessary provisions, fuel, and water, with an adequate margin for emergencies, and a look at the early tide and weather outlook. Because weather forecasts cannot be depended upon for more than three days, for longer trips one can only make recourse to the sailing directions and monthly current and weather charts. However, irrespective of one's objectives, it is essential that an itinerary be flexible; that is, specific plans should be laid for one or more alternatives in case of contingency. It is always provident to expect and prepare for the worst.

Secondly, one's itinerary (and alternates) should be filed with some responsible person, with instructions to contact the Coast Guard for appropriate action in case you do not show up on schedule. This information should include a description of the vessel, her name, number, and passenger manifest, and any other identifying features that might assist an aerial search. All of these features are embodied in the standard aircraft flight plan required (commercial) or recommended (private) by the FAA for all controlled airport traffic. Unfortunately, the Coast Guard is not set up to handle equivalent sail plans, which would doubtless benefit search and rescue (SAR) operations. Short of receiving a distress signal, it has no way of keeping track of vessel movements, unless alerted by a third party.

Lastly, in preparing for sea, a thorough inspection and checkout should be conducted to ensure that everything is in serviceable condition and all mechanical and electrical equipment is functioning normally. In this, one can take a tip from the aircraft pilot, who runs through a detailed checklist before *every* flight. A similar list for a small-boat skipper might include:

1. Inspect all through-hull connections, hoses, and fittings for evidence of leakage. Cycle all shutoff cocks for smooth action. Inspect and pump bilges, and remove any waste accumulation. Operate toilets.
2. Open and/or ventilate the engine compartment. Inspect, start, and operate engine, auxiliary generator, and mechanical pumps; refrigeration system; and automatic steering device, if any. Check all fittings for leakage and for normal gauge readings. Check fire control devices to see if they need filling or are outdated. Check and record fuel and water supply levels.
3. Check battery charge and water level. Test all circuit breakers, lights, and other elements of the electric power system.
4. Inspect sails, standing and running rigging, anchor tackle, and lifelines for evidence of wear or chafe. Particular attention should be given to rudder and engine controls.
5. Inspect all emergency equipment for condition, availability, and operating readiness.

6. Check all navigation aids; these include compass, sextant, binoculars, radio direction finder, fathometer, and radios. Do you have the right charts, signal flags, sailing directions, etc.?

Most of these items also require periodic checking at sea, and a suitable list should be posted and watch assignments made to ensure that they are carried out. If the list has a transparent plastic overlay, it can be annotated, signed, and dated with grease pencil.

Much of the above may seem like nit-picking. There is a great temptation in day- or weekend-sailing just to jump aboard, hoist sail or fire up the engine, and shove off. No airplane pilot would think of doing this—and no seaman should!

26

Tactics

EDUCATION AND PRACTICE

Tactics comprise the operational phase of seamanship, and include all things that go into implementing strategy. While strategy begins in port, from the time the lines are cast off until they are made fast again, strategy and tactics go hand in hand. Because sea conditions can change so rapidly, one's preliminary sailing plan and its alternates should be regarded only as general guidelines, subject to continuous revision. When racing, of course, there is the additional complication of seeking advantage in speed and position.

While it is often said that seamanship is something that cannot be learned from books, I would hold that this is largely a hangover from the days of general illiteracy—else why should we have them?* But few would contest that tactical skill requires long and frequent practice. No one would field a football team or a racing crew fresh out of the classroom and expect to win, but a lot of time can be saved by a few preliminary skull sessions.

Since, in this chapter, we are dealing with the unusual rather than the ordinary, practice is all the more important. At advanced sea states, decisions must be made and carried out almost instantaneously, if catastrophe is to be avoided. You can learn to make knots, bends, and hitches in your living room, but a practiced seaman can do them blindfolded in a few seconds on a wildly pitching deck, that is being swept sporadically by green water. Here, *blindfolded* is not a jest; at night, under really bad conditions, you might as well be. In the days of sail, the first lessons of an apprentice consisted of learning the function and position of every piece of gear—a list numbering some thousands of items. If you believe yourself a seaman, try to do the ordinary, blindfolded, in moderate weather: hoist and take in a sail; take and shake a reef; go below and bring up any specified object. All this can be done with practice.

Fundamental to acquiring skill through practice is to develop a system and adhere to it. This is best done by trial and error, for boats and individual preferences vary greatly, but we can offer a few general guidelines.

Stowage: In discussing construction, I mentioned that everything should have its place and nothing made-do. This applies to everything from fire extinguishers to your bifocals; know where everything is, and always replace each item. If you sail

*My morning paper tells of two young men rescued after drifting 49 days on their third attempt to reach the South Seas in a converted 5,000-gallon fuel storage tank. This book, at least, might have helped discourage them before starting.

with strangers or guests aboard, the essentials should be pointed out. Items having similar functions should be categorized and stowed together. It often helps to label drawers and lockers under general headings, and to post operating instructions and precautions for such items as engines, fire-control equipment, pumps, and toilets. Sails and sail bags should be conspicuously labeled, and preferably hung from hooks. Special hooks at the four corners of a forward hatch, on which to hang a sail bag, greatly facilitate rapid stowage of foresails, which need only be bundled and dropped through the hatch; after cockpit locker bins can be similarly arranged for the main and mizzen (fig. 145). If things are properly arranged, and suitably stowed, you will be able to lay hands on anything without having to think. This is the essence of a tidy and shipshape vessel.

Cruising yachtsmen often develop a host of ingenious laborsaving devices. On boarding the 28-foot, round-the-world ketch *Marco Polo* at Takaroa Atoll, I inquired about dozens of 3-foot lengths of small stuff bent every few feet to the lifelines. Her skippers remarked that in rough weather it was a great convenience to always have a bit of lashing at hand, no matter where one was on deck.

Know Your Boat: Every boat—including quantity-produced one-designs—has its own peculiarities, and handles somewhat differently. Although it is not commonly realized, most boats are right- or left-handed; that is, under otherwise identical conditions, they sail slightly faster on a particular tack. With an off-center screw, this is quite apparent, but small hull, keel, or rudder asymmetries, or a slight rigging eccentricity, can bestow a tendency to yaw or make leeway that requires compensating trim and produces additional drag, such that speed differs by a tenth of a knot or so on alternate headings. These slight differences can only be learned by experiment,

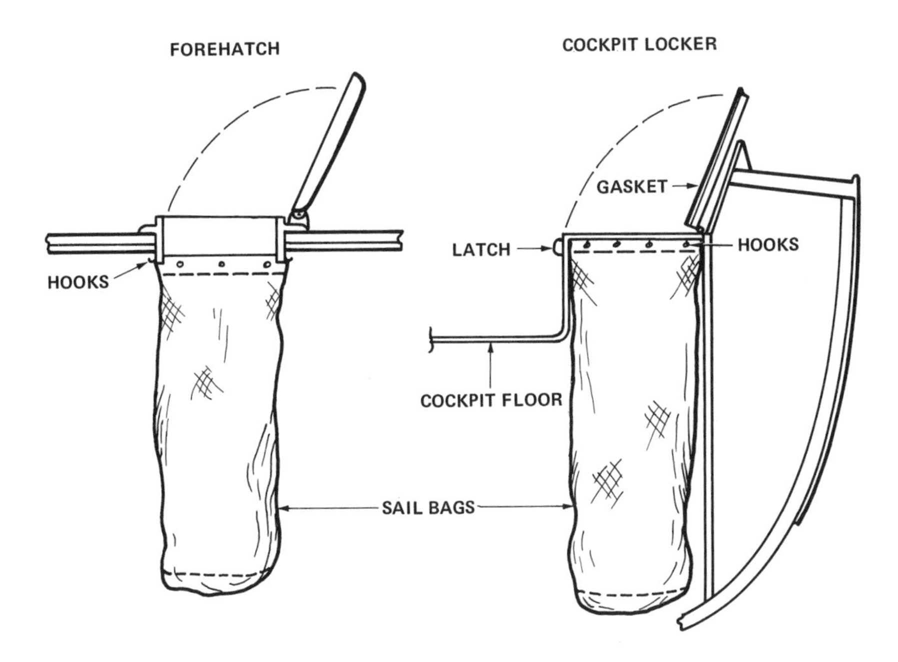

Fig. 145. Sail stowage in rough weather is much facilitated by hanging sail bags from hooks in forehatch or cockpit locker. Lid for either must be watertight.

but a preferred tack may be a deciding factor in a close race, or may save you a day or two on a long cruise.*

An important factor in handling larger vessels is a table of times and stopping distances for standard power settings, and the turning radii for standard speeds, as functions of load displacement. Although perhaps less critical for smaller vessels, these numbers are easily determined, and may be of advantage in competition, or in emergencies.

Safety Habits: Get in the habit of thinking *safety.* Emergencies at sea often develop unexpectedly, and more often in rough weather. Most can be prevented by advance planning and practice. The following may seem obvious, but how many have you mastered?

1. The Coast Guard lists overloading as a primary cause of boating accidents. With full tanks and a rough sea, how many people can you safely accommodate aboard without exceeding your boat's stability limit? The Coast Guard *Boating Guide* gives formulae for calculating safe loads; have you ever bothered to use them?
2. Fire is an ever-present hazard. Even if adequately equipped with current extinguishers, have you ever used one? Start an oil fire in a safe place and try to put it out. You may find it more difficult than you think, particularly in a strong wind.
3. How far can you run your boat upwind in a 30-knot breeze with a specified quantity of fuel? You can answer this by keeping a careful log of engine RPM, fuel consumption, wind, and sea state.
4. With a full battery charge, how long can the engine be cranked without starting? More particularly, when should you stop cranking in order to save power for pumps, lights, radios, etc.?
5. You hit a floating log and begin to make water. With all pumps manned, the cabin sole is awash in 30 minutes. How long can you run toward port before foundering? To answer this, you need to know the bilge capacity below each waterline, the height of your margin line, and the level to which the water can rise before engine and/or pumps cease working.
6. A squall is approaching. As you are taking in sail, your main halyard jams in its sheave half-mast. Now what?**

In succeeding sections, we will discuss a number of emergency and heavy weather tactics, some of which must still be regarded as suggestions. This is all the more reason to practice in moderate weather, since their effective application varies widely, depending upon sea state, relative to the type and size of vessel. A good deal of experimenting may be required to find the best procedure and to carry it out smartly.

*"Hook" Beardsley, longtime Star Champion of the 1930s, was a master at diagnosing a boat's peculiarities. Having won a race, he often offered to trade boats with his closest rival, and would usually win again.

**Hoist the boom with the topping lift, and brail everything to the mast with the halyard fall. It is assumed you have dropped the foresail and started the engine, and are headed upwind when the squall hits.

CLASSIFYING SEA STATE AND PERFORMANCE

One of the interesting things about sea state is that it is scalable; that is, for a given weather situation, as sea state grows, its statistical properties change only as to absolute magnitude and average period. Thus *small, moderate,* and *severe* are only relative terms, depending mainly upon the size of a vessel. For every vessel, there is a limiting sea state in which its average motions will be similar to those of a larger vessel at a correspondingly higher sea state. Thus, for your own boat, it is important to know when a critical sea state is reached—or can be expected—when deciding upon a course of action. Here, "critical" is somewhat subjective, but can usually be taken as that state where most control is lost, the vessel is in jeopardy, and emergency action is indicated. Although part of good seamanship lies in knowing what to do in emergencies, it obviously makes better sense to know your boat's limitations, and to avoid critical conditions wherever possible. This can be done by learning to forecast sea state (Part V), and by a little experimenting with your own boat.

Pick a day when strong offshore winds are forecast for your area with a probability of holding steady for at least twenty-four hours.* This will give you a weather shore, with sea state increasing offshore in accordance with the wind strength and fetch lines on your CSS diagram (fig. 79, p. 189). With some preliminary estimate of critical conditions, you will also know how far out you must go before encountering them, and whether you can plan your departure so as to arrive in daylight. Be sure there is no lee coast offshore, in case you cannot make it back and have to ride it out.

After filing a sail plan, and preferably being accompanied by a larger escort, head out downwind, stopping every few miles to round up and to practice the following tactics:

1. Sail a closed polygon of six or eight legs, centered on the wind direction, and plot a polar speed diagram (fig. 121, p. 290).
2. Practice heaving-to, employing various methods described below under "Heavy Weather Tactics" (p. 359), and note which method provides maximum comfort and safety. Put over a marker staff and drogue, and note drift rate with respect to it.
3. While hove-to, practice abandoning ship, taking all necessary survival equipment. (Leave a crew member aboard with the engine idling, just in case.)
4. Practice recovering a "man" overboard, utilizing a sandbag of body weight, buoyed up with a standard life vest, horseshoe buoy, and marker staff.

Not all of these need be done at every stop; perhaps twice (under moderate and severe conditions) is sufficient for items 3 and 4, and they constitute better drills if the alarm is called unexpectedly.

Continue working your way seaward, repeating these operations at progressively shorter intervals as conditions worsen,** but stop short of the point where you

*This may take some waiting. But, in most continental areas, suitable winds occur at least several times a year. Along a curving coastline, some point can often be selected where the local wind is offshore, and an alternate, crosswind port of refuge is available (fig. 146).

**If sailing, it is important to realize that much more canvas can be carried downwind than up, and sail should be progressively shortened before heading down to your next stop. In fact, it is a general cruising admonition not to carry more sail downwind than your boat can stand up to close-hauled.

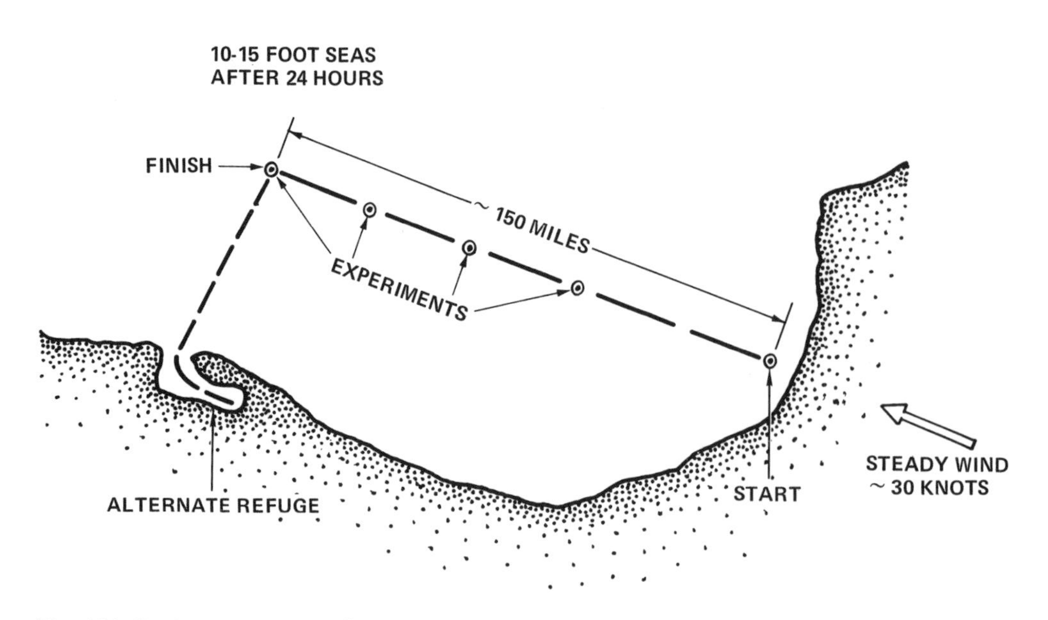

Fig. 146. Optimum situation for practicing rough weather tactics under controlled conditions in a growing fetch.

can no longer make ground to windward under either sail or power. Stop shorter yet if the wind increases. If it is a long way back, you may have to wait until the wind goes down, unless you have wisely selected an alternate refuge on some crosswind heading.

Having successfully completed this exercise, you should have a very good idea of what your boat can stand up to, and how she performs under various conditions—better, perhaps, than might be obtained in several seasons of racing. Being able to quantify performance under known conditions is something that few racers have time for. Secondly, you will be able to look at a weather chart, examine your CSS diagram, and know just what you are up against. Lastly, you will know which emergency procedures work best for your boat at different sea states.

HEAVY WEATHER STRATEGY

As previously noted, *heavy weather* is a relative term that depends upon the type and size of vessel. A large sailing yacht can carry full sail in a wind that would require a smaller one to double-reef; the latter may lie comfortably hove-to under conditions that might capsize a larger power cruiser. However, a seaworthy vessel of any size, if well handled, can ride out all but the worst storms with little damage. This is why we have stressed the importance of knowing how your boat performs at various sea states and what tactics to employ in hazardous situations. For the day sailor or weekend racer, the answer is very simple: if things look bad, don't go out. But anyone who spends much time at sea may get caught out in a weather that will tax the ability to survive.

Avoidance: Unless committed, as in ocean racing, almost without exception the best strategy is storm avoidance. The Coast Guard attributes a large percentage of boating casualties to failure to heed adverse weather signs. Fishermen almost universally prefer the loss of a few days' catch to the risk of loss of equipment or damage to a vessel, and they head for port when the sea is rising and the glass is falling. In the Introduction to Adlard Coles' *Heavy Weather Sailing,* Alasdair Garret sets forth the hallmarks of the seafarer as "humility, prudence, and the recognition that there is no end to learning and to the acquisition of experience." He further defines prudence as "the ability to distinguish between the risk which can reasonably be accepted having regard to prevailing conditions and the risk which must be rejected as unacceptable." After a race in which 30 percent of all entries were disabled, one might well ask whether it was prudence or providence that led to winning.

Avoidance, like racing, requires a little advance planning. Familiarize yourself with local weather patterns. When working up your sailing plan, identify alternate ports of refuge along your route—and points of danger as well. Carry small-scale pilot charts of these places and be sure your sailing directions, tide and current tables, and notices to mariners are up-to-date. Knowing your boat's speed at various sea states and headings, you will then be prepared to rapidly estimate escape courses and sailing times, and to make the most prudent judgment about action.

In this, there is always a time factor. In addition to limiting course and speed options, rising wind and sea state make navigation difficult. In unfamiliar waters, entrance passages may be difficult or impossible to negotiate because of darkness, alteration of tidal currents and depths by wind effects, or heavy wave breaking in anomalous places. Thus, a decision to seek shelter must be made early enough to avoid these hazards.

Even given time, running for port may be unwise, unless a harbor provides sure protection against the elements. Major storms sometimes convert seemingly safe refuges into death traps, and few are immune to hurricanes. Given a stout vessel, confidence in one's own ability, and any serious doubt about making a safe refuge, a prudent skipper may elect to remain at sea.* Again, if the decision is made early enough, there may be several avoidance tactics open. A weather shore offers considerable protection from high seas because of the limited fetch, but thought should be given to possible changes of wind direction. With shifting winds, working around the lee of an (isolated) island is a safe procedure—even in a hurricane—provided that your engine is powerful enough to stem wind and currents. Off a lee coast, one should head out in time to provide plenty of sea room for drifting when headway is no longer possible. In the open sea, if a storm's progress can be gauged, an early decision may make it possible to lay a course that will put you as far out as possible in its navigable sector, where conditions will be more moderate.

In addition to advance planning, all of these avoidance measures require judgment, as well as prudence. These, in turn, rest on experience; and experience is preferably gained by practice, rather than by misadventure.

Storm Preparations: Assuming you elect to remain at sea, and have made adequate preparations in port, stormproofing your boat reduces to buttoning up, securing any loose objects, and readying safety and emergency equipment. To what extent somewhat depends upon the anticipated storm duration and intensity, but in any

*Private power yachts are generally better off in port. Few have adequate fuel reserves, stability, and watertight integrity to withstand storms at sea, and losing one's boat in a poor anchorage is preferable to the risk of foundering.

case it is better done early than late. Remember that, in heavy weather, visibility and communication may be poor or nil; it may be difficult or impossible to move about on deck; the simplest functions will take great effort, and any forethought that will simplify subsequent shiphandling will prove a blessing. As on all large ships, it is a good idea to have a predetermined action plan that includes a checklist grouped into categories that are assigned to various crew members. The checklist will differ, depending on conditions and the type and size of the vessel, but may include any or all of the following:

1. Watertight Integrity

(a) Close, secure, and/or plug *all* exterior hull and cabin openings, except the main companionway hatch and the engine compartment ventilators.*

(b) Remove ventilator funnels, unless you are sure they will survive green water impact.

(c) Open cockpit drains and check for plugging.

(d) Board up all frangible ports and port lights, and reinforce companionway doors if their impact strength is in doubt.

(e) Inspect bilges and bilge pump screens; remove any accumulated waste. Check operation of all pumps; void bilges, toilets, and holding tanks. A great virtue of the latter is that toilets can be used in extreme weather without risk of back flooding.

2. Power and Control

(a) Start the engine and electric power plant, and run through your operating checklist for all mechanical and electrical systems. If you elect to shut down the engine, keep the power plant running, so that batteries remain fully charged. In rough weather sailing, it may be better to keep the engine ticking over, since an occasional power assist may greatly facilitate maneuvering and sail handling.

(b) Check rudder controls and linkage. Ready your emergency tiller, and have warps and drags coiled for streaming if emergency speed or directional control becomes necessary.

(c) Ready storm sails, sheets, and tackle in some accessible location. (Storm jib should have stay shackles at head and tack.) Provide brailing lines for smothering loose sail. Ready boom crutches and lashings. Boom travelers should be secured amidships.

3. Clearing Up and Tightening Down

(a) Clear decks and cockpit of all nonessential items. Double-lash all heavy objects (spinnaker booms are better lashed along the life rails).

(b) Stow all loose objects below deck, padding to fill voids.** Secure all drawers and lockers. Stow all loose clothing, bedding, pillows, and

*If the engine is not to be used, close fuel and exhaust port cocks. Never close engine compartment ventilators unless imperative to avoid flooding. Lee cabin ventilators can be left open as long as weather permits.

**Half-empty iceboxes are a particular problem. The best remedy I know is to stuff them with pillows encased in plastic bags.

anything else that might come adrift and find its way into a flooded bilge.

(c) Prepare and store hot food and drink in insulated containers. Prepare dry snacks in plastic bags. Secure stove gimbals, and shut off fuel or gas cocks. Pass out seasickness pills and disposable bags, so that it will not be necessary for anyone ill to come on deck.

4. Navigation

(a) Take a navigation fix or calculate your DR and note it on the chart. Record barometer and wind data, and estimate drift rate. From weather signs and reports, prepare storm and sea state forecast. Plot storm track and possible course alternatives. Haul in taffrail log. Jot down pertinent radio frequencies, and fuel and water reserve.

(b) Check navigation, deck, and cabin lighting systems. Ready emergency lighting. Deck crew should don exposure suits, flotation vests, and safety harnesses.

5. Emergency Equipment

(a) Hoist radar reflector. Ready signal and distress flares. Slip-reef life buoys to prevent self-launching.

(b) Ready emergency repair gear: shores, fothering blanket, gloves, cable cutter, pump handle, ax, etc.

(c) Ready survival kit for possible abandonment of ship.

(d) Post emergency watch bill, and assign action stations.

This list may seem long, and is intended to cover extreme conditions, but with a little practice all of it can be accomplished in ten or fifteen minutes. You will then be prepared for the worst, and if it does not materialize, you can chalk it off as practice. With a green crew, such a drill is desirable *before* you get into the thick of a storm; the more you can reduce everything to routine, the better your chances of coming through unscathed.

HEAVY WEATHER TACTICS

Storm situations and tactics are discussed at some length in the appended references, and these are recommended reading for all who venture off soundings. In this section, we will review their conclusions, and attempt to interpret them in the light of what we have learned about vessel behavior in advanced sea states. I have also drawn on other sources relevant to the handling of large ships and smaller power craft.*

If shiphandling in storm seas is not the ultimate test of seamanship, it is without doubt the best discussed. Few subjects have evoked more controversy or speculation. Although most of this book is devoted to acquainting the reader with what the sea is like as an operating environment and why vessels behave as they do, when all is said and done, and the chips are down, the best (though certainly not the only) survival tactic *may* be to do nothing at all! Given a stout, well-prepared vessel

*Especially Crenshaw's *Naval Shiphandling*, the *Admiralty Manual of Seamanship*, and Kock and Henderson's *Heavy Weather Guide*.

and plenty of sea room, with her helm lashed to weather, she may take her own head and lie-to in relative safety and comfort. But, then again, she may not.

To wit, the *Admiralty Manual of Seamanship* cites the following accounts, here paraphrased for brevity:

> In a large Bay of Biscay storm, an aircraft carrier making 7 knots into the seas suffered severe pounding and took solid water over her flight deck, while a destroyer holding the same course two cables off her lee quarter rode comfortably. In the same storm, a large freighter foundered in the trough, her engines having broken down. Under similar circumstances, two cargo liners on the same route suffered opposite fates; one, lying disabled in the trough, rode out the gale safely; her sister some miles away foundered while steaming slowly, head to sea.

In light of our previous discussion of ship behavior in random seas, one can draw several inferences from these three accounts:

1. No single tactic is universally successful; whether drifting broadside or heading into the seas, one ship may founder while a similar vessel survives.
2. When drifting, seamanship clearly is not at issue, and a single extraordinary wave could spell the difference. From figures 72 and 73 (pp. 175–177), we can see that the highest 10 percent of all waves in a random sea are about twice as high as the average wave present. But, in a twenty-four hour sample, there is a 5 percent probability of a single wave 3.4 times as high as the average wave.
3. When heading into the seas, it is unlikely that a single wave could produce catastrophic pitching in a large vessel, but it might disable her or break her back. She might be better off running with the sea on her quarter.
4. Vessels of different sizes can behave quite differently, other things being the same, making station-keeping difficult or hazardous.*

If there is no optimum tactic for all ships at a given sea state, nor for a given vessel at all sea states, the best course of action clearly rests upon the skill and judgment of the skipper. As the *Admiralty Manual* puts it, ". . . The captain, knowing his own ship and her stability and handling qualities, forecasting the future trend of the weather and considering the sea room available, must decide for himself which is the best method to adopt in the prevailing circumstances." In fact, the number of

*That the destroyer was making rather better weather of it than the carrier may seem to go against the grain of earlier remarks about ship response. Although the ship specifications are not given, one can make the following estimates of response parameters:

Ship	***Speed*** **(knots)** V_s	***Est. waterline length*** **(ft.)** L_W	V_s/L_W	***Est. wavelength range*** *L* **(feet)**	L/L_W
Carrier	7	500	0.31	500–1,000	1.0–2.0
Destroyer	7	240	0.45	500–1,000	2.1–4.2

On the same heading, exciting forces would be roughly the same for both vessels, and pitching and heaving motions, therefore, would be proportional to their respective response factors. If the latter can be assumed to follow the same general trend as those for the cargo vessel in figure 129 (p. 305), then, anywhere within the wavelength range given above, one would expect the carrier's responses to be higher—rather than lower—than those of the destroyer. Further speculation is pointless without more specifics, but this illustration points up the difficulty of generalizing ship behavior under extreme conditions.

possible elections is rather limited, decreases with increasing sea state, and can be further reduced by eliminating the worst choices.

Review of Forces: Before plunging into tactics, we shall briefly recapitulate the forces acting on a vessel at advanced sea states, since their control is essential to decision making and shiphandling. In general, all forces will be markedly different from those attendant on normal operation:

Wind and Wave Forces: In the early stages of a growing storm sea, wind and waves comprise directional resistances that impede progress on upwind headings and abet it downwind. Force intensity increases roughly as the square of wind speed and average wave height, and varies with relative heading, as shown in figure 114 (p. 277). Note that, in shifting from end-on to 30° off the bow or stern, these forces nearly triple! Their point of application is always such that, unless compensated for by control forces, they act to turn a vessel so that the net resultant force is maximum—usually with beam to wind and sea.

Neither force is steady, and their time scales differ. Wind gusts vary in direction and intensity from minute to minute, and this variability increases with average windspeed and sea state. The average wind direction also shifts from hour to hour as the storm progresses. Wave forces vary from second to second, depending upon the frequency of encounter (effective wave period), and their average intensity varies from minute to minute as a result of wave group interference. Wave directional spread is greater than that of the wind, and the mean wave direction lags behind the mean wind shift.

Both the intensity and the variability of these forces increase with sea state, and, in extreme cases, although the wind may persist from some general quadrant, all sense of wave direction may be lost (plate 17, p. 212). Lastly, an FDS sea state is characterized by heavy breaking of even the highest waves, imposing additional impact forces as high as a ton per square foot.

Inertial Forces: The aforesaid exciting forces are resisted by a vessel's inertia in the form of her various response factors, which, when multiplied by their respective force components, give her average behavior in a seaway. As explained in Part VI, to the extent that wind and sea are directional, certain speed and heading combinations can result in very large synchronous motions even in moderate seas, and optimum shiphandling involves attempting to avoid such combinations while still making headway on some desired course.

At advanced sea states, increasing forces at all spectral wave periods and increasing randomness of direction make it impossible to avoid these combinations, and it is necessary to reduce speed to moderate excessive motion. Speed reduction, in turn, means loss of control and the ability to hold a desired heading, in which case a vessel may have no recourse to avoid drifting into danger, except heaving-to. Survival is then related mainly to the statistical probability of encountering breaking waves large enough to cause structural damage and subsequent flooding, or to flip her bodily over. For a given vessel and sea state, the CP diagram (fig. 135, p. 320) shows that pitchpoling requires higher waves than capsizing (roughly in the ratio of length to beam), implying that a vessel is better off with head or stern to the seas than drifting broadside. However, her chances are still better if the seas are taken at some small angle (10°–20°), rather than end-on, owing to the tendency to slew around and recover.

Control Forces: Assuming the normal functions of rudder, screws, and sails to be well known, we consider here the changes in control forces brought about by high wind and seas. In general, all are greatly reduced, just at a time when much greater control is necessary to overcome abnormally high and variable exciting forces. Rudders, screws, and sails are all fluid reaction devices that are designed to achieve maximum efficiency in steady flow at specified speeds and angles of incidence. Under these conditions, the control forces are proportional to the square of mean flow speed. When a vessel is rolling or pitching heavily at reduced speed in high waves, flow over these surfaces is no longer steady, but fluctuates wildly in magnitude and direction. The result is a great loss of efficiency and the imposition of high, alternating, transverse loading: the rudder whips back and forth; the engine labors and then races; the sails lull and strain. Any of these effects can cause damage and further loss of control, and learning how to minimize them and still maintain control is the essence of heavy weather tactics.

Maneuvering in High Seas: In simplest aspect, maneuvering is the application of control forces to overcome the external and inertial forces acting on a vessel. In high seas, the former are generally smaller—and the latter much higher—than normal, and more time and/or power are required to effect the same maneuver than in calm water. In addition, high, alternating, wave orbital particle velocities can produce strange control reactions that are sometimes temporarily opposite to those normally expected. For example, when running before a following sea, as a vessel teeters on a wave crest, the ahead component of particle velocity may completely blanket the propeller slipstream and cause momentary backflow across the rudder, whose action is thereby reversed. The propeller, now operating above normal incidence (stalled out), slows down and loses up to half its thrust, and the remaining power is diverted into radial flow (the proverbial egg-in-the-fan). Interaction of this flow with the vessel's underbody forces her stern strongly to starboard (right-hand screws). The normal tendency to correct this yawing moment by applying hard right rudder only makes things worse; the vessel yaws to port, and may slide broadside down the crest and broach. The correct tactic here would be to use reverse rudder and (in smaller vessels) reverse power, unless you are sure that an abrupt power increase would restore normal rudder control. More to the point, it would be better to alter course slightly to starboard so that the wind and wave forces would compensate the tendency to swing to port with each overtaking crest.

In a similar sense, the following tactics take advantage of forces and control changes, and can be used to expedite maneuvering:

1. Even in random seas, there is usually some underlying rhythm; groups of high waves are succeeded by groups of low waves *(smooths)*, and turning is more easily and safely accomplished in the latter. The time interval between smooths can be determined by observation, and is often 1–2 minutes. Vessels under 1,000 tons can usually reverse heading in a minute or less.

2. Although the turning radius will be larger, most ships turn fastest at high speed. To avoid braking the vessel and blanketing the screw, allow speed to build up before putting the rudder hard over.
3. Turning off the wind is easier than turning upwind; the opposing forces and period of encounter are both lower, and more time is available to gather headway. If in the trough and desiring to head upwind, it may be easier to bear off and come through 270°, by which time the vessel will have acquired speed, directional stability, and maximum steering control.
4. Most single-screw vessels are right-handed (propeller turns clockwise when viewed from astern), and force the stern to starboard. This tendency is greatest for large, slow-turning screws, and conversely. Therefore, other things being equal, turning to port is easier than to starboard.
5. Because the centers of wind and wave pressure are usually forward of a vessel's CLP, she will always tend to turn stern-to when backing down.
6. Under sail or power, maximum speed made good to windward will be achieved by heading into the wind as closely as possible, so as to minimize wind and wave forces. The ahead component of both forces is maximum at 30° off the wind (fig. 114, p. 277) and decreases to either side. Although a sailboat can gain speed by falling off, her effective $V_m g$ will be less, owing to increasing drift forces.
7. Running before high seas is the most precarious point of sailing: if too slow, there is risk of pooping; if too fast, of surfing, yawing, and broaching; if at wave speed, of loss of steering control. Moreover, there are many possibilities for synchronous pitching (fig. 130, p. 306). Most vessels will handle more comfortably taking the seas at some optimum angle on the quarter, as best determined by experiment.
8. Barring heavy breaking, only sailboats with heavy ballast keels may find running in the troughs tolerable. For lack of adequate roll damping and dynamic stability, this course is proscribed for most other vessels.
9. The maximum sail that can be safely carried should be judged by the peak gust intensity, and not by the average wind. Unless one is racing, it should never be more than the boat can stand up to close-hauled. Trimming sails to minimize weather helm (although never so much as to produce lee helm) will greatly facilitate turning.
10. Freak waves three or four times the average height (and freak hollows equally deep) can occur in any random seaway, and under marginal conditions, a constant watch is necessary to meet them end-on.

Other than the above, we can remark more generally that a straight course is seldom the best course; the sea is best viewed as an ever-changing obstacle course, through which one threads one's way, as Dickens puts it, "with nice discrimination, infinite zeal, and indefatigable assiduity." Again, all things are relative. Large ships with limited maneuvering capability may make better speed on quartering headings, because of reduced pitching and rolling, and should preferably tack on up-, down-, or crosswind courses.* Smaller vessels, being more susceptible to wave forces, will do

*Tacks of as much as 30° from the average course add only 13 percent to the distance traveled, although wind drift will roughly double.

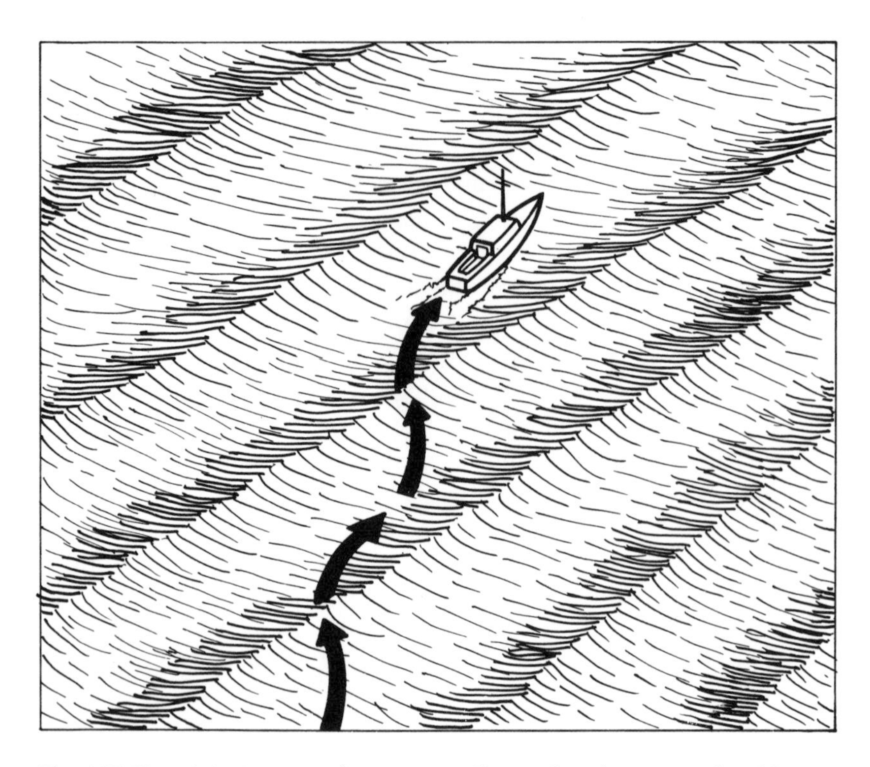

Fig. 147. To minimize wave forces, a small vessel under power should steer a sinuous course, heading up to meet wave crests, and bearing off in the troughs to compensate.

better to lay a sinuous course on oblique headings, meeting the larger seas head-on, or stern-to, and veering off crosswind in the troughs (fig. 147).

Somewhat different tactics are necessary for a sailboat when beating vs. broad-reaching, because of the relative wind shifts produced by wave orbital motions. These effects are shown in figure 148 for a boat beating at 45° and reaching at 135° to a true wind of 20 knots, in a regular swell of height H = 20 feet and period T = 12 seconds. Here, the orbital velocity is $u = \pi H/T$ = 5.2 ft./sec. = 3.1 knots, upwind under the troughs, and downwind under the crests. If her sails are trimmed for optimum incidence (30°) to the relative wind (V_r) when in the trough, they will luff on either heading as the boat rises on a crest, and she must then bear off to keep her sails drawing and maintain her average speed of 8 knots. Note that V_r shifts by a much larger angle (16°) when reaching than when close-hauled (4°), and will be still larger the higher the sea and the farther her course off the wind. If, instead of bearing off, the downwind boat sheets tighter to correct the luff, she runs a risk of jibing when the wind hauls aft in the next trough. All of these effects are aggravated by pitching, rolling, and wind gustiness, which make for very difficult downwind sailing in rough seas.

Lastly, we should mention surfing, a tactic now widely used in light- and medium-displacement sailboat racing, and one that can be employed by high-performance power boats as well—under the right conditions. Here, *right conditions* are the key words; somewhat like downhill ski racing, surfing is a precarious exercise in high-speed control, where the slightest misjudgment can cause disaster. It requires expert knowledge of wave behavior and a boat's handling characteristics, seas that

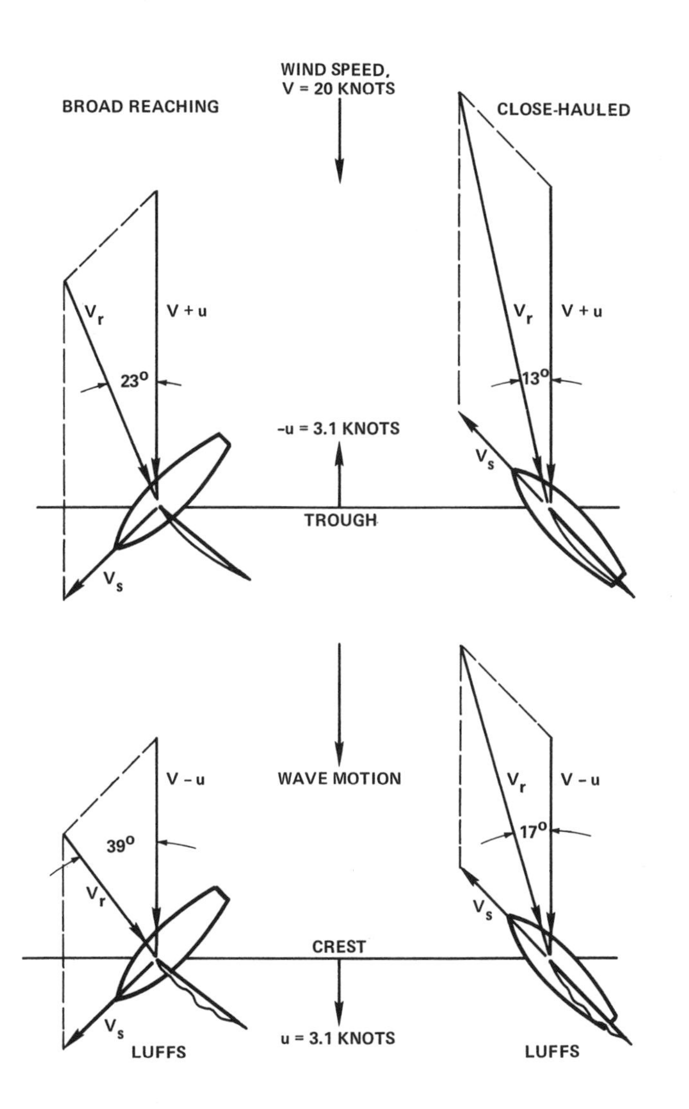

Fig. 148. Whether broad-reaching or close-hauled, a sailboat will tend to luff on the crests of large waves, owing to a forward shift in the relative wind caused by wave-induced speed changes. Sheet or heading changes are required to compensate. Shift is much greater when reaching (16°) than when beating (4°).

are sufficiently steep and regular, and good visibility. Again, like skiing, properly mastered, it gives a matchless sense of exhilaration and control over one's environment, provides a great speed advantage when racing, and can be an important survival asset in outrunning storms or maneuvering in high seas.

As described in Part VI, surfing is a form of controlled, downwind tacking, where temporary bursts of speed are achieved by accelerating to near-wave-velocity on the forward faces of large, near-breaking crests. Once the boat has gathered way, she is turned away from the direction of heaviest breaking, and guided diagonally down the wave slope at an angle best calculated to maintain a relatively constant elevation just beneath the break. In this attitude, good rudder control is possible, because the local wave orbital velocities are directed upslope at about one-third the wave speed (0.9T knots).* The duration of one's "ride" may persist for about 5–6 wave periods (60–75 seconds for long, regular swells), and may be extended by 3–4 more ridable waves in the same group.

Sailboat surfing is most often undertaken while running under spinnaker and main, where, in addition to skillful steering, expert sail control is required to prevent backwinding or jibing, as a result of the large shifts in V_r that accompany speed and heading changes. There is also a strong tendency to yaw and broach as a boat heels in gusts.**

Tactical Alternatives: We consider now tactical situations in marginal sea states where progress can no longer be maintained and future actions are limited to maximizing safety and minimizing discomfort until conditions moderate. Here, the number of possible choices is restricted at most to the following: (1) running under power into wind and seas at a speed and heading that minimizes adverse motions; (2) running off under power, with the additional hazards of pooping or broaching; and (3) heaving-to without power, and attempting to maintain a comfortable attitude by any available means of control.

Which choice, and when to make it, depends upon the vessel and the circumstances, but the decision should be made early enough so that it can be altered if necessary. If delayed too long, some choices may become irreversible; for example, in a rising sea, it may be impossible to turn and head into the seas if running, and conversely.*** Therefore, you should always ask yourself what you will do if things get worse, and try to adopt a course of action that reserves some flexibility. Lastly, where survival is in question, the key criteria—in order of increasing importance—are *control, stability,* and *buoyancy.* Having these, riding out a bad storm is mainly a matter of patient inconvenience.

In the following, we shall discuss these four tactics separately, pointing out their respective advantages and difficulties. As you will see, some of them may be proscribed for a particular vessel, owing to excessive motion or lack of adequate control. But often several alternatives can be employed successively as conditions change during the course of a storm:

Heading the Seas: This tactic has the sole—albeit significant—advantage of minimizing leeward drift. Its disadvantages are the punishment imposed on vessel and crew alike, the power (and fuel consumption) necessary to maintain steerage way in the face of high wind and wave forces, and the difficulties of avoiding dangerous pitching in random seas and of altering course to some more comfortable heading if things get too bad. Despite these objections, heading storm seas is often elected by high-performance military vessels, whose strength and high bows are well-adapted to withstanding wave impact, and, if the seas are not too large, also by larger ships anxious to minimize lost time. Except in dire necessity (as off a lee shore), sailboats

*Speedometer readings in excess of 20 knots have been reported by large surfing sailboats. Such speeds are in error by the astern component of orbital velocity, but still represent a considerable increase over hull speed.

**See *Heavy Weather,* by C. A. Marchaj, *Sail,* March 1973.

***Turning off the wind is generally easier. But I was once aboard a deeply laden 100-foot schooner plowing under power into the teeth of a 50-knot gale. Her skipper would greatly have preferred to turn and run back to port, but dared not, for fear that she would broach in the trough.

and small power craft should avoid this option; they often lack power to keep rudder control, are vulnerable to taking green water over the bow, and usually will be more comfortable on some other heading.

Careful rudder and power control are the keys to driving upwind. Power should be reduced to the bare minimum to provide steerage way, and the precise heading sought that requires least control effort. This heading will generally lie close to the wind's eye, or perhaps fine on the port bow in single-screw vessels, whose sterns tend to walk to starboard. Small speed adjustments may reduce pitching to some extent, but do not expect to be comfortable. Steering in this situation is a continual progression of small corrections; it takes a long time for average response to be felt, and there are many random digressions that must be ignored unless they indicate a persistent trend. Above all, strive to keep the bow from wandering as much as 30°, where greatly increased forces can weathervane you right around beam-to.

If your boat can take the punishment, the chief enemy is fatigue, which is best combated by frequent watch relief—as often as every 15 minutes under really severe conditions. If you can spare them, two hands in the cockpit are good insurance in sudden emergencies.

Running Off: Given ample sea room, running off on some optimum downwind heading is the preferred storm tactic for most small vessels and large ships alike. Not only are all motions much less violent, but the longer period of encounter is less conducive to fatigue. Except for dead off the wind, there are always two equivalent course options, one of which may take you out of the storm's path, or out of a heavily traveled shipping lane.

While optimum course and speed depend upon vessel and sea state, the principal objectives are to minimize adverse motions and wave impact. This usually amounts to finding a compromise between rolling and pitching, while at the same time attempting to avoid two great pitfalls: uncontrolled surfing down the face of steep waves, and being pooped by plunging breakers. These problems are not mutually exclusive; there is no ideal speed that will avoid both. At advanced sea states, the highest waves will have periods of at least 9–10 seconds and crest velocities (and orbital velocities, if breaking) approaching 30 knots, which no vessel can outrun under these conditions. Thus, a vessel will always be overtaken by the larger waves and will run some risk of pooping if they happen to break at the wrong instant. If running fast enough to minimize pooping, she will have enough way on to begin to surf on the steeper crests, and may accelerate so rapidly that she plunges into the trough ahead. Here, the resonance diagram (fig. 130, p. 306) can be consulted for a course and speed that best avoid synchronism with the larger waves,* and small vessels can play the pooping-surfing game on a wave-by-wave basis, briefly applying power and rudder so as to take breaking crests stern-to,** bearing off a little above this course as the vessel accelerates, reducing power and sliding diagonally across the wave until the crest passes, and then resuming the original course. Again, this

*The resonance diagram gives a band of course-speed options, from which one should select the one on which the vessel handles most comfortably. Owing to steering difficulties at low speed dead off, this will more likely be found at slightly higher speed and at some angle between 120° and 160° off the wind.

**There is some evidence that a strong screw wake is also effective in causing premature, less-violent wave breaking.

procedure requires considerable skill, judgment, and knowledge of a vessel's responses. The tendency is usually to overcontrol; an optimum heading will usually lie close to that which she would normally assume if given her head, and only timely corrections are required to properly meet "freak" waves that threaten to overwhelm her.

Large ships, of course, cannot respond rapidly enough to deal with individual waves, but, having much greater freeboard, they run less risk of pooping and will usually ride out a storm most comfortably by running at low speed and taking the seas on the quarter.

Heaving-to without Power: This is another controversial tactic. It minimizes drift and conserves fuel, but is disallowed for heavy rollers. These include warships that are designed to have low roll damping for gun steadiness, and many cargo vessels in the light ballast configuration or those liable to cargo shifting.* Nevertheless, the *Admiralty Manual* states that "authentic cases have been reported in recent years of ships successfully riding through, or very near to, the center of typhoons in the Pacific with engines stopped." Although rolling and pitching heavily, the ships sustained no damage, which is attributed to adequate metacentric height and good watertight integrity.**

Short of encountering breaking waves high enough to risk capsizing or pitchpoling (i.e., waves equal to the beam or length, respectively), and given adequate dynamic stability, watertight integrity, and strength to withstand breaking seas, small craft are often better adapted than large ships to heaving-to in storm seas. Their shorter free periods are farther removed from synchronism with the highest exciting forces, and they can sometimes employ passive control means to maintain a preferred heading with respect to wind and seas. Although motions may be more violent hove-to than when running off, the former is much less demanding of the skipper; having assured oneself that everything has been done to weatherproof one's vessel, there is little to do but wait.***

As in running off, there are several options in heaving-to, the choice of which depends upon the vessel and sea state, and is best determined by experiment, as earlier described. In order of increasing sea state, you can try the following:

(*a*) *Heaving-to under Short Sail*

Below wind speeds at which even storm sails are in danger of blowing

* Insurance investigations also attribute the foundering of several commercial fishermen to attempting to run in the trough when heavily loaded, owing to the tendency for fish packed in crushed ice to shift and wedge against the low side of the bilge.

** Recall (Part VI) that stability, of itself, is not an index of a vessel's rolling tendencies: a circular cylinder has no stability, but (friction neglected) will not roll because waves cannot exert torque on it; a wide, flat plank has high stability, but will not roll more than the wave slope. In general, heavy rolling results from a combination of a limited range of dynamic stability, low damping, and relatively boxlike sections. Under all conditions, maximum rolling will occur at wave synchronism with the vessel's free roll period.

*** Expert opinion seems sharply divided on the advisability of heaving-to in small power craft. Captain Beebe says "absolutely not," yet small fishermen the world over routinely ride out gales in this manner. A famous designer of pleasure boats in the range 20–60 feet has told me that most power yachtsmen, inexperienced with heavy weather, tend to overcontrol and are apt to broach in running downhill, and would be much better off shutting down and letting the boat take its head. The chief problems are cockpit flooding and lack of adequate strength topside to withstand breaker impact. Any means of controlling a boat's heading so as to minimize these effects are worth attempting.

out,* many sailboats will ride quite comfortably to a storm jib and trysail. If steering becomes too arduous, try sheeting the jib to weather and lashing the helm down, in which case the boat will drift to leeward, while slowly fore-reaching or sliding astern. Again, the best combination is a matter of experiment, and some boats do better under trysail or jib alone.

Whatever the combination, because of the time, effort, and danger of working on deck under gale conditions, the advantages of prior practice and early decision cannot be overstressed. Riding out a force 10 gale northwest of Bermuda in a 30-foot sloop, Errol Bruce found that a skilled hand, working at the end of a safety line in pitch darkness, took 40 minutes to set a storm jib. The same sail was lowered and reset three times during the course of the 36-hour storm (*Rudder,* March 1972).

(*b*) *Lying-to with a Sea Anchor*

Bow to the wind, this seems to be a universally recommended tactic for small power craft, and is espoused by the Coast Guard. Although I can find little evidence of practical success among the power boat fraternity with sea anchors in heavy weather, there are a number of accounts of sailboats having used them to advantage—both bow- and stern-to. But there are as many reports of failures.

From a dynamic standpoint, lying-to with a sea anchor is much akin to ordinary anchoring or towing under similar conditions. Because wind and wave forces act ahead of the center of lateral plane, they tend to turn a vessel broadside; unless precisely aligned with their resultant, she is basically unstable in yaw, and, like a bull tethered to a ring in his nose, will sheer wildly from side to side, and be alternately exposed to much greater forces with each swing. As shown later ("Anchors and Anchoring," p. 381), most of these difficulties can be resolved with proper design and deployment; the system must be sufficiently elastic to yield to the seas, but strong enough to withstand average wind and current forces.

Many vessels simply cannot be prevented from sheering without some independent means of controlling yaw (engine, riding sail, etc.). Failing this, a (smaller) sea anchor may be effective in controlling heading and drift in some more natural attitude, as in (c), below.

(*c*) *Doing Nothing*

Lash the helm amidships and see how she lies. Depending on the balance of forces, most vessels will adopt a mean heading at some angle to wind and seas that can be adjusted for optimum comfort by a different helm position, a storm trysail rigged as a weathervane, or a weighted line trailed from bow or stern.** As you will note from figure 114 (p. 277), the force balance does not change much with heading at angles greater than 45° from bow or stern,

* The strength of sails varies greatly, depending on construction and condition. Storm sails usually have bolt ropes all round (often wire luff ropes), and are cut rather full to reduce clew tension when oversheeted. In *Voyaging under Sail,* Eric Hiscock devotes several pages to rigging and management of heavy weather sails.

** An auto tire casing makes an excellent drag weight—if of course, you have providently brought one along. I know an instance of a 10,000-ton freighter that rode out a hurricane with her anchor and 50 fathoms of chain dangling from the hawse.

and a relatively small drag change can have a large effect. If you can strike a tolerable attitude, you have nothing but offing or a possible collision to worry about, unless conditions worsen.

(*d*) *Streaming Warps*

The tactic of drifting downwind without power and streaming warps or bights of hawser to control speed and heading has been employed by many boats under 100 tons, with varying success. As Adlard Coles points out, boats have been broached, pitchpoled, and pooped under these conditions, in which a vessel's stern is most vulnerable to wave impact. Yet, one can argue that breaking waves are always a random hazard in heavy weather, and any way downwind reduces both the probability of encounter and the severity of impact. Broaching may result from streaming too little hawser; pitchpoling, from too much. In any case, a weighted line produces more drag than one trailed free.

Because of bluff transoms, vulnerable deckhouses, or large, open cockpits, some boats are markedly unsuited for this tactic, whereas stoutly built double-enders may fare very well. Assuming that optimum speed can be controlled by adjusting the length or number of lines trailed, steering will be easiest if they come aboard in pairs, port and starboard, roughly athwart the rudder post, so as to equalize their turning moments (fig. 149). The optimum point can be determined in advance by experimenting under power in calm water. It will often be found that the genoa sheet winches are well situated for belaying these lines, making for easy adjustment of length (*A*). If not, they can be passed through quarter chocks, or through snatch blocks secured to life-rail stanchions, and belayed to bitts—or to the base of the mast (*B*). In all cases, once the proper length is established, they should be wrapped in chafing gear. Because survival is often predicated upon encounter with a single "freak" wave, keep a sharp knife handy to free the boat in emergency.

(*e*) *Running at Hull Speed*

While I know of boats that have run off streaming warps for as much as 300 miles in extratropical cyclones, all of the above must be regarded as temporary delaying tactics in a rising sea state. Under extreme conditions, the ultimate hazard is being rolled over or pitchpoled by extraordinary breaking seas, which—unless hopelessly confused—are best avoided by running free before them at *sufficient speed to give positive steering.*

Whether or not this is practicable depends a good deal on the type of boat, and how she is rigged, but Adlard Coles cites a number of cases where skippers have asserted that running at speed was the only possible tactic under the circumstances, and he is inclined to give the benefit of the doubt to those who have tried it.

Extreme sea conditions are ordinarily accompanied by winds strong enough to drive a boat at hull speed, particularly if kept about 30° on the quarter, where the ahead force component is maximum, and sailboats with large rudders and having maximum windage forward under bare poles may be able to maintain adequate steerage without further ado. In other cases cited, steering assistance and roll stability were obtained by sheeting a boom

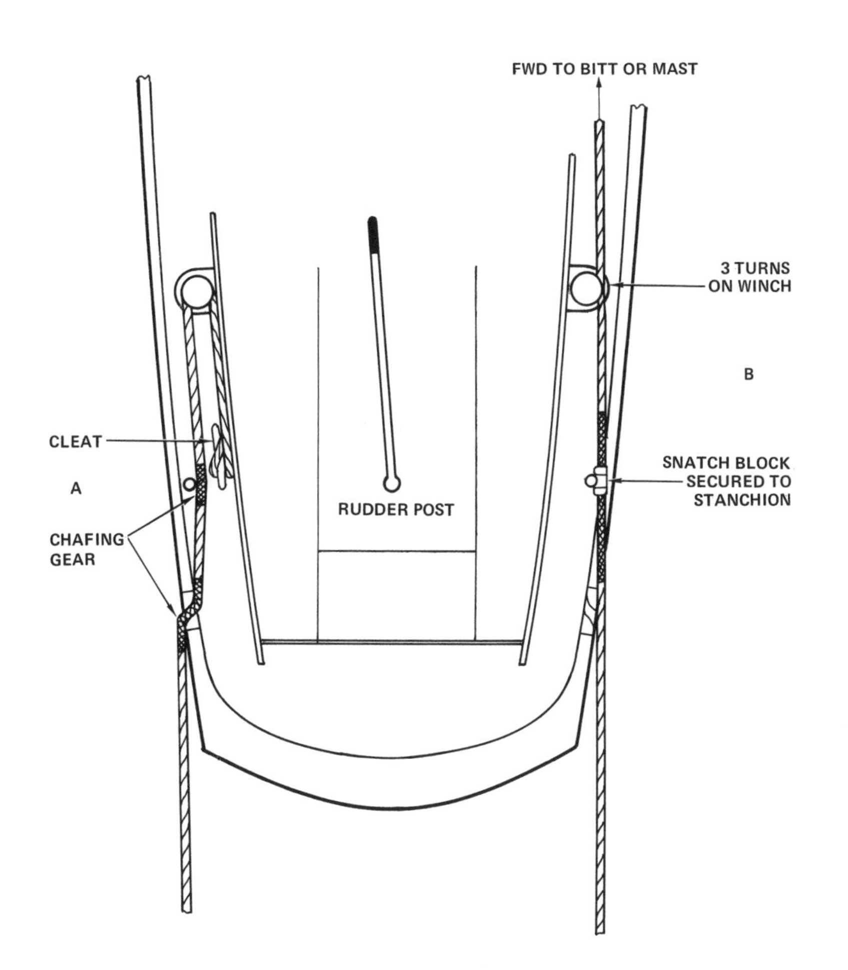

Fig. 149. Lines trailed for speed or heading control should come aboard in pairs abeam the rudderpost for optimum steering, and should be well protected against chafe at all points of motion.

staysail or a storm jib amidships. Both Vito Dumas and Sir Francis Chichester continuously carried one or two sails, suitably reefed, in winds of hurricane force.

The principal problem with running off under bare poles or sails is that the boat may surf so fast that it outruns the waves, only to stall and be pooped in the trough. This problem has been successfully avoided by a number of cruising yachtsmen by towing a small drogue, usually cobbled up out of anything handy. In Part VIII, we analyze such situations, and describe how to design a drogue suitable for your own boat.

In reviewing these many accounts, four points stand out: (1) Hull speeds of 5–8 knots were maintained; (2) seas were taken 15°–20° to the stern; (3) some sail, however small, was found necessary for stability and control; and (4) controlled surfing tactics were employed to prevent broaching.

Coles raises the interesting point that he cannot see why tactics that are successful under the worst conditions should not apply as well to ordinary gales, and there seems to be no sound dynamical argument against this premise.

Thus, we are led to the paradoxical question as to why other tactics should be recommended. There are several answers: great skill and judgment are required to run downhill in high seas; steering is exhausting under these conditions, and a storm may last for several days (one of the reasons why racing crews are redundantly large); lastly, you may not have several hundred miles of sea room.

SQUALL TACTICS

We have talked a lot about heavy weather tactics in which waves are the principal hazard, and wind is a secondary impediment to shiphandling. With squalls, things are just the opposite; waves are usually of little consequence, but the first wind gust (sometimes of hurricane strength) may knock you flat in the water, fill you up, and sink you before you can say "red-rubber-baby-buggy-bumpers," or may drag you on the rocks from a secure anchorage before you can rub the sleep from your eyes.*

Although squalls are generally of short duration, their worldwide distribution, frequency of occurrence, suddenness, and intensity put them in the front ranks of all marine hazards to small craft—particularly to those left unattended at anchor or in marina slips. Unless capsized without hope of early rescue, the threat to life is small, but no one likes to lose a boat unnecessarily, especially when a little alertness and relatively simple measures will go a long way to prevent its loss.

As described more generally in Part II, squalls are intense, local wind systems that accompany steep atmospheric pressure fronts. The latter may occur in clear weather, as the result of pressure buildup behind a mountain ridge to the point where it is suddenly relieved through passes, and air cascades down the valleys *(katabatic squall);* in sultry, tropical weather, under developing thunderheads; or in middle latitudes, as a line squall that precedes a larger cold front. In all cases, the wind comes suddenly, often from an unexpected direction, and with little or no advance barometric warning. Visible signs may consist of the rapid growth of a local cumulus cloud with heavy rainfall beneath, the rapid advance of a solid, dark cloud wall, or by nothing more than a line of dark, ruffled water. Although radio weather may announce the possibility of squall conditions, these disturbances are commonly so localized that they pass unreported, and there is no way to gauge their possible intensity until they hit; at night, there is no sure way even to detect their approach.**

Squall tactics vary according to the circumstances, but the basic problem is always the same: how to deal with an abrupt, short-duration (10–60 minutes) wind of hurricane intensity. Most authorities concur on the following measures:

Advancing Squall at Sea: The chief danger is being caught broadside with sail up. Start your engine immediately and head the advancing front. Get all sail in, smothered, and

*Despite years of sailing in the Bahamas, W. S. Kals's 30-foot sloop, *Boheme,* was thrown high and dry on the rocks by an estimated 110-knot squall gust, while riding to a 35-pound anchor and 15 fathoms of ⅜-inch chain in 6 feet of water. Although having suffered only slight damage, she was wrecked three days later by heavy surf, owing to Kals's lack of a radio, and his inability to attract the attention of the many passing vessels with pencil flares, signal lamp, sun mirror, or even signal fires.

**Intense rain squalls can be seen on radar screens.

lashed; sheet booms tight amidships; close all topside openings to prevent taking water if knocked down.

As the squall gets close, try to analyze and head the surface wind direction; this can usually be determined by alignment of whitecaps, flying spray, etc. The first gust is usually the most intense, and if taken squarely, half the battle is won. If the boat's head falls off, chances are you will not be able to control it, and will be better off continuing the turn to take the wind astern; in the absence of high waves, most boats are more easily controlled stern-to.* Trailing a bight of line will reduce drift, if offing is a problem.

Remember, squall fronts can move at fifty knots, so that there may be little time for preparations; the object is to get the boat squared away and watertight as rapidly as possible—and then hang on. The worst is over.

If drifting into shallow water, lower your heaviest anchor until it strikes, and veer a scope of 8-10. With enough power, you *may* be able to control horsing and relieve anchor strain. If you have prepared for it in advance, this is really the time for emergency anchoring, as described in Part VIII.

Even if your hook fails to catch, or your cable parts and you are blown ashore, stay with the boat. Hitting a reef is no worse than an auto collision at 10 mph. After the boat comes to rest, a second anchor may keep her from pounding heavily. Even if holed and sinking, you are better off aboard than attempting to swim, unless you have fetched up on a nice sandy beach. Squalls seldom last long enough to make abandonment worth the risk.

Advancing Squall at Anchor: Unlike a storm, a squall seldom gives you time to pick up the hook and get to sea. However, you may have time to set a second anchor about 45° from the first and centered on the wind direction.** Veer a scope of 7 on both cables, unless there is danger of fouling with nearby boats, in which case veer enough more to clear them. If you have a third anchor, keep it ready to drop from the stern, so as to arrest swinging. Button up, as when at sea, and use your engine to relieve anchor strain and control heading.

Squalls at Slippage: Unless the slip itself carries away (this happens more often than you might think), a properly secured boat is relatively safe. Given time to move, a leeward berth is preferable to being caught to windward. In either case, shorten and double up all lines, adding breasts and springs where necessary. Peak halyards secured on both sides to any convenient belaying point may prevent a sailboat from being blown over and fouling another's rigging. As before, button up, and then see what you can do to secure other craft whose owners are not aboard.

* Some authorities advise putting out a sea anchor. Maybe so—in a small dinghy—but have you ever tried to carry 10 square feet of canvas forward and deploy it in a 60-knot breeze?

** If the first cable is chain, the second should be nylon.

PART VIII

Emergency Procedures

27

Emergency or Exigency?

Although it may appear that the previous chapter was principally devoted to emergencies,* most old-time seafarers would regard our discussion of storm strategy and tactics as nothing more than standard operating practice, to be expected and taken in stride on any long voyage. If an apology for overstressing the obvious is due, it is that my remarks were addressed to a gentler breed of yachtsmen (your author included), who have not had the experience of being rousted from a soaking berth at midnight, to climb icy ratlines 100 feet above the deck of a square-rigger beating her way through a howling blizzard off Cape Horn, and edging 40 feet out on the footrope of a swaying yard to smother the frozen tatters of a blown-out upper topsail.

Those ships and men are long gone, but the sea is still the same. Even today, it is considered nothing short of heroic to take a single-hander around the Horn, where the usual might seem to most of us to be a continuous emergency.

In the same context, many nautical books discuss common emergencies as if they always occurred under benign conditions. Indeed, quite the reverse is true; accidents at sea seem to pick the worst times to happen. A helmsman is most apt to be swept overboard by a breaking sea at night, when it may be impossible to execute a Williamson turn or, having done so, to find him—let alone to launch a boat, heave him in, and get both back aboard.**

Accordingly, in Part VIII, we take a more realistic look at a few procedures that require immediate action (even if they should not be regarded as unexpected), introduce some new concepts that may be hard for an old salt to swallow, and present arguments against some other measures traditionally accepted.

OIL ON TROUBLED WATERS

The concept that angry seas can, to some extent, be ameliorated by a film of oil appears universally in nautical literature, and seems to have originated in early whaling days, from observations of the apparent calming effects of whale oil liberated during flensing operations. Since then, most marine references have advocated the use of oil without qualification, although the *Admiralty Manual* does question its effectiveness on larger waves.

*My dictionary defines *emergency* as "a sudden, unexpected occurrence demanding immediate action." Perhaps *exigency* would be a better word.

**Allan Villiers accounts of backing down his full-rigged frigate *Joseph Conrad* on a fresh gale southeast of New Zealand, to lower a boat and pick up the ship's cat—certainly the luckiest feline in the world.

Yet, long tradition notwithstanding, there is considerable physical and scientific evidence against any substantial diminution of the *significant* waves at elevated sea states, whether by oils or any other surface additives that can be carried aboard ship in practicable quantities!* Whether or not the instant of breaking of larger waves can be *procrastinated* by oils, etc., is still somewhat open to question, owing to our lack of understanding of the breaking process in deep water.

Damping of Small Waves: The theory of small-wave damping by surface films is well confirmed by experiment, both in the laboratory and—on a limited scale—in the field. I can say this with some assurance, having contributed modestly to the more recent results. Most effective damping agents fall in the category of *dispersants;* that is, their molecules mutually repel one another so that, when poured upon the water, the agent spreads out over the surface to form a very thin film that ultimately may be only a few molecules thick. This dispersing action is so strong in some agents, such as the *dodecanols,* that they can spread *upwind* against a 15-knot breeze.

The damping action of such films arises not from augmentation or diminution of surface tension directly, but from the fact that the work (energy) required to stretch the film as waves pass through it is not recovered when the film shrinks again, but instead is dissipated as heat. Physically, the result of this one-way energy flow is the inhibition of *horizontal* surface particle motion, without much effect on vertical motion. But because ripples and wavelets cannot grow higher without also growing longer, the net effect of films is to prevent the growth of small waves into large waves. Plate 27 shows the dramatic difference in the surface appearance of a 4-acre pond under 35-knot winds before and after the addition of a film of ordinary laundry detergent:** with no film, waves up to 1 foot high and 7 feet long have arisen at the downwind end of the 800-foot fetch; with a monomolecular detergent film, small capillary waves cover the surface, but cannot grow larger.

Like surface tension, the film effect is a function of surface curvature (stretch per unit surface area). The percentage rate of energy dissipation, relative to the average energy of wave motion, is approximately $0.7T^{-2.5}$ percent per second, where T is the wave period: a 1-second wave will lose energy at the rate of 0.7 percent per second; a 10-second wave at 0.002 percent per second. Hence, effective damping is restricted to wavelengths in the capillary range ($T < 1$ second), while longer and larger waves will propagate virtually unaffected through a localized film-covered area.

Effect on Breaking: The effects of films on wave breaking are not so easily quantified. There is no question that a surface film an acre or two in extent brings about a dramatic change in the appearance of the sea surface. All waves shorter than 5 feet or so rapidly disappear. Because the frequency distribution of whitecaps is heavily weighted toward the more numerous small waves (which are also more strongly influenced by the wind), the average number of whitecaps and the amount of

*Oil several inches thick and covering several square miles has a demonstrable damping effect, as also does field ice, where the floe edges are substantially in contact.

**Powdered detergent was applied by hand-scattering at the rate of 200 lbs./hr.

Plate 27. Two views of a 40-acre pond under 35-knot winds. The addition of laundry detergent at the rate of 200 lbs./hr. completely prevents all but the smallest capillary ripples.

wind-whipped spray are considerably reduced. These qualitative features are matters of common observation, but have not been reduced to numbers.

But the breaking of larger waves is attributed to transient instability, resulting from the coincidence of many underlying component waves on an energy scale much too large to be influenced by localized film dissipation, and there is no rational basis for invoking surface tension forces as a moderating agent. Even so, the most specific evidence I know of is still qualitative. First, in the laboratory I have been unable to detect significant modification of the breaking of paddle-generated, periodic waves, 1 to 2 feet high by addition of a variety of filming agents. Second, a Scripps Institution vessel, returning from a recent north Pacific cruise, had been running for two days before 40–50 knot winds and fully developed seas, when it encountered a large oil slick about 100 miles northwest of Point Conception, California. Scientists aboard observed that, while incipient whitecapping was virtually eliminated, the largest waves continued to break with apparently undiminished intensity, even after the wave energy had traveled several miles through the slick.

In summary, and all things considered, the old-time practice of pumping out several hundred gallons of bunker crude (one of the least effective agents) to form a calm lee for small-boat launching may have a strong psychological (and a slight physical) advantage. As regards survival of small vessels in storm seas, I suggest that oil is a waste of time and makes a mess of the boat. However, a pint of dodecanol attached to a life buoy, for dispensing when a rescue vessel is within visual range, might be an important asset in sighting a crew member lost overboard.

28

Anchors and Anchoring

ANCHORING: ART OR SCIENCE?

Although anchoring is not ordinarily regarded as an emergency procedure, it might better be viewed as such. It is a moot point whether more small craft are lost at anchor or at sea, but, as any diver can testify,* the number of anchors festooning many undersea coastlines points up a much-neglected part of seamanship.

Anchoring technology has advanced remarkably within the past twenty years. Massive oil drilling rigs and large oceanographic weather buoys are routinely anchored in some of the stormiest regions of the world's oceans. Despite the fact that the analysis contained in the first edition of this book placed it in the forefront of industrial anchoring technology, most of the suggestions advanced have not filtered down to the operating level of small commercial vessels and ordinary yachtsmen. I take this to be a deficiency in the original presentation, in which undue emphasis was placed on open-sea anchoring under storm conditions—a last resort for the beleaguered yachtsman.

Accordingly, in the following, we present a much more simplified analysis, showing that, wherever protected by a limited fetch from breaking seas large enough to threaten her watertight integrity, a vessel up to 100 tons can safely be anchored under storm conditions, using ordinary ground tackle, with a few simple modifications. As with everything before, we take the position that it is just as important to explain *why it works* as it is to state *what works*.

ANCHORING DYNAMICS

Even where protected from the direct action of large storm waves, a vessel lying to a single anchor is in dynamic equilibrium between her inertial force and the resultant of all external forces acting at any instant. Aside from the wind, wave, and current forces previously considered, the main distinction between the equilibrium of a vessel moving upwind under power in a seaway and one at anchor is that rudder and propulsive forces are replaced by the anchor rode** tension, usually applied near the

* I know a southern California abalone diver who now earns a good part of his livelihood in anchor salvage. Even in a remote Bora Bora lagoon, I counted fourteen large ship anchors within a mile of the entrance pass.

** *Rode* is defined as any line connecting a vessel to its anchor.

bow, and acting diagonally downward at some angle from the horizontal. With steady external forces (static equilibrium), the anchoring equations can be solved without difficulty: the anchor need only be capable of resisting the resultant horizontal drag force; the bow need only support the rode weight; and the rode need only be long (or heavy) enough to keep the anchor shank angle below a safe maximum (about 8°), and strong enough to sustain the tension at its upper end.

However, as we all know, these forces are never steady, but rather random, oscillatory, and often contrary. Under severe conditions, an anchored vessel can perform alarming gyrations, surging back and forth, swinging through a wide arc, and bucking and plunging like a frightened mule. All this plays havoc with anchor tackle: at best, you may spend a sleepless night at anchor watch; at worst, lose your anchor—and possibly your boat.

Nevertheless, because of the acknowledged difficulties of appraising dynamical forces, vessel responses, and anchor holding power in different media, most writers on the subject have confined themselves to the assumption of static forces, multiplied by large safety factors, which leads to weights and strengths of anchor tackle larger than necessary to withstand the actual forces involved. Under most circumstances, one can hardly quarrel with this—if it always works. Aside from extra labor, there is little penalty in heavy tackle, and the extra weight can often be used to advantage as inside ballast.

But, judging from the present record, brute force methods do not always work, so it behooves us to look a little closer at the problem of dynamic anchoring. As we shall see, a knowledge of ship response sometimes allows us to outwit Nature where it is not practicable to overwhelm her.

Motions of an Anchored Vessel: The motions of a vessel drifting freely in a seaway can be described as a random series of wave- and wind-induced oscillations in six degrees of freedom (Part VI), superimposed on a relatively slow average drift. Ideally, the optimum anchoring system would consist of a horizontal restraint just sufficient to overcome the drift forces, but elastic enough to accommodate random motions without transmitting much additional force to the anchor. But elastic restraint (a restoring force) has the effect of converting the vessel's three nonoscillatory modes of motion (surge, sway, and yaw) into oscillatory modes, each of which has a natural free period that may result in large-amplitude oscillations if in synchronism with some exciting force.

Of these (now) six oscillatory modes, roll can be neglected, since it is not appreciably affected by anchoring restraint. The other five can be grouped into three combinations(pitch and heave, surge, and sway and yaw), whose responses are quite different, and must be considered separately (fig. 150):

Coupled Pitch and Heave: These motions are only affected by the vertical component of rode tension, which, in turn, is equal to the weight (in water) of that portion of the rode clear of the bottom. Since anchor tackle is normally stored in the forepeak, an anchored vessel will be relieved of the change in weight of the suspended portion (negligible for nylon, and about 15 percent for chain), plus that of the anchor and whatever portion of the rode is lying on the bottom. This weight loss increases bow freeboard, and allows the bow to lift more readily to seas—a tendency that is

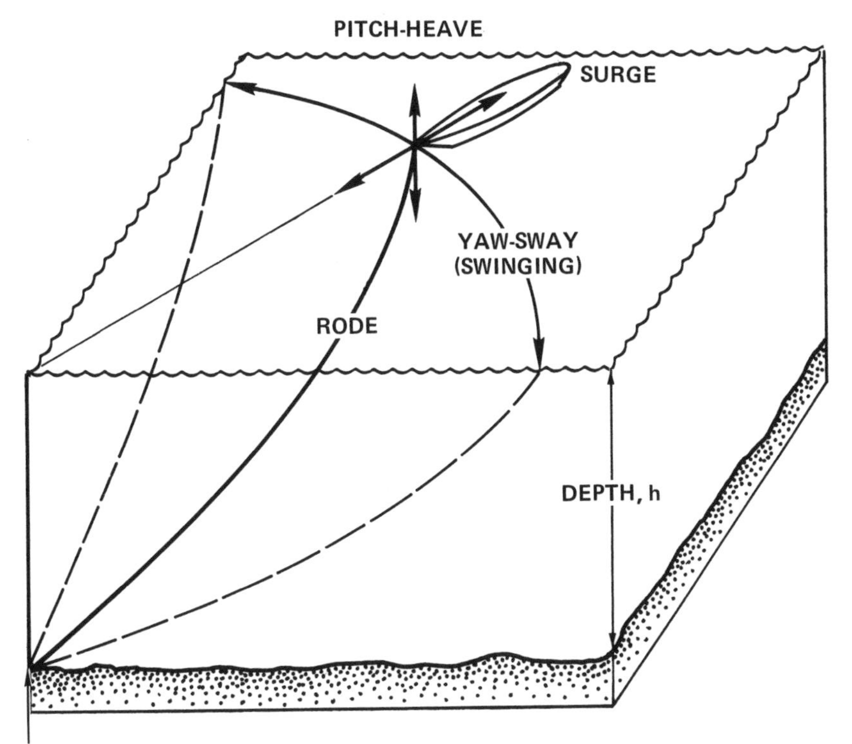

Fig. 150. The three significant modes of motion of an anchored vessel.

somewhat opposed by frictional drag on a taut rode, or by alternately picking up and laying down a loose one. In all events, pitch and heave response are little altered by anchoring, and, for practical purposes, the range of bow motion that must be allowed for can be taken as the maximum probable fetch-limited wave height.

Pitch and heave are of no importance to anchor holding power, provided that the anchor shank angle is lower than about 8° from the horizontal (i.e., the upward component of rode tension must be less than one percent of the horizontal component). While the equilibrium shank angle will be established by the balance of forces later described, due allowance must be made for relative depth changes resulting from both waves and tides.

Surge Response: In a strong wind, a vessel anchored under elastic restraint undergoes fore-and-aft surging oscillations in response to waves and wind gusts. Because the total resistance (damping) is rather high for slow longitudinal motions, it can be shown by rather complicated arguments that resonant surge-amplification only occurs among vessels tightly moored dockside. With elastic anchoring, surge forces will always be much smaller than those imposed by sway and yaw.

Coupled Sway and Yaw: These motions comprise the familiar swinging of an anchored vessel—the exact dynamic analogue of a tailless kite. Because the centers of wind and wave pressure act forward of the center of lateral resistance (fig. 114, p. 277), most boats are dynamically unstable in yaw when anchored by the bow, and will sheer from side to side by as much as 60° from the wind direction. The oscillation period increases with anchor scope (the ratio of rode length to water depth), decreases with increasing wind force, and is roughly independent of vessel displacement; typical periods are 1 to 5 minutes.

Although commonly accepted as a fact of Nature, swinging has manifold disadvantages. It not only requires lots of extra room in a crowded anchorage, but, because wind and wave forces increase rapidly with angle off the bow, it can also treble or quadruple average rode tension. An anchor is basically a unidirectional device, and alternate swinging over a wide arc can snap its shank, break it out, or cause it to saw its way through soft holding ground a hundred feet or so in the course of a night. Lastly, *swinging greatly aggravates the problem of rode chafe where it passes through a bow chock or bears against the stem (or the bobstay).* All these add up to make swinging the primary cause of anchor failure.

A Child's Guide to Swinging: Of the three modes of anchored motion discussed above, swinging is both the most serious problem and the most easily prevented. Nautical books suggest a number of remedies, most of which do not address the basic problem: that of correcting the natural force imbalance that induces swinging, so that a vessel will lie to a single rode and be free to pivot with the wind, while still maintaining enough elasticity to accommodate random motions. For the record, these include (fig. 151):

(a) dropping a second anchor off the stern at either extreme of the swinging arc;
(b) lying to two bow anchors set 45° apart and centered on the wind direction;
(c) riding midway between two bow anchors set 180° apart and aligned with the wind direction; and
(d) riding to three bow anchors set 120° apart.

Each of these methods has its pros and cons, but all involve extra tackle that is difficult to deploy and retrieve in heavy weather. The problems of fouling and suitable holding ground are multiplied in proportion to the number of anchors. Only method (b) provides the necessary elasticity, but if the wind shifts, you are back again to one anchor.

Sea conditions and the type of boat permitting, the preferred way is to anchor by the stern, or stern quarter (powerboats). Sheer heresy? I think not. Any vessel that swings to a bow anchor will usually ride stably by the stern; the resultant of yawing forces no longer acts behind the CLR, and the vessel becomes self-stabilizing. Stern anchoring admittedly raises problems of wind and wave forces from a direction never considered by the designer, but most cruising sailboats and many power craft have relatively soft waterlines aft, and some stern overhang. Given enough flexibility in the anchor system, they will rise to stern seas, and remain drier and more comfortable than if yawing and swinging to a bow anchor.

Deferring the mechanics till later, I can assert that, in 40 years of chartering, I have habitually anchored by the stern—not by heaving an anchor off the fantail, but by bending a dockline to the bow anchor rode with a rolling hitch, and hauling it aft to a stern mooring cleat. Not only does the boat lie comfortably while other boats are horsing around, but cabin ventilation is greatly improved. And, if the wind gets up and some chafing occurs, I've only lost a dock line, and not an expensive rode.

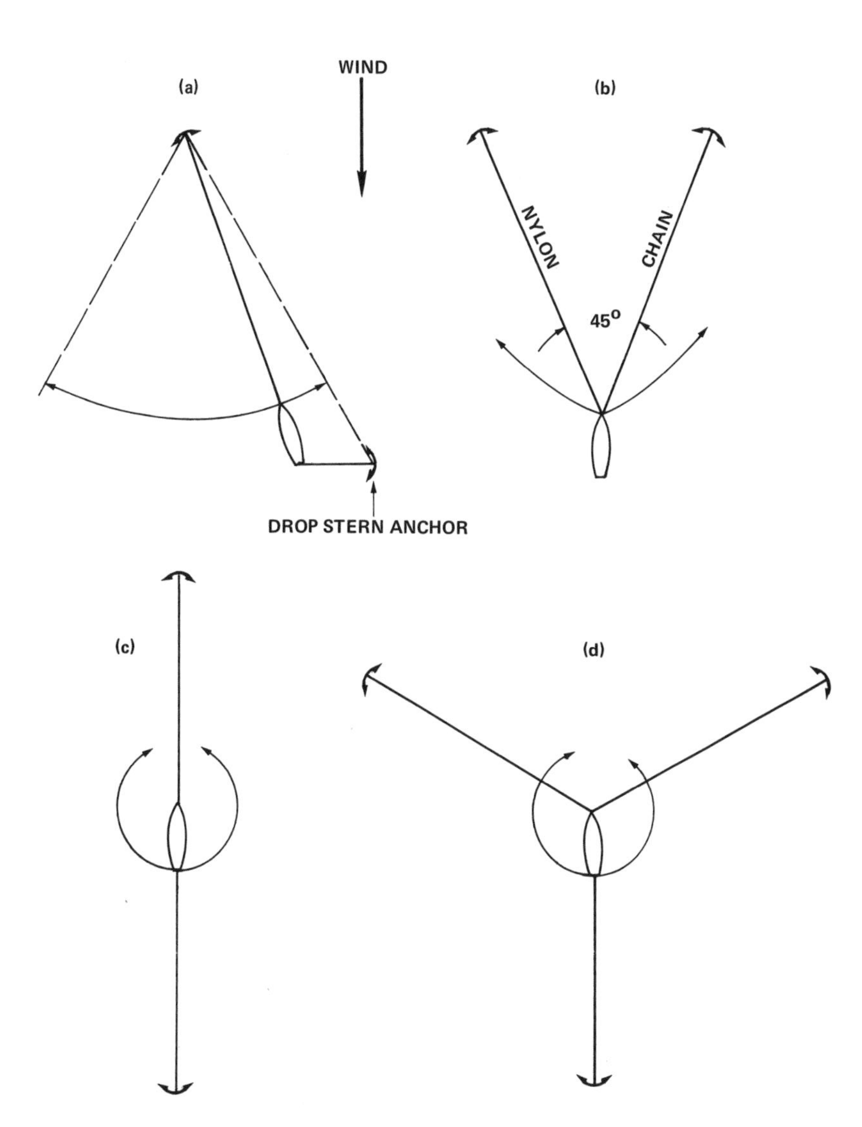

Fig. 151. Four multiple-anchor methods of preventing swinging. Method *(b)* is most practical for strong winds in shallow water.

For boats that do not take kindly to stern seas, swinging to a bow anchor can be minimized by any means that resists a vessel's tendency to fall off and forereach across the wind:

1. If the wind is not too strong, a storm staysail rigged as far aft as possible, and sheeted tight amidships, may keep an unruly boat's head up—particularly if assisted by corrective rudder during passage of wave crests.
2. A small sea anchor or a barely buoyant tire casing trailed from the stern on a long nylon line will also provide a restoring moment that reduces swinging radius. The average effect will be optimum if the drag is half a wavelength astern, since maximum drag then occurs when the vessel is in a wave trough and most likely to sheer off. Nylon is essential for stern drags; an inelastic line is apt to part from sudden strains, or to slack and foul propeller or rudder.

3. A forward bridle on the bow rode also may provide restoring torque, and allows the bow to lift more easily to steep seas.

But, irrespective of which end of the boat you elect to anchor from, the ensuing discussion of elastic anchoring makes due allowance for the worst conditions, including wave and current forces to be expected in high-wind, limited-fetch situations. Lastly, I would like to emphasize that storm anchoring is only a modest variation of ordinary anchoring, requiring a few additional precautions, a different anchor, and perhaps a little advance practice—just to make sure it works.

ELASTIC ANCHORING

While the force-displacement equilibria of a catenary chain have been recognized for centuries, the useful working stretch of a chain loaded near to its design tension is extremely limited. In fact, with a scope of 7 (a commonly recommended value), the difference in surface displacement between a chain with half its length lying on the bottom and one pulled bar-taut is *less than one percent!* Thus, in a sudden wind gust, or even moderate seas, a vessel is apt to come to the end of its tether with a jerk that parts the chain or breaks the anchor out (plate 28).

True elastic anchoring is a concept that perforce awaited the development of nylon rope, which can be stretched by as much as 50 percent without breaking, and by 25 percent without permanent elongations.* Today, most authorities agree that only chain-nylon combinations are worth considering for anchoring small vessels in strong winds, and that chain is best regarded as a flexible, abrasion-resistant weight that permits the nylon-chain junction to be kept clear of the bottom, while at the same time insuring that the anchor stock angle never exceeds about 8°. The object of this section is to show how best to proportion these elements for optimum versatility: as we shall see, it is perfectly feasible to design an anchoring system adequate for everyday use that can also be configured for storm anchoring.

Static Rode Equilibrium: Aside from anchor type and holding power (discussed later), the essence of elastic anchoring is proper rode selection, relative to vessel displacement, water depth, and the prevailing wind and wave forces. In this section we show how a rode behaves under steady forces, and particularly how much safe stretch is available to accommodate random motions as the static loading is changed.

Figure 152 shows the schematic steady-force equilibrium for chain and nylon rodes, to which the following definitions apply:

d = diameter of nylon rope (in.)
d_c = rated diameter of chain (in.)
h = water depth (ft.)
l = equilibrium (stretched) length of nylon rode (ft.)
l_o = initial (unstretched) length of nylon rode (ft.)
l_c = constant length of chain rode (ft.)
w = unit weight (in water) of chain rode (lbs./ft.—nylon is considered weightless)

*Polypropylene rope has nearly comparable stretch properties, but its lower strength, poor chafe resistance, and the difficulty of making secure hitches and splices make it an inferior alternative to nylon.

Plate 28. The limited elasticity of a catenary chain is dramatically illustrated by this photograph of a Coast Guard cutter storm-anchored off Cape Cod. Her starboard chain has already snapped, and she shortly dragged her port anchor aground.

F_h = average horizontal wind force on vessel = anchor-holding force (lbs.)
P_m = $F_h/\cos\theta_m$ = rode tension at top (lbs.)
T_b = breaking tension of nylon rope (lbs.)
T_{bc} = breaking tension of chain (lbs.)
W_c = wl_c = total weight of chain rode (lbs.); also the anchor stock hold-down weight for a nylon rode
θ_m = angle between horizontal and tangent to rode at top (degrees)
θ_o = lead = angle between bottom and anchor stock (degrees)

In addition to which we will define a few other quantities as needed. If this list seems long, it includes the minimum number of variables required to describe rode equilibrium, which shows that even static anchoring is a rather complicated business.

For any nylon-chain combination, there is a very simple geometric relation between the ratio of (chain) weight to horizontal force and the angles θ_m and θ_o that the rode makes with the surface and bottom, respectively:

(1) W_c/F_h = tangent θ_m − tangent θ_o

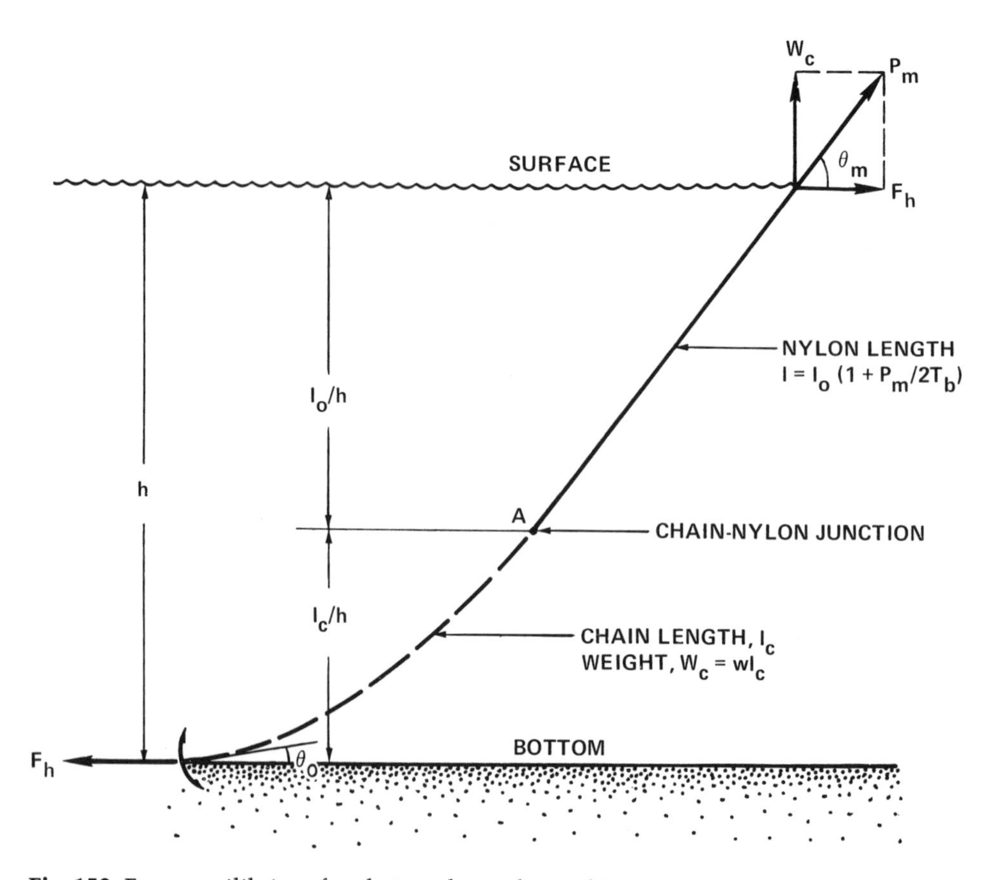

Fig. 152. Force equilibrium for chain-nylon rode combinations.

The bottom angle, θ_o, is always smaller than the surface angle because of the curvature of the catenary chain. Thus, if we assume as a design condition that the surface angle shall never exceed the safe anchor stock angle (θ_o = 8°) under the maximum design surface force F_h, then we can forget about the precise chain curvature; once properly set, the anchor will hold, and may in fact continue to bury itself in soft bottom.

Under maximum load (small chain curvature, and θ_m = 8°), the geometry of figure 152 also leads to a second, approximate, relation,

(2) $\text{sine}\ \theta_m \approx h\ /\ [l_c + l_o(1 + P_m/T_b)]$

where the quantity in parentheses is the amount the nylon rode can stretch beyond its original length l_o. Recalling the trigonometric law for small angles, tangent 8° ≈ sine 8° ≈ 0.14, and taking, as a limiting design load for nylon rope, P_m/T_b = ¼ (half the nonrecoverable stretch), Equation (2) can be reduced to a simple relation between nylon and chain lengths, and water depth:

(3) $l_o + 0.8l_c = 4.8h$

Rode Selection: While Equation (3) applies generally to steady forces, it is subject to certain limitations on chain length and anchoring depth, in order to accommodate surge and heave motions. Figure 153 is an expanded version of the lower left corner of the CSS diagram (fig. 79, p. 189), which is intended to cover practical storm-an-

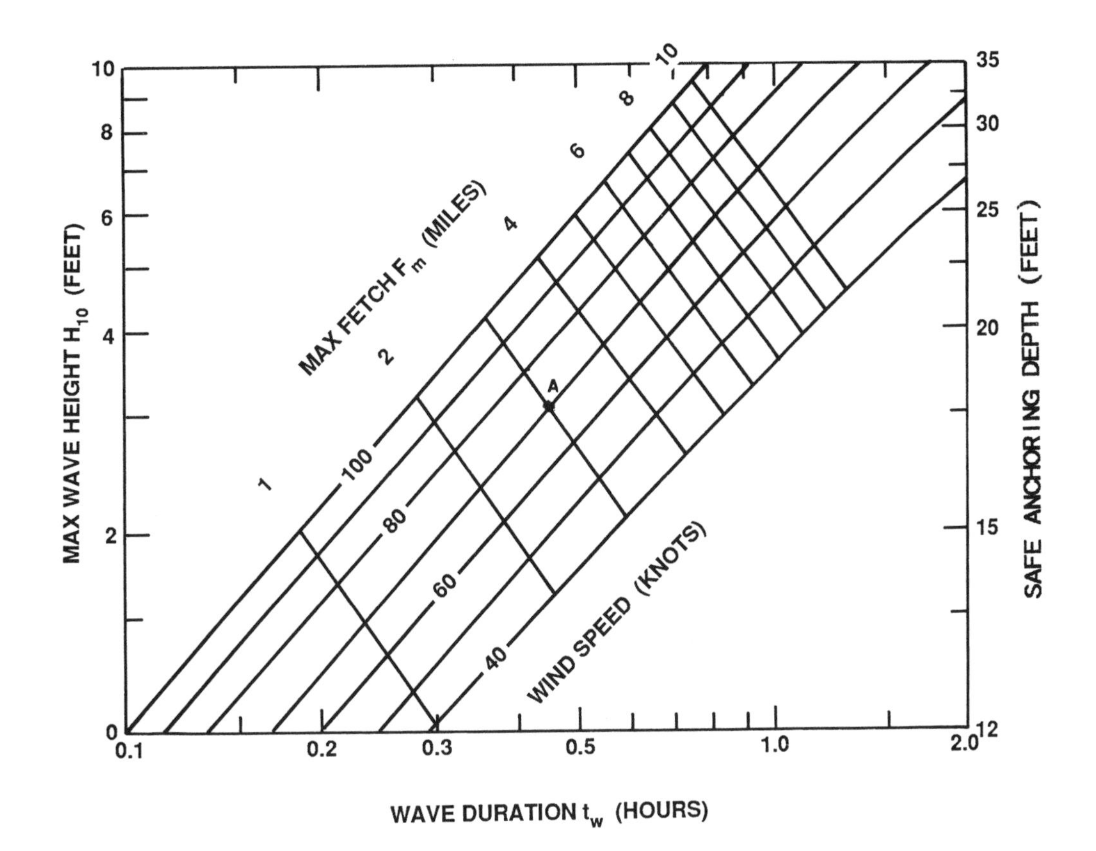

Fig. 153. CSS diagram for small fetches and high winds. Safe anchoring depths are free from wave enhancement by shoaling.

choring situations. It covers wind speeds of 40–100 knots blowing over fetches of 1–10 (nautical) miles, which produce maximum wave heights in the range of 1–10 feet, corresponding to minimum safe anchoring depths of 12–35 feet, respectively. The latter depths refer to mean low water, and include a 10-foot provision for keel depth. As an example, point (*A*) shows that, when limited by an upwind fetch of 3 miles, an 80-knot wind can only produce 3.5-foot breaking waves, if you are anchored in at least 17.5 feet of water. In this water depth, horizontal wave surge will also be limited to the wave height.

Equation (3) is quite flexible as regards the respective lengths of chain and nylon; the anchor scope diminishes from 7.2 with an all-chain rode (no elasticity) to 4.8 with all-nylon (no abrasion protection). Since you are not depending upon the chain weight to restrict anchor stock angle, the chain need only be long enough to provide a convenient, abrasion-resistant connection to the anchor. But the shorter the chain, the longer the nylon you will need to satisfy Equation (3).*

If you have a windlass, and normally anchor with a chain, you can put a false link at a convenient distance from the anchor, which can be broken to splice in the proper

*For boats without a chain winch, making the chain length equal to freeboard height plus keel depth lets you know you are free to get underway as soon as the nylon-chain junction appears over the bow roller.

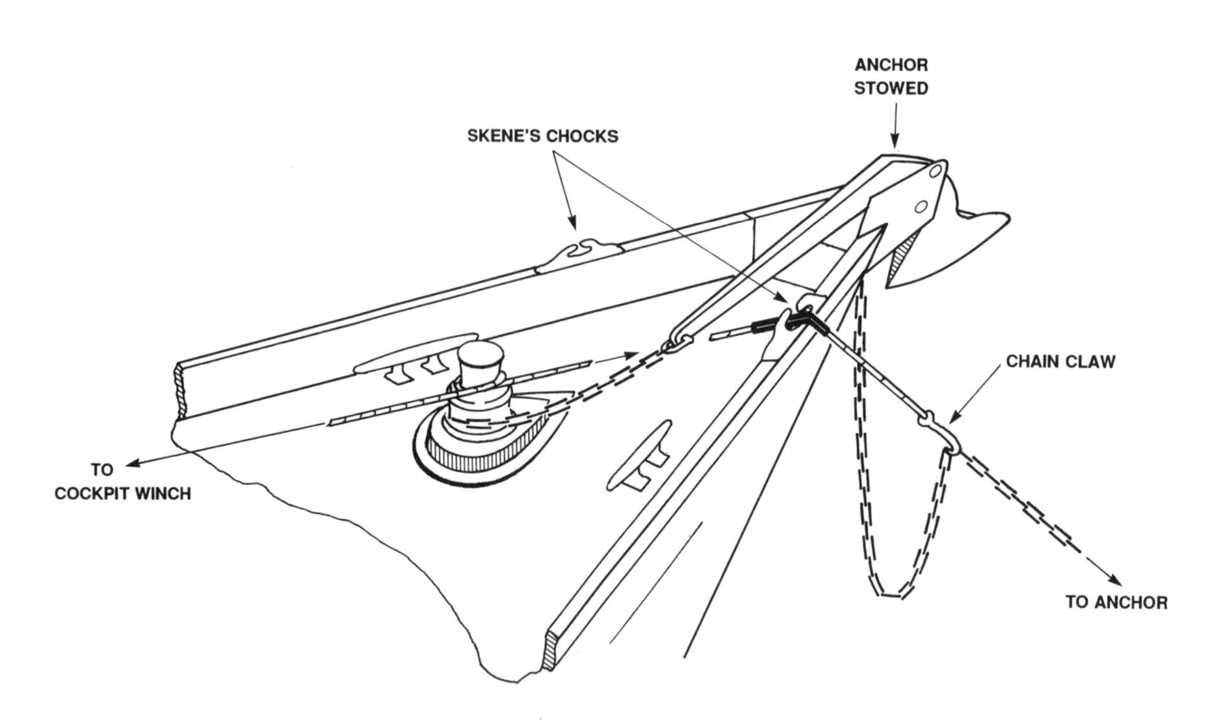

Fig. 154. Compliance of all-chain rode in daily anchoring can be increased with nylon snubber. Note chafing gear where nylon rode passes through metal chock.

length of nylon for storm anchoring. In such cases, rode compliance can be increased by a nylon snubber (fig. 154), whose length can be adjusted to suit current conditions.

So far, we have said nothing about rode strength, which is determined only by maximum wind and current forces. Allowing for a ± 30° swinging arc (fig. 113, p. 275), and adding 10 percent for induced current drag, we take the design horizontal force to be

(4) $F_h = 0.20V^2W^{2/3}$ lbs.

for powerboats, which gives an extra 15 percent margin over sailboats under bare poles. This load factor is shown on the left vertical axis of figure 155, corresponding to wind speeds of 40–100 knots, and boat displacements of 1–100 tons. Assuming, as above, that the maximum rode tension never exceeds 25 percent of breaking, the right-hand scales show the respective nylon and chain sizes necessary to withstand forces on the left-hand scale. These sizes reflect current manufacturer's tables for three-strand, marine-grade nylon Gold Line, and galvanized BBB-proofed chain. While somewhat stronger, braided nylon will not meet our required specifications for abrasion resistance or elasticity, and once the braid chafes through, the line is worthless.

Comparison of the rode sizes in figure 155 with those recommended by the American Boat and Yacht Council reveals that the latter are about twice as strong for the same boat size and wind speed. Fine! Their recommendations are based upon 12 percent of breaking strength, instead of the 25 percent assumed here. If your boat is already equipped with heavier chain and nylon, go ahead and use them. There is no weight penalty for heavy chain—except for the effort of hauling it back aboard—but you should still obey the rode length rule for storm anchoring. There is at least one well-documented case of a boat that survived 100-knot winds in a limited fetch, while anchored by almost exactly the tackle suggested here (see page 393).

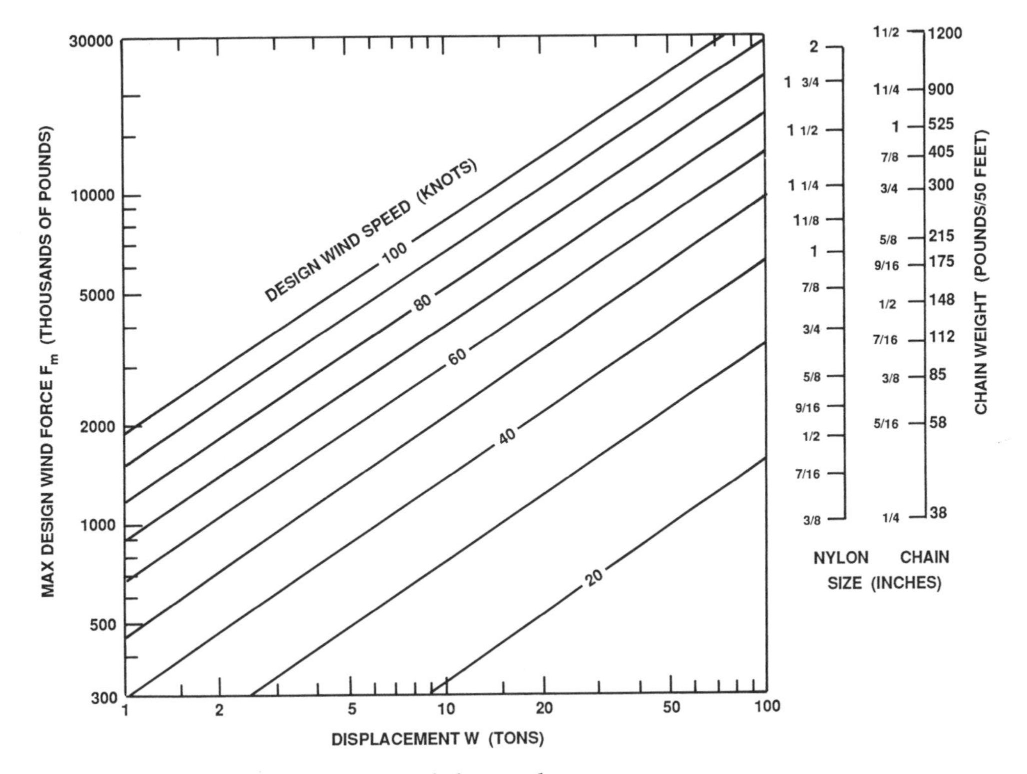

Fig. 155. Strength chart for nylon and chain rodes.

Anchor Selection: Storm anchoring compounds most of the problems of ordinary anchoring. You are looking for good holding ground in wave-sheltered water, deep enough to minimize wave enhancement by shoaling and with as much room to leeward as possible. If you are familiar with the area, you may know what the bottom is like; otherwise, your charts and depth recorder may provide clues. Lacking these, it may be best to rig two different types of anchors in tandem (see page 395) in order to increase your chances of a safe hookup. Even with known bottom conditions, and safe loads determined from figure 155, the best choice of anchors remains controversial. Earl Hinz's excellent book* reviews the current (light-wind) choices, but provides little analytical help. If yours is "not to reason why," you can refer to his convenient table. Otherwise, you will want to read the new and much broader attempt to fathom the murky depths below.

First, it is convenient to divide all anchors into two general classes, which I will call the *spades* and the *hooks*, according to whether their holding power depends upon the cohesive shear resistance of sediments, or upon hooking up on something solid. Of the spades, for storm anchoring at least, there are three serious contenders: the American *Danforth*, the British *CQR*,** and the Scottish *Bruce*.

Pound for pound, the Danforth has the largest area, and, once well-embedded, it has unquestionably the highest holding force. But, because of its large area-to-

* *The Complete Book of Anchoring and Mooring* (Centreville, Md.: Cornell Maritime Press, 1986.)

** The CQR anchor was invented by the renowned British scientist, Sir Geoffrey Taylor, a man with little boating experience, showing that application of sound basic principles can sometimes supersede many generations of practical experience.

weight ratio and limited fluke travel, it tends to kite from a moving vessel (a serious disadvantage in emergency anchoring), and may glide some distance over hard bottom without digging in—particularly if fouled by grass or weed. The CQR does not kite, and is renowned for its ability to roll and bury, even at relatively large shank angles. The Bruce is similarly good at self-engagement, and holds well in hard mud and sand, but pulls out rather easily at high shank angles. In clay, it tends to cake up, and must be cleaned before resetting. All things considered, I would give the edge to the CQR when anchoring blind under duress. But, given some knowledge of bottom conditions, one can do much better than groping in the dark.

Among the hooks, the best choice is undoubtedly the yachtsman, or *kedge*. It will hang up on any ledge or crevice, bury through weeds and engage against large cobbles, and even give fair holding in hard sand. It is particularly good in coral, although it may be difficult to dislodge if it falls into a hole. But a hole may be the best place to drop the hood in shifting wind. I have personally planted many an anchor in holes by diving, when working around coral reefs in the presence of high waves and strong currents. Thought of as a hook, rather than a spade, a kedge's holding power depends only upon its bending strength, for which recommended sizes usually suffice.

Returning to spade selection, the object is to equate a vessel's maximum rode tension (Equation (4), above) with anchor holding force. The latter is well demonstrated to be proportional to fluke area (i.e. to the two-thirds-power of its weight, w), multiplied by an appropriate sediment cohesion coefficient, C_s,

(5) $0.2V^2W^{2/3} = C_sK_aw^{2/3}$

where the fluke-area proportionality coefficient K_a depends upon the type of anchor. Tentative values for both coefficients are given in table 11. For example, anchor coefficients are graded according to fluke area per unit weight. Lighter versions of Danforth and plow anchors might be accorded slightly higher coefficients, if suitably tested.

Table 11. Bottom sediment (C_s) and anchor fluke (K_a) coefficients.

Type of sediment	C_s	Type of sediment	C_s	Type of anchor	K_a
Silty clay	1	Sand with clay	6	Danforth HT	10
Silt with shells	2	Sticky clay	8	CQR	7
Coral sand	4	Stiff clay & sand	10	Bruce	6

Sources: C_s: US NAV DOCKS DM-26 (11/59); K_a: E. Hinz, *Anchors and Anchoring,* Table 6-1.

Rearranging (5), and taking the product of the coefficients C_s and K_a to be a single "anchor coefficient," C_s, we obtain an anchoring equation relating all relevant variables:

(6) $w/W = V^3/10C_a^{3/2}$

We note immediately that, other things equal, anchor weight is proportional to the *cube* of wind speed; if the wind doubles, you must increase the anchor weight by *eight!* This immediately dispels the usual notion that a storm anchor is "one size larger" than a working anchor; it is more apt to be three times as heavy.

Equation (6) is plotted in figure 156, which might be called an anchor analysis diagram (AAD), since it relates anchor and vessel weight to wind speed, anchor type, and sediment shearing stress. In this form, it is far more versatile than an anchor selection chart. Knowing any four variables in Equation (6), the fifth can be found by inspection, as shown by the following examples:

1. You enter an anchorage where the bottom sediment type is noted on a chart, and want to know safe anchoring wind speed. Divide your anchor weight by your boat displacement, and project a horizontal line from the left scale; look up anchor and sediment coefficients in table 11, multiply them together, and project a vertical line up from the bottom scale. The intersection of these lines gives desired wind speed.

2. You are outfitting your 45-foot, 10-ton ketch for a South Pacific cruise, and wish to buy a storm anchor suitable for coral sand in 40-knot winds. As a trial, pick a Danforth Hi Tensile, so the $C_a = K_a \times C_s = 10 \times 4 = 40$. Project a line up from bottom to the 40-knot wind curve, and then left, to read $w/W = 5.5$. The required anchor weight is $w = 5.5 \times 10 = 55$ lbs. Corresponding working anchor (30 knots) and lunch hook (20 knots) weights would be 25 and 7.5 lbs., respectively. However, if your working anchor is a CQR, it would have to weigh about 40 lbs. ($C_a = 7 \times 4 = 28$).

3. In 1983, Jack Ronalter rode out a 100-knot typhoon in Tahiti lagoon. His 30-foot, 3.5-ton sloop was anchored in 45-feet of water on a 35-lb. CQR and a 22-lb. Bruce, deployed in parallel. Both rodes satisfy the rode length and strength equations, (3) and (4). Is his experience consistent with Equation (6)?

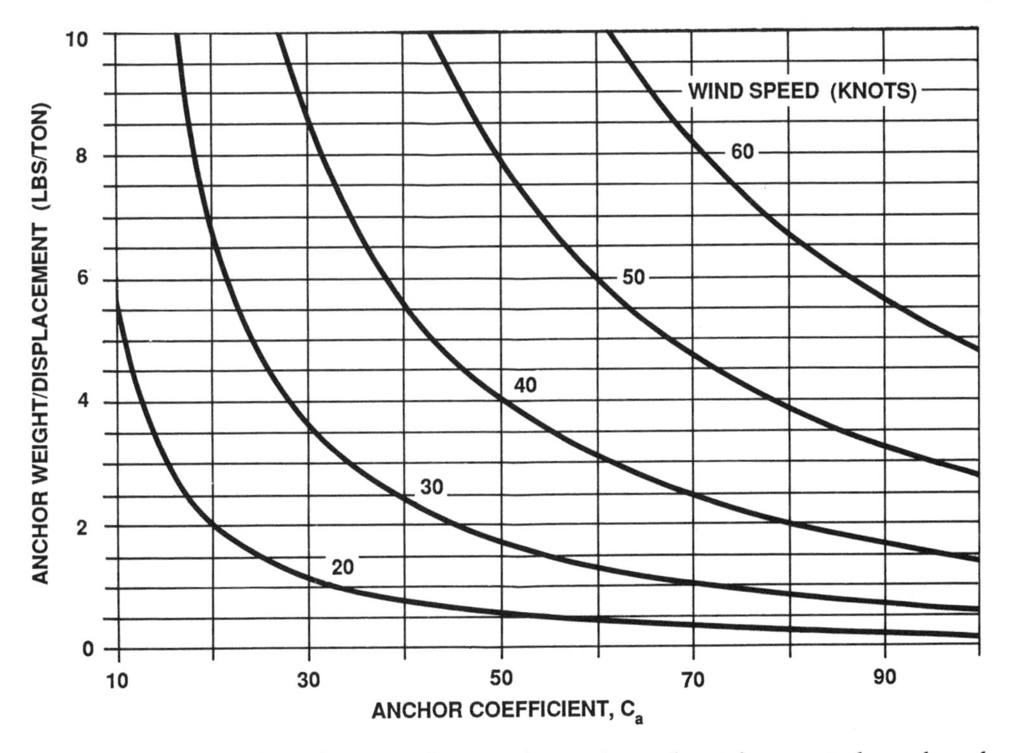

Fig. 156. Anchor analysis diagram relates anchor and vessel weights to wind speed, anchor type, and soil strength.

For parallel (or tandem) anchors, we must find the wind speed for each separately; the resulting wind speed is then the square root of the sums of the individual speeds. Taking $C_a = 6$ (sandy clay), we have:

Anchor	w/W	K_a	C_a	V (knots)	$V_t = \sqrt{(V_1^2 + V_2^2)}$ (knots)
CQR	10.0	7	42	49	
Bruce	6.3	6	36	39	63 knots

But the 100-knot windspeed was measured at Maeva airport, 33 feet above ground, whereas Ronalter's sloop's center of effort was about 5 feet. Wind speeds characteristically increase as the logarithm of altitude; i.e., as $63 \times \log (33/5) = 121$ knots, which is close to the maximum gust velocity (132 knots) reported at the airport. Thus, the anchor equation appears to work well enough for all practical purposes.

Special Provisions: Most boats today are simply not suited to bringing synthetic rodes aboard under high tension in rough water. Instead, they come equipped with small mooring chocks and cleats, and with a stemhead weldment that carries one or two anchors resting on rollers wide enough to pass chain, rope, or connecting fittings. This arrangement serves well enough for ordinary duty, but stories abound about stemhead fittings being bent sideways or torn clear off in storms, and of ropes chafed through wherever they rub against a sharp corner. Thus, it behooves those who put frequent—and occasionally heavy—demands upon their anchor tackle to make the few inexpensive modifications necessary for effective storm anchoring.

Whether anchored by bow or stern, the most appropriate interconnection between your boat and its anchor rode is a mooring bridle, preferably of braided dacron, having twice the breaking strength of the rode. At its seaward end, the bridle is made into an eye by seizing and whipping around the thimble of a Nylite connector*, which is shackled, via a swivel, to a similar eye in the anchor rode (fig. 157). At its upper end, the bridle is preferably hitched directly to large horn cleats let into the boat's toe rail. The cleats can replace your mooring chocks, but should be through-bolted to backing plates, so as to carry the same load as your stemhead fitting. As in ordinary mooring, the bridle eyes should be chafe-dipped, or otherwise protected by stitched leather wrapping, where they contact the cleats. With such a system, there will be no relative abrading motions in any part, and ordinary anchoring can be achieved simply by removing the bridle.

Under moderate conditions, even bluff-transom power craft can be comfortably anchored by the stern quarter, using a bridle, but with unequal parts that can be adjusted for optimum stability under prevailing conditions. With powerboats that cannot anchor safely by the stern, there is always danger that a forward bridle may chafe against the boat's cutwater when it is plunging and sheering. In such cases there may be no practical means of bringing a rode aboard, other than by shackling it to a short length of anchor chain that passes over the stemhead bow roller. In such cases, the chain *must* be constrained to prevent its springing out of place, because it can do a great deal of damage by thrashing about.

* Available from Samson Ocean Systems, Ferndale, WA 98248.

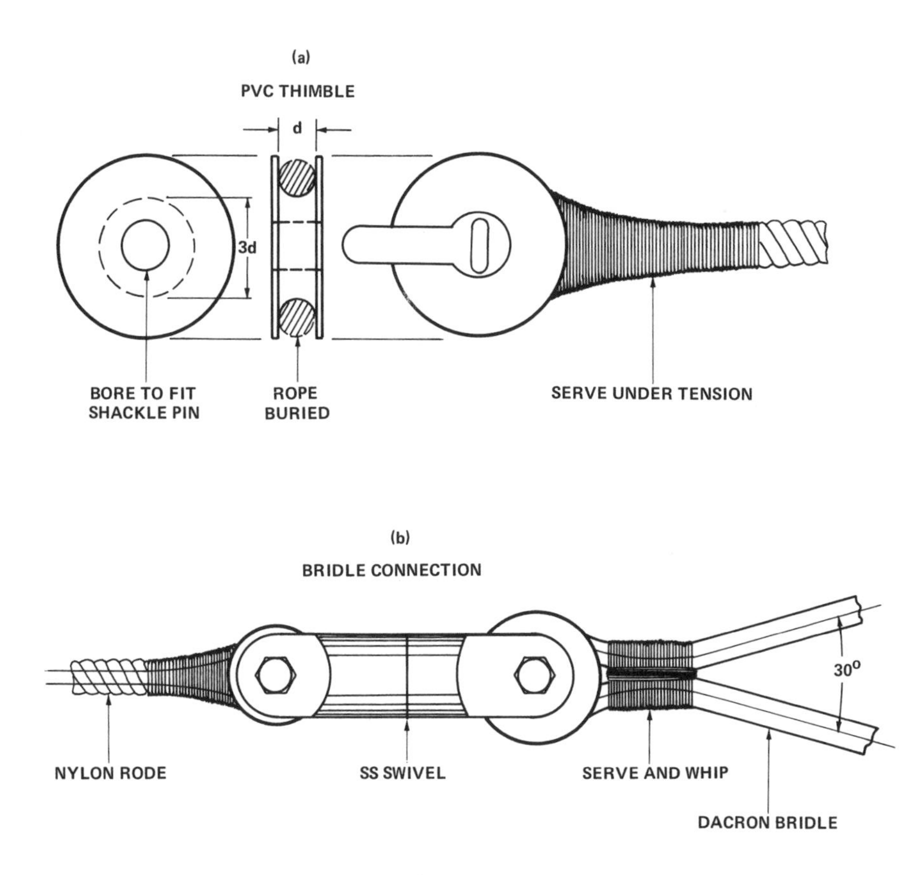

Fig. 157. *(a)* Special Nylite thimbles for splicing a nylon rode to develop full breaking tension. *(b)* Method of joining rode to a Dacron bridle with an interconnecting swivel.

But whatever system you employ, it should be practiced in advance. By mooring the boat to an offshore buoy in a moderate sea, you can easily make adjustments to obtain optimum riding behavior and freedom from fouling.

Where holding ground is uncertain, it is desirable to rig two anchors in tandem. This is best accomplished by shackling the second anchor to 5 fathoms of chain, and the latter to a ring or eye welded to the crown of the first anchor (Danforth or kedge), or to the shank at the point where it bends down (CQR).* The plane of this eye should be vertical, and in line with the anchor shank, so as not to inhibit normal hinge action of the fluke(s). Since all rode combinations are designed for an 8° shank angle, proper holding of the first anchor will in no way be affected by the second. If using a spade and a hook, the spade should come last; if two spades, the smaller last. There is no advantage in using two hooks.

Anchoring Precautions: Given a limited wave fetch and the proper tackle, storm anchoring reduces to proper leeway from obstacles and other boats, and a good bottom hookup.

*A crown eye is a general convenience on any anchor, not only for tandem rigging, but also for attaching a trip line, or for catting the anchor aboard.

With an effective scope of 7 (stretched nylon), you will need a clearance circle of at least eight water depths around you, and, hopefully, no one else anchored to weather.* Cyclonic storms usually imply at least 90° wind shifts, and most of the damage from storm anchoring results from interboat collisions, or from fouling one another's anchor lines by dragging. If at all possible, it is better to relocate than to be fouled.

The anchor should be set by lowering it slowly to the bottom, backing down as you pay out the rode, and then applying full engine power until cross-bearings indicate you are not moving. If you drag, and need to relocate or change anchors, you can always overhaul the bridle and rode by hand. If conditions permit, it is always desirable to inspect the hookup, either by dinghy and lookbox, or preferably with a faceplate and snorkel. In charter work, I carry a scuba regulator, and rent an air tank, solely for the purpose of hand-setting and -recovering anchors. In an unfamiliar location, there is nothing so comforting as to know that you are securely anchored. Even so, one should always maintain an anchor watch in storm situations. If you drag, you may have to put out an extra anchor by conventional means, and then stand by to prevent chafing. We assume you have anticipated this problem and have a second (and third?) system ready.**

SEA ANCHORS AND DROGUES

Sea anchors and *drogues* are collapsible, parachutelike devices that can be deployed to reduce the drift speed of a vessel. The distinction between these terms is largely one of size, relative to boat length. Sea anchors—usually over-age, surplus military parachutes—are commonly utilized by commercial fishermen to remain overnight in the vicinity of fish schools without burning expensive fuel. Drogues are smaller devices that can be towed astern to moderate speed and maintain steering control in heavy weather. In common with anchoring, there is very little engineering guidance in the literature regarding appropriate sizes and deployment procedures.

Drogue Analysis: Once deployed, a vessel is in equilibrium between the wind force and the combined drags of its underbody and the drogue, all of which are interrelated. While the drogue drag depends only upon its shape relative to the water, wind and hull drag can vary by a factor of five between relative wind headings of 0° and 90° (fig. 114, p. 277). Thus, a small drogue, which might have little effect on a vessel's broadside drift rate, may exert enough bow (or stern) torque to swing her end-to, and allow her to run off in this fashion at a controlled rate. This technique has been used repeatedly by cruising yachtsmen, and the only new factor is the ability to design a drogue for this particular purpose.

Because there are so many variations of hull form, one cannot readily generalize drift rate on random headings. We can, however, make reasonable drift estimates

*Stern-anchored in an inlet of Roatan, Honduras, during a 40-knot norther, I came on deck for a midnight anchor check, only to see a 200-ton shrimper dragging silently past, so close that his trawling boom missed our shrouds by a foot. My yell woke up the only crewman aboard, who had been asleep the whole time.

**A good way to do this is to have the rode(s) already made up and stowed in a plastic bucket with the chain on top, so that one need only shackle on the desired anchor and lower it over the side. In calm weather, you can put the bucket and anchor into a dinghy, row it over to an auxiliary site, lower the anchor by hand over the transom, and row the bitter end back to the boat.

for vessels riding to a drogue large enough to keep a heading within, say, 10° off the wind. Assuming, in addition, that the vessel's drift speed with respect to the water is slow enough that all drag forces vary as the square of velocity ($V_s / \sqrt{L_w} < 1.0$), and the above- and below-water hull drags scale similarly with displacement, my formula for the ratio of a drogue's inflated diameter D to vessel waterline length L_w is

$$(7) \qquad \frac{D}{L_w} = \frac{1}{17}\left[0.021\left(\frac{V - V_d}{V_d - V/30}\right)^2 - 1\right]^{1/2} \geq \frac{1}{17}$$

where V is the wind speed, and V_d the vessel speed over the bottom.

Because of similitude, this formula contains no term for vessel displacement. It says that, if we proportion drogue diameter to waterline length, all vessels will drift at the same rate for the same wind speed. This can obviously be so only if the drogue contributes most of the drag, which is the reason for the inequality on the right end of the above equation.

The drogue equation is graphed in figure 158, for three different wind speeds. A horizontal line through any point on these curves indicates the drogue size necessary to slow a vessel to the true drift speed read vertically below the same point. Vessel speed V_s, relative to the water, will be less than V_d by the magnitude of the wind current, V_c taken here as one-thirtieth the wind speed (small arrows at lower left). If extended upward infinitely far above this diagram, each wind speed curve would become tangent to its respective wind current arrow, showing that, no matter

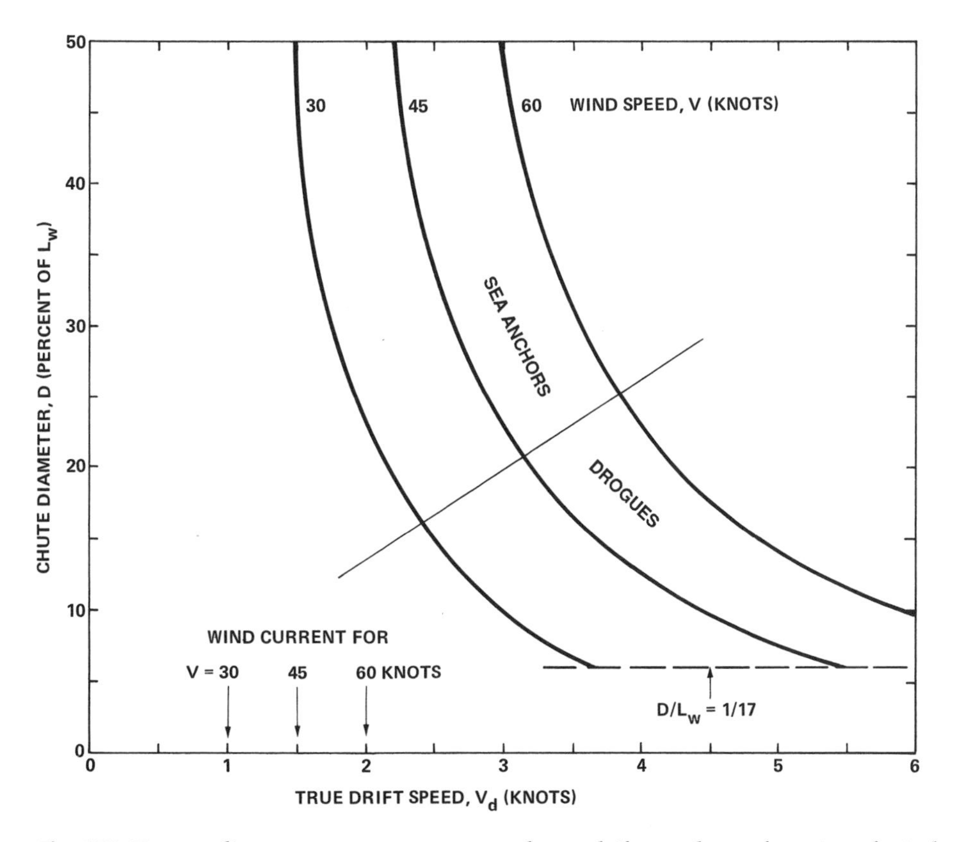

Fig. 158. Drogue diameter necessary to control true drift speed, as a function of wind speed.

how large the drogue, it cannot slow a vessel below the surface drift speed V_c. To the lower right, these curves terminate in a dashed line giving the minimum drogue size (D/L_W = 1⁄17) for which hull resistance begins to dominate drift equilibrium. The diagonal line is a compromise between these extremes, effectively dividing the diagram into two regions: sea anchors, intended for drifting restraint in light and moderate winds, and drogues, for heading control and speed moderation under storm conditions.

Because of their larger size, sea anchors are effectively restricted to parachutes. Drogues up to about 3 feet in diameter can be purchased from marine suppliers. For intermediate sizes (3–8 ft.), I suggest the do-it-yourself designs shown in figure 159, which can be made up by your sailmaker from the same material as your storm trysail. All dimensions are in units of *D*, the nominal inflated diameter (fig. 158). The design folds flat, opens by itself, and should be deployed with a bridle on your nylon anchor rode.

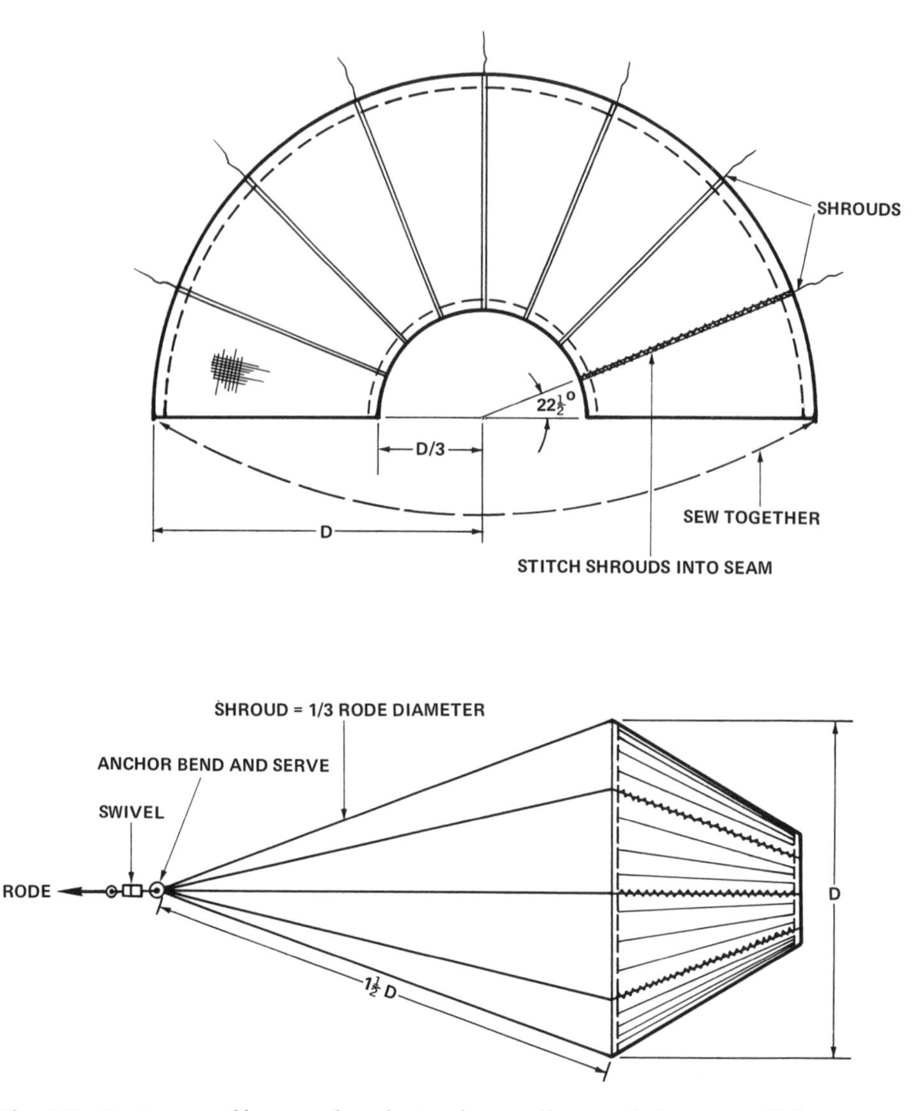

Fig. 159. Do-it-yourself sea-anchor design for small control drogues. All dimensions are in units of *D*, the nominal drogue diameter, obtained from figure 158.

Deploying Parachute Drogues: Given a choice, parachute sea anchors should also be sized according to figure 158. Buy only nylon chutes; canvas ones tend to mildew in a damp environment. For storage, remove the chute from its pack, and discard everything except the main chute, shrouds, and D-rings. Stretch the chute out on a level surface, and rig a ⅜-in. Dacron trip line *internally* from the crown straps to either D-ring. Pass the line through the ring, and tie it back on itself with a rolling hitch, leaving about a 6-foot excess. Chain-gather the shrouds into a compact bundle, and stuff the chute, crown-first, into a small nylon duffle bag.

To deploy the chute, rig your bridle and nylon rode on deck, and shackle the bitter end, via a swivel, to the chute D-rings. Feed the chute, crown-first, off the weather side, holding the rode near the D-rings until the chute fills, and then paying the rode out as the boat drifts.

To recover the chute, overhaul the bridle by hand until you can get the rode on your genoa sheet winch, then haul it in until you can reach the trip line. Untie the trip line from the D-ring, and cleat it astern. Pay off the rode a little, and the chute will trip itself, so that it can be hauled aboard by the trip line.

29

"Man Overboard"

SIC TRANSIT HOMO IN MARE

Losing someone overboard in the open sea is surely one of the greatest calamities—particularly if the victim cannot be found after a reasonable search, and is suspected to be paddling aimlessly in vain hope of rescue. Yet, the annals of the sea are sprinkled with just such happenings, as testified by the lucky few who have survived. In *Invisible Horizons,* Vincent Gaddis cites several graphic accounts, here paraphrased for brevity:

1. *(July 1908)* Seaman James Wilkinson was torn from the lifelines on the deck of the battleship USS *Minnesota* by a huge wave; some minutes later, another wave deposited him safely on the deck of the following USS *Vermont.*
2. *(? 1958)* Chief Officer Francis Schremp of the freighter SS *John Lykes* was washed overside off Bermuda; the next wave washed him back aboard.
3. *(November 1956)* Seaman John Craig was swept from the trawler *Dorileen* off Aberdeen; washed back aboard a half-hour later, he found the skipper was out looking for him in the ship's launch.

Not all were rescued so promptly:

4. *(October 1962)* Swedish sailor Per Svahlin disappeared from forward lookout aboard the freighter *Horn Crusader* off Santa Barbara in a heavy fog. Thirty-five minutes later, her skipper turned to look for him, and took as long again to reach the same dead-reckoned position. Seven minutes later—after altering course by 7° *on pure impulse*—he was heard shouting.
5. *(December 1965)* Seaman Arne Nicolaysen slipped on a wet deck and fell unnoticed from the SS *Hoegh Silver Spray* into the Gulf of Mexico. He swam for 30 hours, and was miraculously rescued by the tanker *British Surveyor.*

Such accidents seem less common aboard small boats—perhaps because of closer interdependence of crewmen. In July 1957, after fifteen eventless Transpac races, the cutter *L'Apache* was midway to Honolulu. Crewman Ted Sierks, leaning outboard against the boat's lifeline to repair a broken boom tackle, fell overboard when the line parted. Sierks caught the log line astern, but it broke. He was seen to reach one of two life rings thrown him, but *L'Apache,* under spinnaker in a 30-knot wind, took 11 minutes to jibe and tack back. They found the second ring—but not Ted, who drifted for 30 hours in heavy seas, unable to attract the attention of any of a multitude of yachts, ships, and aircraft that were combing the ocean around him. He was finally sighted by the navy escort U.S.S. *Munro* on her last leg before scheduled search curtailment. He was estimated to have drifted about 50 miles.* This widely publicized rescue deserves special attention because it bears on some key points of safety and survival:

1. Sierks wore no safety harness—else he probably would have been recovered quickly.
2. *L'Apache* took 11 minutes to jibe and turn back; I have seen a spinnaker doused in 2.
3. They found the empty ring but could not sight Ted, who could not have drifted more than 100–200 yards away in that length of time.
4. A number of search vessels passed within a quarter-mile without seeing him.
5. The search was due to be abandoned when he was found.

Calamity has a way of raising embarrassing questions. But, without pointing a finger, one can ask how much different it would be if Ted Sierks' accident happened today. The answer is, much better, but still more could be done. Specifically, Sierks might expect to be thrown (and be able to reach) a flagged marker pole with attached horseshoe buoy, drogue, and self-actuating strobe flasher. If the crew of *L'Apache* practiced *quick-stop* procedures now recommended by the USYRU, they might be able to turn the boat within a half-mile, which is barely within the down-sun visibility of a strobe in daylight, and Sierks might be plucked out of the water within a half-hour or so. If they took a mile to stop the boat, they could easily miss him, and would be forced to adopt some sort of area search of the type discussed below. With 5-mile strobe visibility at night, he still would have a good chance of being sighted by dawn the next morning. But if Sierks were thrown a buoy equipped with an EPIRB, he would be located by patrol aircraft within a few hours; with a portable VHF transceiver, he would be found by *L'Apache* within a few minutes (see below).

Statistics of "man overboard" accidents are hard to come by, but such accidents are undoubtedly more common than one might suppose. Who has spent much time at sea without falling in the drink a few times, or witnessing a dozen others? At the end of a week's cruise aboard a 50-foot schooner, my 3-year-old daughter, never permitted on deck without a life jacket, slipped overside unnoticed as we were motoring back up the bay to dockside. She was fished out by a following boat, and handed to us, still unwitting, as we were tying up. Surely the vast majority of cases where the victim was quickly recovered pass unrecorded; those where he or she was not are often lacking in details. Section 37 of the Federal Boat Safety Act provides for a Casualty Reporting System, which hopefully pinpoints equipment and procedures needing improvement. But safety at sea is mostly a matter of common sense, and it is a skipper's responsibility to educate himself or herself and to see that proper safety measures are always carried out. As always, this involves preparation and practice.**

*This report implies some navigational confusion; 50 mi./30 hrs. = an average drift of 1.7 knots; at a wind speed of 30 knots, a maximum drift rate of 1 knot would seem more likely at his latitude (~ 25°N).

**Practice can sometimes be carried too far. A recent newspaper story tells of a man who conducted weekly overboard drills in which his wife was always the unwilling victim. After several months, she rebelled, swam ashore, and filed for divorce.

CAUSE AND PREVENTION

Most "man overboard" accidents fall into one or more of the three categories discussed below. The remedies suggested may seem obvious, but are often neglected. Some of these have been mentioned previously, but bear reiterating in present context.

Slipping and Falling: Slipping is undoubtedly the commonest cause of falling overboard. Even in calm, dry weather, boat decks are apt to be slippery; salt incrustation is hygroscopic and attracts atmospheric moisture. Plastic decks are notoriously dangerous because they are smooth and do not absorb moisture. The common, molded antiskid patterns may actually aggravate matters by trapping evaporated salt in minute indentations. In contrast, natural, unfinished teak decking is one of the best traction surfaces—wet or dry. There are a number of antiskid preparations designed to make decks less slippery, in the form of either abrasive coatings or stick-on strips. Both are a little rough on bare feet, but crew members should be wearing deck shoes anyway.

Although intended to prevent falling overboard, most lifelines seem better designed to promote it. Marine catalogs list the standard rail stanchion height as 26 inches, which catches the average adult slightly above the knee. Small boats often have lower rails for purely aesthetic reasons. Human engineers assert that 30 inches is the minimum practical height for safety. Here, there is an obvious conflict between function and appearance that only judgment or regulation can alter. While additional protection in heavy weather can be obtained by rigging higher lines between shrouds or stays, these may interfere with working the boat, and a more practical solution is to wear a safety harness that is clipped to a fore-and-aft line amidships. In all events, a second lifeline should be provided about 8 inches above the toe rail, to catch anyone in danger of sliding overside beneath the top line. Needless to say, netting is still better, particularly where children are concerned.

Equipment Failure: Understrength or overage safety gear may be more dangerous than no gear at all, since it gives a false sense of security (Ted Sierks' accident was caused by lifeline failure). We have previously discussed the strength advantage of continuous-welded-tube guardrails, as opposed to stanchions and wire-rope lifelines. The former are much stronger, better looking, and less susceptible to local corrosion. They have found wide acceptance among power craft, but few sailboat designers seem inclined toward the obvious. The reasons are obscure: higher initial cost is soon offset by the necessity for lifeline replacement, and the wind drag of a 1-inch tube is less than that of a sheathed wire rope or twisted Dacron line of *half* this diameter.

However, things being as they are, all lifelines, harness lines, fittings, and snaps should be inspected regularly for wear or corrosion, and proof-tested to withstand a 1,000-pound outboard load (about the pull that can be developed by a strong man operating a sheet winch with a 12:1 mechanical advantage, and applied through a two-part block).

Underestimating Hazard: Most of the accidents cited at the beginning of this section are attributable to failure to recognize that occasional extraordinary seas can sweep the decks of even large ships in heavy weather. The extreme probability curves of figure 73 (p. 177) rise very steeply for expectancies of less than 1 hour. On a 4-hour watch, you can reasonably expect one wave to be at least 1.5 times higher than the average of the highest 10 percent of all waves encountered. Thus, if your lee rail is dipping under every few minutes, there is a good chance that the deck will be swept once or twice on that watch. Because hard driving in heavy weather is often the key to winning ocean races, a very close line is threaded between safety and disaster. Plenty of crewmembers fall overboard on races; the fact that more are not lost is perhaps because racing crews are attuned to emergencies and quick to respond.

RESCUE PROCEDURES

Most popular articles treat "man overboard" emergencies under the assumption that visual contact with the victim can always be maintained. This being the case, rescue is mainly a matter of proper procedure and quick action. However, it is not commonly realized how quickly someone in the water can disappear, so that rescue must be preceded by a search—with the odds heavily weighted against finding the person! There are a number of well-documented cases in which search vessels have passed within 150 yards of people floating in life preservers—even in rubber boats—without their being sighted, despite good visibility and moderate seas. At 10 knots, a boat will travel 150 yards in 27 seconds, and few boats of any size can be stopped and turned within this radius and time frame (*L'Apache* took 11 minutes). Therefore, it is important to recognize that, unless the sea is very flat, you will *probably* lose sight of the victim, and should gear your preparations and action plan accordingly.

Search is a subject in itself. There is practically nothing written on ship-sea search, to which I have devoted the closing section of this book. Prompt *rescue* procedures, on the other hand, have received considerable expert study in recent years, resulting in drastic revision of traditional methods for sailing yachts and small power craft.

Quick-Stop Rescue: Recognizing the importance of always keeping the overboard victim in sight, the objective is to turn the boat upwind and slow to maneuvering speed as quickly as possible, with someone detailed to keep watching. A buoyant life sling is then trailed astern on 150 feet of line, and the boat spirals in on the victim until he or she can grasp the line and don the sling. The boat is then headed up, the sails dropped, and the victim hauled in to weather amidships, where he or she can be lifted aboard by the sling. At night, or if the victim is lost sight of, an auxiliary, strobe-lighted, personal flotation device is thrown at the earliest possible time to mark the spot, and to give the victim a target to swim to.

The specifics of this technique depend upon the type and size of boat, and the number of crew available to assist. But, with practice, crewed sailing yachts have wrestled "man overboard" victims aboard in 2–6 minutes; small, two-hand teams (wife hauling husband on multipart tackle) in 8–30 minutes. Powerboats should take less time.

Bagged life-sling kits, complete with trailing line, are available commercially, and the most apt and detailed description of the whole procedure I have seen appeared in *Practical Sailor*, August 1988.

Delayed Rescue: But what if some misadventure prevents the boat from returning promptly and the MOB cannot be located and is believed to be out there, seeing but unseen? This is the time to deploy your marker pole, with attached horsehoe buoy, sea anchor, and strobe light, carried as required equipment by racing boats, and optionally by most cruising boats. You have then several options:

1. If you have adopted my suggestion in the first edition of this book to attach a survival kit containing a portable 5-watt VHF transceiver to your horseshoe buoy, you can simply call the MOB on your master unit, and get directions. Unofficially, the Coast Guard search and rescue (SAR) Directorate heartily recommends this procedure. It is always quicker and easier for everybody if you can find the victim yourself.
2. If you are one of the 14 percent of all offshore yachts that carries an emergency position indicating radio beacon (EPIRB) attached to your buoy, you can call the Coast Guard on Channel 16/15 and ask them to direct you to the victim's position via your SATNAV. If you do not have a SATNAV, they will still be glad to learn that it is a real emergency, and not just one of the 98 percent of all EPIRB calls that are false alarms (one-quarter of which they cannot dispel and must respond to). In this connection, the new, self-actuating Class 1 EPIRB is registered to your boat, and provides the Coast Guard with useful auxiliary data.* Class 1 EPIRBs are now required on all commercial fishing vessels.
3. With no EPIRB, you can call the Coast Guard, report your position, and stand by the marker pole. Like the MOB, it is anchored to the water, and will be the focal point for their search—and they must find you before they start. Meanwhile, you can start preparing for your own search (see below).
4. If you cannot reach the Coast Guard, your only recourse is to commence your own search, according to procedures outlined in the last section of this book.

COLD-WATER SURVIVAL

"Man overboard" search procedures necessarily involve the question of survival expectancy in cold water, the answer to which has hardly changed from shipwreck survivor data collected after World War II. Seeking to improve the Coast Guard's simplistic survival guide published in the first edition of this book, I spent several months exploring the mysterious world of cold-water physiology and persuaded myself that the rate at which a human body cools down is more easily approximated by a thermodynamic model than by a physiological one, provided that appropriate values are used for insulation and heat production, etc.**

*Even EPIRBs are not infallible. Originally outgrowths of aircraft crash-locator beacons, marine Class A and B EPIRBs broadcast a homing signal on commercial/military distress frequencies (121.5/243 MHz), and are now received by four SARSAT polar-orbiting satellites and retransmitted to 20 (mostly) northern hemisphere ground stations, where the beacon's location is determined by Doppler shift, and retransmitted to the appropriate Coast Guard SAR Coordination Center. But, to be acted upon, the signal must pass a reliability test, be within range of Coast Guard patrol limits (1200 nm), and be within range of a satellite ground station. This excludes all of the Southern Hemisphere (except Australia) and wide expanses of the Northern Hemisphere oceans, including Hawaii.

The new Class 1 (406/121.5 MHz) EPIRBs transmit a pulsed signal that can be interpreted by SARSAT as a position location, which is transmitted more rapidly to the Coast Guard's computer. Because of their unreliability, Class A and B units will be phased out, probably by 1995. Class C units, having no satellite capability, are already virtually obsolescent.

**For those interested, the relevant paper, "A Thermodynamic Model for Cold Water Survival," is published in the *Proceedings of the Antarctic Diving Symposium*, Smithsonian Institution, Washington DC, July 1990.

In simplest terms, *hypothermia* is the medical name for the body's response to tissue cooling, leading progressively to numbness, muscular paralysis, mental disorientation, coma, and death from cardiac or respiratory arrest. Initially, the body responds to cooler water by generating internal heat, either involuntarily (by shivering) or by external exercise. It can stabilize itself after a small drop in core temperature as long as it can continue to produce heat. But if the rate of heat loss exceeds heat production, the core will continue to drop until the ability to produce heat ceases, after which the body cools off exponentially, like a bag of water surrounded by an insulating jacket of fat and muscle.

The key to survival is to provide more insulation or more heat, so as to keep one's core temperature within the critical range. But the more you exercise, the faster the blood flows, cooling the extremities. Expert swimmers can generate enough heat to keep going for many hours at water temperatures as low as 50–60 degrees, but the only sure solution is more insulation. Thus, body fat is an advantage for survival, but we ordinary mortals must make do with neoprene wet suits, or bulky exposure suits in which it is difficult to navigate.

Figure 160 is a set of curves, generated from my thermodynamic equation,* showing the time it takes for the body's core temperature to drop to 86° F (hypothermic death) when immersed in water at temperatures shown on the lower scale. The curves are given in 6-mm (¼-in.) increments of body fat, corresponding to 2-mm increments of closed-cell neoprene wet-suit foam. The only assumption made here is that the victim is capable of producing metabolic and muscular heat at the rate of 0.8 watts per pound of body weight for the duration of immersion (roughly equivalent to dog-paddling or treading water).

One can see at a glance that a man of average weight (6 mm of fat) would be expected to survive 6 hours in water at 60° F; adding a 2-mm wet suit would increase his survival time to 22 hours. No wonder surfers can sit out there all day in 60° water!

How do you estimate your own fat thickness? Figure 161 is my representation of standard Navy tables. Locate your height at the left, and your circumference index in inches (men: waist – neck; women: waist + hips – neck) at the bottom, and then find the nearest diagonal line. Follow it up to read fat thickness. Values between lines should be interpolated by eye; I have left out intervening lines to avoid confusing the figure. If you, as a skipper, have each crew member jot down his or her own fat thickness, you can assist any Coast Guard search by reporting the water temperature and survival expectancy of an overboard victim. In this, add 1-mm of equivalent fat thickness for each layer of cotton or wool clothing worn.

There are times when it is useful to know how long a person can work in water of a given temperature without falling below the stabilization tolerance limit (core temperature about 92° F), or, conversely, how much supplemental insulation a person needs to do light work for a desired period of time. Figure 162 shows a set of curves similar to those of figure 160, except calculated for an end-point temperature of 92° F, and for shorter practical work periods. Locate the desired work time and water temperature, and then find the required equivalent fat thickness. Subtract your own fat thickness, and make up the difference with neoprene foam, in the following ratio: 1 mm neoprene = 3 mm fat. For example, if you

*These curves have yet to be accepted by the Coast Guard, or by any public agency. But they are the only representation that fits all known survival data.

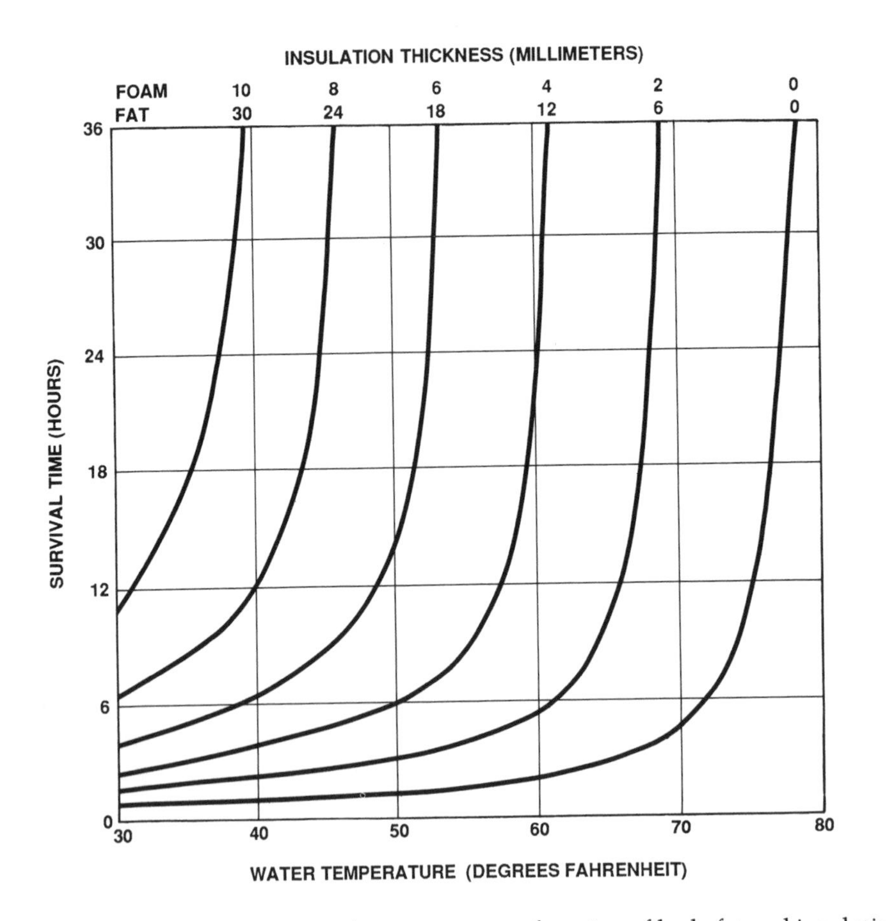

Fig. 160. Cold-water survival time curve as a function of body fat and insulation thickness.

have 6 mm fat, you will need 24/3 = 8 mm of neoprene to work for 1 hour at 30°. This, in fact, is about what Scripps Institution divers wear for shallow work in Antarctica. But, if you want to dive deeper, you will need to double the neoprene thickness for each 40 feet of depth, to compensate for foam compression and partial loss of insulation.

SEARCH PROCEDURES

Searching for a person overboard, as mentioned above, usually results from some circumstance that defeats your prompt rescue plan—anything that results in losing sight of the victim, and subsequent inability to find him or her within a few minutes. At a wind speed of 30 knots, even while drifting without sails, you will be separating from a swimming person at 60–90 feet per minute, and will catch only occasional glimpses of the victim's head after one or two minutes. This is why it is imperative to throw over your lighted marker pole and drogued horseshoe buoy at the earliest possible time; it will not drift away from the swimmer, and, if you can keep it in sight, you will have a little time to ponder the best course of action.

Coast Guard Procedures: If the Coast Guard is involved in your action plan, it is useful to know how they go about searching, so that your efforts can at least be

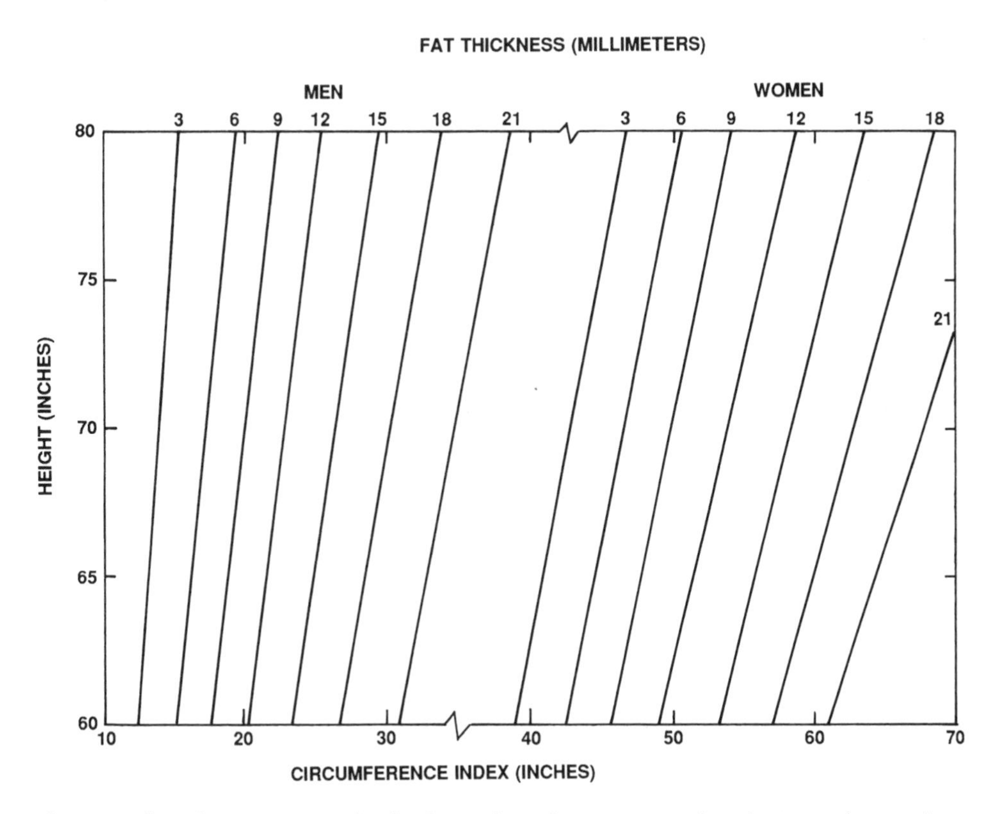

Fig. 161. Chart for estimating fat thickness from hip, waist, and neck circumference (see text).

coordinated. Figure 163 is a block diagram of the Coast Guard's search and rescue (SAR) system. Basically, the oceanic and inland waters from the mid-Atlantic to the mid-Pacific are divided into 18 regions, or rescue coordination centers (RCCs), all under the direction of a western or eastern SAR coordinator, who defines the scope of effort and delegates appropriate RCC participation for each action alert. The RCC selects a mission coordinator (MC), who actually runs the show.

Whether you send a Mayday to the Coast Guard, your EPIRB starts beeping, or the *Exxon Valdez* runs on a rock and starts spewing oil into Prince William Sound, the pipeline is the same: the message alert comes into the local RCC, which first evaluates it as an emergency (2 percent) or a false alarm, and then estimates the scope of effort required to deal with it. The mission coordinator has many options: there will be a number of search and rescue units (SRUs) in house, and, if more muscle is required, he or she can go upstairs and request help from other RCCs, and from military or even civilian sources. Ordinarily, searches are handled at the lowest practical level, but the message traffic goes clear to the top.

The mission coordinator performs many functions simultaneously. Communications are maintained with the reporting source, in order to collect information about the accident and advise about search progress. Air-sea-weather agencies prepare search aids (visibility, drift, water temperature, etc.). Aircraft and ship mission plans are drawn up, and the SRUs are activated and briefed. If the search area is remote, operational control may be shifted to an on-the-scene coordination unit especially established for the purpose. Despite the complexity of even the simplest search operation, which may involve the participation of hundreds of

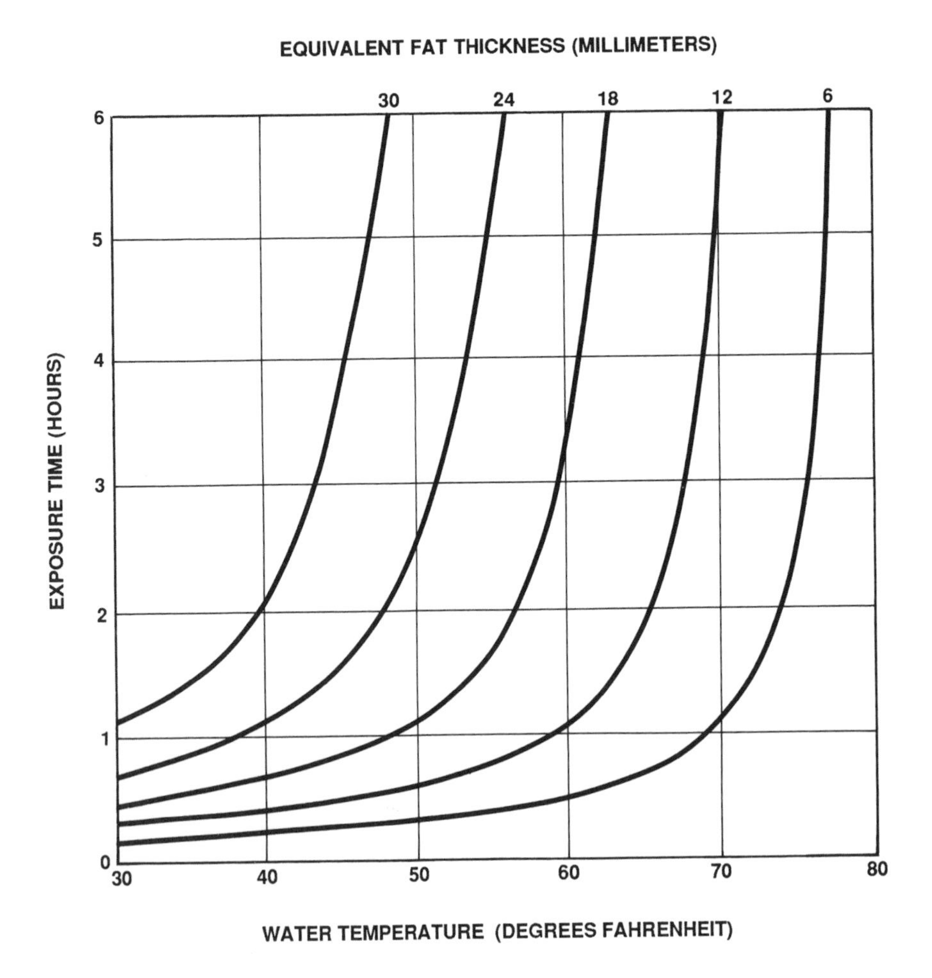

Fig. 162. Working exposure-time curves for estimating equivalent fat thickness as a function of water temperature.

personnel, it is not unusual for the first SRUs to arrive on the scene and commence searching within an hour or two in coastal waters. So, if you're testing your EPIRB in a pail of water in the cockpit, be sure you do it within the proper window (1-sec. transmission within first 5 min. after any hour), or a lot of people may be scrambled unnecessarily.

Searching, itself, may take many forms, of which I will describe only two (fig. 164). All searches start from the best position reference—usually your boat, located by your SATNAV report, by radar, or by RDF cross-bearings from shore stations. But if your report says "This guy fell over 2 hours ago, and I've been sailing south at 6 knots ever since," it may start at some other point precalculated to best cover the area of uncertainty, after figuring in wind and tidal shift, and such factors as daylight, sun angle, fuel consumption, and survival expectancy.

The upper pattern in the figure is a *creeping-line*, coordinated air-sea search. A boat moves at a predefined speed down the center of the search area, and a plane (or helicopter) flies a transverse pattern above it, so arranged that it crosses the boat on each leg. All pattern dimensions are controlled by a navigation plot designed to maintain a predetermined visual sweep width (area scanned to each side).

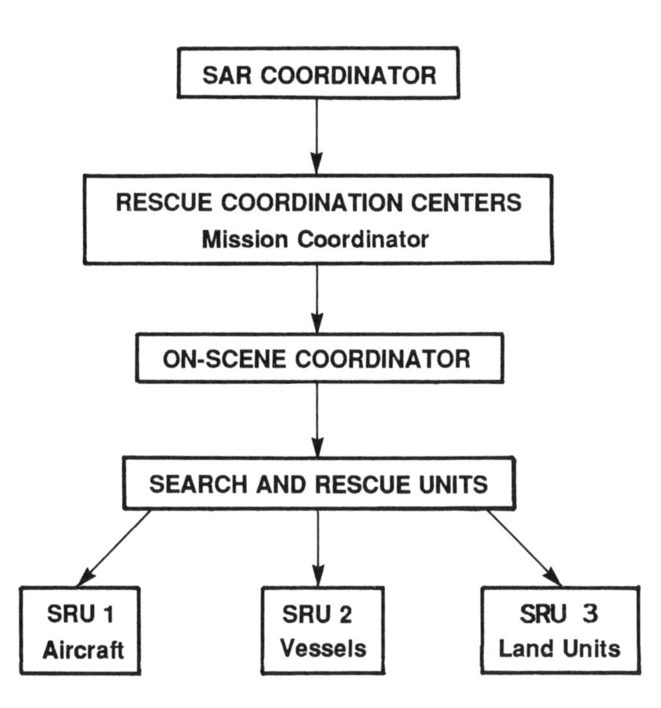

Fig. 163. Block diagram of U.S. Coast Guard Search and Rescue (SAR) infrastructure.

The lower pattern is an *expanding square* search, and is usually intended for rather dense coverage of a limited area. It can be performed by either ship or aircraft, and is controlled by Loran or SATNAV navigation, without reference to visual aids.

As might be expected, aircraft cover greater areas more quickly than do surface units, and have the advantage of greater visual range, but at the expense of limited flight duration. All this notwithstanding, the Coast Guard reckons the probability of finding an unassisted swimmer, separated by more than an hour from a drifting boat, at about 1 percent in daylight, and *zero* at night. The chances are improved by anything that enhances visual or radar detection, and, if there is an EPIRB floating within a mile of the person's position, the probability approaches the victim's cold-water survival potential.

SEARCHING ALONE

This section is devoted to the improvident 86 percent of all yachtsmen who go to sea without some sort of radio-communication aid that can be thrown overboard at the best DR location of a person lost at sea. Who would start an overnight race, or even an overnight trip, without an anchor? Without fuel? Without water? Plenty of people do. And many cruising yachtsmen undertake long voyages without a radio transceiver. The Coast Guard rescues them by the hundreds every year, and the only thing most of them never overlook is enough booze to maintain a happy ship.

Joking aside, even if you have lost sight of an MOB, have thrown over a marker pole with an EPIRB attached, and have called the Coast Guard, with a little advance

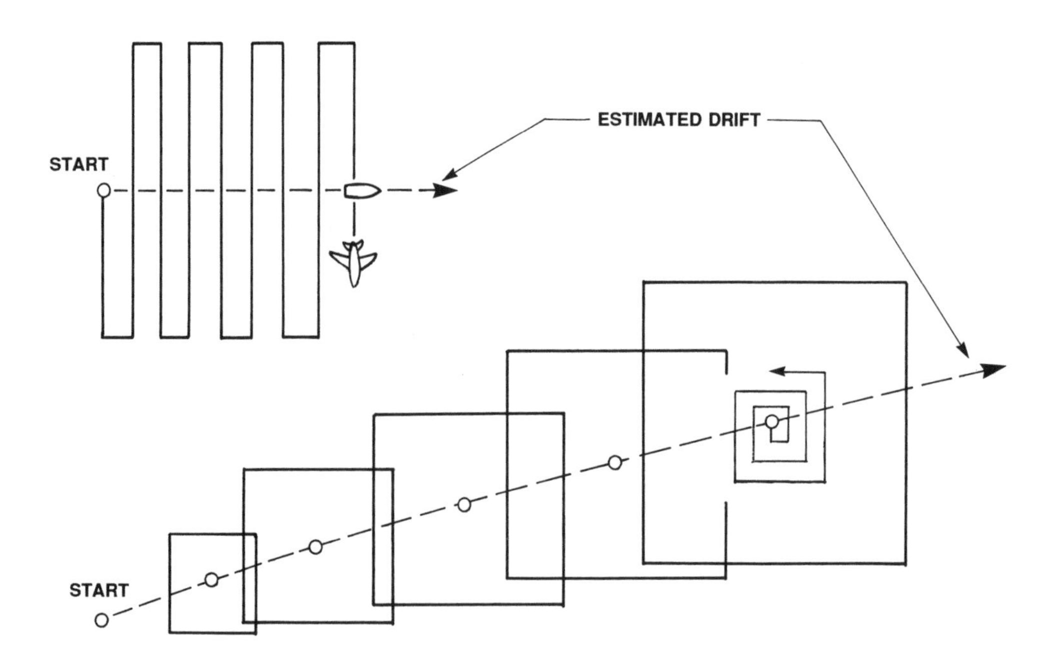

Fig. 164. Search patterns used by the U.S. Coast Guard: *(top)* creeping line; *(bottom)* overlapping squares.

preparation you can make a local, high-density search of the region surrounding the marker pole with a relatively good chance of finding the victim before the Coast Guard can get there. With no Coast Guard available, it may be your only chance.

Search Philosophy: It is important to realize at the outset that random searching has little hope of success; an effective search always follows a systematic pattern, whose character depends upon the search craft and the prevailing circumstances. Nevertheless, if the accident was witnessed, unless you badly goof your DR plot, there is a high probability that the victim is within a mile or so of your marker pole, and you will have a good chance of finding the person if you start soon and go about it systematically.

In this regard, the ordinary MOB pole required for offshore racing is inadequate for search operations because of marginal visibility. The search marker *beacon,* as we shall hereafter refer to it, should have a binocular visibility of at least 1 mile in 10-foot seas, which is about the highest sea state (20-knot FDS) at which it is practical to search. To save you the trouble of carrying an extra pole, the standard pole can be adapted for search by securing a 1-foot-diameter Day-Glo plastic sphere or biplane in place of (or in addition to) the existing flag. The details are left up to experiment, but the best system is demountable for storage, leaving the flag available to fulfill the qualifying regulations.

As described more fully below, a search pattern consists of an array of legs, connected by shorter traverses, along which the search vessel proceeds on fixed magnetic bearings for specified running times. The transverse distance between adjacent legs is called the *sweep width,* which we can otherwise define as the maximum distance abeam at which there is a high probability of sighting the victim from a moving vessel. From a small vessel, sweep width depends upon a number of factors: the victim's visual cross-section, the observer's eye height, sea state, sun position, atmospheric visibility, vessel speed, and the angle of binocular vision, to cite a few.

Figure 165 gives some perspective of what you are up against. With good visibility and 20/20 vision, the human eye can resolve an object subtending about 2 minutes (1⁄30 of a degree) of arc. Thus, under ideal conditions, an 8-inch human head can be seen at a range of 1,150 feet—about like looking for a flea at a distance of 8 feet. The angular range of vision with this acuity is fairly narrow, but the eye can sweep rapidly. Within the few seconds during which a swimmer may be visible on the crest of a wave, he or she can reasonably be detected over an angle of about 30°. While standard 7 × 50 binoculars can extend the visual range to about 8,100 feet, the visual angle is then narrowed to around 7°. This angle increases to as much as 15° in special wide-angle glasses—an important asset in searching.*

Even in moderate seas, unless a swimmer is very close to the search vessel, he or she will intermittently be hidden by intervening wave crests (points A and A'), and only clearly visible when at the top of the highest crest in his or her vicinity (B and B'). The percentage of time the swimmer is visible is related to the *product* of three time-dependent probabilities.

1. The probability that the swimmer is near the crest of a high wave: From figure 72 (p. 175), we can see that waves having average heights within the upper 10 percent of all waves present occur only about 2.5 percent of the time. Thus, the swimmer will be visible for only a few seconds out of forty average wave periods. This probability decreases as sea state rises, and can be increased by raising a flag or some other visual target above head height.
2. The probability that the searcher's eye height is always higher than that of all intervening waves: For zero eye height (two swimmers trying to see each other), the individual probability of being on a wave crest is the same as that given above, and the *joint* probability is a few seconds in 40 × 40 = 1,600 periods, i.e., practically zero. Unless you are fortunate enough to pass within a wavelength of the swimmer, practical searching requires that you should be able to see the horizon at all times. This is why radio search is so much more efficient; the antenna can be atop the mast.
3. The probability that the searcher's binocular angle intersects the swimmer's position at the instant the latter is visible: Continuous sweeping with binoculars is useless. The only practical way to increase this small probability is to look on a constant bearing long enough so as to be sure of seeing the swimmer if he or she is within the 7° field of view, and within binocular range.

Further discussion of probability is pointless, because there are always too many variables to calculate it with certainty. It should be clear, however, that random searching is a waste of time. Systematic search involves experimentally establishing one's sweep width and setting up a search pattern, as described below. If the victim is anywhere within the pattern, you will have a reasonably good chance of sighting the person.

*The standard 7 × 50 binocular was developed during World War II to optimize night search. Seven-power is a compromise between weight and eye fatigue; the 50-mm objective lens gives a visual image just the size of an expanded pupil, thus maximizing light-gathering power. Neither power nor lens size has anything to do with angular aperture, which is close to 7° on all except wide-angle glasses. For daylight looking, 6 × 30 binoculars are just as good, and much easier on the eyes and arms.

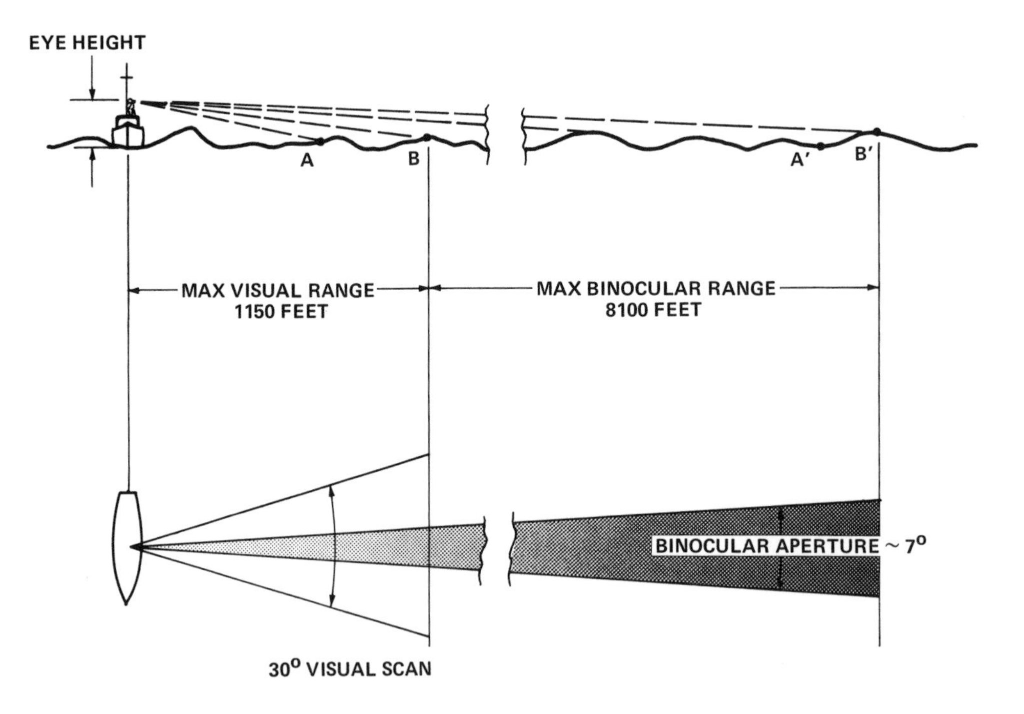

Fig. 165. Swimmer sighting limitations include eye height, binocular aperture, and the probability of shadowing by intervening waves.

Sweep Geometry: Because time is so important, the search objective is to maximize the area swept in unit time (*sweep rate*). As will be shown, this is accomplished by maximizing sweep speed at the expense of sweep width. These factors are geometrically related, and both are limited by sea state and orientation of the search pattern (which may also be affected by sun position and wind direction): sweep speed is limited by the difficulty of holding a steady course, and sweep width is constrained because sea state limits, binocular range, and adverse motions enhance eye fatigue.

Figure 166 is a generalized sweep diagram, whose specifics can be adjusted to maximize sweep rate in a particular situation, and which accounts for all pertinent physical and environmental factors. It assumes that four persons are available for optimum searching:*

(a) a navigator, who lays out the search pattern, keeps track of the ship's position, and gives course and speed instructions;
(b) a helmsman, who cons the vessel, and periodically scans the entire horizon for signals (flashes, flares, smoke, etc.) that might otherwise not be detected;
(c) a visual (naked-eye) observer, who scans only the near field (within 1,150 ft.**) over a 30° sector abeam the vessel; and

*With only three men available, the navigator can double as visual observer; with only two, you are necessarily restricted to visual sweeping, unless the vessel has autopilot.

**If searching for something larger than a human head, the effective visual range will be increased by about 1,300 feet for each foot of target width.

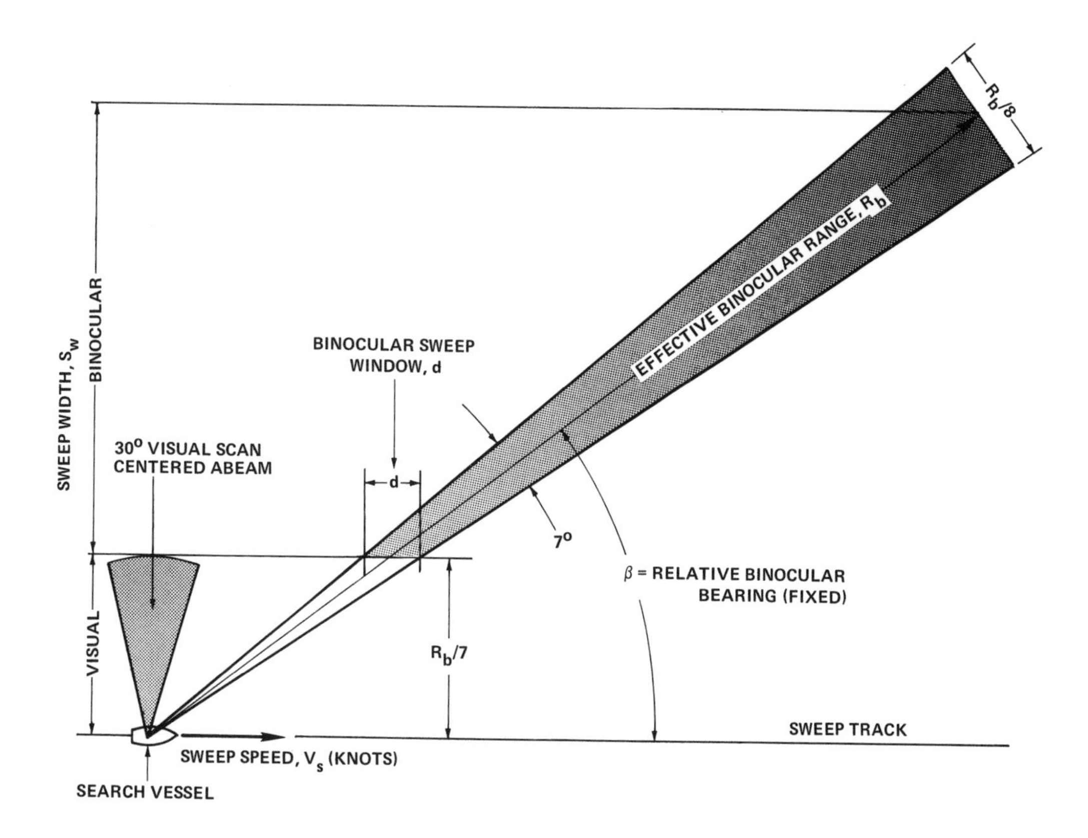

Fig. 166. Schematic search geometry: as ship proceeds along sweep track, visual observer looks abeam; binocular observer looks at constant angle β to heading.

(d) a binocular observer, who looks in a fixed direction making an angle β with the vessel's search track.*

The philosophy behind this arrangement is that, as the vessel moves along each leg of its search track, the visual observer will sweep out a path whose width equals his or her visual detection range. If this observer limits the scan cycle to 4–5 seconds (about the longest interval a target will remain near the top of a large wave crest), he or she will have a high probability of seeing anything within the designated scan sector. At the same time, the *binocular* observer is sweeping a path of width $S = R_b \sin\beta$, where R_b is the effective binocular range. As we have said, R_b varies with sea state, but it can be determined by simple experiment (described below) on the first leg of your search pattern. The reason for looking at an angle β, rather than abeam, is because you want to maximize sweep speed and yet always ensure that any target traversing the binoculars' narrow (7°) field of view remains within it long enough to be seen at the prevailing sea state. Since the binoculars' near field is also covered by the visual observer's sweep width, the critical target transit duration t_b equals the oblique distance d, subtended across the binocular field by its intersection with the

* Holding a constant binocular bearing is best accomplished with compass-bearing binoculars. With ordinary binoculars, you must reference your sighting angle to the ship's heading by keeping a shroud or something else in the field of vision.

margin of the visual sweep (here, called the *sweep window*), divided by the vessel's sweep speed: $t_b = d/V_s$.

Without boring you with the details, from the geometry of figure 166 the following key relations are easily established

(a) $\sin \beta = (R_b / V_s t_b)^{1/2}$ = relative binocular bearing
(b) $S_w = R_b \sin \beta$ = sweep width (miles)
(c) $S_r \times V_s = V_s R_b \sin \beta$ = sweep rate (sq. mi./hr)

Here, t_b is in minutes, R_b in miles, and V_s in knots. These equations assume you are using 7-power binoculars with a 7° field, and that the visual sighting distance is always one- seventh the binocular distance. In high seas, or if using wide-angle glasses, the actual sweep widths will be larger, and you will have some overlap between legs. So much the better.

To reduce these equations to useful curves, we need to establish a functional relationship between t_b and R_b. Recalling that t_b is the length of time one needs to look to be sure of seeing a target if it is within binocular range, there are two ways of doing this. The hard way is to take R_b as the actual range of binocular resolution (8,100 feet, for a human head), and to calculate t_b from wave statistics as a function of sea state. So far, this has not been done, and the results would be in doubt because the character of a random sea cannot be defined well enough for all cases of interest. The easy way is to pick t_b arbitrarily (but judiciously), and to determine R_b experimentally. This can be done in a few minutes on your first search leg, using only your binoculars and a stopwatch. After running this experiment a few times aboard one of our ships, I have selected t_b = 1 minute as a practical compromise at moderate search speeds, and at sea states in which searching is practical for small vessels; it has the added advantage that t_b disappears from the above equations.

Having chosen t_b = 1 minute, proceed as follows to find R_b:

1. Prepare a dummy target having roughly the same dimensions as the search target. This can be a boat fender, weighted at one end with a piece of chain to prevent drift. Rig the dummy on the fantail for heaving on command. The crewmember on the fantail should have a wristwatch, stopwatch, pencil, and paper.
2. Station a binocular observer at the point that will later be occupied when searching, i.e., a position affording the highest possible eye height. When the vessel is on a steady course at her maximum practical search speed (usually limited by the observer's inability to hold the glasses on a steady bearing), the fantail observer should record vessel speed and wristwatch time to the nearest second, and throw the dummy over.
3. The observer then tracks the dummy, calling out "lost" when it disappears behind waves, and "found" when it reappears.
4. When the lost–found interval is judged to be approaching 1 minute, the fantail person should commence starting the stopwatch on "lost" and stopping it again on "found," noting the elapsed time, and resetting the watch.
5. The first time the dummy disappears for more than 1 minute, he or she should note wristwatch time and stop the stopwatch simultaneously. This ends the experiment.

6. By subtracting the last stopwatch interval from elapsed wristwatch time, you obtain the total running time before the dummy was lost for 1 minute. Multiplying this time by vessel speed gives R_b, which should be converted to miles.

Knowing R_b and V_s, you can now refer to figure 167 to determine the correct binocular bearing (right-hand scale), sweep width, and sweep rate (left-hand scale). The two sets of R_b curves are given in quarter-mile increments (500 yards), and can be interpolated, using the small tick marks along the left margin. You will note that, for any fixed binocular range, increasing search speed requires that the binocular bearing angle be decreased, in order to expand the search window so that a target will still traverse it in 1 minute. This has the effect of decreasing sweep width and increasing sweep rate. However, there are practical limits to the advantage gained by higher search speed and sweep rate: total track length, the number of pattern legs, and fuel consumption also increase; more time is wasted in turning; and position accuracy deteriorates. For these reasons, the optimum binocular angle will be about $\beta = 30°$ for small boats, and I have adopted this value in the following numerical example.

Suppose that your dummy-sighting experiment indicates an effective binocular range of 1 mile. Locating the intersection of the horizontal line $\beta = 30°$ with the binocular bearing curve for $R_b = 1.0$, project a vertical downward and read sweep speed, $V_s = 4.0$ knots. At intermediate intersections with the lines labeled $R_b = 1.00$, read sweep rate = 2.0 sq. mi./hr., and sweep width = 0.5 mile on the left-hand scale.

Search Pattern: Having optimized sweep rate for local conditions, it remains to establish a search pattern that makes maximum advantage of our peculiar sweep geometry. The choice of pattern also involves the following considerations:

(a) It should be centered on the marker beacon, because it is the only position reference, and identifies the most probable vicinity for the victim.
(b) It should be navigationally simple, with provision for frequent position checks and maximum overlap near the center.
(c) The observers should be always looking down-sun during the day.
(d) Drift errors should tend to compensate, rather than accumulate.

There are many possible patterns, and you may have fun trying to better the solution I have come up with for this interesting geometric problem, which is shown in figure 168. It satisfies all of the above conditions and assumes only visual bearings on the marker beacon. While the pattern is designed for daylight searching for a passive visual target, the method is perfectly general, and can be expanded to work in darkness at greater ranges, if the victim and dummy target are equipped with similar lights, and the marker beacon with a distinguishing light.

The search pattern consists of a circle of radius R_b (initially swept out with binoculars while the search vessel is holding station on the marker beacon) surrounded by a polar array of trapezoidal lobes. Each lobe is bounded laterally by equal inbound and outbound radial legs (labeled S_L for *sweep length*), and terminally by a traverse of one sweep width, S_W. Each lobe is connected to its neighbor by a

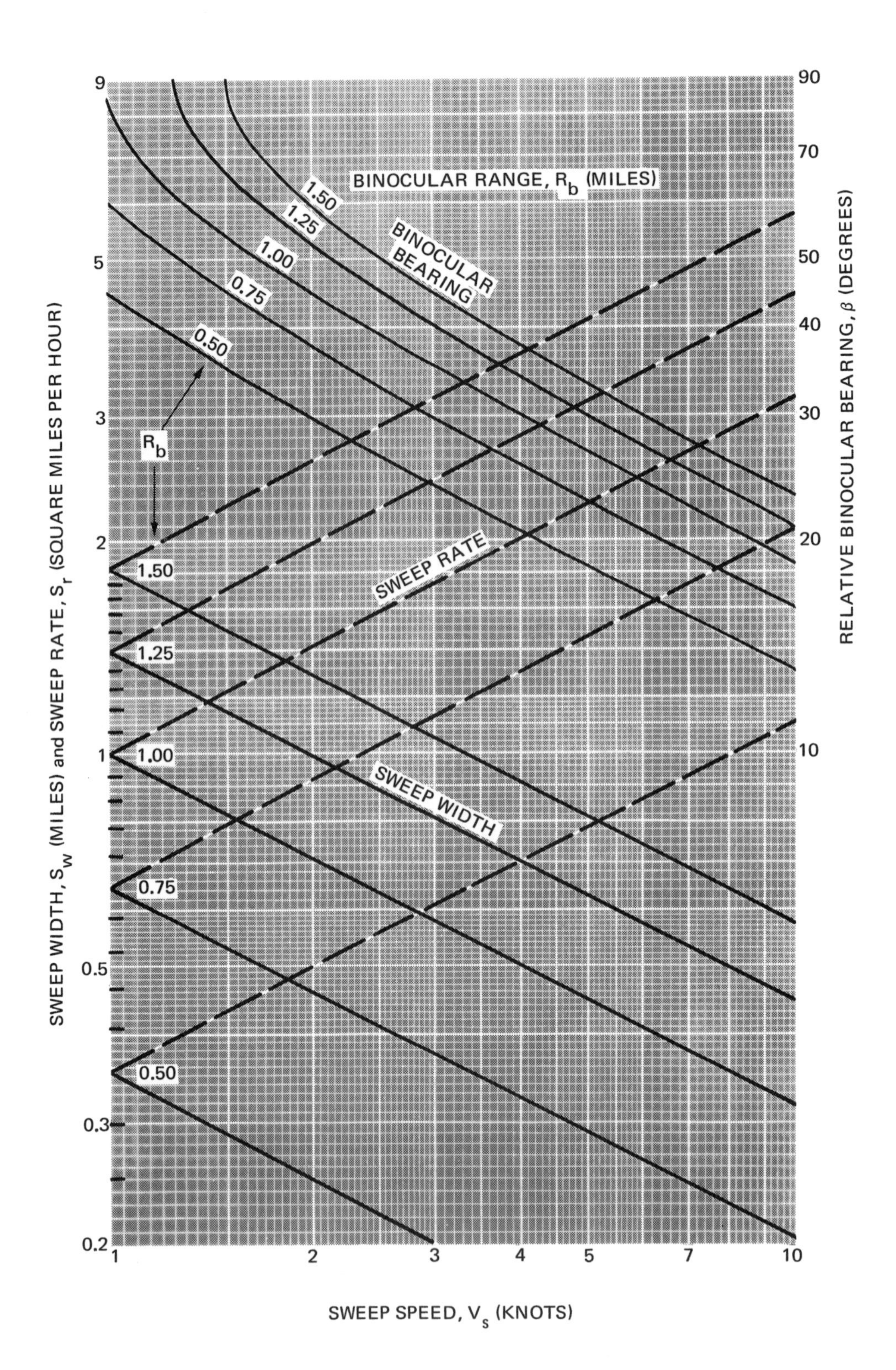

Fig. 167. Sweep parameters as a function of sweep speed V_S and relative binocular bearing.

shorter traverse (such as *a–b*). The arrows diverging midway from each leg or traverse indicate the binocular observer's search direction, always making a constant angle β with the local track segment. Note that these arrows point in a nearly common direction that rotates clockwise as the vessel proceeds along the track. To accomplish this, it is necessary for the observer to shift his or her relative bearing from side to side, or from bow to stern, each time the vessel changes heading.

Although the average rate of rotation of the sighting vector depends upon pattern size and vessel speed, by proper orientation of the initial track leg with respect to the sun's position (θ), the sun can be kept out of the observer's line of vision. In the Southern Hemisphere, of course, the sun rotates counterclockwise, and the entire pattern should be reversed accordingly.

In addition to being sun-synchronized, the above search pattern has other advantages. It is circular, which maximizes search density. Because the search legs converge inward, there is considerable sweep overlap toward the center. The initial center sweep, with the vessel stationary with respect to the marker beacon, is the fastest means of scanning the most probable location of the victim. All pattern legs are run on straight azimuths toward or away from the marker beacon, which is visually relocated at the end of each lobe, thus providing recurrent position checks. Because the entire pattern is referenced to the marker beacon, which drifts at the same rate as the victim, cumulative drift errors are largely avoided. Distance errors are minimized by running alternate legs in opposite directions, and reckoning all distances from the ship's log, or from running times referenced to her speedometer. Lastly, none of the complicated heading and sighting angles need be calculated; they can be lifted directly from a running maneuvering-board plot that is laid out knowing only binocular range (dummy experiment), sweep width (from figure 167), and sweep length (as shown below). All this can be done on the first search leg, and the plot corrected if necessary at the end of each pattern lobe.

All dimensions of the schematic search pattern (fig. 168) are given in units of R_b, the effective binocular range. In a real search situation, the actual dimensions and the number of pattern lobes will most likely be governed by total search time, which may be limited by present and prospective weather, the survival expectancy of the victim, or the number of hours of daylight. The procedure set forth below is geared to a maximum of 28 hours of daylight search (about 36 hours of real time), during which a maximum search radius R_s = 5.25 miles can be swept,* encompassing about 86 square miles of ocean. Barring gross navigational errors, a victim of a witnessed accident should be found well within this radius, and 36 hours is ample time for other help to arrive, if it is forthcoming. However, there is always a possibility of an unwitnessed accident, where the victim's position uncertainty may be much larger than 5.2 miles, and for which contingency I have included the following equations necessary to expand the pattern to any practicable size:

*Assuming good visibility, moderate seas, and maximum binocular range (R_b = 1.5 miles). With a lighted target, R_b may be considerably larger at night.

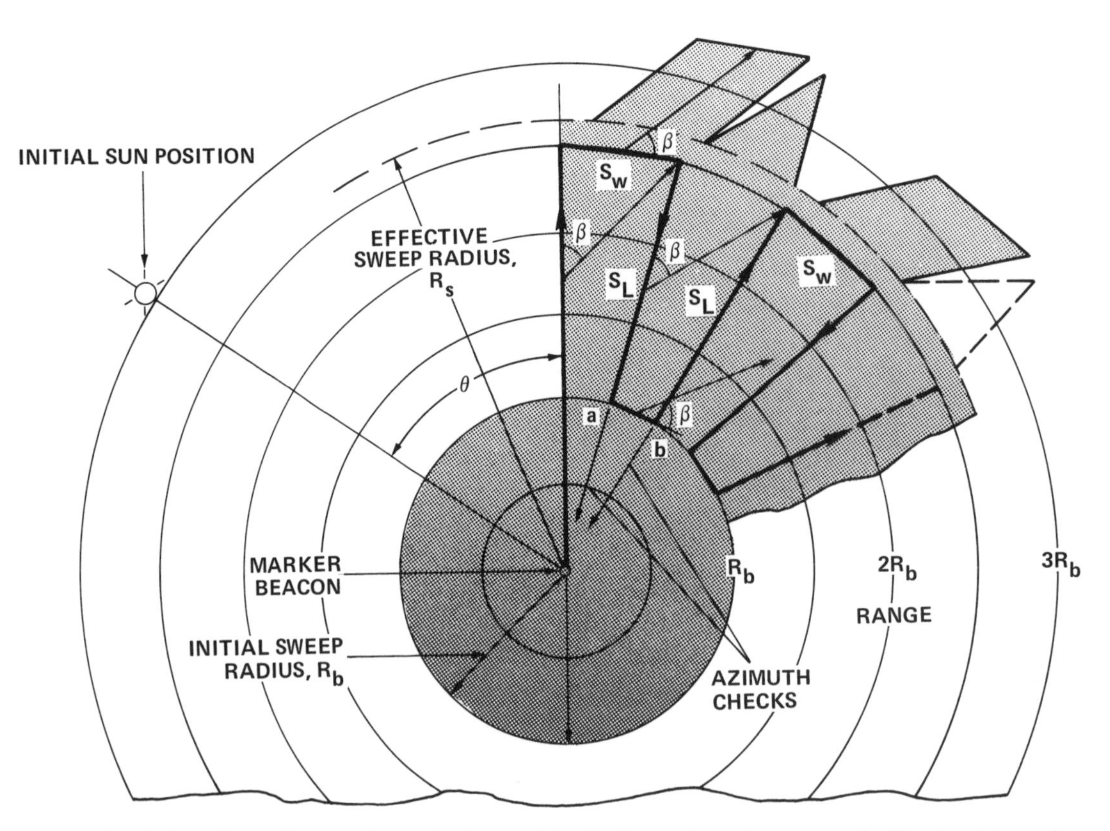

Fig. 168. Schematic search track plot, with all dimensions in terms of effective binocular range R_b.

Total sweep radius (mi.):	$R_s = S_L + 1.2\ R_b$
Total area swept (sq. mi.):	$A_s = \pi R_s^2$
Total sweep time (hrs.):	$t_s = 1 + \sin^2\beta \left[2 + \frac{S_L}{R_b}\left(1 + \frac{2 + 2S_L/R_b}{\sin\beta}\right)\right]$
Total track length (mi.):	$L_s = R_b\ (1 + 2t_s)$
Number of pattern lobes:	$N_L = \frac{\pi\ (1 + S_L\ /R_b)}{\sin\beta}$

These equations assume that the searcher is using 7-power binoculars, with a 7° field, in which case the optimum binocular bearing angle is β = 30° (sin β = 0.50), the optimum search speed is $V_s = 4R_b$ knots, and the sweep width is $S_w = 0.50R_b$ miles. With these values, all other pattern parameters are given in figure 169, as functions of total sweep time. The diagram can be used either of two ways: given t_s, project a vertical to intersect the curve, locate the tick mark nearest to the left, and from that point read S_L/R_b and R_s/R_b horizontally to the left and right, respectively; or, assuming R_s/R_b, do just the reverse. In either case, the tick mark gives the number of pattern lobes. It also indicates the optimum angle θ between the initial leg of the search pattern and the sun's azimuth that minimizes sun

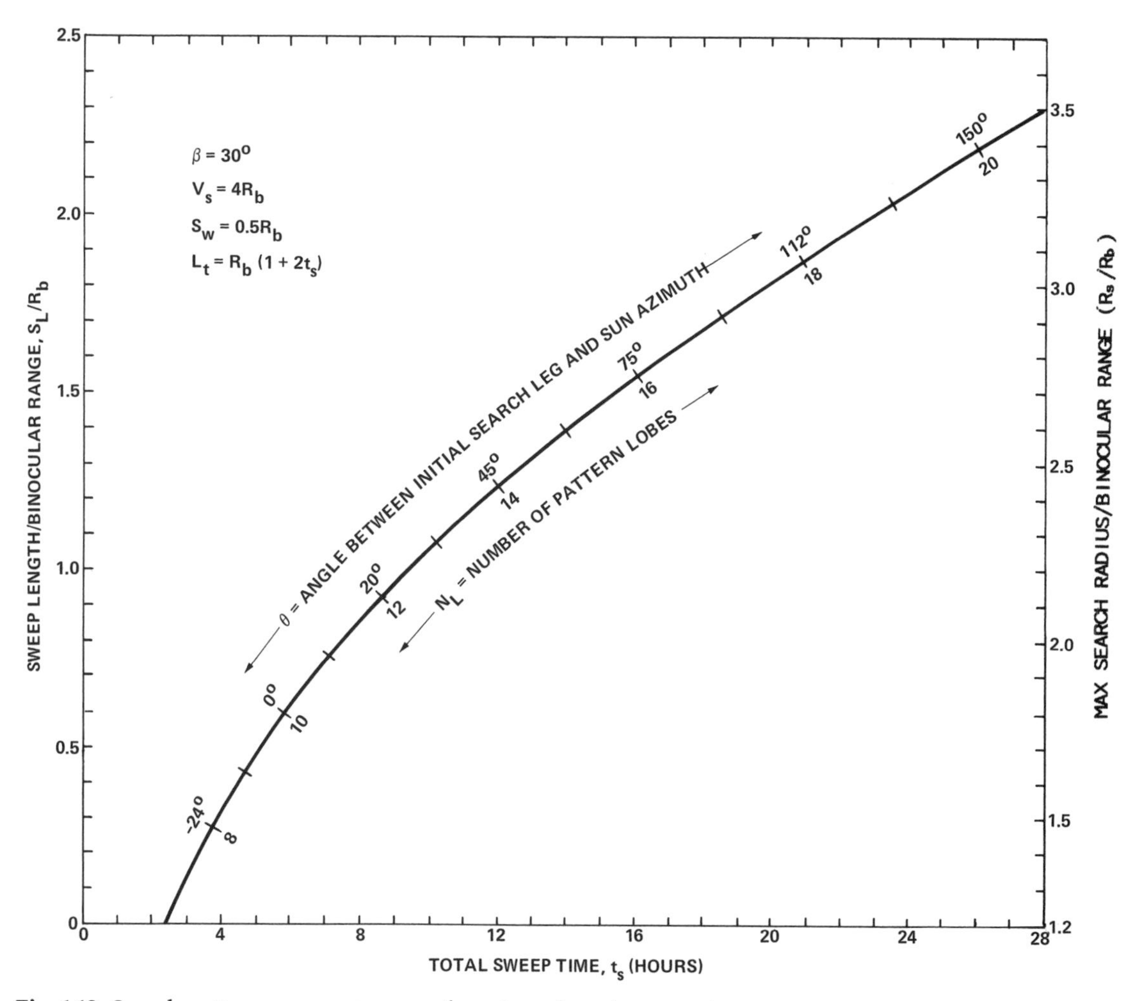

Fig. 169. Search-pattern parameters as a function of total sweep time.

interference during the search. For search durations less than 6 hours, θ is negative, indicating a sun azimuth between the initial leg and the observer's line of sight. Shorter search durations suffer also from impractically small sweep lengths. The curve intersects the time axis at $t_S = 2.5$ hours, which is just the time required for the vessel to sail around a circle of radius R_b.

Search Procedure: We now have everything in hand needed to run through a practical search exercise. Let us assume that you are aboard your 46-foot ketch, *Dandelion,* which is close-reaching south southeast through the Gulf Stream en route from Boston to Bermuda. The weather is unsettled with occasional squalls, and a weak low is forecast to the northeast. The wind is generally southwest at 12 knots, with moderate seas. There is no moon.

At 0230 hours you run into a sharp squall, and call all hands to shorten sail. While getting in the main, the halyard jams midway, and the boom jibes suddenly, knocking your favorite drinking buddy over the side, clad only in undershorts and sweatshirt. He surfaces, waving his arms to signal no injury, and you heave the trailing buoy, but he fails to reach it. You cast it off, but in darkness, and beyond the range illuminated by the spreader lights. You are uncertain whether or not he eventually picks it up.

Owing to the jammed halyard, it takes you 11 minutes (à la Sierks) to get shipshape, start your engine, and run back to your best DR estimate of his position. Finding nothing, you broadcast a radio alert, giving your position and the circumstances. As rapidly as possible, you assemble your marker beacon (lighted, but with no radio) attaching it by a pendant to your second horseshoe buoy, to which you seize a net bag containing his wet suit and survival pouch. This done (10 more minutes), you put the beacon over, post a lookout and a radio watch, and set someone to work making up a dummy target. Pending dawn, there is little more you can do, other than to circle the buoy slowly, spreader lights on, and scanning visually with a portable searchlight in hopes of sighting him afloat in the first buoy.*

Meanwhile, you take stock of the situation. The squall has blown by, and the wind has veered west, freshening to 20 knots, with a prospect of clearing weather as the low passes to the north. The victim is a good swimmer, and, even though he is unprotected and has failed to reach the first buoy thrown, he has a reasonable expectancy of staying afloat through the coming day in the warm Gulf Stream. It is questionable whether he could survive an additional 24 hours. Your radio alert was picked up by the Coast Guard, who is dispatching a cutter from Norfolk, but it cannot hope to arrive before midnight. At this point, you discover that someone has left a crescent wrench lying in a tray near the binnacle, and realize that you have been running about 10° off course all night, and that the combination of compass error and squall drift might add up to as much as a 1-mile position uncertainty with respect to the victim, who is probably floating somewhere to your west. Of your crew of six, you have five left to work with.

This combination of circumstances calls for a systematic search over the largest area that can be swept during the approaching day. Your almanac indicates that sunrise will occur at 0455, and that you can expect about 15 hours of good visibility. From figure 169, you find that you can complete a 15-lobe sweep in 14 hours, using a sweep length $S_L = 1.4R_b$, and will sweep out a circle of radius $R_s = 2.6R_b$. Your ensuing search procedure (fig. 170), identified by the letters *a, b, . . . h* along the search track, includes the following steps:

1. At first daylight (0430), and while holding station on the marker beacon *(a)*, you sweep the entire horizon clockwise, visually and with binoculars. Although your effective binocular range (R_b) is yet unknown, by looking on a constant bearing for 1 minute, and then shifting the bearing clockwise by 6° for another minute, and so on,** you will complete a 360° sweep in one hour, and automatically scan everything within the area encompassed by the subsequent search grid. Because the sun will impair looking eastward as soon as it peeps above the horizon, you should commence the sweep about 30° south of the sun, and will have swept the 60° up-sun sector before

*If he has been rehearsed in your rescue drills, and is within swimming distance, he should make for the marker beacon, knowing that there will be another buoy there, equipped for signaling. He also knows that you will return to the beacon at dawn to commence daylight search.

** With 7° binoculars, this gives a 1° overlap. Using wide-angle glasses, you can increase the angle between consecutive bearings accordingly.

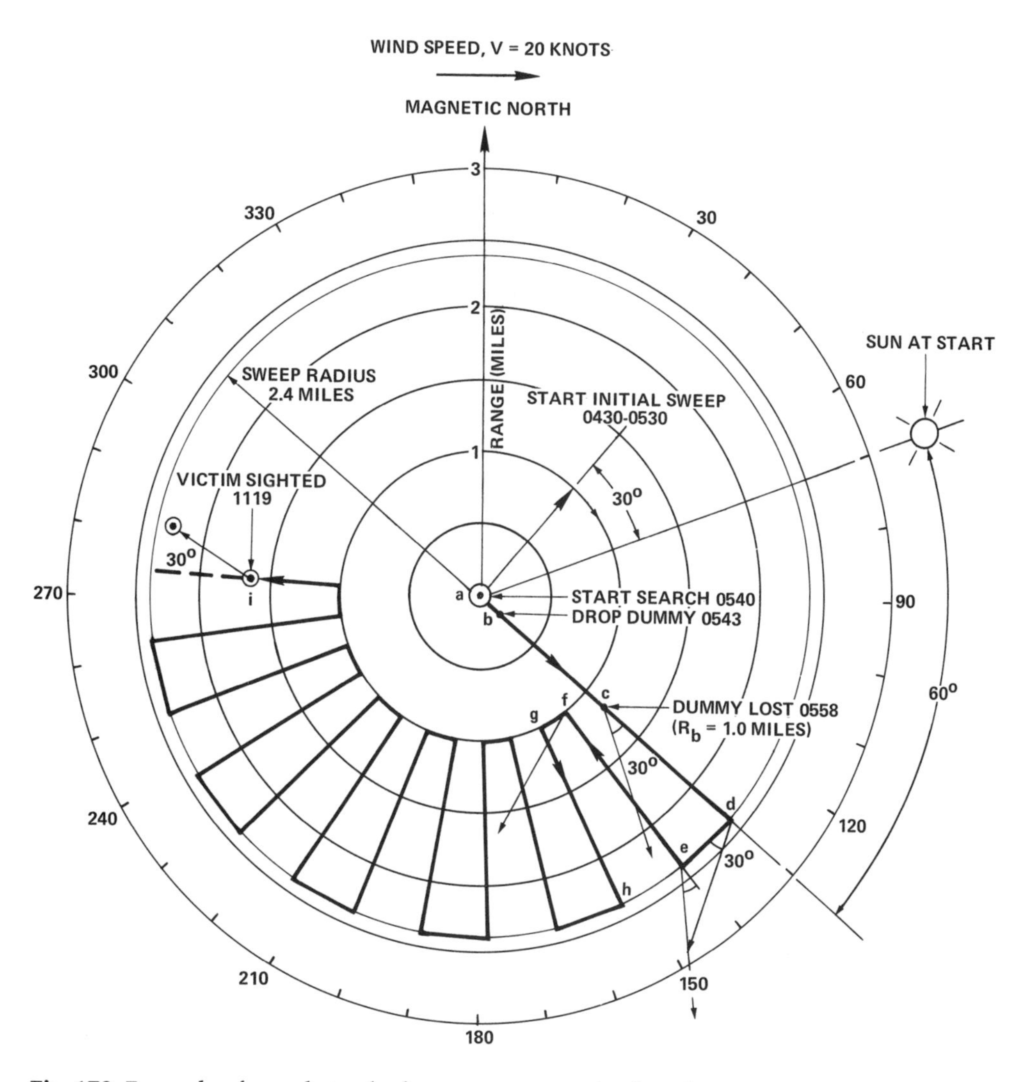

Fig. 170. Example of search-track plot on a maneuvering board.

sunrise. Failing this, you can always come back to the beacon later, on some inbound search leg, after the sun has moved around.

This initial sweep is very important; it covers the most likely position of the victim at an early time, and is accomplished most rapidly and with the least effort. (Had *L'Apache* made such a sweep, there is little question that Ted Sierks would have been sighted.)

2. Having sighted nothing during your initial sweep, the next step is to lay out the direction of the initial leg (*a–d*) of your search track on a maneuvering board, with the marker beacon at the center, and magnetic North at the top. It is now 0530; the sun is up and bearing 070°. From figure 169, the optimum heading angle is 60° clockwise of the sun's position, or 130°.
3. Getting underway on this bearing, as soon as you have achieved a comfortable, steady speed, heave the target dummy over (point *b*), and commence the tracking experiment previously described (p. 414). Let's suppose that you run for 15 minutes at 4 knots before the target is lost for 1 minute (point *c*),

thus obtaining R_b = 0.25 hr. × 4 knots = 1.0 mi.* This number immediately establishes the scale of your pattern, giving you:

V_s = 4.0 knots = track speed
S_W = 0.50 mi. = sweep width
S_L = 1.4 mi. = sweep length

Now, lay out a distance $R_b + S_L$ = 2.4 miles along your initial heading, thus establishing point *d*. With a compass, draw a circle of this radius, and a smaller circle of radius R_b, centered on the marker beacon. With dividers set at S_W = 0.50 miles, start at point *d* and prick off equal distances along the larger circle. Draw radii (such as *e–f*) through all these prick points, and connect them with straight traverses (such as *d–e*, and *f–g*, respectively). You have now completed the track plot, and, if you are not trailing a log, you can compute running times along S_W and S_L at 4 knots. All this should easily be completed with the twenty minutes or so it takes to reach point *d*.

4. Upon arriving at point *d*, as reckoned by log or DR plot, the navigator gives the helmsman the new course (lifted from the maneuvering board), and also gives the binocular observer the new search bearing (30° to port of traverse *d–e*). The visual observer always looks abeam on the same side.
5. At point *e*, the vessel turns back on an inbound radial leg, with the binocular observer looking 30° to port on the stern quarter. As point *f* is neared, the marker beacon should easily be located with binoculars, since it is an elevated, bright orange sphere, whose bearing should be well established.** At point *f*, the navigator takes the beacon's bearing, corrects the maneuvering plot, if necessary, and gives new heading and looking instructions for traverse *f–g*, which is then run off, and point *g* checked by a second bearing. This completes one pattern lobe.
6. The search is continued along successive lobes until you sight your drinking buddy midway on the first leg of the seventh lobe, alter course to pick him up, and cancel your radio alert. If you have not deviated too far from your plot schedule, he will have been in the water a little less than 9 hours, and will be getting a little thirsty.

If you fail to find him after completing the entire pattern, there is no recourse, except to repeat the search the following day—assisted, hopefully, by the Coast Guard. Any subsequent search should be coordinated with them, after you have explained your procedure and defined the area already covered. By attaching a radar reflector to the marker beacon, the cutter can cover an extensive region exterior to this area, while you go over it again. However, as we have said, the Coast Guard has its own methods, and you should assist them in any way possible.

The Person in the Water: If all crew members are acquainted with your "man overboard" search and rescue procedures, there are several ways in which a future

*At point *c*, the binocular observer should immediately turn and focus his or her attention 30° off the starboard bow, with the visual observer scanning a 30° sector centered abeam (fig. 166). They will have missed part of the first leg, but will pick it up on the last.

**If the beacon is not sighted from point *f*, leg *e–f* should be extended until it is located, since the search cannot reasonably proceed without it. The track is then recommenced along the next leg (*g–h*).

accident victim can enhance the chances of early rescue. As already noted, the MOB will know that the marker beacon is the search focus, that the vessel will hold station on it for an hour before commencing its radial sweep, and that it will sight the beacon at the end of every pattern lobe. Thus, a good swimmer can determine the beacon's position—either visually, or from observations of the vessel's movements—and may be able to reach it, and hope to attract attention with flares or strobe flash in the survival kit attached thereto, or may be able to swim so as to better intercept the search pattern on some earlier leg. In either case, the victim should not abandon the life buoy, but should tie the trailing line to his or her waist and tow it behind.

If already equipped with a survival kit, the MOB will know from the vessel's movements where search attention is directed, and should only signal at times when he or she is likely to be within the observer's visual sector. Further, if the helmsman, who periodically scans the entire horizon, does so every time the vessel makes a turn to a new leg, the victim can observe this, and act accordingly. Alternatively, since most sailors now wear waterproof watches, if some sort of schedule is set up when briefing crewmen on procedures, horizon scans and rescue signals can be synchronized.

All else failing, if the victim is anywhere within visual range of either the marker beacon or the search vessel, he or she will have the comfort of knowing that a systematic procedure is being followed, and that there is a good chance of being sighted, if the victim remains calm and conserves energy.

Epilogue

There are many other emergencies not touched on here that are common enough at sea: collision, grounding, fire, flooding, and capsize all exact their toll of the unprepared, the unaware, or the unlucky. But most of these are treated well enough elsewhere. Originally, I had planned to say much more, but this book is already long enough—perhaps too long—and one has to stop somewhere. Parts of it may seem foreboding, but none of it (I hope) pessimistic. As noted, in discussing shark attacks, I am a great believer in calculated risk. Knowing the sea, the weather, and your boat is the best way to minimize this risk. That is what this book is all about. Happy sailing!

References

It would be impractical here to list the several hundred books and articles reviewed in preparing this book. Accordingly, I have included only principal references. Most authors cited anecdotally are appended under "Recommended Reading."

Oceanography

Bascom, Willard. *Waves and Beaches.* New York: Doubleday & Co., 1964.

Defant, A. *Physical Oceanography.* New York: Pergamon, 1961.

Dietrich, G. *General Oceanography: An Introduction.* New York: Interscience Publishers (John Wiley and Sons), 1963.

Halstead, Bruce W. *Dangerous Marine Animals.* Cambridge, Md.: Cornell Maritime Press, 1959.

———. *Poisonous and Venomous Marine Animals of the World.* 3 vols. Princeton, NJ.: Darwin Press, 1972.

Hill, M. H., ed. *The Sea.* New York: Interscience Publishers (John Wiley and Sons), 1963.

Hogben, N., and F. E. Lumb. *Ocean Wave Statistics.* London: Her Majesty's Stationery Office, 1967.

Kinsman, Blair. *Wind Waves.* Morristown, N. J.: Prentice Hall, 1964.

Marine Fouling and Its Prevention. Annapolis, Md.: U.S. Naval Institute, 1952.

Neumann, G. *Ocean Currents.* Amsterdam: Elsevier Publishing Co., 1968.

Sverdrup, H. V., Martin W. Johnson and Richard H. Fleming. *The Oceans.* New York: Prentice Hall, 1946.

Wiegel, Robert L. *Oceanographical Engineering.* Morristown, N. J.: Prentice Hall, 1964.

Meteorology

Flöhn, Hermann. *Climate and Weather.* New York: McGraw-Hill Book Co., 1968.

Heavy Weather Guide. Annapolis, Md.: U.S. Naval Institute, 1984.

Meteorology for Mariners. London: Her Majesty's Stationery Office, 1967.

Taylor, George F. *Elementary Meteorology.* Morristown, N. J.: Prentice Hall, 1963.

U.S. Naval Marine Climatic Atlas. Vol. 3, The World. Washington, D. C.: U.S. Government Printing Office, 1970.

Ship Dynamics

Barnaby, Kenneth C. *Basic Naval Architecture.* Tiptree, Essex (England): Anchor Press, 1948.

Beebe, Robert T., *Voyaging under Power.* New York: South Seas Press, 1975.

Comstock, John P., ed. *Principles of Naval Architecture.* New York: Society of Naval Architects and Marine Engineers, 1967.

Gilmer, Thomas C. *Modern Ship Design.* 2d ed. Annapolis, Md.: U.S. Naval Institute Press, 1977.

Marchaj, C. A. *Sailing Theory and Practice.* New York: Dodd, Mead & Co., 1964.

Saunders, Harold E., *Hydrodynamics in Ship Design.* New York: SNAME, 1957.

St. Denis, Manley, and W. J. Pierson, Jr. "On the Motions of Ships in Confused Seas." *SNAME Transactions,* 1954, pp. 280-357.

Yacht Design and Construction

CCA Tech. Committee. *Desirable and Undesirable Characteristics of Offshore Yachts.* Edited by John Rousmaniere. New York : W. W. Norton, 1987.

Fox, Uffa. *Sail and Power.* London: Peter Davies, 1936.

Gutelle, Pierre. *The Design of Sailing Yachts.* Camden, Me.: International Marine Publishing Co., 1986.

Henry, Robert G., and Richard T. Miller. *Sailing Yacht Design.* Cambridge, Md.: Cornell Maritime Press, 1965.

Herreshoff, L. Francis. *The Common Sense of Yacht Design.* New York: Rudder Publishing Co., 1934.

Johnson, Peter. *Ocean Racing and Offshore Yachts.* New York: Dodd Mead & Co., 1970.

Kinney, Francis S. *Skene's Elements of Yacht Design.* New York: Dodd, Mead & Co., 1962.

Lord, Lindsay. *Naval Architecture of Planning Hulls.* Cambridge, Md.: Cornell Maritime Press, 1954.

Marchaj, C. A. *Seaworthiness: The Forgotten Factor.* Camden, Me.: International Marine Publishing Co., 1986.

West, Gordon, and Freeman Pittman. *Marine Electronics.* Camden, Me.: International Marine Publishing Co., 1985.

Seamanship

Admiralty Manual of Seamanship. Vol. 3. London: Her Majesty's Stationery Office, 1964.

Coles, K. Adlard. *Heavy Weather Sailing.* New York: John DeGraff, 1967.

Crenshaw, Capt. R. S., Jr. *Naval Shiphandling.* Annapolis, Md.: U.S. Naval Institute, 1965.

Hiscock, Eric C. *Cruising Under Sail.* New York: Oxford Univ. Press, 1963.

Kals, W. S. *Practical Boating.* New York: Doubleday & Co., 1969.

Kotsch, William J., and Richard Henderson. *Heavy Weather Guide.* 2d. Ed. Annapolis, Md.: Naval Institute Press, 1984.

Rousmaniere, John. *The Annapolis Book of Seamanship.* New York: Simon and Schuster, 1989.

Voyaging Under Sail, New York: Oxford Univ. Press, 1972.

Recommended Reading

Bruce, Errol. *Deep Sea Sailing.* London: Stanley Paul & Co., 1961.

Chichester, Sir Francis. *Gypsy Moth Circles the World.* New York: Pocket Books, 1969.

Dumas, Vito. *Alone Through the Roaring Forties.* New York: John DeGraff, 1960.

Gaddis, Vincent. *Invisible Horizons.* New York: Ace Books, 1972.

Gerbault, Alain. *In Quest of the Sun.* New York: John DeGraff, 1955.

Hinz, Earl R. *The Complete Book of Anchoring and Mooring.* Centreville, Md.: Cornell Maritime Press, 1986.

Hiscock, Eric C. *Around the World in Wanderer III.* New York: Oxford Univ. Press, 1964.

Johnson, Electa, and Irving Johnson. *Yankee Sails Across Europe.* New York: W. W. Norton & Co., 1962.

Knox-Johnston, Robin. *A World of My Own.* New York: William Morrow & Co., 1970.

Lane, Frank W. *The Elements Rage.* New York: Chilton Publishing Co., 1965.

National Search and Rescue Manual. Vol. 1, Washington. D.C.: U.S. Govt. Printing Office, 1986.

Pidgeon, Harry. *Around the World Single-Handed.* New York: John DeGraff, 1955.

Robinson, W. A. *To the Great Southern Sea.* London: Peter Davies, 1966.

Rousmaniere, John. *Fastnet, Force 10.* New York: W.W. Norton, 1980.

Slocum, Joshua. *Sailing Alone Around the World.* New York: Sheridan House, 1954.

Villiers, Alan. *Cruise of the Conrad.* New York: Charles Scribners Sons, 1937.

Index

I

J

K

L

M

N

O